Achieving sustainable cultivation of temperate zone tree fruits and berries

Volume 2: Case studies

It is widely recognised that agriculture is a significant contributor to global warming and climate change. Agriculture needs to reduce its environmental impact and adapt to current climate change whilst still feeding a growing population, i.e. become more 'climate-smart'. Burleigh Dodds Science Publishing is playing its part in achieving this by bringing together key research on making the production of the world's most important crops and livestock products more sustainable.

Based on extensive research, our publications specifically target the challenge of climate-smart agriculture. In this way we are using 'smart publishing' to help achieve climate-smart agriculture.

Burleigh Dodds Science Publishing is an independent and innovative publisher delivering high quality customer-focused agricultural science content in both print and online formats for the academic and research communities. Our aim is to build a foundation of knowledge on which researchers can build to meet the challenge of climate-smart agriculture.

For more information about Burleigh Dodds Science Publishing simply call us on +44 (0) 1223 839365, email info@bdspublishing.com or alternatively please visit our website at www.bdspublishing.com.

Related titles:

Achieving sustainable cultivation of temperate zone tree fruits and berries Volume 1: Physiology, genetics and cultivation
Print (ISBN 978-1-78676-208-5) Online (ISBN 978-1-78676-210-8; 978-1-78676-211-5)

Achieving sustainable cultivation of apples
Print (ISBN 978-1-78676-032-6) Online (ISBN 978-1-78676-034-0; 978-1-78676-035-7)

Achieving sustainable cultivation of mangoes
Print (ISBN 978-1-78676-132-3) Online (ISBN 978-1-78676-134-7; 978-1-78676-135-4)

Achieving sustainable cultivation of tomatoes
Print (ISBN 978-1-78676-040-1) Online (ISBN 978-1-78676-042-5; 978-1-78676-043-2)

Achieving sustainable cultivation of tree nuts
Print (ISBN 978-1-78676-224-5) Online (ISBN 978-1-78676-226-9; 978-1-78676-227-6)

Chapters are available individually from our online bookshop:
https://shop.bdspublishing.com

climate•SMART•publishing
IN AGRICULTURAL SCIENCE

BURLEIGH DODDS SERIES IN AGRICULTURAL SCIENCE
NUMBER 54

Achieving sustainable cultivation of temperate zone tree fruits and berries

Volume 2: Case studies

Edited by Professor Gregory A. Lang,
Michigan State University, USA

burleigh dodds
SCIENCE PUBLISHING

Published by Burleigh Dodds Science Publishing Limited
82 High Street, Sawston, Cambridge CB22 3HJ, UK
www.bdspublishing.com

Burleigh Dodds Science Publishing, 1518 Walnut Street, Suite 900, Philadelphia, PA 19102-3406, USA

First published 2019 by Burleigh Dodds Science Publishing Limited

Library of Congress Control Number: 2019941166

British Library Cataloguing in Publication Data
A catalogue record for this book is available from the British Library

ISBN 978-1-78676-212-2 (Print)
ISBN 978-1-78676-215-3 (PDF)
ISBN 978-1-78676-214-6 (ePub)
ISSN 2059-6936 (print)
ISSN 2059-6944 (online)

DOI 10.19103/AS.2018.0040.2

Typeset by Deanta Global Publishing Services, Chennai, India

Contents

Series list

Title	Series number
Achieving sustainable cultivation of maize - Vol 1 From improved varieties to local applications Edited by: Dr Dave Watson, CGIAR Maize Research Program Manager, CIMMYT, Mexico	001
Achieving sustainable cultivation of maize - Vol 2 Cultivation techniques, pest and disease control Edited by: Dr Dave Watson, CGIAR Maize Research Program Manager, CIMMYT, Mexico	002
Achieving sustainable cultivation of rice - Vol 1 Breeding for higher yield and quality Edited by: Prof. Takuji Sasaki, Tokyo University of Agriculture, Japan	003
Achieving sustainable cultivation of rice - Vol 2 Cultivation, pest and disease management Edited by: Prof. Takuji Sasaki, Tokyo University of Agriculture, Japan	004
Achieving sustainable cultivation of wheat - Vol 1 Breeding, quality traits, pests and diseases Edited by: Prof. Peter Langridge, The University of Adelaide, Australia	005
Achieving sustainable cultivation of wheat - Vol 2 Cultivation techniques Edited by: Prof. Peter Langridge, The University of Adelaide, Australia	006
Achieving sustainable cultivation of tomatoes Edited by: Dr Autar Mattoo, USDA-ARS, USA & Prof. Avtar Handa, Purdue University, USA	007
Achieving sustainable production of milk - Vol 1 Milk composition, genetics and breeding Edited by: Dr Nico van Belzen, International Dairy Federation (IDF), Belgium	008
Achieving sustainable production of milk - Vol 2 Safety, quality and sustainability Edited by: Dr Nico van Belzen, International Dairy Federation (IDF), Belgium	009
Achieving sustainable production of milk - Vol 3 Dairy herd management and welfare Edited by: Prof. John Webster, University of Bristol, UK	010
Ensuring safety and quality in the production of beef - Vol 1 Safety Edited by: Prof. Gary Acuff, Texas A&M University, USA & Prof. James Dickson, Iowa State University, USA	011
Ensuring safety and quality in the production of beef - Vol 2 Quality Edited by: Prof. Michael Dikeman, Kansas State University, USA	012
Achieving sustainable production of poultry meat - Vol 1 Safety, quality and sustainability Edited by: Prof. Steven C. Ricke, University of Arkansas, USA	013
Achieving sustainable production of poultry meat - Vol 2 Breeding and nutrition Edited by: Prof. Todd Applegate, University of Georgia, USA	014

Acknowledgements

We wish to acknowledge the following for their help in reviewing particular chapters:

- Chapter 3: Dr Geza Bujdoso, NARIC Fruitculture Research Institute (FRI), Hungary
- Chapter 7: Prof. James Luby, University of Minnesota, USA
- Chapter 10: Emeritus Prof. Craig Chandler, University of Florida, USA
- Chapter 12: Dr Chad Finn, USDA-ARS, USA; and Dr John R. Clark, University of Arkansas, USA

Introduction

Temperate fruits include stone/drupe fruits (such as peach), pome fruits (such as apple) and berries (such as strawberries). Like other crops, cultivation of these fruits faces a number of challenges. These include the need to optimize yields, sensory and nutritional quality; the dynamic threats from biotic and abiotic stresses in a changing climate; the increasing cost and decreasing availability of labour; and the need for more efficient use of resources to minimise environmental impact. This second volume of *Achieving sustainable cultivation of temperate zone tree fruits and berries* includes case studies of individual fruits that illuminate the specific challenges they face and ways these are being addressed.

Part 1 Stone and pome fruits

The first part of the volume summarises research in improving the production of stone and pome fruits. Chapter 1 addresses advances and challenges in peach breeding. Characteristics such as a relatively short juvenile period, self-pollination and small genome size have made peach a model fruit for breeding and genetics research. The chapter reviews the moderate chill peach variety development program involving the USDA-ARS, the University of Georgia and the University of Florida. It shows how the program addressed the challenge of developing new early season varieties using low chilling genotypes with the appropriate sensory quality and firmness required for long-distance shipping. As the chapter explains, the improved firmness realized by the shift to non-melting germplasm has made it possible to leave ripening fruit on the tree longer, resulting in improved fruit size and appearance as well as eating quality. The chapter also reviews ways of improving disease resistance, focusing on peach fungal gummosis (PFG). It describes the steps involved in developing interspecific hybrids with improved resistance, as well as progress in understanding the genetics of PFG resistance and linkage mapping for marker-assisted selection.

Chapter 2 shifts the focus from breeding to improving sustainability in peach production. By adopting precision management techniques, growers can reduce the environmental impact of fruit growing without sacrificing quality and yields, while maintaining income. The chapter discusses how improvements in sustainable peach production can be achieved by adopting an interdisciplinary approach. Research on plant physiology provides the foundations for mechanistic models to predict crop performance in real time, allowing growers to fine-tune their orchard management. The chapter illustrates this by describing the importance of vascular flows to peach fruit growth, including factors such as vapour pressure deficit (VPD), and how this can be used to develop deficit irrigation programs, as well as mulches to reduce soil evaporation, to optimize both fruit development and water use. The chapter also reviews the latest research on the role of photosynthesis in peach production, including the effects of temperature, light and water availability. It shows how this understanding can be used to develop methods to modulate the light environment of the orchard (e.g. via shading hail nets) in order to reduce evapotranspiration, photoinhibition and heat stress which reduce yields. The chapter also discusses how recent advances in sensor technologies can be used, for example, to improve water stress management in peach crops.

Moving from peaches to cherries, the subject of Chapter 3 is advances and challenges in cherry breeding. Cherry breeding is currently carried out in many countries, by public and private programs, and sweet cherry cultivars are continuously being released. However, classification into clear-cut groups of existing cultivars is difficult, because there is a vast continuum of morphological diversity and many traits are influenced by differences in environmental factors among growing locations, including climate and soil characteristics as well as cultural practices. Despite the high number of available commercial cultivars, both sweet and sour cherry cultivation are still based on a small number of cultivars. After offering a historical overview of cherry breeding, the chapter describes the main achievements in conventional breeding, before considering methodologies and the latest advances and key cultivars. The chapter considers new approaches and phenotyping protocols.

Continuing to examine cherries, the subject of Chapter 4 is advances and challenges in sustainable cherry cultivation. Sweet cherries are an inherently challenging crop to produce sustainably, given the significant risks of crop loss from weather events, birds, insects and diseases, and requiring extensive manual labour due to large tree canopies and small delicate fruits. Nevertheless, cherry production has increased dramatically worldwide for the past two decades, driven by strong consumer demand and innovations in plant materials, efficient orchard training systems, orchard microclimate modification technologies, and improved physiological knowledge. The chapter examines the optimization of orchard tree development and fruit yields/quality by developing a foundational understanding of cherry morphology, growth, fruiting, cultivars and rootstocks, and how to utilize these components to re-design canopy architecture and orchard production systems for more labour-efficient, sustainable production.

Moving from cherries to pears, Chapter 5 considers advances and challenges in pear breeding. Although pear (*Pyrus* spp.) is an economically important fruit worldwide, pear cultivars and production practices have been among the slowest of the temperate fruits to change to meet modern consumer demands and labour-efficient orchard production. In the USA, the majority of the pear orchards are low-density plantings with large trees of long-standing cultivars that can reach up to 15 feet in height. Large vigorous trees require labour-intensive management and are relatively inefficient in terms of application of inputs such as water, pesticides, and bio-regulators. The chapter reviews the range of pear cultivars and pear rootstocks, and includes a discussion on germplasm resources. Breeding techniques and improvement of particular traits are considered, including dwarfing, precocity, cold hardiness, fire blight resistance, tree architecture and self-incompatibility.

Staying with the theme of pears, Chapter 6 examines advances and challenges in sustainable pear cultivation. Excessive vigour of European pear varieties and a dearth of dwarfing rootstocks pose significant challenges to the establishment and management of modern high-density orchards. The pronounced negative relationship between pear tree vigour and precocity requires intensive horticultural intervention to expedite a return on investment and to achieve maximum yield potential. Plantings of low to moderate tree densities are no longer economically sustainable given their characteristically inconsistent fruit quality and suboptimal yields. Nascent technologies and novel horticultural strategies have the potential to balance reproductive and vegetative development of pear trees and facilitate the cultural management of high-density orchards. The chapter provides a review of pear floral biology, and fruit setting habits and their complex interaction with environmental factors, along with practical horticultural strategies to promote balanced canopies.

Passing from pears to apples, Chapter 7 examines advances and challenges in apple breeding. Breeding new apple varieties is costly and time-consuming, often selecting for consumer-preference traits at the expense of other traits of agronomic importance. However, combining both sustainable cultivation with market acceptability would benefit growers, consumers and the environment. The chapter summarises the current status of apple breeding and genomics research, taking a forward look at the key factors that may improve selection efficiency within apple breeding programmes to simultaneously enhance both resource use efficiency traits and resilience to biotic and abiotic stress. The chapter discusses how coupling enhanced automated phenotyping, rapid cycling through generations, genome-assisted selection and genome editing using CRISPR-Cas9 can improve breeding programme productivity. The chapter also covers advances in genetic characterization of key rootstock traits.

Chapter 8 deals with advances and challenges in sustainable apple cultivation. Developing sustainable apple cultivation is dependent on both a better knowledge of tree architecture and physiology in relation to fruiting, and on how the tree interacts with its abiotic and biotic environments. Improving knowledge in these areas is crucial to take into account for more sustainable, low-input production. The chapter provides an overview of apple tree growth and fruiting, and explores methods for more sustainable apple training and pruning management.

The final chapter of the section, Chapter 9, addresses advances and challenges in sustainable plum and apricot cultivation. Sustainable fruit cultivation may depend on the combination of a number of factors, including optimal growing conditions, correct selection of cultivars and rootstocks, and application of modern crop cultivation techniques. Sustainability also implies promotion of biodiversity, improvement of microbiological processes in the soil, and protection of the environment (e.g. by avoiding pollution from fertilizer and pesticide run-off). It has been estimated that fruit production may need to increase two- to three-fold to meet future demand, but this needs to be achieved in a sustainable way. The chapter explores ways this might be achieved for plum and apricot production. The chapter covers the genetic resources available for cultivation of these fruits, and the environmental factors affecting plum and apricot cultivation.

Part 2 Berry fruits

The second part of the volume summarises challenges for sustainably cultivating berry fruits. Chapter 10 is focussed on advances and challenges in strawberry breeding. The commercial strawberry (*Fragaria* × *ananassa*) is a popular temperate fruit that is both nutritious and widely appreciated for flavour. The chapter highlights some of the newest innovations in strawberry production, with particular emphasis on genetic improvement of the crop. The trend for developing more robust and sustainable strawberry cultivated varieties via genetics is discussed in detail and useful technologies are reviewed, including high-throughput genotyping and quantitative trait locus (QTL) analysis, targeted sequence capture, third generation sequencing and expression QTL studies. High-throughput phenotyping is also covered, which is an increasingly important area of interest, both to improve breeding through traditional selection and for integration with genomics data for discovery of novel traits.

Going into greater depth on strawberries, Chapter 11 is a case study of how evolving market expectations impact strawberry sustainability. The strawberry industry is facing the reality that it has been built on an unsustainable foundation. First, water is increasingly scarce in some of the key regions where strawberries are grown. Aquifers continue to be depleted and land is subsiding in some areas as water is withdrawn. The chapter offers a case study of the impact of market expectations on the US strawberry industry. After an introductory survey of the history of strawberry production in the USA, the chapter examines the development of an annual strawberry production system. This includes looking at the impact of shifting markets on the sustainability of strawberry production and the challenge of increasing inputs with expansion of annual production in favorable locations. Finally, the chapter assesses the sustainability of the current model of strawberry production.

The subject of Chapter 12 is advances and challenges in raspberry and blackberry breeding. Raspberries and blackberries (*Rubus* spp.) are important fruit crops with increasing levels of production. The chapter focuses on key challenges in achieving more sustainable production, the tools available to breeders, and the future of breeding for sustainability of raspberry and blackberry crops. The chapter examines desirable traits for sustainability, molecular tools and resources.

Continuing to focus on brambles, Chapter 13 considers advances and challenges in sustainable raspberry/blackberry cultivation. *Rubus* crops are important for human health and for rural economies. As demand for these berry crops increases at a time of changing climate and increasing consumer concerns about sustainability, new breeding strategies and cultivation practices are needed. The chapter addresses some of the challenges and solutions to continued sustainable growth, including managing pest and disease stresses and the effects of climate change, as well as reducing environmental impacts.

The volume's concluding chapter, Chapter 14, deals with advances and challenges in blueberry breeding, There is a growing body of research on the biological processes underlying key physiological traits in blueberry (*Vaccinium corymbosum*). Studies have shown high levels of genetic diversity are present within this species, much of which remains to be harnessed. The chapter introduces both the recent advances and current challenges in the breeding of blueberries, with particular focus on demand and production in the UK. Key cultivars currently used in the industry are listed and their advantages and disadvantages are discussed. The genetic material available and its use in breeding programmes is covered, including crossing with other species within the genus *Vaccinium* to obtain desirable traits. The chapter includes sections on phenotyping and marker-assisted breeding, and an extensive discussion on the improvement of flavour in blueberry is also provided.

Part 1

Stone and pome fruits

Chapter 1

Advances and challenges in peach breeding

Dario J. Chavez and Rachel A. Itle, University of Georgia, USA; Daniel Mancero-Castillo, Universidad Agraria del Ecuador, Ecuador; Jose X. Chaparro, University of Florida, USA; and Thomas G. Beckman, USDA-ARS, USA

1 Introduction

The genus *Prunus* belongs to the family Rosaceae and contains approximately 230 species divided into five subgenera: *Prunophora* (*Prunus*), *Amygdalus*, *Cerasus*, *Padus* and *Laurocerasus* (Rehder, 1940), which includes peaches and nectarines (*Prunus persica*), plums (*P. salacina* and *P. domestica*), cherries (*P. avium* and *P. cerasus*), almonds (*P. dulcis*) and apricots (*P. armeniaca*). In 2015, the production of *Prunus* edible fruits and seeds surpassed 43 million metric tons with production found across Asia, Europe, North America, South America, Australia and South Africa (FAOSTAT, 2016).

Peach is a member of subgenus *Amygdalus*, which includes almond, Gansu peach (*P. kansuensis*), Tibetan peach (*P. mira*), mountain peach (*P. davidiana*) and *P. ferganensis*. These species are sexually compatible with each other and produce viable and fertile hybrids (Martínez-Gómez et al., 2003). *Prunus* breeding programmes have used hybridization among these species for improvement of genetic resistance to insect, nematodes and pathogens (Gradziel et al., 2001).

The centre of origin and domestication for peach and almond is southeastern and eastern Asia (Hedrick, 1911). Peach domestication dates back to more than 4000 years with peach genetics evolving by inbreeding, random

http://dx.doi.org/10.19103/AS.2018.0040.15

drift from a reduced number of founders in the breeding programmes and heterosis (Faust and Timon, 2010; Li et al., 2013). During peach domestication in China, three main groups were recognized (Scorza and Okie, 1991; Wang, 1985). The southern group of peaches are found along the Yangtze River in the provinces of Jiangsu, Zhejiang, Jiangxi, Hubei, Hunan and Sichuan, and are characterized by their adaptation to mild winters and hot wet summers, similar to the climate of the southeastern United States. The northern group is found along the Yellow River in Shandong, Hebei, Henan, Shanxi, Shaanxi and Gansu provinces, and are characterized by their adaptation to cold winters and hot dry summers. The third group is adapted to arid northwest China.

From China, peaches were spread through Persia to Europe by the Romans (Scorza and Okie, 1991; Wang, 1985). Peaches made their way to North America via Spanish explorers, first through St. Augustine, Florida, United States, and Mexico (Scorza and Okie, 1991), and then were spread by Native Americans. These peaches were early ripening yellow non-melting flesh types. Later, an additional introduction occurred in the 1800s by the French and English, who brought white melting flesh peaches to the United States. However, all these materials lacked commercial quality (Scorza and Okie, 1991; Sharpe et al., 1954; Sherman et al., 1996; Wang, 1985). The first superior peach varieties in the United States came in the late 1800s and early 1900s from peach producers who identified and selected chance seedlings recovered from open pollinated (OP) seed of plants grown on their farms (Floyd, 1920; Layne and Bassi, 2008). Once these first selections were made and their superior benefits during production were observed, breeding and selection started in state and federal breeding programmes using hybridization between desired parents. However, cultivation and commercialization in the United States did not change until 1850, with the introduction of a superior cultivar, 'Chinese Cling', from China. After that, other superior cultivars followed 'Chinese Cling', such as its OP seedlings 'Georgia Belle' and 'Elberta' (Scorza and Okie, 1991). 'Elberta' had been widely used as a breeding parent in peach breeding programmes across the United States and can be found in the pedigree of most commercial peaches (Layne and Bassi, 2008; Scorza and Okie, 1991; Wang, 1985).

The diverse climates and growing regions in which peaches are grown successfully, plus its relatively short juvenile period (2–3 years), self-pollinated behaviour, small genome size and the identification of important Mendelian traits, have made peach a model fruit for breeding and genetics research (Abbott et al., 2002; Bielenberg et al., 2009). The assembled peach scaffolds cover nearly 99% of the peach genome and over 92% of its orientation has been confirmed. Furthermore, 74 757 *Prunus* ESTs have been queried against the genome at 90% identity and 85% coverage. The peach reference genome contains 27 852 predicted protein-coding genes with a mean percentage of between about 60% A + T bases and 40% G + C [(A+T)/(G+C): 1.5] (Verde et al.,

2013). This GC content is slightly higher than that observed in other species such as *Arabidopsis thaliana* (about 36%), indicating relatively high stability and high percentage of protein-coding genes (Verde et al., 2013).

There are a number of public and private peach breeding programmes in the United States, and their geographic locations and chilling requirements generally are classified as northern (high-chill), western (mid- to high-chill), southern-central (mid-chill) and southern (low-chill) (Layne and Bassi, 2008). In this context, few breeding programmes have focused on mid- and low-chill variety development in the United States until recently.

2 History of the cooperative regional moderate chill peach variety development project

A moderate chill peach breeding project commenced in 1986 as a cooperative regional effort involving the U.S. Department of Agriculture-Agricultural Research Service (USDA-ARS, Byron, Georgia), the University of Georgia (Tifton and, later, Griffin, Georgia) and the University of Florida (Gainesville, Florida). Originally, the project's evaluation site was located near Quitman, Georgia, but was moved in 1991 to its current location at the University of Georgia Research and Education Center outside of Attapulgus in southwest Georgia. The main goal of this collaboration is to develop new peach and nectarine varieties adapted to the lower coastal plain shipping industry of the southeastern United States. Potentially this geographical area is broad, running from coastal South Carolina southward and then westward across South Georgia and north Florida following the Gulf coast west into Texas and then south to the Mexican border. At present, the primary production centre is in South Georgia along its border with Florida, with the largest production volume typically coming from the Quitman area in Brooks county. At the time the project started, this production area was usually the first long-distance shipping industry to send peaches to market in the spring, typically commencing in late April or early May. This industry, though small, representing at most ~5% of Georgia's peach acreage in the 1990s (Hubbard et al., 1998), has, nevertheless, significant economic impact due to the premium paid for early season fruit. Nonetheless, there was little breeding support in terms of new variety development, and the industry relied primarily on two varieties, 'Flordaking' and 'June Gold', for much of its production volume. Both of these varieties were produced as spin-offs by breeding programmes (the University of Florida and Armstrong Nurseries in California, respectively) focused on distinctively different production areas and markets.

2.1 Breeding priorities and approach

The cooperative regional breeding project is focused on the development of early season fresh market varieties. Hence, they must meet minimum size and appearance requirements, and also possess sufficient firmness for long-distance shipping. Quality is important, but the varieties this industry relied upon initially, like most early season varieties in other production areas, presented a decidedly 'low bar' with respect to quality. Initially, fruit smaller than 2 inches were occasionally accepted in the market (though at a much-reduced price); however, over time, fruit smaller than 2.25 inches were no longer marketed due to consumer preferences for larger fruit. Breeding to improve size, appearance and quality is straightforward in its approach, relying primarily on careful selection of breeding parents that offer improvements in one or more desired characteristics, but whose weak points can be offset by the other parent used in the cross.

The market window that this industry targets is both narrow, typically just six weeks in duration, and early, starting in late April and running through the end of May when the main production areas in central Georgia and South Carolina typically begin to ship in volume. Addressing these two issues requires careful manipulation of chilling requirement (which in turn largely determines bloom date) and fruit development period (days from bloom to harvest). Appropriate chilling requirement genetics are important in order to minimize spring frost risk and to adequately break endodormancy for a strong bloom and vegetative bud break. Low chilling requirement genotypes can ensure routine fulfilment of chilling requirement in this climatic region (nominally USDA zone 8b-9), but also are at risk of blooming too early and suffering frequent crop losses to spring frosts. Conversely, while higher chilling requirement genotypes might delay bloom and thereby ensure little or no spring frost damage in most years, they risk the possibility of inadequate breaking of endodormancy in low-chill years, resulting in delayed and extended bloom during warmer weather and possibly incompletely formed flowers that result in low fruit set and poor fruit shape. Additionally, a substantial bloom delay typically translates into a delayed harvest date, thereby potentially creating a gap in the production stream, possibly missing the variety's market window altogether.

The general rule for selecting varieties appropriate to a production area is to choose those with a chilling requirement no higher than 75% of the long-term (i.e. 50 years) chilling average (to ensure regular breaking of endodormancy) and no lower than 50% of the long-term chilling average (to avoid spring frost injury). In practice this is not always as easy as it sounds; recent winters have varied significantly from the long-term average (~675 chill hours) at the project's evaluation site in Attapulgus (T.G. Beckman, *unpublished data*). The long-term chilling average suggests that a chilling requirement between 350 and 500 chill hours is desirable; however, over the last 10 years, the average chill accumulation has declined to 590 chill hours, suggesting a range of 300-450 h might be more prudent.

As a rule, early ripening fruit (with a typically short fruit development period) softens more quickly than does mid- or late-season fruit (Beckman and Krewer, 1999). This issue was initially a major problem for the moderate chill peach breeding project. The first crosses were made with conventional melting flesh-type germplasm like that used in most fresh market peach breeding programmes. However, when the fruit development period was shortened to hit the needed harvest window, it resulted in progeny that often were too soft for shipping and were, therefore, discarded. The resultant heavy culling of the hybrid seedling populations profoundly slowed progress and nearly caused the project to collapse (Rahn, 1997). The search for an alternative strategy focused on other flesh types, such as stonyhard (a novel 'non-ripening' type) and non-melting (a 'slow to soften' type typically used in the development of processing peaches). Both were tried initially, but non-melting germplasm proved more useful and adaptable to the development of fresh market types, which thereby moved the project forward. It is now being augmented by slow ripening 'crispy' types (as exemplified by 'Big Top' nectarine) which provide excellent firmness as fruit approach ripeness. Crispy types are superior to the typical melting type, but then ultimately soften in the hands of consumers to a product similar to that provided by main season melting-type varieties.

Disease resistance is also a breeding priority. Bacterial spot (*Xanthomonas campestris* pv. *pruni*) is of primary importance in most Eastern U.S. breeding programmes, given its potential to significantly degrade the marketability of a crop. Unfortunately, test years for bacterial spot resistance are quite variable for screening germplasm. Nonetheless, the programme utilizes several highly susceptible selections as sentinels for high disease pressure. Peach fungal gummosis (PFG), incited by *Botryosphaeria dothidea* is another disease endemic to the southeastern U.S. peach industry with the potential to cause yield losses of 25–40% with susceptible cultivars (Beckman et al., 2003; Ezra et al., 2017). This disease pressure is particularly strong in the hot and humid lower coastal plain. Periodic evaluation of cultivar releases and advanced breeding lines for susceptibility to PFG has been the standard methodology in breeding programmes (Beckman and Reilly, 2005; Beckman et al., 2011). However, the recent identification of potential markers for resistance to this debilitating disease (Mancero-Castillo et al., 2018) should allow more rapid elimination of selections and potential breeding lines having unacceptable susceptibility.

Tree architecture is a recent, increasingly interesting genetic selection priority in this programme, which was prompted by the discovery of a distinctive spur-type growth habit in a hybrid population that presented a much more open canopy form that required substantially less pruning than normal types (T.G. Beckman, *personal observation*). Although the inheritance of this trait is still not fully characterized, it appears to be a suite of traits under the control of several genes. Observation of their breeding behaviour has revealed that the most useful forms involve not only the production of relatively long-lived spurs, but also an overall reduction in lateral branching and the presence of large diameter shoots (Porter et al., 2002).

2.2 Progress

The moderate chill peach breeding project has an extraordinarily broad germplasm base (Chaparro et al., 2011). Perhaps as a result, several new traits have been discovered in the project's hybrid populations. These include two single gene traits providing either the complete absence of red blush (Beckman et al., 2005b) or its totality in the fruit (Beckman and Sherman, 2003); a single gene trait controlling early defoliation (Chaparro and Beckman, 2011): and a likely polygenic (T.G. Beckman, *unpublished data*) trait providing a semi-freestone phenotype in non-melting flesh types (Beckman and Sherman, 1996).

To date, the chilling range targeted in this programme (300–450 chilling hours) has resulted in a significant improvement in cropping reliability of every non-melting cultivar release compared to the previous standard varieties produced for each market window (Beckman et al., 2008).

The improved firmness realized by the shift to non-melting germplasm has made it possible to leave ripening fruit on the tree longer, resulting in improved fruit size and appearance as well as eating quality, most notably higher soluble solids content (SSC) and lower titratable acidity (TA), achieving a balanced SSC:TA that is more acceptable to consumers (Beckman et al., 2008). An unanticipated benefit also associated with the shift to non-melting germplasm was a profound reduction in split and shattered pits (Beckman et al., 2008), which had been a major shortcoming of the previously dominant melting flesh varieties, 'Flordaking' and 'June Gold', in this market window.

To date, this project has released eight varieties (Table 1), six of which are non-melting. Of these, four have achieved significant commercialization. 'Gulfprince', the first non-melting flesh cultivar, was released in 1999 (Sherman et al., 2000) and was intended only to provide growers with a demonstration of the potential of non-melting-type fruits for the fresh market. However, it has been extraordinarily reliable in this climate in spite of spring frosts and variable winter chilling from year to year. 'Gulfprince' requires ca. 400 chill hours and typically ripens ca. 110 days from bloom, which is slightly too late for the ideal shipping window for this industry due to competition with the beginning of high volume production in central Georgia and South Carolina. Nonetheless, 'Gulfprince' has found a niche for local and roadside sales in this industry.

'Gulfprince' typically ripens with 'Juneprince' peach. The 1995 tree census for Georgia (Hubbard et al., 1998) noted that there were ca. 17000 trees of 'Juneprince' planted in South Georgia. As of 2017, nearly 13000 trees of

Table 1 Peach and nectarine variety releases from the cooperative (USDA-ARS, University of Georgia and University of Florida) regional moderate chill peach development programme for the lower coastal plain production area (1986–2018)

Cultivar	Year	Type	Flesh type	Flesh colour	Chill	Ripe	Reference
'Sunsplash'	1994	Nectarine	Melting	Yellow	400	Mid-May	Krewer et al. (1994)
'White Robin'	1997	Peach	Melting	White	500	Mid-late May	Beckman et al. (2000)
'Gulfprince'	1999	Peach	Non-melting	Yellow	400	Early June	Sherman et al. (2000)
'Gulfking'	2003	Peach	Non-melting	Yellow	350	Early May	Beckman et al. (2005a)
'Gulfcrest'	2003	Peach	Non-melting	Yellow	525	Mid-May	Krewer et al. (2005)
'Gulfcrimson'	2007	Peach	Non-melting	Yellow	400	Mid-late May	Krewer et al. (2008)
'Gulfsnow'	2012	Peach	Non-melting	White	400	Early June	Beckman et al. (2013)
'Gulfatlas'	2013	Peach	Non-melting	Yellow	400	Mid-June	Chaparro et al. (2014)

'Gulfprince' had been propagated by nurseries (J. Watson, *pers. comm.*). If one assumes that the total number of trees in production in this maturity window is essentially static, then 'Gulfprince' may have displaced an equal number of trees of 'Juneprince' and would now be the dominant variety in this maturity window.

'Gulfking' was released in 2003 as an alternative to 'Flordaking' (Beckman et al., 2005a). 'Gulfking' requires ca. 350 chill hours and ripens ca. 75 days from bloom. Although not quite as large as 'Flordaking', 'Gulfking' has cropped more reliably and provides a much better appearance in combination with a substantial improvement in firmness for shipping. As of 2017, over 44 000 trees of 'Gulfking' had been propagated. There were over 69 000 trees of 'Flordaking' in production in 1995 (Hubbard et al., 1998); using similar assumptions as above, 'Gulfking' appears to have overtaken 'Flordaking' as the dominant variety in this maturity window.

'Gulfcrest' was released in 2003 as a non-melting alternative to 'Flordacrest' to fill a small production gap between 'Flordaking' and 'June Gold' (Krewer et al., 2005). 'Gulfcrest' requires ca. 525 chill hours and ripens ca. 75 days from bloom. Despite its slightly smaller size, 'Gulfcrest' has displaced 'Flordacrest', having a better appearance and superior firmness due to its non-melting flesh. There were ca. 6500 trees of 'Flordacrest' in production in 1995. As of 2017, nearly 12 000 trees of 'Gulfcrest' had been propagated, thereby relegating 'Flordacrest' to a very minor role in this ripening window.

'Gulfcrimson' (Krewer et al., 2008) was released in 2007 and has been very well received by growers as a replacement for 'June Gold', which in recent years often produced small, late ripening crops of poorly shaped fruit due to inadequate chilling. 'Gulfcrimson' requires ca. 400 chill hours and ripens ca. 90 days from bloom, and produces fruit that are larger, more attractive, firmer and of higher eating quality than 'June Gold'. There were ca. 45 000 trees of 'June Gold' in production in 1995 (Hubbard et al., 1998). As of 2017, nearly 19 000 trees of 'Gulfcrimson' had been propagated, and it is expected to displace 'June Gold' as the second most widely planted cultivar in this production area within the next few years.

The first white flesh, non-melting release, 'Gulfsnow' (Beckman et al., 2013), was made available for grower trial in 2012. This is the first product of an effort to develop a portfolio of white flesh non-melting cultivars for the lower coastal plain of the southeastern United States. 'Gulfsnow' requires ca. 400 chill hours and ripens ca. 100 days from bloom. There are no other suitably adapted, shipping quality white flesh peaches currently available for this production area. 'Gulfsnow' is a very large and attractive fruit.

The programme's most recent release, 'Gulfatlas' (Chaparro et al., 2014), was made available for grower trial in 2013. 'Gulfatlas' requires ca. 400 chill hours and ripens ca. 120 days from bloom, producing an exceptionally large and well-coloured fruit. It ripens in a market window that previously had no regionally adapted varieties. 'Gulfatlas' extends this industry's potential market window into June, when growers may have a potential advantage in fruit size and quality over early season production from areas to the north. Whether the moderate chill industry will capitalize on this advantage remains to be seen.

The rapid adoption of 'Gulfprince', 'Gulfking', 'Gulfcrest' and 'Gulfcrimson' by growers in the targeted production area is the most compelling evidence that the project's novel use of non-melting germplasm for the fresh market is successful. As this is a continuing project, additional cultivar releases are anticipated.

3 Case study: breeding for resistance to PFG

Peach domestication and breeding for market requirements has resulted in the development of hundreds of new cultivars with horticultural traits beneficial to consumers and the market chain, but often lacking tolerance to biotic and abiotic stresses. Public and private peach breeding programmes have focused on a vast number of traits, such as chilling requirement, frost tolerance, higher fruit quality and improved shelf life, as well as pest and disease resistance (Bielenberg et al., 2009). Breeding for peach pest and disease resistance has been complex, usually relying upon closely related species as sources of useful genes (Gradziel et al., 2001).

The University of Florida (UF) stone fruit breeding and genetics programme began developing peaches for low-chill areas in 1952 (Layne and Bassi, 2008; Scorza and Okie, 1991; Sharpe et al., 1954; Sharpe, 1961, 1969; Sherman et al., 1984). This programme was created to produce varieties with exceptional fruit quality, a low-chill requirement for adaptation to tropical-subtropical climates, and disease resistance suitable for areas with high disease pressure (Andersen et al., 2001; Sharpe et al., 1954; Sherman et al., 1984, 1996). Over 40 peach and nectarine varieties have been released by UF and are currently being grown and tested around the world (Layne and Bassi, 2008).

Within the UF stone fruit breeding programme, resistance to fungal gummosis (*Botryosphaeria dothidea*), also known as peach fungal gummosis (PFG), has been a major goal for several decades. PFG is characterized by deposits of gum exuded through the bark of trunks, limbs and twigs (Weaver, 1974), and can result in dieback of limbs, scaffolds and even entire trees. Peach trees can be infected through wounds and/or lenticels, where gum exudate will typically accumulate (Weaver, 1979). PFG is easily spread by water splash, wind and tools that came in contact with infected wood in trees or from infected prunings left in or near orchards (Pusey, 1989). *Botryosphaeria* conidia and spores can infect lenticels and wounds (Sutton, 1981). Infection through the lenticels typically results in blistering, followed by necrosis of the inner bark, outer xylem and cambial layers (Britton and Hendrix, 1984; Britton et al., 1990). The first external symptom of infection is swollen lenticels, followed by gum exudation on the trunk, scaffolds and limbs during periods of high moisture. Young shoots may present some blistering before the appearance of gum. Gum exudation is reduced during the dry winter months and infection scars are visible as a coarse periderm layer of black colour, compared to a healthy smooth periderm of grey-silver colour. Often, the first symptoms of this disease can be seen as early as the first season after planting the orchard (Fig. 1).

Species in the family Botryosphaeriaceae have a wide range of hosts and geographic distribution (Brown and Britton, 1986; Smith, 1934; Slippers et al., 2017). Examples of other fruit plant diseases linked to *Botryosphaeria* species include apple white rot, apple black rot, mango stem-end rot, cane canker in blackberry, grapevine canker, blueberry stem blight and shoot blight of pistachio. In the peach production areas of the southeastern United States, PFG is widespread and a major concern for growers.

Several management practices have been proposed to reduce PFG. Dead wood and prunings can be removed from the field or flail-mowed to speed decomposition, thereby reducing the sources of new inoculum. Chemical control with $CaCO_3$ and latex paint with $CuSO_4$ has been used to some effect in Asia, but provided no control in a U.S. study (Beckman et al., 2003). Captafol and, to a lesser extent, Captan, provided control, but the number of sprays required to obtain only partial control (particularly with Captan) appear uneconomic and likely to promote fungicide resistance. This is especially an issue for Captan, an important component of the spray programme for the Southeast U.S. industry's spray programme. Moreover, the registration of Captafol has now been cancelled for peach. It is hard to quantify how much damage peach gummosis causes in

Figure 1 Peach fungal gummosis (PFG) incited by *Botryosphaeria dothidea*. (a) Close up of a highly susceptible genotype with high incidence of gummosis. (b) Susceptible (left) and resistant (right) full-sib progeny from a *P. persica* × *P. dulcis* interspecific cross. Photo credit: José X. Chaparro.

the southeastern United States. Pusey et al. (1986) reported that peach trees inoculated with *Botryosphaeria* can have a mortality rate of 17–46% 1 year after inoculation. Similarly, Beckman et al. (2011) reported that in field conditions, infected peach trees can experience dieback and yield reduction of 40–60%.

The negative impact of PFG on peach production (growth and yield), and the lack of cost-effective chemicals and/or cultural management practices, created a need for fundamental research to identify sources of genetic resistance to PFG and utilize them to breed resistant peach scion and rootstock cultivars. In addition, *Botryosphaeria* diseases have gained worldwide interest as a problem in *Prunus* production due to the high species diversity, a global commerce and diverse environmental conditions (Slippers et al., 2017; Marsberg et al., 2017). Preliminary research indicated that variation in the level of resistance to *B. dothidea* exists in commercial peach germplasm. Okie and Pusey (1996) identified resistance in the genotypes 'Harbrite', 'Eagle Beak' (PI43289), 'Shau Thai Tao' (PI65821) and NRL1 (Nemaguard × Rutgers RL F_2). More recent work (Beckman and Reilly, 2005; Beckman et al., 2011) identified additional commercial peach cultivars with PFG resistance. In almond (closely related to peach), previous evaluations of germplasm from France and the former U.S.S.R. indicated high levels of resistance to *Monilinia* and *Fussicocum* (teleomorph *Botryosphaeria*) (Wood, 1925; Kester and Asay, 1975).

3.1 Breeding and PFG resistance evaluation methodology

Introgression of resistance to PFG is a priority for the UF stone fruit breeding programme. The following information is a summary of the work published by Mancero-Castillo et al. (2018). Interspecific hybrids were produced between three species within subgenus *Amygdalus*: Kansu peach (*P. kansuensis*) genotype A (PK), 'Tardy Nonpareil' (TNP) almond and peach selections FL97-47C, 'Flordaguard' (FG) and 'Okinawa' (OK). Interspecific hybrid populations of FG × PK, OK × PK, FG × TNP, FL97-47C × TNP and FL97-47C × PK were screened for resistance/susceptibility to gummosis in field conditions in Gainesville, Florida, USA, where natural presence of the inoculum *Botryosphaeria* occurs. Genotypes that did not present gum exudate, swollen lenticels and limb dieback symptoms (Fig. 1) were classified as resistant (Beckman and Reilly, 2005). Lines with resistant genotypes were classified as potentially segregating for resistance to gummosis (Table 2). Out of the hybrids produced, resistant FG × TNP1260, susceptible FG × TNP1258, FG × PK3, FG × PK6 and FL97-47C × PK1 F_1 hybrids were selected as potential parents to be used for test crosses with susceptible and resistant individuals to characterize the inheritance of the resistance. These hybrids and four peach genotypes (AP00-30WBS, 'UFSharp', FL97-47C and 'Flordaguard') were re-propagated by grafting onto 'Flordaguard' rootstock six times to confirm its resistance/susceptibility using field evaluation techniques.

Table 2 F_1 interspecific hybrids evaluated for resistance to peach fungal gummosis (PFG) in field conditions at the University of Florida, Gainesville, Florida, USA (Mancero-Castillo et al., 2018)

F_1 interspecific hybrids	Number of hybrids	Phenotype[a]	
		Susceptible	Resistant
'Flordaguard' × 'Tardy Nonpareil'	5	3	2
'Flordaguard' × *Prunus kansuensis*	10	10	0
FL97-47C × 'Tardy Nonpareil'	8	5	3
FL97-47C × *Prunus kansuensis*	1	0	1
'Okinawa' × *Prunus kansuensis*	31	31	0

'Flordaguard', Okinawa and FL97-47C are *P. persica* selections; 'Tardy Nonpareil' is a *P. dulcis* selection.
[a] Resistant = trees with no visual symptoms, Susceptible = trees with clear disease symptoms.

Additional grafted trees of resistant FG × TNP1260 and susceptible FG × TNP1258 were screened using field evaluation techniques with a trellis system and artificial inoculation (Okie and Pusey, 1996) at the USDA-ARS Southeastern Fruit and Tree Nut Research Laboratory, Byron, Georgia, USA. The trellis structure was formed by T-shaped wooden supports, with a 1.2-m-long horizontal piece attached 2 m above ground, with supports 10 m apart. Welded wire fencing and a misting system were attached at the top of the trellis (Fig. 2). Peach genotypes were planted below the trellis system in one row at 0.3-m spacing between plants. Freshly cut branches and limbs with symptoms of PFG were collected, placed and maintained on top of the welded wire fencing throughout the season. The sprinkler system was turned on for 10 min each hour from 7 am to 7 pm from April to June (4–6 weeks). Conidia and ascospores spread from the dead branches and limbs to inoculate the test trees, creating optimal environmental conditions for infection.

Visual evaluations were made on the trunk and scaffolds during fall, and genotypes were rated for gummosis based on a disease scale (Beckman et al., 2003) as follows: 0 = no gum sites or lesions, 1 = 1 or 2 gum sites or lesions, 2 = 3–10 gum sites or lesions, 3 = 11–25 gum sites or lesions, 4 = 26–50 gum sites or lesions and 5 = 51 or more gum sites or lesions. This test identified that all replicates of FG × TNP1258 were consistently susceptible and all replicates of FG × TNP1260 were resistant (Table 3).

3.2 Genetics of PFG resistance

Confirmation of susceptible and resistant genotypes constituted an important step to perform the test crosses used to characterize the inheritance of peach resistance to *Botryosphaeria*. Once peach genotypes and F_1 hybrids were classified as resistant or susceptible to gummosis (Table 3), test crosses were performed with the F_1 hybrids being backcrossed to peach selections

Figure 2 Trellis system used to screen peach germplasm for peach fungal gummosis (PFG). (a) Trellis structure and field layout. (b) Welded wire fence, sprinkler system and infected limbs. (c) Mist on for field inoculation. Photo credit: T. G. Beckman.

AP00-30WBS and FL97-47C (Table 4). The objective of these crosses was to produce segregating populations between susceptible × resistant genotypes and susceptible × susceptible genotypes.

The BC_1F_1 populations were grown at UF in field conditions with natural inoculation of *Botryosphaeria*. To aid with disease progression and inoculation, overhead irrigation was turned on weekly. Populations were equally divided to create four pseudo-replicates. In each population, clonal replicates of parent genotypes were included as controls. These populations and clonal replicates were rated twice a year from 2012 until 2015 using the gummosis

Table 3 Phenotypic data for peach fungal gummosis (PFG) rating, stem diameter and leaf colour of peach genotypes and interspecific hybrids at the USDA-ARS, Southeastern Fruit and Tree Nut Research Laboratory, Byron, Georgia, USA (Mancero-Castillo et al., 2018)

Genotypes	Mean gummosis rating[a]	Mean diameter	Leaf colour	Phenotype
Peach				
AP00-30WBS	3.40a	10.17a	Green	Susceptible
'UFSharp'	2.41a	9.95a	Green	Susceptible
FL97-47C	1.91ab	8.75a	Green	Intermediate resistance
'Flordaguard'	3.45a	11.29a	Red	Susceptible
Interspecific hybrids				
FG × PK3	2.72a	13.49a	Red	Susceptible
FG × PK6	2.91a	13.77a	Red	Susceptible
FG × TNP1258	3.79a	10.23a	Red	Susceptible
FG × TNP1260	0.23b	10.27a	Red	Resistant
FL97-47C × PK6	1.61ab	13.91a	Green	Intermediate resistance

Different letters indicate significant difference ($P < 0.001$).
FG = 'Flordaguard', TNP = 'Tardy Nonpareil', PK = *Prunus kansuensis*.
All means calculated from at least six clonal replicates measured twice a year from 2012 to 2015.
[a] Gummosis rating: 0 = no lesion, 1 = 1 or 2, 2 = 3-10, 3 = 11-25, 4 = 26-50, 5= >51 lesions.

rate (Beckman et al., 2003). Backcross populations segregated for resistance and susceptibility to *Botryosphaeria* as identified using gummosis rates from natural inoculations (Fig. 3). Populations of 'UFSharp' × (FG × TNP1260) and AP00-30WBS × (FG × TNP1260) were subsequently used to determine the inheritance of the resistance to PFG. A binary scale was used to separate the backcross populations rating with genotypes with a mean gum rate ≥2 classified as susceptible and genotypes with a mean gum rate of <2 classified as resistant. Segregation was determined to be 69:81 resistant to susceptible, suggesting the presence of a monogenic resistance coming from almond.

Table 4 Backcross segregating population means for peach fungal gummosis (PFG) rating and trunk diameter at the USDA-ARS, Southeastern Fruit and Tree Nut Research Laboratory, Byron, Georgia, USA (Mancero-Castillo et al., 2018)

Backcross populations	PFG phenotype	Gummosis rating	Diameter (cm)
'UFSharp' × (FG × TNP1260) (n = 113)	Sus × Res	2.32a	9.69ab
'UFSharp' × (FG × PK6) (n = 100)	Sus × Sus	2.09a	11.06a
FL97-47C × (FL97-47c × PK) (n = 50)	Sus × Sus	2.17a	9.97ab
AP00-30WBS × (FG × TNP1260) (n = 37)	Sus × Res	2.37a	8.82b

Gummosis rating: 0 = no lesion, 1 = 1 or 2, 2 = 3-10, 3 = 11-25, 4 = 26-50, 5= > 51 lesions. Data followed by a different letter represents significant differences within a column ($P > 0.01$).

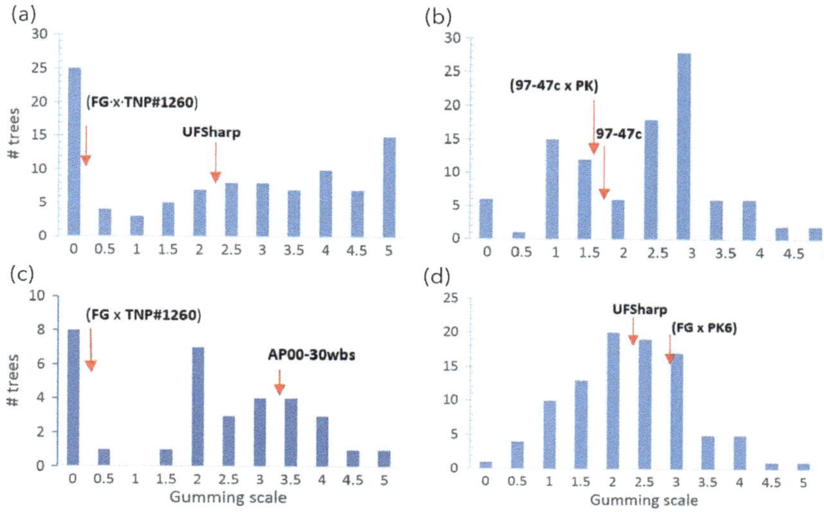

Figure 3 Distribution of gumming scores for segregating backcross populations evaluated from 2012 to 2016. Parts (a) and (c) represent populations generated by backcrossing the resistant hybrid ('Flordaguard' peach × 'Tardy Nonpareil' 1260) to 'UFSharp' and AP00-30WBS, respectively. Parts (b) and (d) represent distributions generated by the backcross populations FL97-47C × (FL97-47C peach × *Prunus kansuensis*) and 'UFSharp' × ('Flordaguard' peach × *Prunus kansuensis*), respectively. Red arrows indicate the mean gummosis ratings of the parents. Source: Adapted from Mancero-Castillo et al. (2018).

3.3 Validation and linkage mapping

The characterization of a trait of interest can be affected by the screening methods and locations used during breeding and selection. To confirm the stability of PFG resistance from segregating populations, three clonal replicates were propagated from 113 trees from 'UFSharp' × (FG × TNP1260) and 36 trees from AP00-30WBS × (FG × TNP1260) onto 1-year-old 'Flordaguard' rootstock plants. These trees were planted under a trellis structure for artificial inoculation, as described previously, at the USDA-ARS Southeastern Fruit and Tree Nut Lab in Byron, Georgia, USA. The results obtained from gummosis ratings from the trellis inoculations and the field observations in Gainesville, Florida, USA, were compared. Strong correlations, R^2 values of 0.51–0.71 for AP00-30WBS × (FG × TNP1260) and 'UFSharp' × (FG × TNP1260) populations, respectively, were observed. These results validated the techniques being used and the classification of the phenotypes as resistant/susceptible to PFG. In addition, it was determined that the use of a susceptible rootstock ('Flordaguard') for propagation did not affect the overall rating of the susceptible/resistant genotypes after being artificially inoculated and maintained under the trellis.

DNA of peach, F_1 hybrids and BC_1F_1 progeny for all families were extracted using a modified CTAB method (Doyle, 1991; Chavez and Chaparro, 2011). A total of 40 SSR markers distributed across the peach genome (~10.5 cM between markers) were selected from the *Prunus* 'Texas' almond × 'Earlygold' peach (T × E) reference map (Dirlewanger et al., 2004; Jung et al., 2008) as described by Carrillo-Mendoza et al. (2010). Forward markers were fluorescently labelled using FAM or HEX labels. PCR parameters and

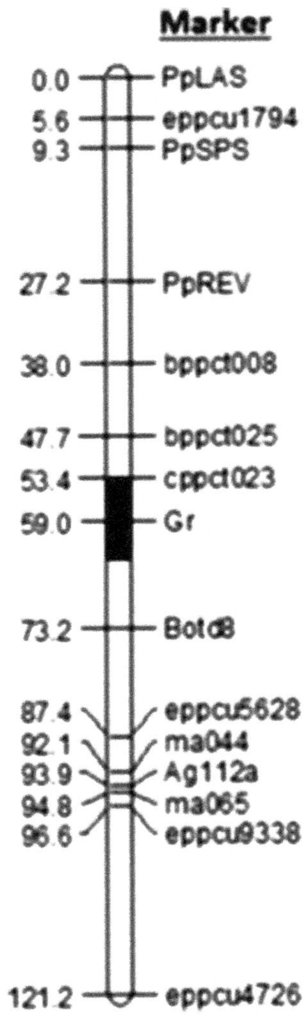

Figure 4 Chimeric chromosome 6/8 generated by the reciprocal translocation between chromosomes 6 and 8 from red leaf peach 'Flordaguard'. Darker region denotes reciprocal translocation site reported by Jáuregui et al. (2001) and Lambert and Pascal (2011). Source: Adapted from Mancero-Castillo et al. (2018).

fragment analysis procedures were done as described in Carrillo-Mendoza et al. (2010). Mapmaker/Exp 3.0 (Lander et al., 1987) software was used to construct the linkage maps. These maps were previously developed for the same populations in studies of tree architecture characteristics as reported by Carrillo-Mendoza et al. (2010). Phenotypes obtained from screening procedures for PFG resistance for BC_1F_1 progeny 'UFSharp' × (FG × TNP1260), FL97-47C × (FL97-47C × PK) and 'UFSharp' × (FG × PK 6) were used to perform segregation analyses.

Quantitative trait locus (QTL) analyses and identification were performed using R/QTL statistical software using the composite interval mapping function (Zeng, 1994). Logarithm of odds (LOD) thresholds were calculated using 1000 permutations. Once a linkage map and location of the PFG resistance locus was identified, additional markers were run to saturate the physical location. All the populations using FG × TNP1260 were used to construct a genetic linkage map. A total of 48 SSRs and the PFG resistance locus were mapped to seven linkage groups (instead of eight) due to the 6/8 reciprocal translocation present in red leaf peaches (Blake, 1937; Lambert and Pascal, 2011). The PFG resistance locus *Botd8* was found in the 6/8 linkage map from 'UFSharp' × (FG × TNP1260) using the binary data as previously described to classify resistant and susceptible genotypes (Fig. 4). Previously, a single dominant locus for red leaf (*Gr*) was mapped near the reciprocal translocation between chromosomes 6 and 8. All populations produced with 'Flordaguard' segregated for *Gr*. PFG resistance co-segregated with green leaves in backcross populations using FG × TNP1260 (Table 5). In addition, QTL analyses using gummosis ratings found the PFG resistance locus *Botd8* in the 6/8 linkage group, with resistant tree gummosis ratings ranging from 0 to 0.7. *Prunus kansuensis* populations did not show a QTL for PFG resistance (Fig. 5).

Table 5 Segregation for tolerance/susceptibility to peach fungal gummosis (PFG) and leaf colour in peach × (peach × almond) and peach × (peach × *Prunus kansuensis*) backcross populations (Mancero-Castillo et al., 2018)

	Progeny phenotype						
	Red	Red	Green	Green			
Backcross populations	Tol	Sus	Tol	Sus	Test ratio	χ^2 (1 df)	
FL97-47C × (FL97-47C × *PK*) (*n* = 50)	na	na	19	31	0:0:1:1	2.88	ns
UFSharp × (FG × *PK*) (*n* = 100)	18	23	24	35	1:1:1:1	0.17	ns
UFSharp × (FG × TNP1260) (*n* = 113)	9	59	43	2	1:1:1:1	68.29	***
AP00-30WBS × (FG × TNP1260) (*n* = 36)	3	19	14	1	1:1:1:1	31.92	***

FG = 'Flordaguard', TNP = 'Tardy Nonpareil', PK = *Prunus kansuensis*.
*, **, *** = significant at the 0.05, 0.01, 0.001 levels, respectively.
χ^2 = chi-square, ns = not significant ($P > 0.01$), na = not applicable, Tol = Tolerant to PFG, Sus = Susceptible to PFG.

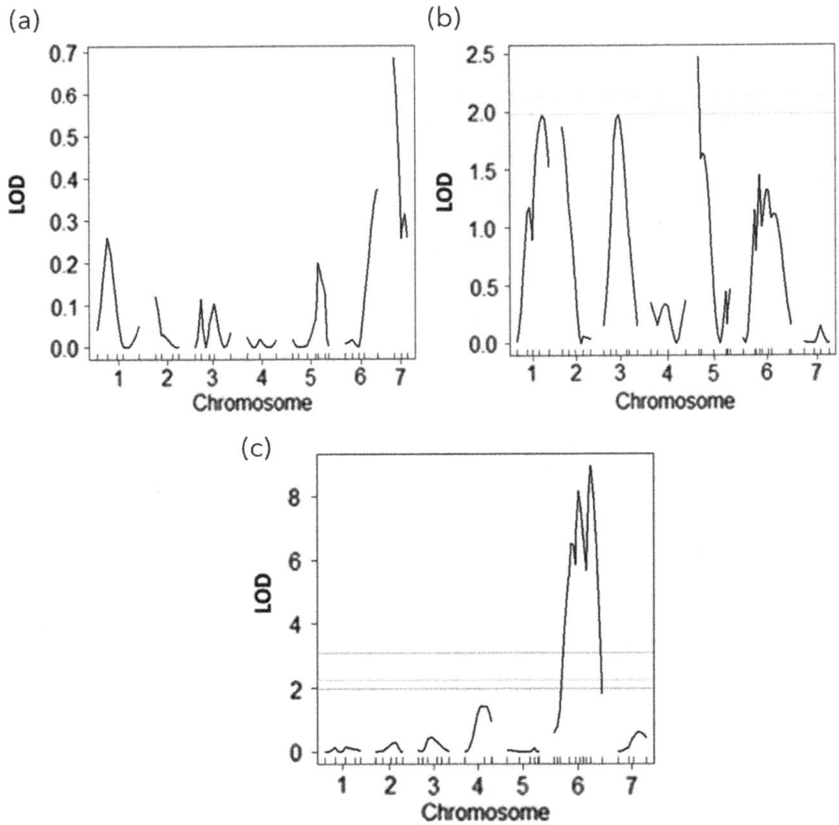

Figure 5 Graphical results of quantitative trait locus (QTL) analysis for peach fungal gummosis (PFG) resistance in backcross populations. (a) FL97-47C × (FL97-47C peach × *Prunus kansuensis*), (b) 'UFSharp' × ('Flordaguard' peach × *Prunus kansuensis*) and (c) 'UFSharp' × ('Flordaguard' × 'Tardy Nonpareil'1260). LOD = logarithm of odds. Colour dotted lines indicate LOD thresholds: red α = 0.01, green α = 0.05, blue α = 0.1. Source: Adapted from Mancero-Castillo et al. (2018).

4 Future trends and conclusion

The identification of PFG resistance from 'Tardy Nonpareil' almond provides a source for breeding resistance into cultivars of peach, nectarine and almond. The UF programme is continuing to increase the marker density within the region linked with PFG resistance, with the goal of identifying the candidate gene(s) associated with this resistance. The availability of high-density marker systems in Rosaceae and the optimized protocols to screen for PFG resistance provide a unique opportunity for marker-assisted selection. In addition, comparative genomics studies can be done with other species within the family Rosaceae and other plant species to identify regions that can harbour resistance genes for diseases caused by *Botryosphaeria* species.

5 Where to look for further information

A well-known reference for peach, production and cultural management, insect and pest, genetics and research programmes around the world can be found in *The Peach: Botany, Production and Uses* by Layne, D.R. and Bassi, D. (see Layne and Bassi, 2008).

The most complete genome database for Rosaceae species, including peach, can be found in the Genome Database for Rosaceae (GDR) (http://www.rosaceae .org/). GDR was established by USDA Specialty Crop Research Initiative, the USDA National Research Support Project 10, the NSF Plant Genome Research Program and the Washington Tree Fruit Research Commission.

6 References

Abbott, A., Lecouls, A., Wang, Y. and Georgi, L. (2002). Peach: The model genome for Rosaceae genomics. *V Int. Peach Symp.*, 8–11 July 2001, Davis, CA, pp. 199–209.

Andersen, P. C., Sherman, W. B. and Williamson, J. G. (2001). Low chill peach and nectarine cultivars from the University of Florida breeding program: 50 years of progress. *Proc. Fla. State Hort. Soc.* 114:33–6.

Beckman, T. G. and Krewer, G. W. (1999). Postharvest characteristics of moderate-chill peach varieties. *Proc. Fla. State Hort. Soc.* 112:236–41.

Beckman, T. G. and Reilly, C. C. (2005). Relative susceptibility of peach cultivars to fungal gummosis (*Botryosphaeria dothidea*). *J. Am. Pom. Soc.* 59:111–16.

Beckman, T. G. and Sherman, W. B. (1996). The non-melting semi-freestone peach. *Fruit Var. J.* 50:189–93.

Beckman, T. G. and Sherman, W. B. (2003). Probable qualitative inheritance of full red skin color in peach. *HortScience* 38:1184–5.

Beckman, T. G., Krewer, G. W. and Sherman, W. B. (2000). 'White Robin' peach. *HortScience* 35:958–9.

Beckman, T. G., Pusey, P. L. and Bertrand, P. F. (2003). Impact of fungal gummosis on peach trees. *HortScience* 38:1141–3.

Beckman, T. G., Krewer, G. W. and Sherman, W. B. (2005a). 'Gulfking' peach. *J. Am. Pom. Soc.* 59:94–6.

Beckman, T. G., Rodriguez Alcazar, J., Sherman, W. B. and Werner, D. J. (2005b). Evidence for qualitative suppression of red skin color in peach. *HortScience* 40:523–4.

Beckman, T. G., Krewer, G. W., Chaparro, J. X. and Sherman, W. B. (2008). Potential of non-melting flesh peaches for the early season fresh market. *J. Am. Pom. Soc.* 62(2):52–7.

Beckman, T. G., Reilly, C. C., Pusey, P. L. and Hotchkiss, M. W. (2011). Progress in the management of peach fungal gummosis (*Botryosphaeria dothidea*) in the southeastern U.S. peach industry. *J. Am. Pom. Soc.* 65(4):192–200.

Beckman, T. G., Chaparro, J. X. and Conner, P. J. (2013). 'GulfSnow' peach. *HortScience* 48:126–7.

Bielenberg, D., Gasic, K. and Chaparro, J. X. (2009). An introduction to peach (*Prunus persica*). In: K. M. Folta and S. E. Gardiner (Eds), *Genetics and Genomics of Rosaceae*. Springer New York, New York, NY, pp. 223–34.

Blake, M. A. (1937). Progress in peach breeding. *Proc. Am. Soc. Hort. Sci.* 35:49–53.

Britton, K. O. and Hendrix, F. F. (1984). Studies toward the biology and control of peach tree gummosis. *Stone Fruit Tree Decline Workshop Proceedings: Proceedings of a Workshop Held 30 October-1 November 1984*, at the Appalachian Fruit Research Station, Kearneysville, WV.

Britton, K. O., Hendrix, F. F., Pusey, P. L., Okie, W. R., Reilly, C. C. and Daniell, J. W. (1990). Evaluating the reaction of peach cultivars to infection by three *Botryosphaeria* species. *HortScience* 25:468–70.

Brown, E. A. and Britton, K. O. (1986). *Botryosphaeria* diseases of apple and peach in the Southeastern United States. *Plant Dis.* 70:480–4.

Carrillo-Mendoza, O., Sherman, W. B. and Chaparro, J. X. (2010). Development of a branching index for evaluation of peach seedlings using interspecific hybrids. *HortScience* 45:852–6.

Chaparro, J. X. and Beckman, T. G. (2011). Evidence of a new single gene trait controlling pre-mature defoliation in peach. *Acta Hort.* 962:147–9.

Chaparro, J. X., Sherman, W. B. and Beckman, T. G. (2011). Genetic diversity of low and mid-chill peach cultivars. *Acta Hort.* 962:35–8.

Chaparro, J. X., Conner, P. J. and Beckman, T. G. (2014). 'GulfAtlas' peach. *HortScience* 49(8):1093–4.

Chavez, D. J. and Chaparro, J. X. (2011). Identification of markers linked to seedlessness in *Citrus kinokuni* hort. ex Tanaka and its progeny using bulked segregant analysis. *HortScience* 46:693–7.

Dirlewanger, E., Graziano, E., Joobeur, T., Garriga-Caldere, F., Cosson, P., Howad, W. and Arus, P. (2004). Comparative mapping and marker-assisted selection in *Rosaceae* fruit crops. *PNAS* 101:9891–6.

Doyle, J. (1991). DNA protocols for plants. In: G. M. Hewitt, A. W. B. Johnston and J. P. W. Young (Eds), *Molecular Techniques in Taxonomy*. NATO ASI Series (Series H: Cell Biology), vol. 57. Springer, Berlin, Heidelberg, pp. 283–93. doi:10.1007/978-3-642-83962-7_18.

Ezra, D., Hershcovich, M. and Shtienberg, D. (2017). Insights into the etiology of gummosis syndrome of deciduous fruit trees in Israel and its impact on tree productivity. *Plant Dis.* 101:1354–61.

FAOSTAT data. (2016). http://www.fao.org/faostat/en/?#search/peach.

Faust, M. and Timon, B. (2010). Origin and dissemination of peach. In: M. Faust and B. Timon (Eds), *Horticultural Reviews*. John Wiley & Sons, Inc., Oxford, UK, pp. 331–79.

Floyd, W. F. (1920). Peach growing in Florida. *Fla. Coop. Ext. Bul.* 27:5–8.

Gradziel, T. M., Martinez-Gomez, P., Dicenta, F. and Kester, D. E. (2001). The utilization of related *Prunus* species for almond variety improvement. *J. Am. Pom. Soc.* 55:100–8.

Hedrick, U. P. (1911). The plums of New York. *N. Y. Dept. Agr. 18th Ann. Rpt.* v. 3, pt. 2.

Hubbard, E. E., Florkowski, W. J., Park, T. A. and Witt, H. J. (1998). Commercial peach tree inventory and prospectus, Georgia 1995. Univ. of Georgia Research Report, No. 650.

Jáuregui, B., de Vicente, M. C. and Messeguer, R. (2001). A reciprocal translocation between 'Garfi' almond and 'Nemared' peach. *Theor. Appl. Genet.* 102:1169–76.

Jung, S., Staton, M., Lee, T., Blenda, A., Svancara, R., Abbott, A. and Main, D. (2008). GDR (genome database for Rosaceae): Integrated web-database for Rosaceae genomics and genetic data. *Nucleic Acids Res.* 36:1034–40.

Kester, D. E. and Asay, R. Y. (1975). Almond breeding. In: P. Janick and J. N. Moore (Eds), *Advances in Fruit Breeding*. Purdue University Press, Lafayette, IN, pp. 382–419.

Krewer, G., Beckman, T. and Sherman, W. (1994). 'Sunsplash' nectarine. *HortScience* 29:339–40.

Krewer, G. W., Sherman, W. B. and Beckman, T. G. (2005). 'Gulfcrest' peach. *J. Am. Pom. Soc.* 59:91–3.

Krewer, G. W., Beckman, T. G., Chaparro, J. X. and Sherman, W. B. (2008). 'Gulfcrimson' peach. *HortScience* 43(5):1596–7.

Lambert, P. and Pascal, T. (2011). Mapping *Rm2* gene conferring resistance to the green peach aphid (*Myzus persicae* Sulzer) in the peach cultivar 'Rubira®.' *Tree Genet. Genom.* 7:1057–68. doi:10.1007/s11295-011-0394-2.

Lander, E. S., Green, P., Abrahamson, J., Barlow, A., Daly, M. J., Lincoln, S. E. and Newberg, L. A. (1987). MAPMAKER: An interactive computer package for constructing primary genetic linkage maps of experimental and natural populations. *Genomics* 1:174–81.

Layne, D. R. and Bassi, D. (2008). *The Peach: Botany, Production and Uses*. CABI International, Cambridge, MA, 615pp.

Li, X., Meng, X. and Jia, H. (2013). Peach genetic resources: Diversity, population structure and linkage disequilibrium. *BMC Genet.* 14:84. doi:10.1186/1471-2156-14-84.

Mancero-Castillo, D., Beckman, T. G., Harmon, P. F. and Chaparro, J. X. (2018). A major locus for resistance to *Botryosphaeria dothidea* in *Prunus*. *Tree Genet. Genom.* 14:26.

Marsberg, A., Kemler, M., Jami, F., Nagel, J. H., Postma-Smidt, A., Naidoo, S., Wingfield, M. J., Crous, P. W., Spatafora, J. W., Hesse, C. N., et al. (2017). *Botryosphaeria dothidea* : A latent pathogen of global importance to woody plant health. *Mol. Plant Pathol.* 18:477–88. doi:10.1111/mpp.12495,

Martínez-Gómez, P., Arulsekar, S., Potter, D. and Gradziel, T. M. (2003). Relationships among peach, almond, and related species as detected by simple sequence repeat markers. *J. Am. Soc. Hort. Sci.* 128:667–71.

Okie, W. R. and Pusey, P. L. (1996). USDA peach breeding in Georgia: Current status and breeding for resistance to *Botryosphaeria*. *Acta Hort.* 374:151–8. doi:10.17660/ActaHortic.1996.374.19.

Porter, G. W., Sherman, W. B., Beckman, T. G. and Krewer, G. W. (2002). Fruit weight and shoot diameter relationship in early ripening peaches. *J. Am. Pom. Soc.* 56(1):30–3.

Pusey, P. L. (1989). Availability and dispersal of ascospores and conidia of *Botryosphaeria* in peach orchards. *Phytopathology* 79:635–9.

Pusey, P. L., Reilly, C. C. and Okie, W. R. (1986). Symptomatic responses of peach trees to various isolates of *Botryosphaeria dothidea*. *Plant Dis.* 70:568–72.

Rahn, D. (1997). The dawning of the perfect peach. UGA Research Reporter, Summer, 1997.

Rehder, A. (1940). *Manual of Cultivated Trees and Shrubs Hardy in North America*. 2nd ed. Macmillan, New York, NY.

Scorza, R. and Okie, W. R. (1991). Peaches (*Prunus*). *Acta Hort.* 290:177–234.

Sharpe, R. H. (1961). Developing new peach varieties for Florida. *Proc. Fla. State Hort. Soc.* 74:348–52.

Sharpe, R. H. (1969). Sub-tropical peaches and nectarines. *Proc. Fla. State Hort. Soc.* 82:302–6.

Sharpe, R. H., Webb, T. E. and Lundy, H. W. (1954). Peach variety tests. *Proc. Fla. State Hort. Soc.* 67:245–7.

Sherman, W. B., Rodriguez, J. and Miller, E. P. (1984). Progress in low-chill peaches and nectarines from Florida. *Proc. Fla. State Hort. Soc.* 97:320–2.

Sherman, W. B., Lyrene, P. M. and Sharpe, R. H. (1996). Low-chill peach and nectarine breeding at the University of Florida. *Proc. Fla. State Hort. Soc.* 109:222–3.

Sherman, W. B., Beckman, T. G. and Krewer, G. W. (2000). 'Gulfprince' peach. *J. Am. Pom. Soc.* 54:82–3.

Slippers, B., Crous, P. W., Jami, F., Groenewald, J. Z. and Wingfield, M. J. (2017). Diversity in the Botryosphaeriales : Looking back, looking forward. *Fungal Biol.* 121:307–21. doi:10.1016/j.funbio.2017.02.002

Smith, C. O. (1934). Inoculations showing the wide host range of *Botryosphaeria ribis. J. Agric. Res.* 49:467–76.

Sutton, T. B. (1981). Production and dispersal of ascospores and conidia by *Physalospora obtuse* and *Botryosphaeria dothidea* in apple orchards. *Phytopathology* 71:584–9.

Verde, I., Abbott, A. G., Scalabrin, S., Jung, S., Shu, S., Marroni, F., Zhebentyayeva, T., Dettori, M. T., Grimwood, J., Cattonaro, F., et al. (2013). The high-quality draft genome of peach (*Prunus persica*) identifies unique patterns of genetic diversity, domestication and genome evolution. *Nat. Genet.* 45:487–94. doi:10.1038/ng.2586.

Wang, Y. (1985). Peach growing and germplasm in China. *Acta Hort.* 173:51–5.

Weaver, D. J. (1974). A gummosis disease of peach trees caused by *Botryosphaeria dothidea. Phytopathology* 64:1429–32. doi:10.1094/Phyto-64-1429.

Weaver, D. J. (1979). Role of conidia of *Botryosphaeria dothidea* in the natural spread of peach tree gummosis. *Phytopathology* 69:330–40.

Wood, M. N. (1925). Almond varieties in the United States. *USDA Bull.* 1282:1–42.

Zeng, Z.-B. (1994). Precision mapping of quantitative trait loci. *Genetics* 136:1457–68.

Chapter 2

Advances and challenges in sustainable peach production

Luca Corelli Grappadelli, Brunella Morandi and Luigi Manfrini, University of Bologna, Italy; and Pasquale Losciale, University of Bari, Italy

1 Introduction

Current trends in separate disciplines (physiology, modelling, precision agriculture) are converging towards the realization of peach (*Prunus persica*) orchards of increasing sustainability. By adopting precise management techniques, growers already can, and will in the future, be more able to reduce the environmental impact of fruit growing, without sacrificing quality and yields, while maintaining income. This interdisciplinary approach builds on physiological knowledge that provides the foundations for mechanistic models that predict crop performance in real time, allowing growers to fine-tune their orchard management. It can be easily foreseen that, in the near future, time- and space-resolved management practices will become mainstream, ushering in the era of precision fruit growing (PFG). The timing and intensity of irrigation, fertigation and so on will vary across the orchard, during the day and along the season, to boost tree performance while reducing losses of unused water, fertilizer, labour and so on. While some data also indicates a positive carbon footprint for peach, in line with that of apple (*Malus domestica*) (Scandellari et al., 2016), improving the efficient use of water, fertilizers and other resources will greatly add to the sustainability and profitability of peach orchards. This is an exciting time for these studies, as scientific and technological advances are opening new fields of research in this endeavour.

http://dx.doi.org/10.19103/AS.2018.0040.16

1.1 Sustainability and resilience

Sustainable fruit production involves economic, environmental and social aspects. It is widely accepted that yield or gross margin alone can no longer be the metric used to evaluate horticultural practices. Assessment of carbon footprint, energy use, respect of biodiversity and life cycle analysis are commonly carried out in studies addressing production systems and their impact on the environment. Separation of economic and environmental sustainability is undesirable, as economic activity inevitably affects the environment and those impacts are currently paid by society at large. These environmental impacts raise consumer concerns over standards of production (i.e. food safety) and process standards (i.e. low carbon and water footprints in production system). The rising interest in good farm practices (e.g. GlobalGAP) is a response to this societal concern for process standards. As fruit production is embedded in a wider agricultural context, it must play its part in ensuring that management practices do not subtract from, but rather add to, resource preservation and land stewardship.

1.2 Climate change

Increased constraints for horticultural production are expected because of climate change, as higher average daily temperature and/or less precipitation throughout the growing season have the potential to increase the degree of stress to which trees are subjected, thereby reducing their efficiency and yields. Higher temperatures threaten productivity via increased plant organ respiration and reduced plant water use efficiency (WUE), although the effects on fruit quality may be more difficult to predict. Higher evapotranspiration and reduced precipitation can be compensated through irrigation but current water management approaches, based on maintaining/restoring hydrological conditions in the orchard, may need to be revisited. As societal demands to reduce water consumption are already present in many agricultural areas of the world, and will likely increase in the future, a drastic decline in irrigation water consumption is needed. Therefore, the implementation of effective strategies to improve farm resilience is necessary to preserve/increase the quality of production and yields.

1.3 Precision fruit growing

In this chapter, the expression 'precision fruit growing' (PFG) is used in place of 'precision horticulture' to refer to practices related to orchard management. We propose this distinction in light of the specific characteristics of fruit trees that make them quite different from broad-acre crops such as vegetables, for example tomato (*Solanum lycopersicum*). Specific issues typical of tree fruit production include three-dimensional canopies, uneven distribution of fruit within the canopy itself and irregularity of tree structure and size within the orchard. Sustainable orchards and mitigation of climate change can go together in a context of precise crop management by gauging tree needs before inputs of energy, water, nutrients and labour are delivered, quality and productivity can be optimized, while waste and inefficient use of resources can be avoided. Precision fruit management is a distinct interdisciplinary research area where diverse knowledge and expertise converge, including tree physiology, horticulture, engineering, modelling and artificial intelligence, to name a few. The technical challenge of collecting data is, however, still daunting (as common methodologies/sensors are currently somewhat lacking), and it is of great importance that these efforts be corroborated by the knowledge that the data being collected are 'horticulturally important'.

At present, orchard management is on the verge of a revolution, where robust mechanistic models of fruit/tree/orchard performance are coming together with technologies able to measure very accurately key physiological parameters that provide insight on the potential of the fruit to grow to its highest quality while maintaining good yields. This chapter thus considers key aspects related to tree physiological efficiency and productive performance and addresses their potential to be included in PFG approaches. Examples of potential integration of this diverse knowledge into operational scenarios will then be provided.

2 Peach fruit growth and vascular flows

Fruit development is a complex mechanism involving biochemical and biophysical processes leading to the seasonal increase in fruit fresh and dry weight. Water, carbohydrates and mineral elements accumulate in the fruit via phloem and xylem inflows. Fruit xylem flow depends on both the water potential gradients ($\Delta\Psi$) and the hydraulic conductance (K) of the xylem-to-fruit pathway. On the other hand, phloem unloading relies on symplasmic or apoplasmic mechanisms, depending on whether carbohydrate transport to the fruit occurs via the symplast, due to turgor pressure and/or concentration gradients, or via the apoplast, due to specific carbohydrate transporters (Patrick, 1997). However, fruit can also lose water via epidermal transpiration and/or xylem backflows (Lang, 1990). Fruit transpiration depends on both the anatomical features of the fruit epidermis (fruit surface conductance – g_c) and the environmental conditions (i.e. vapour pressure deficit – VPD) (Jones and Higgs, 1982). In addition, xylem backflows can occur when leaves transpire water at rates so high that stem water potential drops to a value more negative than that of the fruit. Fruit water losses (by either transpiration or xylem backflows) decrease fruit turgor and potentially increase stem-to-fruit water potential gradients (ΔΨ), with a positive impact on the fruit capacity to attract xylem and phloem flows (Morandi et al., 2010). Therefore, fruit vascular flows are highly related to tree performance (i.e. carbon assimilation, water relations), environmental conditions (i.e. VPD) and fruit anatomical features (i.e. xylem K and epidermis g_c).

2.1 Peach fruit growth during the season

Peach fruit growth is described by a double sigmoid model which divides development into three different stages: an initial phase of exponential growth (stage I), when cytokinesis occurs; a second phase of pit hardening, when fruit diameter increase significantly slows (stage II); and a final exponential phase, when fruit growth is by cell expansion (stage III) (Weinberger, 1941).

Throughout fruit development, sorbitol is the most translocated assimilate in peach phloem, but it maintains very low concentrations (about 4%) in the fruit flesh (Génard and Souty, 1996). At stage I, fructose and glucose are the predominant sugars in the fruit tissue, although their concentrations decrease from pit hardening to harvest (Moriguchi et al., 1990; Vizzotto et al., 1996; Lo Bianco and Rieger, 2002). Sucrose is the main form of carbohydrate accumulated during the last stages of fruit development, rising rapidly at the onset of stage III and reaching values around 70% of total sugar content.

As fruits develop, growth is characterized by an increasing amount of water exchanges from tree to fruit, and from fruit to atmosphere. This is due to increasing epidermal permeability and xylem function (Morandi et al., 2007a). In fact, following the formation of microcracks in the fruit epidermis and the increase in fruit total surface, water losses by transpiration rise as the fruit grow (Lescourret et al., 2001). Fruit water losses induce a decrease in fruit water potential, via an increase in the osmotic concentration and a decrease in the turgor pressure (McFadyen et al., 1996). In this physiological model, as more water is lost by a fruit, more water can be drawn from the phloem (which contains osmotically active carbohydrates) and subsequently from the xylem streams into the fruit itself as sketched in Fig. 1. Nectarine and peach show a similar behaviour in terms of transpiration and vascular flows (Morandi et al., 2007a): xylem and phloem account for about 70% and 30% of fruit total inflows, respectively, while more than 55% of daily water import is lost by transpiration at both stage I and stage III. For example, a 100 g peach at stage III receives about 8 g of sap daily by xylem and phloem flows but half is lost by transpiration (Morandi et al., 2007a).

2.2 Peach fruit growth during the day

The availability of automatic fruit growth gauges (Fig. 2) to continuously monitor fruit diameter variation during the day (Morandi et al., 2007b) has

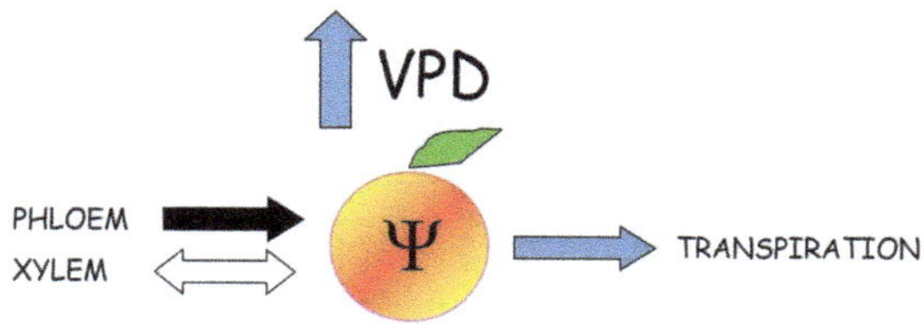

Figure 1 Fruit growth as a biophysical balance between fruit inflows (phloem and xylem) and outflows (fruit transpiration and xylem backflow). As VPD increases the fruit loses water via transpiration and its water potential decreases. Lower fruit water potentials increase the force with which the fruit attracts water and carbohydrates through phloem and xylem inflows.

facilitated study of fruit growth mechanisms on a short timescale (daily) basis. Such studies have shown how, unlike other species, peach fruit exhibit similar daily growth dynamics between their initial and last stages of development (Morandi et al., 2007a). At sunrise all vascular and transpiration flows to/from the fruit are usually in equilibrium after the overnight rehydration of fruit tissue and the relatively low VPDs typical of this time of day (Fig. 3). Afterwards, as temperature increases, fruits start to lose water by transpiration, but these losses are not balanced by xylem inflows to the fruit, which are low (probably due to the high amount of water directed to transpiring leaves, which reach lower water potentials) (McFadyen et al., 1996). The negative balance between inflows and transpiration losses causes the typical midday shrinkage of peach fruit (Morandi et al., 2007a) and decreases their turgor pressure (McFadyen et al., 1996). This may facilitate translocation and bulk flow phloem unloading into the fruit tissues (Patrick, 1997), linking fruit transpiration to phloem import. Subsequently, following the typical decrease in leaf stomatal conductance that occurs in the afternoon, leaf and stem water potential increase, and this causes the stem-to-fruit water potential gradient to increase. Xylem flow to the fruit

Figure 2 Custom-built fruit gauges for accurate and continuous monitoring of peach fruit growth.

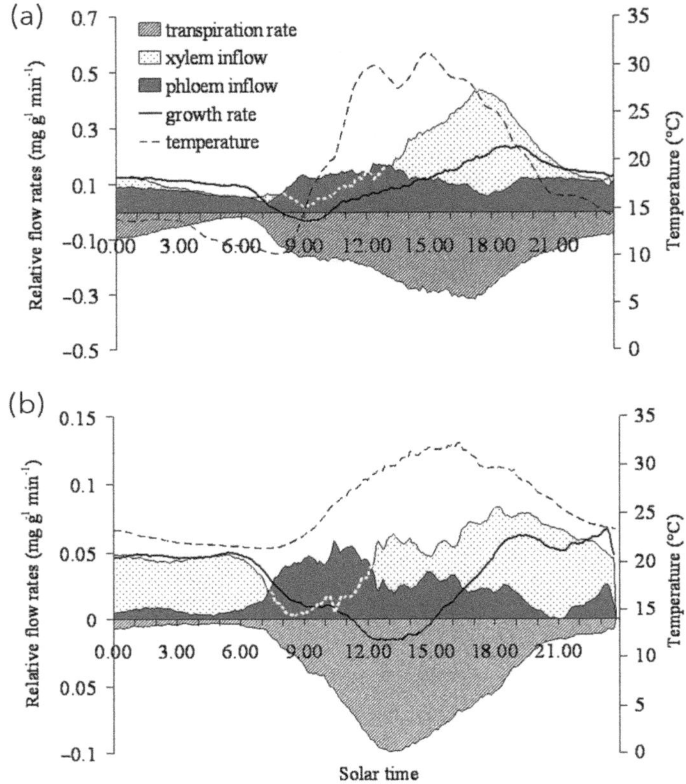

Figure 3 Diurnal courses of fruit relative growth rate (RGR), specific phloem, xylem, transpiration flow rates (mg g^{-1} min^{-1}) and temperature for 'Redhaven' peaches, during stage I (a) and stage III (b) of fruit development. Maximum SE at stage I were 0.051, 0.046, 0.056 and 0.070 for RGR, phloem and xylem inflows, and transpiration rates respectively, while at stage III they were 0.011, 0.025, 0.023 and 0.020 for RGR, phloem and xylem inflows, and transpiration rates, respectively.

responds to this gradient and increases during the afternoon and night, first fully rehydrating the fruit and then causing an increase in volume (Fig. 3).

2.3 Factors influencing peach fruit growth and vascular flows

Proper functioning of the peach growth mechanism depends on resource availability, environmental conditions and genetic background, and is fundamental to peach fruit development. Variations in both orchard management and environmental factors, therefore, may deeply and readily affect the mechanisms described so far and, consequently, orchard productivity.

2.3.1 The orchard microclimate

Results from several trials show that, absent water limitations, the high transpiration rate that characterizes peach and nectarine fruit (Lescourret et al., 2001) is a positive feature for fruit growth. High transpiration decreases fruit water potential and enhances the strength of import phloem flow with consequently positive effects on fruit fresh and dry matter gain (Fig. 4).

For example, it has been found that peach fruit growth and final quality in terms of dry matter and soluble solids content can be increased by reflective mulches that positively affect whole tree photosynthesis (Costa et al., 2003) and increase fruit transpiration. This, in turn, improves their potential to import fresh matter passively from phloem and xylem tissues (Fig. 5) (Morandi et al., 2012). In these conditions, it is of utmost importance to guarantee appropriate water supply to peach trees as, in the presence of very high VPDs, transpiratory losses may exceed xylem import thus limiting daily fruit growth (Fig. 4). On the other hand, fruit bagging (to limit transpiration) resulted in significant reductions in fruit growth and dry matter accumulation (Li et al., 2002; Morandi et al., 2010). It can be concluded that all conditions that reduce light distribution and potential fruit transpiration within the canopy, such as shading nets and/or dense canopies resulting from inaccurate pruning, may cause negative effects on peach fruit growth and quality. These findings open a wide range of possibilities to optimize the orchard microclimate through specific management practices that increase VPD and light distribution within the orchard, such as mulching with reflective materials, appropriate pruning and training systems, photoselective nets and so on.

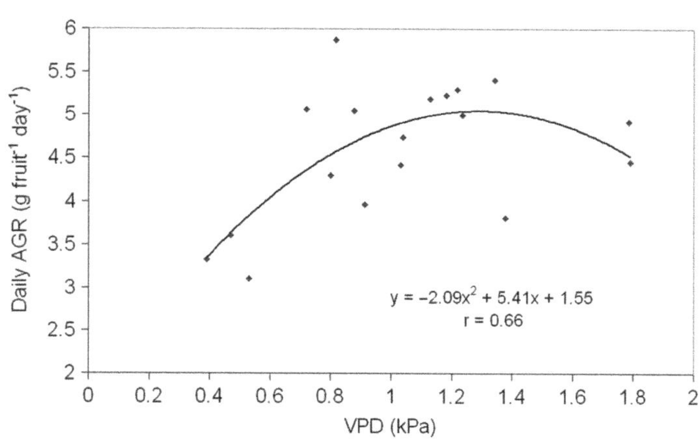

Figure 4 Relationship between air vapour pressure deficit (VPD) and daily absolute fruit growth rate (AGR). Peach fruit growth is directly related to VPD until an optimal environmental condition of 1.2 kPa. Higher VPDs may be too stressing for peach fruit growth.

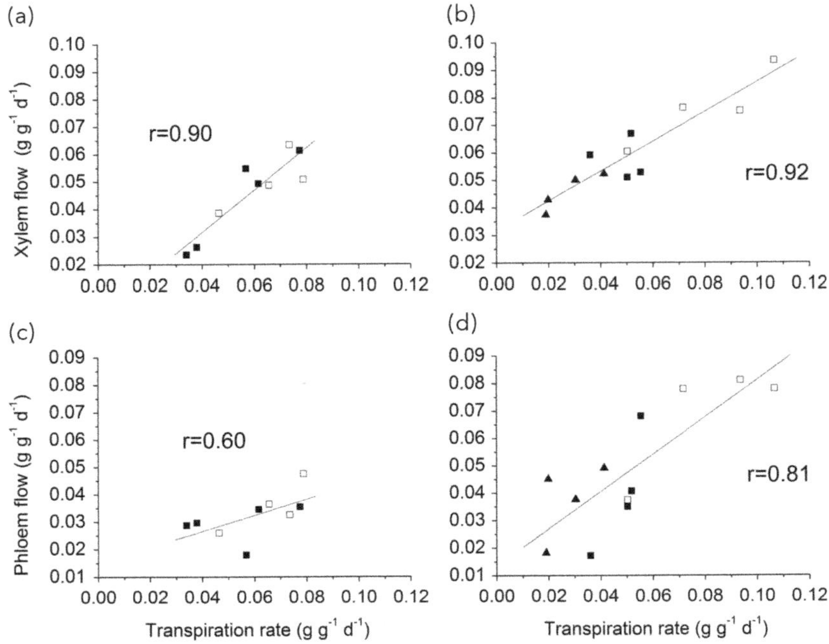

Figure 5 Relationships between specific transpiration rate (g g^{-1} day^{-1}) and xylem (a,b) and phloem (c,d) flows (g g^{-1} day^{-1}) in 'Red Gold' (a,c) and 'Alice Col' (b,d) nectarine fruit on Extenday® reflective mulches (open squares), in control conditions (closed squares) and under shading nets (closed triangles).

2.3.2 Water availability

Due to the high tree-to-atmosphere water exchanges typical of peach, this species is particularly sensitive to water stress. In fact, high water volumes are necessary to replenish the transpiratory losses of peach orchards. Although not many studies have been carried out on this subject, negative relationships have been found at harvest between stem water potential, fruit size and production (Naor et al., 1995). This can be explained by the fact that, as vascular flows respond to water potential gradients, the lower the stem water potential (due to water scarcity) the smaller will be the fruit-to-stem $\Delta\Psi$ with consequent reduction in phloem and xylem flows to the fruit. However, strategies to reduce orchard water use may risk compromising fruit quality: heavy shading, for example, has been shown to reduce fruit growth and fruit dry matter concentration due to a negative effect on fruit transpiration and on fruit passive phloem unloading (Morandi et al., 2012). Therefore, it is important to develop alternative strategies to face water scarcity challenge, such as, for example, the application of regulated deficit irrigation protocols, which are widely used in peach production.

2.3.3 Source/sink relationships

Crop load management represents one of the most important factors influencing peach fruit growth as it strongly affects fruit-to-fruit competition and resource availability in terms of water and dry matter (Pavel and DeJong, 1993; Grossman and DeJong, 1995).

Early in the season, peach fruit import via the phloem and xylem appears to be sink-limited throughout the day, with no or very slight differences in fruit growth velocity among trees with different crop loads. In contrast, during cell expansion, phloem flow and fruit growth rates are clearly subjected to source limitations due to fruit-to-fruit competition. However, Morandi and Corelli Grappadelli (2009) report how peach fruit in high cropping conditions seem able to 'sink strengthen' their phloem and xylem flows due to an increase in their specific transpiration rate. Being smaller, fruit on high crop load trees expose more skin surface per unit of fresh weight with consequent increases in specific transpiration, which brings their midday fruit water potential to lower values by loss of turgor. This strengthens the stem-to-fruit hydrostatic pressure gradients and passively drives both xylem and phloem fluxes to their tissue. These findings, which derive from simple biophysical considerations, suggest how sinks developed under conditions of high competition may be more 'active' in attracting water and assimilates, than fruit on low-cropping trees. However, the latter are characterized by a higher carbon gain in absolute terms and by a higher water balance between xylem gains and transpiration losses. The combination of these factors determines the higher growth rates typical of fruit on low-cropping trees.

From this discussion, several points can be noted: (1) to ensure proper fruit growth, peaches must spend part of the day under moderately high VPD, as this boosts their capacity to function as passive phloem downloaders, increasing their dry matter content which, later in the day, functions as a driver of water imports through the xylem; (2) higher VPDs result in higher water requirements to balance the increase in orchard evapotranspiration; therefore, if reflective mulches were applied to increase fruit growth, the water supply should be increased to balance fruit water losses by transpiration; (3) techniques to save irrigation water can be devised for peaches, but they must not focus on the reduction of VPD; and (4) the sink strength of the fruit might be increased by subjecting the tree to 'micro' water stresses during the warmer part of the day, with positive effects on fruit performance.

3 Photosynthesis: the engine of productivity

Photosynthesis drives tree productivity; it is dependent on abiotic and biotic factors such as light intensity, water availability, temperature, nutrition, pest damage and the partitioning of resources within the plant. A deep understanding of this fundamental process is essential to boost sustainable fruit production; from this knowledge innovative management practices can be derived to maintain or improve yields of high-quality fruit, even under reduced availability of resources such as water and fertilizers.

Several reviews on tree photosynthesis have been published and a specific chapter on photosynthesis in peach has been written by DeJong and Moing (2008). In this chapter the photosynthetic process is described briefly with the goal of understanding or assessing new strategies able to improve peach production efficiency.

3.1 The photosynthetic process

Photosynthesis occurs in the leaf chloroplasts. While general plant physiology textbooks provide a detailed description of the events involved in photosynthesis, a brief recap of the main elements is provided here to facilitate the discussion of tree-light relationships.

When visible light is absorbed by a leaf, the chlorophyll molecules forming the light harvesting complexes (LHC) of photosystems II (PSII) and I (PSI) transition to a high-energy and unstable state called *singlet state*. Neighbouring chlorophyll molecules in the LHCs exchange this vibrational energy between one another until the excitation reaches the chlorophyll molecules of the core complexes of PSII and PSI (P680 and P700, respectively). These two molecules return to the *fundamental state* by first losing one electron (forming $P680^+$ and $P700^+$). This event initiates the electron transport chain. The $P680^+$ in the PSII core complex is one of the most powerful organic oxidizing molecules, able to oxidize water to H^+ and O_2, thus returning to its *fundamental state*. The electron lost by PSI ($P700^+$) reduces ferredoxin and then the electron carrier NADPH. $P700^+$ returns to the *fundamental state* in a non-ionic (P700) form by receiving the electron lost by $P680^+$. The photons intercepted by the photosystems are thus converted into reducing power (NADPH[H^+]) and biochemical energy (ATP). These products fuel the activity of the key enzyme of photosynthesis, ribulose bis-phosphate carboxylase-oxygenase (RuBisCO).

RuBisCO is an unusual enzyme in that it is able to catalyse both carboxylation (photosynthesis) and decarboxylation (photorespiration). Water and/or heat stresses predispose RuBisCO activity towards photorespiration, as these two reactions are in competition and the prevalence of one or the other is dependent on temperature and CO_2 and O_2 concentrations at the stomata level. Water stress that causes stomatal closure will favour photorespiration, since less CO_2 and more O_2 will be present in the mesophyll, and the leaf will tend to become warmer due to reduced transpiration. In peach, the optimum temperature for photosynthesis is 27–30°C, while photorespiration increases with temperature (Crews et al., 1975). These biochemical pathways co-exist in

C3 plants, and there is also a continuous release of CO_2 due to cellular dark respiration. In eco physiological studies, the assessment of leaf gas exchanges is commonly based on infrared gas analysis which measures the concentrations of CO_2 and H_2O in the airstream entering and leaving the assimilation chamber of the instrument. Therefore, what is measured and reported is *net photosynthesis* (P_n, µmol m^{-2} s^{-1}), that is the difference between the CO_2 sequestered by gross photosynthesis and the CO_2 released by respiration and photorespiration. P_n is the amount of carbon that potentially can increase the dry matter amount within the tree, including fruit.

3.2 Light

Light drives photosynthesis, in particular photons with wavelengths of 400 nm and 700 nm (blue and red, respectively). Net photosynthesis can be described in terms of a few functional variables. The *saturation point* is the light intensity above which increasing light no longer increases P_n. This intensity for peach is around 900–1200 µmol m^{-2} s^{-1} (Rosati et al., 1999; Losciale et al., 2010). At this irradiance, P_n can reach 20–23 µmol m^{-2} s^{-1} of CO_2, according to the cultivar and the pedo-climatic conditions. The *compensation point* is the irradiance at which the gain of gross photosynthesis balances the losses by mitochondrial respiration and photorespiration, and P_n equals zero. As with the saturation point, the compensation point also depends on cultivar and pedo-climatic conditions and can vary from 20 to 70 µmol m^{-2} s^{-1}. The light environment at which a leaf develops also can affect these two key points. A leaf developed in a high light environment will have higher saturation and compensation points than a shade leaf. A shade leaf, in turn, will be more efficient at low light, quickly reaching its saturation point but with a maximum photosynthetic activity lower than light leaves (Sams and Flore, 1983).

Since tree canopies are comprised of a population of leaves, it would be interesting to know the relationship between irradiance and photosynthesis at the whole canopy level. The energetic basis of orchard productivity lies in the interaction between the tree and the sunlight. The amount of dry matter produced by a tree is linearly related to the amount of light it intercepts (Monteith, 1977; Lakso, 1994). This concept guided the development of intensive plantation systems that aim to achieve high light interception early in the life of the orchard while maintaining good light distribution, thereby improving productivity (Jackson and Palmer, 1972; Jackson et al., 1971; Palmer, 1980). Orchard light interception is related to two main factors: light intensity and planting density. Theoretically, the same orchard light interception could be obtained by decreasing the number of plants and increasing the irradiance, with an expected change in orchard yield. Several authors found a direct relation between planting density and fruit yield (t ha^{-1}) as a consequence of increased leaf area and orchard light interception (Caruso et al., 2000; Nuzzo et al., 2002; Guerriero et al., 1980; Loreti et al., 1989; Hutton et al., 1987; Chalmers and van den Ende, 1989). However, a direct relation between light intensity and productivity has not always been found. 'Ross' peach orchard yields did not differ when trained as cordon or Kearney Agricultural Centre Perpendicular V (KAC-V) canopies at 1196 trees ha^{-1}; nevertheless, the cordon training system intercepted more light than the KAC-V (Grossman and DeJong, 1998). At the single tree level, a similar yield efficiency (expressed as fruit dry weight per trunk diameter) was found for the open vase (299 trees ha^{-1}), cordon and KAC-V training systems, even though the open vase intercepted less light than the KAC-V and the cordon systems, respectively (Grossman and DeJong, 1998).

Whole tree canopy photosynthesis, in 'Valley Red' genetically dwarf peach, didn't reach a steady state when irradiance increased and more light reached the interior leaves (Corelli Grappadelli et al., 1996). On the other hand, Giuliani et al. (1998) and Losciale et al. (2010) observed, on two different peach cultivars and experimental set up, that the maximum whole canopy net photosynthesis was achieved when irradiance was moderate and not at its maximum level (Fig. 6). Excessive PAR which is absorbed by the leaf but cannot be used for photosynthesis is mainly dissipated by photoprotective mechanisms. However, a part of these excess photons still drives

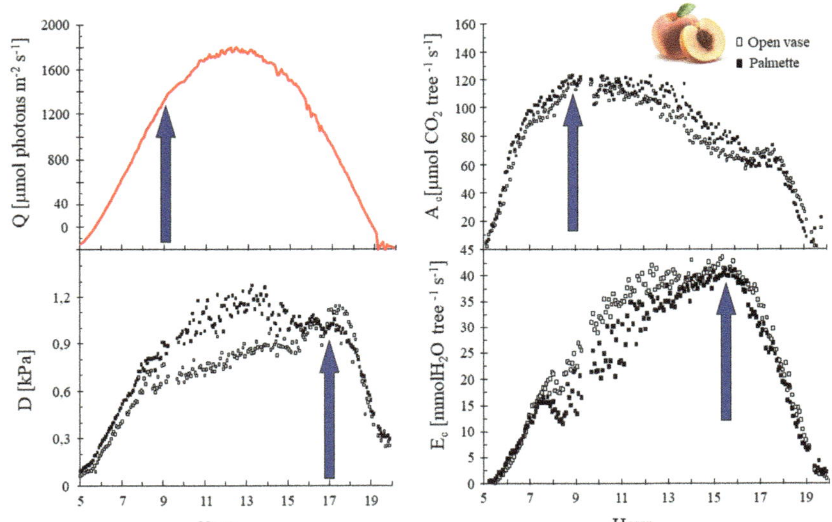

Figure 6 Daily pattern of incoming light (Q, top left), vapour pressure deficit (D, bottom left), net photosynthesis (A_c, top right) and transpiration (E_c, bottom right) measured on the whole canopy of trees trained as open vase (open squares) and palmette (closed squares). Arrows highlight the time of the day when the maximum net photosynthesis and transpiration is reached. Source: Adapted from Giuliani et al. (1998).

photoinhibition and reduces the activity of the photosynthetic apparatus. As a result, a fraction of the dry matter produced is used for photoprotection and to repair photodamage to PSII. In a standard orchard tree, during a clear summer day, almost the entire pool of PSII centres is destroyed and immediately repaired, at the expense of a considerable fraction of the dry matter produced (Losciale et al., 2010, 2011).

3.3 Light dependence of temperature

Absorbed light also drives the increase in temperature, thus photorespiration, VPD and consequently transpiration (Fig. 7). In peach, this behaviour can be observed at both the single leaf and whole canopy levels (Fig. 6). These findings prompted research into ways to reduce incoming light and to

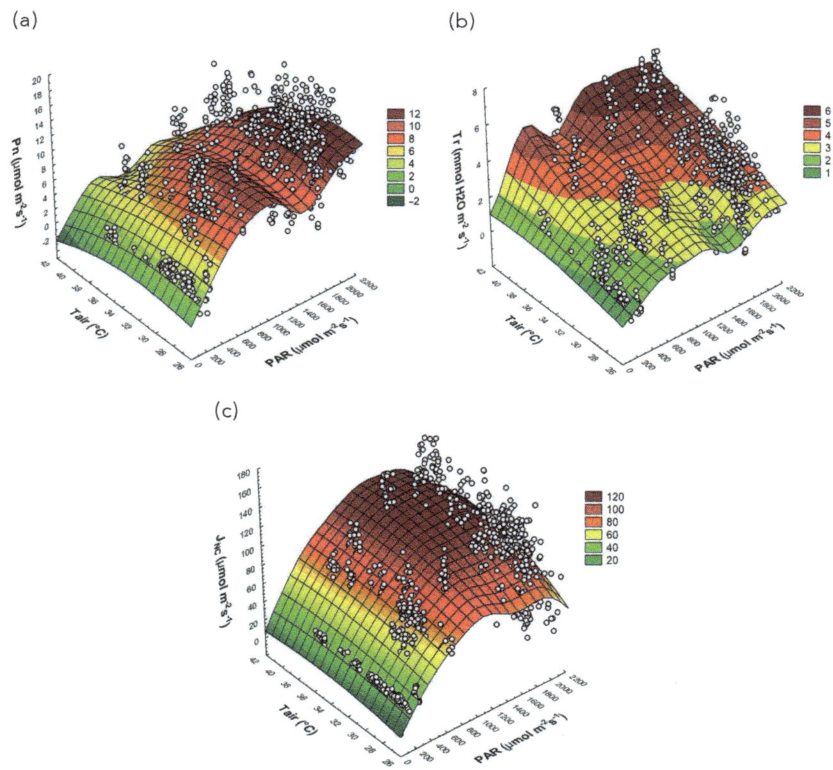

Figure 7 Effect of intercepted light and air temperature (T_{air}) on net photosynthesis (a), transpiration (b) and the electron transport rate of the non-net carboxylative transports (JNC), that is the alternative electron transports and photorespiration (c), measured at single leaf level on peach cv. RedCal. The surface shape and ranking has been obtained using the distance-weighted least squares fitting procedure. Source: Adapted from Losciale (*unpublished*).

increase uniformity of its distribution within the canopy. These objectives can be reached in different ways. The use of reflective kaolin particle film reduced absorbed light and temperature, improving plant performance (Jifon and Syvertsen, 2003; Maletsika and Nanos, 2015). Light interception can be modulated by means of an appropriate training system (Grossman and DeJong, 1998). Another approach is to design the orchard to manage the daily canopy light interception profile. At the University of Bologna, a trial was established with tree rows planted with an orientation of 330°–150° and a canopy inclination of 35° from vertical towards East. With this row configuration, trees exhibited a more constant photosynthetic response during the day (Losciale et al., 2010) due to a more efficient interplay among light intensity, time of day and portions of the canopy that were exposed to incoming direct light.

Another way to reduce the incoming light intensity while increasing the diffuse light within the canopy is the use of shade nets above, and/or reflective materials below the tree canopy. On the columnar peach cultivar 'Alice Col', reducing incoming light by 40% did not reduce net photosynthesis but decreased air temperature and transpiration (Losciale et al., 2011) with a

Table 1 Irradiance (*PPFD*, μmol m^{-2} s^{-1}), leaf net photosynthesis (P_n, μmol m^{-2} s^{-1}), stomatal conductance (g_s, mol m^{-2} s^{-1}), leaf transpiration (*Tr*, mmol m^{-2} s^{-1}), air temperature (T_{air}, °C) and water use efficiency (*WUE*, μmol mmol^{-1}) measured on peach leaves facing either East or West acclimated to full light (CTRL) and 40% light reduction (SHD)

Time	Trt	*PPFD* (μmol m^{-2} s^{-1})	P_n (μmol m^{-2} s^{-1})	g_s (mol m^{-2} s^{-1})	*Tr* (mmol m^{-2} s^{-1})	T_{air} (°C)	*WUE* (μmol mmol^{-1})
East side							
10.00	CTRL	1801	18.43	0.33	6.65	31.1	2.77
	SHD	1049*	18.38	0.25*	5.98*	30.5*	3.07*
13.30	CTRL	1599	10.20	0.17	3.80	30.8	2.69
	SHD	850*	9.23	0.11*	2.74*	29.2*	3.44*
16.30	CTRL	201	3.14	0.07	1.94	28.2	1.74
	SHD	100*	2.43	0.08	1.79	28.2	1.48
West side							
10.00	CTRL	151	5.65	0.13	3.15	26.7	1.87
	SHD	101*	3.87	0.11	2.58	27.5*	1.85
13.30	CTRL	1450	13.02	0.19	5.49	32.8	2.35
	SHD	901*	15.97	0.22	6.64	32.1	2.41
16.30	CTRL	1650	11.89	0.23	5.54	32.8	2.14
	SHD	801*	9.32	0.13*	3.15*	32.1*	2.94*

* Mean separation by *t*-test, $P = 0.05$.
Source: Adapted from Losciale et al. (2011b).

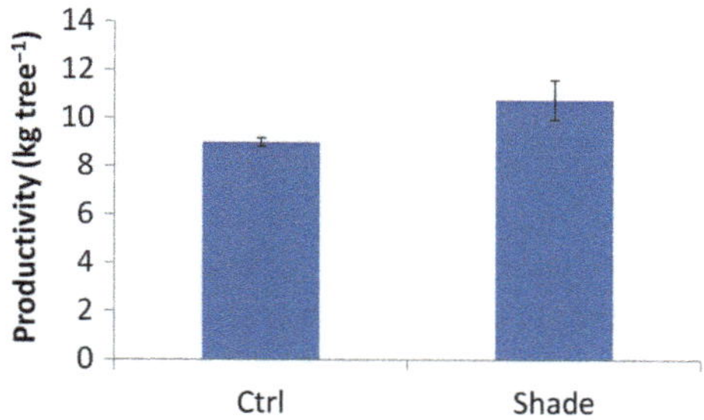

Figure 8 Productivity of 'Alice Col' nectarine (kg $tree^{-1}$) in full light and under shading nets (s.e.: n = 3). Trees were spaced 1.2 × 5 m, at a density of 1666 tree ha^{-1}.

positive effect on WUE (Table 1) and productivity (Fig. 8). Photoprotection and photoinhibition also were reduced and more dry matter was produced. While this extra carbon was potentially available for fruit, whether it can be transported into the fruit remains to be seen, given the biophysical mechanism of peach fruit growth that relies on moderately high VPD to drive fruit transpiration and the subsequent inflows from phloem and xylem. Other authors have demonstrated that the use of reflective mulch materials on the orchard floor can increase the diffuse light fraction through the canopy, thus allowing inner leaves to increase their photosynthetic activity, and to increase production of good quality fruit (Costa et al., 2003; Schmidt et al., 2014).

Shade nets could modulate the incoming light level to the saturation point, but this might cause excessive shading of inner canopy leaves. Heavily shaded leaves act as carbon sinks rather than sources; even worse, they would continue to transpire without any or little gain in carbon assimilation. On top of this, the negative effects of shading on fruit growth that have been discussed above indicate that moderate shading, possibly associated with summer pruning (that also reduces the transpirational surface), is an innovative and sustainable practice in environments with high VPD and irradiance, such as the Mediterranean basin, California, South Africa and much of Australia.

Temperature affects photosynthesis and dry matter accumulation in several ways. First, the carboxylating/oxygenating activity of RuBisCO is dependent on temperature; above an optimal threshold for carboxylation high temperatures enhance photorespiration and dry matter consumption. Second, high temperature, when associated with a high VPD, exposes plants to high rates of transpiration. If the rate of transpiration is equalled by the rate of water uptake then the consequence is only the high water loss and the related reduction of WUE. Under water scarcity and/or with a very high VPD (i.e. >4.0 kPa), plants counter water losses by closing the stomata, thus reducing the internal CO_2 concentration and P_n and, again, increasing leaf temperature and photorespiration.

3.4 Water

In addition to light, water is the main factor affecting plant productivity. As previously explained, water stress reduces net photosynthesis via stomatal closure (stomatal limitation) and via the increase of photorespiration and photoinhibition (non-stomatal limitation). Orchard water management is not easy to improve and become more sustainable due to a lack of improved methods to detect plant water status in a reliable and easy way. In addition, deeper knowledge about the relationship between water use and carbon assimilation is needed to be able to implement sustainable irrigation strategies.

The most widely accepted method to assess tree water status is via the measurement of the midday stem water potential Ψ_s (Shackel et al., 1997; Naor, 2006). Ψ_s values higher than −2.0 MPa (e.g. −1.2 or −1.5 MPa) are considered acceptable for peach productivity (Girona et al., 1993). However, it is still unclear whether peach leaves, when subjected to moderate water stress, close their stomata to reduce water losses and prevent embolism by maintaining Ψ_s values at a safety threshold. This behaviour would prevent embolism but reduce carbon assimilation (Gollan et al., 1985; Socías et al., 1997). Flexas and Medrano (2002) proposed leaf stomatal conductance (g_s) as the best indicator for detecting and monitoring plant water status. As recorded for apple and pear (*Pyrus communis*) (Losciale et al., 2015; Massonet et al., 2007; Galmés et al., 2005), the relation between P_n and g_s is not always linear, and above a saturation point any increase of g_s does not generate a further gain of P_n (Cheng et al., 1996), but may cause excessive water consumption (Fig. 9). Determining stomatal conductance would allow the evaluation of whether the tree is under water stress, but it would not provide information about possible water overuse.

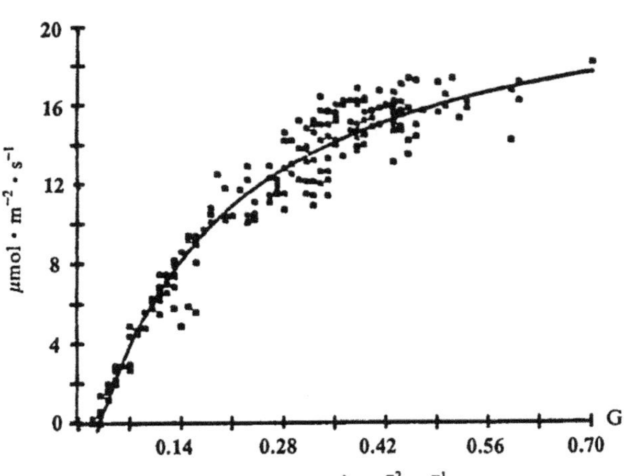

Figure 9 Relation between stomatal conductance and net photosynthesis of leaves of 'Okubo' peach. Source: Adapted from Cheng et al. (1996).

Since water status affects carbon assimilation, another recent approach is the direct detection of net photosynthesis by measuring chlorophyll fluorescence and other derived indices (Losciale et al., 2015). During high irradiance and temperature on summer days, CO_2 assimilation reaches a saturation point when exposure of tree canopies to incoming light is no longer limiting (around 09:00 am) and is maintained at this level until midday. Transpiration from the canopy continues to increase as the irradiance and VPD increase, causing stomatal opening and water loss (Jarvis, 1976). A moderate water restriction during this part of the day could trigger stomatal closure at levels not limiting for P_n but able to reduce excessive water consumption. In the afternoon, the relation between carbon assimilation and transpiration is quite linear and transpiration rate is strictly linked to VPD (reaching its maximum in the early afternoon). Under restricted water availability, peach trees limit water losses by closing their stomata and reducing carbon assimilation (Fig. 10). It is reasonable to hypothesize that while water restriction can reduce CO_2 assimilation in the afternoon, it does not affect photosynthesis in the morning, so a time-resolved precise irrigation would be desirable. If properly timed to assist the fruit in reaching more negative osmotic potentials, this might be an added strategy to improve fruit quality by facilitating fruit growth.

Sustainability of water management in peach cultivation can be increased by maximizing plant carbon assimilation and avoiding any excessive light absorption and evapotranspiration. The accurate and reliable detection of plant water status and canopy functionality can become the driver of water supply management to maximize assimilation, while minimizing transpiration. Reduction of excessive evapotranspiration also can be reached by moderately reducing incoming light (as explained in the previous section), as well as decreasing soil evaporation, using artificial (Wang et al., 2015; Yaghi et al., 2013) or natural (Diacono et al., 2016) mulching materials.

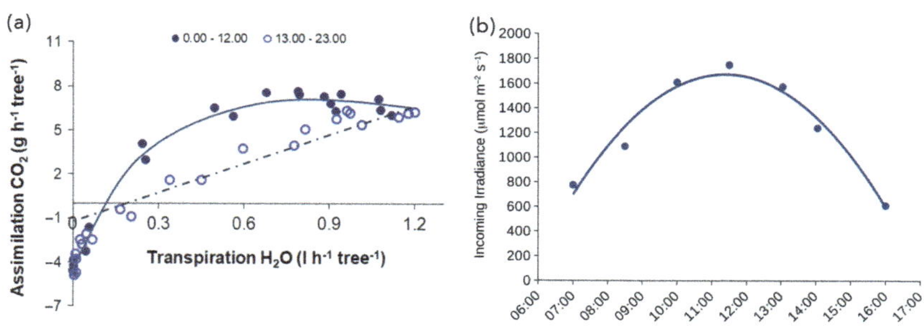

Figure 10 (a) Relation between whole tree CO_2 assimilation and transpiration measured on trees of 'Alice Col' nectarine, during a clear summer day, measured in the first hours of the day (0:00–12:00, blue dots) and in the second part of the day (13:00–23:00, blue circles). (b) Panel illustrates the daily profile of incoming irradiance. Source: Adapted from Losciale (*unpublished*).

The above discussion points out several important features: (1) it is necessary to decrease the absorption of excessive direct light and to increase its distribution within the canopy; (2) there are physiological parameters that can be monitored in real time in orchards, providing information on the 'functional' status of the tree, and these can be used to drive irrigation or other agro-practices; and (3) it may be possible to improve productivity by specific timing of water applications during the day, with the synergistic goals of supporting P_n as well as improving fruit sink strength.

4 Precision fruit growing applications

Sustainable production systems will greatly benefit from the implementation of precision orchard management. Zude-Sasse et al. (2016) pointed out how precision approaches integrated into orchard practice – including in peach cultivation – would significantly improve orchard sustainability. Recent advances in sensor technologies, such as yield monitors (Rahnemoonfar and Sheppard, 2017), fruit growth (Figs. 2 and 11) (Morandi et al., 2007b; Seifert et al., 2015) and canopy sensors (Zarco-Tejeda et al., 2009; Gago et al., 2015), and non-destructive fruit quality testers (Ziosi et al., 2008; Seifert et al., 2015; Guo et al., 2016; Zhang et al., 2017) provide many different data sets to help derive management strategies on the basis of within-orchard variability. Fruit growers often already have basic information on crop production or its quality, but they do not apply it either spatially or temporally to crop management strategies (Taylor et al., 2007; Manfrini et al., 2009; Torres-Ruiz et al., 2016). Examples of basic peach orchard information easily recordable through the growing season include fruit diameter and crop load (Morandi et al., 2005; Manfrini et al., 2012).

Figure 11 Example of a sensor embedded on a fruit, which detects the levels of chlorophyll and anthocyanin pigments during ripening. Data are fed to an algorithm that computes developmental stage (Seifert et al., 2015). Image from https://www.youtube.com/watch?v=2eogq_x_xlA.

From these data, algorithms can be derived to predict fruit size, class distribution and yield from 30 to 20 days before harvest (Morandi et al., 2005).

Real-time monitoring of fruit development to assess its potential to reach a satisfactory size at harvest allows the grower to fine-tune tree management and maximize yields. Since 1990, Monestiez et al. (1990) proposed a within-tree geostatistical approach to assess spatial dependence among fruit, to choose the most appropriate sampling design for calculating tree crop load. Simple- and multi-level systematic sampling, two of the geostatistical designs used most to perform crop load estimation, have been studied in apple and represent an interesting option to estimate the number of fruit to forecast yields in peach orchards (Wulfsohn et al., 2012; Manfrini et al., 2015). More recently, a geostatistical analysis used NDVI (Normalized Difference Vegetation Index) aerial images to define where, within the orchard, samples should be collected. This approach allowed better delimitation of sectors for sampling the crop load within nectarine orchards (Miranda et al., 2015), although it did not appreciably reduce sample size compared to a simpler random sampling.

Looking for alternative solutions, Arnó et al. (2017) studied how multispectral airborne imagery and soil apparent electrical conductivity (ECa) maps can be used as ancillary information to detect spatial variability, with the goal of increasing sampling efficiency in peach orchards. They concluded that NDVI images or ECa surveys could be used to stratify data layers. Currently, however, the best results for determining fruit load and estimating yield early in the season, when fruit are small and green (i.e. difficult to see), have been obtained from manual, direct counting of fruit or flowers (Maldonado and Barbosa, 2016). This practice has an estimated error of around 10% but it is time-consuming and expensive, and is not often suitable for large orchards. To overcome this issue, the new generation of automatic yield estimation based on robotic computer vision provides a viable solution. Recently, Rahnemoonfar and Sheppard (2017) used a trained network based on deep learning techniques to efficiently count fruit with 91% accuracy, which included fruit partly or totally shaded, occluded by foliage or branches, or partly overlapping other fruit.

4.1 Water use efficiency

Although information on crop development is certainly of major importance, preservation and better management of water resources play a key role in improving orchard sustainability, and is thus subject to a great deal of research. For example, the European Commission introduced the requirement to sustainably 'produce more with less' as one of the main themes of its 'Horizon 2020' Framework Program (Geoghegan-Quin, 2013). Characterizing within-orchard spatial heterogeneity in water requirements would assist in improving irrigation WUE and in conserving water. To date, the crop water stress index

(CWSI) has been used successfully as a crop water status indicator in several fruit tree species (Wang and Gartung, 2010; Gonzalez-Dugo et al., 2013). Bellvert et al. (2016) developed the CWSI in three *Prunus persica* cultivars at different phenological stages using canopy temperature measurements (Fig. 12). High-resolution thermal imagery was acquired from an airborne platform and related to leaf water potential (Ψ_l) readings taken throughout the season. This study indicated that CWSI is a feasible method to assess the spatial variability of tree water status in heterogeneous orchards, and could be used to derive Ψ_l maps throughout a complete growing season (Fig. 12).

Unmanned aerial vehicles (UAVs) present an exciting opportunity to monitor orchards with high spatial and temporal resolution; this type of remote sensing promises to be capable of improving water stress management in peach crops. Gago et al. (2015) presented a review on several types of UAVs carrying different remote sensors for peach and other species. Several reflectance indices, such as NDVI, TCARI/OSAVI (transformed chlorophyll absorption in reflectance index/optimized soil-adjusted vegetation index), PRInorm (photochemical reflectance index normalized by considering the chlorophyll content) and thermal imagery obtained from UAVs have shown positive correlations with water stress indicators such as (Ψ_l) and g_s. Chlorophyll fluorescence also could be a good indicator of plant photosynthesis and WUE under water stress. Losciale et al. (2015) and Kalaji et al. (2017) reported on new indices and data interpretation methods that might prove useful in

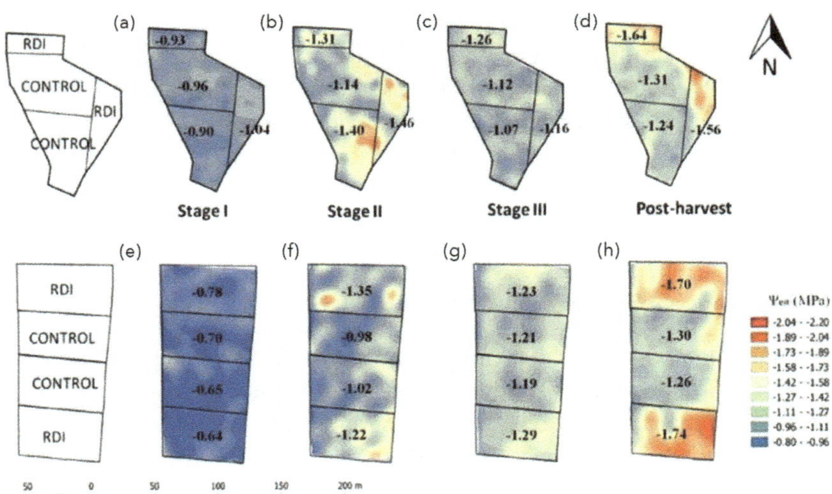

Figure 12 Maps of remotely sensed leaf water potential (Ψ_{est}) at different phenological stages for a 2 ha peach (a-d) and a 2.2 ha nectarine orchard (e-g). Irrigation treatments were full irrigation (CONTROL) and regulated deficit irrigation (RDI). Source: Adapted from Bellvert et al. (2016).

breeding programmes. For selection purposes, availability of high-throughput phenotyping platforms would be quite useful, particularly for application in the field. The authors have proposed novel approaches to detect water stress that require less time and increase resolution scales for the determination of important crop traits as yield or stress tolerance.

Another important area of improvement in sustainable fruit production is that of pesticide application, which is increasingly precise, not only in terms of timing of application relative to infection risk or damage thresholds, but also in terms of 'smart spray application'. The latter techniques are based on sprayers that can detect leaf health during application and turn nozzles on and off accordingly, all while measuring wind direction and intensity to minimize drift and maintaining optimum crop coverage. These sprayers also carry GPS sensors to recognize their proximity to sensitive targets, such as ponds, ditches, roads and so on. However, these are mainly developed to date for apple, and literature about use in peaches is scarce. Balsari et al. (2009) presented a thorough discussion on these sprayers.

In summary, (1) PFG is an emerging science that aims to produce knowledge to facilitate integrated, precise management of the orchard; (2) integrating fruit monitoring into forecasting models of growth is a powerful tool not only to improve fruit quality but also to improve irrigation WUE; (3) traditional plant status indices (e.g. NDVI) are not yet sufficiently accurate for fruit crops; and (4) specifically derived indices of canopy photosynthesis such as the I_{PL} (Losciale et al., 2015, 2017) appear promising, but need further validation.

5 Case study

Bringing together the various aspects discussed so far, and imagining a peach orchard designed for and managed in a context of PFG, the following would/could (in our opinion) be included:

a) Methods to modulate the light environment of the orchard (e.g. via shading hail nets) to achieve light intensities up to the saturation point in order to reduce evapotranspiration, photoinhibition and heat stress, which reduce yields. These covers should not drastically reduce VPD, as this needs to remain sufficiently high to favour fruit growth and dry matter accumulation. Reflective mulches could be coupled to shade hail nets to improve light diffusion and maintain appropriate VPD in the inner part of the canopy, while reducing soil evaporation.
b) Precise diurnal irrigation scheduling aimed at creating a moderate stress during the central part of the day, with the goal of enhancing fruit sink strength in the passive phloem unloading phases, without negatively impacting leaf gas exchange.

c) A decision support system to manage the orchard (irrigation, fertigation, crop load management) equipped with a set of low-cost and accurate plant-based and/or proximal sensors, mounted on self-steering robotic devices. Examples of plant-based sensors include fruit growth gauges (for accurate and automatic monitoring of fruit diameter variation), sap flow sensors, trunk dendrometers, leaf I_{PL} measurements and so on; proximal sensors may include multispectral and infrared cameras for the estimation of crop load, canopy temperature and structure.

d) Variable rate application (both spatially and temporally) of fertilizers, water and so on based on information derived from real-time plant status and aimed at optimizing fruit development.

The integration of the above factors will contribute to the improvement of both orchard production sustainability and crop quality. The collection of this information with 'big data' analytics and artificial intelligence (AI) approaches will in the medium term reduce the number/type of sensors used, as AI will indicate which parameters will be necessary for the system to achieve optimum orchard performance.

6 Conclusion and future trends

Improving the sustainability of peach production requires a high degree of interdisciplinarity as very diverse disciplines must come together to produce the necessary new knowledge. Plant science, physics, engineering, modelling and statistics are among the first that come to mind. One challenge is how to usefully integrate the contribution of each discipline. For example, if image analysis is developed in a laboratory environment, with static devices, it may appear relatively easy to provide horticulturally relevant information, such as fruit number per meter of row, and their size, with the necessary accuracy. However, even the best designed orchard will challenge such a system, as terrain, size and shape of the tree, and tree location in the orchard must be accounted for. A simple bump in the alleyway, tilting the robot might wreak havoc in its accuracy of measurement, as distance from the subject would be frequently changed, due to the uneven orchard floor.

Peach production systems must evolve to accommodate new technologies, while improving quality and yields, if they are to remain profitable. Doing more with less is a familiar theme to growers worldwide, and this, under conditions of climate change, will mean achieving profitability with less water and fewer fertilizer applications, yet with increased evapotranspiration.

PFG approaches founded on sound science will provide growers with a toolbox containing an array of new strategies to be adopted either alone or in synergy to raise their orchard performance to much higher limits. This toolbox will include an orchard design more suited to the application of precise management techniques/technologies (i.e. fruit thinning, spraying, pruning, harvesting etc.), including closer spacing of tree rows, to improve soil use and canopy light interception per ha, and cordon-based, inclined narrow canopies to better expose the fruit to optimal light levels and for easier counting by humans or machine vision/sensors.

These orchards, having more complex configurations, will require a very high level of professionalism to fine-tune management of the various inputs, not unlike a racing car, where small adjustments in management techniques will have profound effects on the outcome. Of course, there will be an equal potential for failure if these adjustments are inaccurate or incorrect! However, as the volume of information and precise management tools available is increasing almost daily, the possibility of adapting orchard management to ever-changing conditions, thus optimizing its performance is increasingly real.

7 Where to look for further information

The relations of plants and microclimate are best described in Jones (2014) and the reader is referred to this text for an in-depth discussion of ecophysiological aspects of tree growing. More in-depth discussions on photosynthesis and photoinhibition can be found in von Caemmerer (2000) and Long et al. (1994).

Fruit growth and its relations to the environment have been modelled by Allen et al. (2005) and Lescourret et al. (2011) using Functional Structural Plant Models (FSPM).

Zhang (2018) provides a reference textbook covering specific aspects of tree management and the possibility to automate them. The review by He and Schupp (2018) discusses automation and robotics in orchard management. There are a number of annual events on precision agriculture, which provide online access to the proceedings, in whose context one can find increasing reference to PFG, such as the International Congress on Precision Agriculture (ICPA) and the European Congress on Precision Agriculture (ECPA).

The readers are further referred to the following websites:

https://www.ispag.org/Proceedings
https://www.ispag.org/publications/ECPA_Proceedings

8 References

Allen, M. T., Prusinkiewicz, P., DeJong, T. M. (2005), 'Using L-systems for modeling source-sink interactions, architecture and physiology of growing trees: The L-PEACH model', *New Phytol.*, 166(3), 869–80.

Arnó, J., Martínez-Casasnovas, J. A., Uribeetxebarria, A., Escolà, A. and Rosell-Polo, J. R. (2017), 'Comparing efficiency of different sampling schemes to estimate yield and quality parameters in fruit orchards', *Adv. Anim. Biosci.*, 8, 471–6.

Balsari, P., Marucco, P. and Tamagnone, M. (2009), 'A crop identification system (CIS) to optimise pesticide applications in orchards', *J. Hortic. Sci. Biotechnol.*, 84(113), 113-16.

Bellvert, J., Marsal, J., Girona, J., Gonzalez-Dugo, V., Fereres, E., Ustin, S. L. and Zarco-Tejada, P. J. (2016), 'Airborne thermal imagery to detect the seasonal evolution of crop water status in peach, nectarine and Saturn peach orchards', *Remote Sens.*, 8, 1-17.

Caruso, T., Barone, E. and Di Vaio, C. (2000), 'Factors affecting tree crop efficiency in young peach trees: Rootstock vigour and training system', *Acta Hortic.*, 557, 193-7.

Chalmers, D. J. and van den Ende, B. (1989), 'Tatura trellis peaches: Productivity over fifteen years', *Acta Hortic.*, 254, 303-6.

Cheng, L., Cheng, S., Shu, H. and Luo, X. (1996), 'Effect of mild water stress in CO_2 assimilation and water use efficiency of field-grown peach trees', *Acta Hortic.*, 374, 121-5.

Corelli Grappadelli, L., Giovannini, D. and Ravaglia, R. (1996), 'Tipi di foglie e loro ruolo nella crescita dei frutti e nella produttività in peschi a diverso habitus vegetativo', *Proceedings of the 'Progetto Finalizzato Frutticoltura - Agro. Bio. Fruit'*, Cesena (Italy), 10-11 May 1996, pp. 148-9.

Costa, G., Corelli Grappadelli, L., Noferini, M. and Fiori, G. (2003), 'Use of light reflective mulch to affect yield and fruit quality', *Acta Hortic.*, 610, 139-44.

Crews, C. E., Williams, S. L. and Vines, H. M. (1975), 'Characteristics of photosynthesis in peach leaves', *Planta*, 126, 97-104.

DeJong, T. M. and Moing, A. (2008), 'Carbon assimilation and budget modelling', in: *The Peach: Botany, Production and Uses*, D. R. Layne, D. Bassi (Eds), CABI, Wallingford, UK, pp. 244-63.

Diacono, M., Fiore, A., Farina, R., Canali, S., Di Bene, C., Testani, E. and Montemurro, F. (2016), 'Combined agro-ecological strategies for adaptation of organic horticultural systems to climate change in Mediterranean environment', *Ital. J. Agron.*, 11(2), 85-91.

Flexas, J. and Medrano, H. (2002), 'Drought inhibition of photosynthesis in C3 plants: Stomatal and non-stomatal limitations revisited', *Ann. Bot.*, 89, 183-9.

Gago, J., Douthe, C., Coopman, R. E., Gallego, P. P., Ribas-Carbo, M., Flexas, J., Escalona, J. and Medrano, H. (2015), 'UAVs challenge to assess water stress for sustainable agriculture', *Agric. Water Manag.*, 153, 9-19.

Galmés, J., Flexas, J., Keys, A. J., Cifre, J., Mitchell, R. A.C., Madgwick, P. J., Haslam, R. P., Medrano, H. and Parry, M. A. J. (2005), 'RuBisCo specificity factor tends to be larger in plant species from drier habitats and in species with persistent leaves', *Plant Cell Environ.*, 28, 571-9.

Génard, M. and Souty, M. (1996), 'Modeling the peach sugar contents in relation to fruit growth', *J. Am. Soc. Hortic. Sci.*, 121(6), 1122-31.

Geoghegan-Quin, M., (2013), 'Role of Research & Innovation in Agriculture. European Commission-SPEECH/13/505', http://europa.eu/rapid/press-release SPEECH-13-505 en.htm

Girona, J., Mata, M., Goldhamer, D. A., Johnson, R. S. and DeJong, T. M. (1993), 'Patterns of soil and tree water status and leaf functioning during regulated deficit irrigation scheduling in peach', *J. Am. Soc. Hortic. Sci.*, 118, 580-6.

Giuliani, R., Magnanini, E. and Corelli Grappadelli, L. (1998), 'Whole canopy gas exchanges and light interception of three peach training systems', *Acta Hortic.*, 465, 309-18.

Gollan, T., Turner, N. C. and Schulze, E. D. (1985), 'The responses of stomata and leaf gas exchange to vapour pressure deficits and soil water content', *Oecol.*, 65, 356–62.
Gonzalez-Dugo, V., Zarco-Tejada, P., Nicolás, E., Nortes, P. A., Alarcón, J. J., Intrigliolo, D. S. and Fereres, E. (2013), 'Using high resolution UAV thermal imagery to assess the variability in the water status of five fruit tree species within a commercial orchard' *Precis. Agric.*, 14, 660–78.
Grossman, Y. L. and DeJong, T. M. (1995), 'Maximum fruit growth potential and seasonal patterns of resource dynamics during peach growth', *Ann. Bot.*, 75, 553–60.
Grossman, Y. L. and DeJong, T. M. (1998), 'Training and pruning system effects on vegetative growth potential, light interception, and cropping efficiency in peach trees', *J. Am. Soc. Hortic. Sci.*, 123, 1058–64.
Guerriero, R., Loreti, F., Morini, S. and Natali, S. (1980), 'Eight years of observations on a peach double-row planted orchard', *Acta Hortic.*, 114, 362–82.
Guo, W., Gu, J., Liu, D. and Shang, L. (2016), 'Peach variety identification using near-infrared diffuse reflectance spectroscopy', *Comput. Electron. Agric.*, 123, 297–303.
He, L. and Schupp, J. (2018), 'Sensing and automation in pruning of apple trees: A review', *Agronomy*, 8(10), 211.
Hutton, R. J., Mc Fadyen, L. M. and Lill, W. J. (1987), 'Relative productivity and yield efficiency of canning peach trees in three intensive growing systems', *HortSci.*, 22, 552–60.
Jackson, J. E. and Palmer, J. W. (1972), 'Interception of light by model hedgerow orchards in relation to latitude, time of year and hedgerow configuration and orientation', *J. Appl. Ecol.*, 9, 341–57.
Jackson, J. E., Sharples, R. O. and Palmer, J. W. (1971), 'The influence of shade and within-tree position on apple fruit size, colour and storage quality', *J. Hortic. Sci.*, 46, 227–87.
Jarvis, P. G. (1976), 'The interpretation of the variations in leaf water potential and stomatal conductance found in canopies in the field', *Phil. Trans. R. Soc. Lond.*, 273, 593–610.
Jifon, J. L. and Syvertsen J. P. (2003), 'Kaolin particle film application can increase photosynthesis and water use efficiency of "Ruby Red" grapefruit leaves', *J. Am. Soc. Hortic. Sci.*, 128, 107–12.
Jones, H. (2014), *Plants and Microclimate: A Quantitative Approach to Environmental Plant Physiology*, 3rd ed., Cambridge University Press, p. 423. ISBN: 978-0521279598.
Jones, H. G. and Higgs, K. H. (1982), 'Surface conductance and water balance of developing apple (*Malus Pumila* Mill.) fruits', *J. Exp. Bot.*, 33, 67–77.
Kalaji, H. M., Schansker, G., Brestic, M., Bussotti, F., Calatayud, A., Ferroni, L., Goltsev, V., Guidi, L., Jajoo, A., Li, P., Losciale, P., Mishra, V. K., Misra, A. N., Nebauer, S. G., Pancaldi, S., Penella, C., Pollastrini, M., Suresh, K., Tambussi, E., Yanniccari, M., Zivcak, M., Cetner, M. D., Samborska, I. A., Stirbet, A., Olsovska, K., Kunderlikova, K., Shelonzek, H., Rusinowski, S. and Bąba, W. (2017), 'Frequently asked questions about chlorophyll fluorescence, the sequel', *Photosynth. Res.*, 132, 13–66.
Lakso, A. N. (1994), 'Apple', in: Environmental Physiology of Fruit Crops. Vol. I. Temperate Fruits, B. Schaffer and P. C. Andersen (Eds), CRC Press, Boca Raton, FL, pp. 3–42.
Lang, A. (1990), 'Xylem, phloem and transpiration flows in developing apple fruits', *J. Exp. Bot.*, 41(227), 645–51.
Lescourret, F., Génard, M., Habib, R. and Fishman, S. (2001), 'Variation in surface conductance to water vapor diffusion in peach fruit and its effects on fruit growth assessed by a simulation model', *Tree Physiol.*, 21(11), 735–41.

Lescourret, F., Moitrier, N., Valsesia, P. and Génard, M. (2011), 'QualiTree, a virtual fruit tree to study the management of fruit quality. I. Model development', *Trees - Structure and Function*, 25(3), 519.

Li, S. H., Génard, M., Bussi, C., Lescourret, F., Laurent, R., Besset, J. and Habib, R. (2002), 'Preliminary study on transpiration of peaches and nectarines', *Gartenbauwissenschaft*, 67(1), 39-43.

Lo Bianco, R. and Rieger, M. (2002), 'Partitioning of sorbitol and sucrose catabolism within peach fruit', *J. Am. Soc. Hortic. Sci.*, 127(1), 114-21.

Long S. P., Humphries, S. and Falkowski, P. G. (1994), 'Photoinhibition of photosynthesis in nature', *Annu. Rev. Plant Mol. Biol.*, 45, 633-62.

Loreti, F., Massai, R. and Morini S., (1989), 'Further observations on high density nectarine plantings', *Acta Hortic.*, 243, 353-60.

Losciale, P., Chow, W. S. and Corelli Grappadelli, L. (2010), 'Modulating the light environment with the peach "asymmetric orchard": Effects on gas exchange performances, photoprotection, and photoinhibition', *J. Exp. Bot.*, 61, 1177-92.

Losciale, P., Zibordi, M., Manfrini, L., Morandi, B. and Corelli Grappadelli, L. (2011), 'Effect of moderate light reduction on absorbed energy management, water use, photoprotection and photo-damage in peach', *Acta Hortic.*, 907, 169-74.

Losciale, P., Manfrini, L., Morandi, B., Pierpaoli, E., Zibordi, M., Stellacci, A. M., Salvati, L. and Corelli Grappadelli L. (2015), 'A multivariate approach for assessing leaf photo-assimilation performance using the IPL index', *Physiol. Plant.*, 154, 609-20.

Losciale, P., Manfrini, L., Morandi, B., Pierpaoli, E., Zibordi, M., Lauri, P. E., Reignard, J. L. and Corelli Grappadelli, L. (2017), 'Fast and reliable phenotyping of leaf functions: A tool for water stress tolerance evaluation', *Acta Hortic.*, 1172, 399-404.

Maldonado, W. and Barbosa, J. C. (2016), 'Automatic green fruit counting in orange trees using digital images', *Comput. Electron. Agric.*, 127, 572-81.

Maletsika, P. A. and Nanos, G. D. (2015), 'Kaolin particle film on peach leaf physiology', *Acta Hortic.*, 1084, 327-34

Manfrini, L., Taylor, J. A. and Corelli Grappadelli, L. (2009), 'Spatial analysis of the effect of fruit thinning on apple crop load', *Eur. J. Hortic. Sci.*, 74, 54-60.

Manfrini, L., Pierpaoli, E., Taylor, J. A., Morandi, B., Losciale, P., Zibordi, M., Corelli Grappadelli, L. and Bastías, R. M. (2012), 'Precision fruit growing: How to collect and interpret data on seasonal variation in apple orchards', *Acta Hortic.*, 932, 461-70.

Manfrini, L., Pierpaoli, E., Zibordi, M., Morandi, B., Muzzi, E., Losciale, P. and Corelli Grappadelli, L. (2015), 'Monitoring strategies for precise production of high quality fruit and yield in apple in Emilia-Romagna', *Chem. Eng. Trans.*, 44, 301-6.

Massonnet, C., Costes, E., Rambal, S., Dreyer, E., Regnard, J. L. (2007), 'Stomatal regulation of photosynthesis in apple leaves: Evidence for different water-use strategies between two cultivars', *Ann. Bot.*, 100, 1347-56.

McFadyen, L. M., Hutton, R. J. and Barlow, E. W. R. (1996), 'Effects of crop load in fruit water relations and crop load in peach', *J. Hortic. Sci.*, 71(3), 469-80.

Miranda, C., Urretavizcaya, I., Santesteban, L. G. and Royo, J. B. (2015), 'Sampling stratification using aerial imagery to estimate fruit load and hail damage in nectarine trees', in *Precision Agriculture'15, Proceedings of the 10th European Conference on Precision Agriculture*, J. V. Stafford (Ed.), Wageningen Academic Publishers, Wageningen, The Netherlands, pp. 541-6.

Monestiez, P., Audergon, J. M. and Habib, R. (1990), 'Spatial dependences and sampling in a fruit tree: A geostatistical approach', Institut National de la Recherche Agronomique, *Technical Report*, 163, p. 30.

Monteith, J. L. (1977), 'Climate and the efficiency of crop production in Britain', *Phil. Trans. R. Soc. Lond.*, 281, 277–94.

Morandi, B. and Corelli Grappadelli, L. (2009), 'Source and sink limitations in vascular flows in peach fruit', *J. Hortic. Sci. Biotech.*, ISAFRUIT Special Issue, 150–6.

Morandi, B., Manfrini, L. and Corelli Grappadelli, L. (2005), 'Innovazione tecnologica nel controllo della crescita del frutto per una peschicoltura di qualità', *Atti XXV Convegno Peschicolo*, Faenza, 23–24 Settembre 2004, pp. 117–21.

Morandi, B., Rieger, M. W. and Corelli Grappadelli, L. (2007a), 'Vascular flows and transpiration affect peach (*Prunus Persica* Batsch.) fruit daily growth', *J. Exp. Bot.*, 58, 3941–7.

Morandi, B., Manfrini, L., Zibordi, M., Noferini, M., Fiori, G. and Corelli Grappadelli, L. (2007b), 'A low-cost device for accurate and continuous measurement of fruit growth', *HortSci.*, 42(6), 1380–2.

Morandi, B., Manfrini, L., Losciale, P., Zibordi M. and Corelli Grappadelli, L. (2010), 'The positive effect of skin transpiration in peach fruit growth', *J. Plant Physiol.*, 167(13), 1033–7.

Morandi, B., Losciale, P., Manfrini, L., Zibordi, M. and Corelli Grappadelli, L. (2012), 'Variations in the orchard environmental conditions affect vascular and transpiration flows to/from peach fruit', *Acta Hortic.*, 962, 395–401.

Moriguchi, T., Sanada, T. and Yamaki, S. (1990), 'Seasonal fluctuation of some enzymes relating to sucrose and sorbitol metabolism in peach fruit', *J. Am. Soc. Hortic. Sci.*, 115, 278–81.

Naor, A. (2006), 'Irrigation scheduling and evaluation of tree water status in deciduous orchards', *Hortic. Rev.*, 32, 111–65.

Naor, A., Klein, I. and Doron, I. (1995), 'Stem water potential and apple size', *J. Am. Soc. Hortic. Sci.*, 120(4), 577–82.

Nuzzo, V., Dichio, B. and Xiloyannis, C. (2002), 'Canopy development and light interception in peach trees trained to transverse Y and delayed vase in the first four years after planting', *Acta Hortic.*, 592, 405–12.

Palmer, J. W. (1980), 'Computed effects of spacing on light interception and distribution within hedgerow trees in relation to productivity', *Acta Hortic.*, 114, 80–8.

Patrick, J. W. (1997), 'Phloem unloading: Sieve element unloading and post-sieve element transport', *Annu. Rev. Plant Physiol. Plant Mol. Biol.*, 48, 191–222.

Pavel, E. W. and DeJong, T. M. (1993), 'Seasonal CO_2 exchange patterns of developing peach (*Prunus Persica*) fruits in response to temperature, light and CO_2 concentration', *Physiol. Plant.*, 88, 322–30.

Rahnemoonfar, M. and Sheppard, C. (2017), 'Deep count: Fruit counting based on deep simulated learning', *Sensors*, 17, 1–12.

Rosati, A., Esparza, G., DeJong, T. M. and Pearcy, R. W. (1999), 'Influence of canopy light environment and nitrogen availability on leaf photosynthetic characteristics and photosynthetic nitrogen use efficiency of field grown nectarine trees', *Tree Physiol.*, 19, 173–80.

Sams, C. E. and Flore, J. A. (1983), 'Net photosynthetic rate of sour cherry (*Prunus cerasus* L. 'Montmorency') during the growing season with particular reference to fruiting', *Photosynth. Res.*, 4(4), 307–16.

Scandellari, F, Caruso, G, Liguori, G, Meggio, F, Palese, AM, Zanotelli, D, Celano, G, Gucci, R, Inglese, P, Pitacco, A and Tagliavini, M. (2016), 'A survey of carbon sequestration potential of orchards and vineyards in Italy', *Eur. J. Hortic. Sci.*, 2, 106–14.

Schmidt, T., Hanrahan, I., Castillo, F. and McFerson, J. (2014), 'Reflective ground covers increase yield of fruit trees', *Acta Hortic.*, 1058, 313–20.

Seifert, B., Zude, M., Spinelli, L. and Torricelli, A. (2015), 'Optical properties of developing pip and stone fruit reveal underlying structural changes', *Physiol. Plant.*, 153, 327–36.

Shackel, K. A., Ahmadi, H., Biasi, W., Buchner, R., Goldhamer, D., Gurusinghe, S., Hasey, J., Kester, D., Krueger, B., Lampinen, B., McGourty, G., Micke, W., Mitcham, E., Olson, B., Pelletrau, K., Philips, H., Ramos, D., Schwankl, L., Sibbett, S., Snyder, R., Southwick, S., Stevenson, M., Thorpe, M., Weinbaum, S. and Yeager, J. (1997), 'Plant water status as an index of irrigation need in deciduous fruit trees', *HortTech.*, 7, 23–9.

Socías, X., Correia, M. J., Chaves, M. and Medrano, H. (1997), 'The role of abscisic acid and water relations in drought responses of subterranean clover', *J. Exp. Bot.*, 48, 1281–8.

Taylor, J. A., McBratney, A. B. and Whelan, B. M. (2007), 'Establishing management classes for broadacre agricultural production', *Agron. J.*, 99, 1366–76.

Torres-Ruiz, J. M., Perulli, G. D., Manfrini, L., Zibordi, M., Lopez Velascoo, G., Anconelli, S., Pierpaoli, E., Corelli Grappadelli, L. and Morandi, B. (2016), 'Time of irrigation affects vine water relations and the daily patterns of leaf gas exchanges and vascular flows to kiwifruit (*Actinidia deliciosa* Chev.)', *Agric. Water Manag.*, 166, 101–10.

Vizzotto, G., Pinton, R., Varanini, Z. and Costa, G. (1996), 'Sucrose accumulation in developing peach fruit', *Physiol. Plant.*, 96, 225–30.

von Caemmerer, S. (2000), 'Biochemical models of leaf photosynthesis', Techniques in plant sciences No 2. CSIRO Publishing (Australia). ISBN: 64306379X.

Wang, D. and Gartung, J. (2010), 'Infrared canopy temperature of early-ripening peach trees under postharvest deficit irrigation', *Agric. Water Manag.*, 97, 1787–94.

Wang, C., Wang, H., Zhao, X., Chen, B. and Wang, F. (2015), 'Mulching affects photosynthetic and chlorophyll fluorescence characteristics during stage III of peach fruit growth on the rain-fed semiarid Loess Plateau of China', *Sci. Hortic.*, 194, 246–54.

Weinberger, J. H. (1941), 'Studies on time of peach thinning from blossoming to maturity', *Proc. Amer. Soc. Hort. Sci.*, 38, 137–40.

Wulfsohn, D., Zamora, F. A., Téllez, C. P., Lagos, I. Z. and García-Fiñana, M. (2012), 'Multilevel systematic sampling to estimate total fruit number for yield forecasts', *Precis. Agric.*, 13, 256–75.

Yaghi, T., Arslan, A. and Naoum, F. (2013), 'Cucumber (*Cucumis sativus*, L.) water use efficiency (WUE) under plastic mulch and drip irrigation', *Agric. Water Manag.*, 128, 149–57.

Zarco-Tejada, P. J., Berni, J. A. J., Suárez, L., Sepulcre-Cantó, G., Morales, F. and Miller, J. R. (2009), 'Imaging chlorophyll fluorescence with an airborne narrow-band multispectral camera for vegetation stress detection', *Remote Sens. Environ.*, 113, 1262–75.

Zhang, Q. (2018), *Automation in Tree Fruit Production: Principles and Practice*. doi:10.1079/9781780648507.0000.

Zhang, B., Peng, B., Zhang, C., Song, Z. and Ma, R. (2017), 'Determination of fruit maturity and its prediction model based on the pericarp index of absorbance difference (IAD) for peaches', *PLoS One*, 12, 1–11.

Ziosi, V., Noferini, M., Fiori, G., Tadiello, A., Trainotti, L., Casadoro, G. and Costa, G. (2008), 'A new index based on vis spectroscopy to characterize the progression of ripening in peach fruit', *Postharvest Biol. Technol.*, 49, 319–29.

Zude-Sasse, M., Fountas, S., Gemtos, T. A. and Abu-Khalaf, N. (2016), 'Applications of precision agriculture in horticultural crops', *Eur. J. Hortic. Sci.*, *81*, 78–90.

Chapter 3

Advances and challenges in cherry breeding

José Quero-García, INRA, University of Bordeaux, France; Amy Iezzoni, Michigan State University, USA; Gregorio López-Ortega, IMIDA, Spain; Cameron Peace, Washington State University, USA; Mathieu Fouché and Elisabeth Dirlewanger, INRA, University of Bordeaux, France; and Mirko Schuster, Julius Kühn-Institut, Germany

1 Introduction

Cherries belong to the *Rosaceae* family, *Spiraeoideae* subfamily, *Amygdaleae* tribe and genus *Prunus*. Within this genus, two subgenera are considered: *Cerasus* and *Padus*. Cultivated cherries belong to the former and comprise the diploid sweet cherry (2n = 2x = 16, *Prunus avium* L.) and the tetraploid sour cherry (2n = 4x = 32, *Prunus cerasus* L.). Other cherry species such as *Prunus tomentosa*, *Prunus pseudocerasus*, *Prunus serotina* and *Prunus laudocerasus* are also grown for their fruit. A complete list of species within these two subgenera can be found in Iezzoni et al. (2017).

It is believed that both sweet and sour cherries may have originated within a region around the Caspian Sea and the Black Sea (Hedrick, 1915). They would have been later spread by birds across Europe (Webster, 1996). Sweet cherry occurs naturally in Europe, from Sweden to Greece, Italy and Spain and into areas of northern Africa (Faust and Surányi, 1997; Hedrick, 1915). Sour cherry grows natively from central and south Europe to north India, Iran and Kurdistan (Küssmann, 1962; Hedrick, 1915). It is a segmental allopolyploid species resulting from a natural hybridization between ground cherry, *P. fruticosa*,

http://dx.doi.org/10.19103/AS.2018.0040.17

and unreduced pollen of sweet cherry, *P. avium* (Olden and Nybom, 1968). Isozyme analyses, genomic *in situ* hybridization and karyotype analysis further confirmed the hybrid origin of *P. cerasus* and identified the presence of both disomic and tetrasomic inheritance (Tavaud et al., 2004; Schuster and Schreiber, 2000; Brettin et al., 2000; Beaver and Iezzoni, 1993). Hence, it was hypothesized that sour cherry originated in regions where the distribution of its progenitor species overlap (Zhukovsky, 1965), resulting in continual gene flow between sour cherry and its two progenitor species.

It is known that cherries were consumed by early inhabitants of Europe during 4000–5000 BC (Webster, 1996), which has been supported by several archaeological findings of cherry pits in Europe from the Neolithic period to Bronze Age (Faust and Surányi, 1997). It is believed that Albanians cultivated cherries before the Greeks and seeds of cultivated cherries from the Roman period have been found in several areas. Theophrastus in 300 BC gave the first written reference of cherry cultivation (Brown et al., 1996). The first genetic improvement of cherries may have occurred in Middle Europe because Faust and Surányi (1997) reported fewer findings of cultivated cherries in the northern and most southern areas of Europe. Cherry cultivation increased from the sixteenth century, with most intensity in central Europe (Watkins, 1976) and was introduced during the nineteenth and twentieth centuries in North America, Far East Asia and later in Southern hemisphere countries such as Chile and Australia. As for sour cherry, it was likely spread by Slavic people from West Asia to East and South-eastern Europe during the sixth- to eighth-century migrations. The major sour cherry production regions in Europe are still primarily the former settlement areas of Slavic people; hence, Germany is the western frontier of the growing area in Europe.

The adaptation of cherries across Europe evolved into a wide pool of genetic diversity with the development of ecotypes adapted to the different regions. High levels of genetic diversity among landraces have been detected in genetic diversity studies carried out in recent years (Campoy et al., 2016; Cachi and Wünsch, 2014a; Ercisli et al., 2011; Ganopoulos et al., 2011; Avramidou et al., 2010; Demir et al., 2009; Guarino et al., 2009; Lacis et al., 2009; Pérez-Sánchez et al., 2008; Rodrigues et al., 2008; Stanys et al., 2008; Schuster et al., 2007; Kaçar et al., 2006; Joublan et al., 2005; Wünsch and Hormaza, 2004a; Tavaud et al., 2001). However, these studies consistently showed a reduction in genetic diversity from wild to landrace to modern bred sweet cherry cultivars (Mariette et al., 2010).

Although cherries have been cultivated for more than 2000 years, cherry breeding started around the early 1800s, but some modern cultivars are just a few generations away from early ancestors (Iezzoni et al., 1990). Cherry breeding is currently carried out in many countries, by public and private programs, and sweet cherry cultivars are continuously being released. A detailed description of the history of sweet and sour cherry breeding, along with the characteristics of the most important current public breeding programs, is given in Quero-García et al. (2017a) and Schuster et al. (2017), respectively.

Hundreds of sweet cherry cultivars are currently available for growers and a large diversity of landraces are preserved, some of them being used in modern breeding programs. From a morphological point of view, Zwitzscher (1961), according to a previous classification by Truchseß (1819), divided sweet cherries into two groups, the soft-flesh Heart cherries and the firm-flesh Bigarreau cherries. Both groups were further subdivided depending on their juice colour. However, classification into clear-cut groups of existing cultivars is difficult, because there is a vast continuum of morphological diversity and many traits are influenced by differences in environmental factors among growing locations, including climate and soil characteristics as well as cultural practices. The variability in tree morphology and fruit characteristics is higher in sour cherry, probably due to its more complex phylogenetic origin. Truchseß (1819) and Hedrick (1915) divided the sour cherry into two groups that vary more or less in both tree habit and fruit characteristics but have a constant difference only in a single, very easily distinguished character, the colour of juice. Sour cherries with red to dark red coloured juice are described as Morellos (Griottes, Weichsel). The sour cherries with colourless juice are the Amarelles (Kentish). Morello fruit are very dark red with spherical or cordate shape. Amarelles are pale red fruits and more or less flattened at the end. As an additional division, Hedrick (1915) described the Marasca cherry. This cherry is a native of Dalmatia in Croatia, where the tree grows wild and is now sparingly cultivated. The fruits are much smaller, are deep red or almost black in colour and have intensively red flesh and juice. The tree and fruit characteristics of the Danish local cultivar 'Stevnsbaer' are very similar. It is possible that 'Stevnsbaer' originated from the Marasca cherry (Stainer, 1975).

Despite the high number of available commercial cultivars, as well as local landraces, both sweet and sour cherry cultivation are still based on a small number of cultivars. It is the case for very old sweet cherry selections such as 'Bing' or 'Burlat' or even old cultivars of unknown origin, such as '0900 Ziraat', in Turkey. The sour cherry cultivar 'Schattenmorelle' (with many local synonyms, for example 'Łutovka' in Poland, 'Griotte du Nord' or 'Griotte Noir Tardive' in France) is dominant in Middle Europe, whereas in the United States sour cherry production is still based on the 400-year-old cultivar from France, 'Montmorency'. The landrace cultivar 'Pandy' (syn. 'Crisana', 'Köröser') and related cultivars are also very popular in Hungary and Romania.

As cherry has a long juvenile period (4–6 years or longer), the introduction of a desired attribute to improved cultivars is a long process. This factor may have promoted the use of the same progenitor in breeding programmes and one well-known example is the continuous use of the sweet cherry cultivar

'Stella' and its descendants in breeding for self-compatibility. As a result, the genetic diversity of sweet cherry cultivars is limited (Choi and Kappel, 2004) and several breeding programmes have recently started to incorporate new sources of diversity, including from the supposed area of origin of sweet and sour cherries (López-Ortega, 2015).

2 Main achievements in conventional breeding

In this section sweet and sour cherry are treated separately, although some breeding objectives and selection criteria are common to both crops. The main breeding goals in sweet and sour cherry have been described previously (Schuster et al., 2017; Quero-García et al., 2017a; Kappel et al., 2012; Sansavini and Lugli, 2008; Bargioni, 1996). Herein, a synthesis of the most important current breeding goals is provided for both crops.

2.1 Sweet cherry

2.1.1 Tree, fruiting structure and flower characteristics

Intensive cultivation of sweet cherry has been traditionally hampered by excessive tree vigour. However, with the advent of a new generation of dwarfing and semi-dwarfing rootstocks, breeding for scion dwarf types has been gradually abandoned. Nevertheless, an excess of vigour is always avoided. In order to establish modern intensive orchards, equipped with complex protection structures against biotic or abiotic stresses, growers have to make increasingly higher investments. For this reason, yield precocity, productivity and regularity of production are nowadays key selection criteria, although all of them may be highly influenced by the rootstock, the planting and the training systems. Sweet cherry cultivars are extremely dependent on climatic conditions for a consistent and sufficient fruit set; hence, regularity of production is becoming increasingly important, even more given the impacts of climate change. One of the major achievements in sweet cherry breeding was the development of self-compatibility, which is currently a major breeding goal in almost every programme. In order to avoid exclusive use of the 'Stella' source of self-compatibility, other sources, such as landraces 'Cristobalina' or 'Kronio', are being used as parents in several breeding programmes.

2.1.2 Tolerance to abiotic and biotic stresses

Rain-induced fruit cracking is one of the most important agronomic problems for sweet cherry growers (Knoche and Winkler, 2017). No reliable field or lab phenotyping protocol exists to clearly evaluate genotypic tolerance to cracking. Hence, only repeated multi-year field observations can help to discard the most sensitive materials, but only those programmes located in rainy regions can easily achieve this evaluation. Very few sources of tolerance to cracking exist; among modern cultivars we can cite 'Regina' and 'Fermina'. Resistance to winter frost has been a major breeding target in countries at the margins of traditional production areas, such as Latvia and Russia, or in very cold continental areas of central European countries. A review on studies aimed at evaluating cultivars' frost resistance, both in field conditions or after artificial freezing test, can be found in Wenden et al. (2017). There is a growing interest, in the last decades, to adapt sweet cherry growing to regions characterized by mild winters, such as south-east Spain, California, central parts of Chile, central and western regions of China and even North African countries. This becomes even more urgent in the current context of global warming. Although several commercial cultivars, such as 'Lapins', 'Brooks' or 'Rainier', are regularly productive, very few cultivars are clearly low-chilling. Very few sources of low-chilling requirements exist, the most well-known being the Spanish landrace 'Cristobalina'. Recently, private breeders from California (Zaiger Genetics and International Fruit Genetics) have released several cultivars which exhibit a very early blooming time, and presumably, low-chilling requirements. Ideal cultivars should have low-chilling requirements for flowering but with sufficient heat requirements in order not to flower too early and hence avoid the risk of frost damage. To our knowledge, this 'ideotype' has not yet been achieved for sweet cherry. Another abiotic stress that is becoming increasingly serious due to global warming is the formation of double fruits. Although marked differences exist in terms of cultivar susceptibility, no studies have yet been carried out in order to elucidate the genetic basis of this trait.

Among biotic stresses, one of the most serious diseases in sweet cherry is bacterial canker caused by *Pseudomonas* spp. (Puławska et al., 2017). The breeding programme carried at John Innes Institute (UK) released a number of bacterial canker-resistant cultivars but unfortunately, most of these cultivars showed some susceptibility to canker, due to the infection of new more virulent bacterial strains (Bargioni, 1996). However, tolerant cultivars exist (e.g. 'Colney', 'Hertford', 'Vittoria') and could be integrated into breeders' pools of genitors. Brown rot, caused by *Monilinia* spp., is one of the most damaging fungal diseases in sweet cherry, causing severe damage on flowers and fruits (Børve et al., 2017). Although several studies have reported different levels of cultivar susceptibility, in no case could resistance be confirmed (Kappel and Sholberg, 2008; Brown and Wilcox, 1989). Cultivars 'Regina', 'Early Korvik', 'Melitopolska chorna' and 'Valerij Chkalov' are reported to have good levels of tolerance to brown rot. Among sweet cherry pests (Papadopoulos et al., 2017), breeding research has only been carried out at East Malling (UK) on resistance to black cherry aphid (*Myzus cerasi* Fab.). Interspecific hybridization was attempted between sweet cherry cultivar 'Napoleon' and clones from the species *Prunus canescens*, *P. incisa*, *P. kurilensis* and *P. nipponica* which all showed resistance to colonization. Some of the hybrids proved to be tolerant but not fully resistant to colonization (Bargioni, 1996).

2.1.3 Fruit-related traits

The most important fruit quality traits for breeders are fruit size, fruit firmness, skin and flesh colour, sugar content and flavour. Additional morphological or biochemical traits can be evaluated in more advanced stages of the selection process. Fruit size is one of the most critical criteria in sweet cherry breeding and the majority of breeders will systematically discard hybrids producing fruits of less than 10 g. Hence, tremendous progress has been achieved and many recently released cultivars produce very large and firm fruits.

Mechanical harvesting of sweet cherries has been mainly developed for the processing industry but interest in a market of stemless sweet cherries for the fresh market is growing (Whiting and Perry, 2017; Kappel et al., 2012). In order to prevent juice loss, oxidation and pathogens attacks, stemless cultivars require an abscission zone between the pedicel and the fruit that produces a hardened scar (Sansavini and Lugli, 2008). Several cultivars suitable for this type of harvest already exist (e.g. 'Ambrunés', 'Cristalina', 'Fermina', 'Linda', 'Sumste' and 'Vittoria', as cited in Quero-García et al., 2017a, Bargioni, 1996).

2.1.4 Extension of harvest period

Many breeding programmes seek to develop extra-early and extra-late cultivars, in order to extend a harvest period which is particularly short for sweet cherries, in comparison to other *Prunus* species. Hence, for instance in Europe, one of the objectives is to breed cultivars ripening before the benchmark cultivar 'Burlat'. To date, no cultivar with a ripening date close to or earlier than 'Burlat' has proven significantly better than 'Burlat', although some new cultivars, such as 'Pacific Red' or 'Nimba', are highly promising. Although very early blooming cultivars are now available, breeders have not yet managed to combine such early flowering trait with a short fruit ripening phase such as the one from 'Burlat', and premium fruit quality traits. Extra-late ripening cultivars such as 'Fertard', 'Penny' or 'Staccato' have been released but progress can still be expected in terms of firmness and cracking tolerance.

2.2 Sour cherry

Because most sour cherries are processed, high productivity, fruit quality, suitability for mechanical harvesting and the extension of harvest period are the main objectives in sour cherry breeding. In the last decades, tolerance to biotic and abiotic stress becomes more important in the selection.

2.2.1 Fertility

A high fertility is important for a reasonable fruit set in sour cherry, because numerous cultivars may have a reduced self-compatibility reaction. Sour cherries are frequently considered to be self-compatible, although self-incompatible and partially self-compatible cultivars exist. Redalen (1984) regarded cultivars with a final fruit set more than 15% as self-compatible. Self-incompatible cultivars may sometimes have low fruit set. Cultivars with an intermediate final fruit set have been characterized as partly self-compatible. Certain pairs of cultivars are cross-incompatible, reciprocally or unilaterally (Bošković et al., 2006). Similar results were obtained in progenies of cross populations in sour cherry breeding. The reasons are unknown for the partially low fertility in sour cherry. Recent investigations demonstrated that a gametophytic self-incompatibility (GSI) system exists in sour cherry (Tobutt et al., 2004; Yamane et al., 2001). This GSI illustrated the occurrence of self-compatible and self-incompatible cultivars in sour cherry, like in sweet cherry. Self-compatibility in sour cherry requires the loss of function for a minimum of two *S*-haplotypes specificity components (Hauck et al., 2006). The reason for the partially self-compatible cultivars is unknown. Low fruit set could also be due to meiotic instability caused by intra- or interspecific crosses or inbreeding effects, resulting in ovule or zygote abortion. Whatever the cause for the low fruit set in sour cherry, any successful cultivar would need to have high yields and be self-compatible.

2.2.2 Tree and fruiting structure

The variability in tree and fruiting structure is very high in sour cherry. The tree sizes vary from an upright and vigorous growth like sweet cherry to dwarf or bushy tree types.

Flower buds can develop on 1-year-old wood and on spurs formed on older wood. Most cultivars form fruits mainly on 1-year-old wood and at a smaller scale on spurs. The production of bare wood is typical of sour cherry cultivars which produce flower buds primarily on the 1-year-old wood. A good example of this fruiting habit is 'Schattenmorelle', a Morello cherry. The tendency to produce blind wood is low in sour cherries where the fruits are preferentially borne on spurs or bouquet spurs on older wood, such as in cultivar 'Achat', for example (Schuster, 2011). The selection of new cultivars should be focused on trees with an upright growth and fruit set primarily on spurs. This tree habit is advantageous for most of the harvest techniques and requires less pruning.

2.2.3 Tolerance to biotic and abiotic stress

Blossom blight and brown rot, caused by *Monilinia laxa* (Aderh. & Ruhl.) Honey, and leaf spot, caused by *Blumeriella jaapii* (Rehm) Arx., are the main fungal diseases in sour cherry. They can significantly damage the trees and reduce the yield in sour cherry orchards. The symptoms caused by *M. laxa* and the degree of susceptibility depend on the climatic conditions and the virulence of specific local races (Budan et al., 2005a). Cultivars or wild cherry *Prunus* species resistant to *M. laxa* are unknown. However, some wild cherry *Prunus* species and sour cherry cultivars show a high tolerance to blossom blight and can be used in resistance breeding programmes. Leaf spot is common in the cherry growing areas in North America and Europe. Only a few sour cherry cultivars are tolerant to leaf spot infections. The *Prunus* species *P. maackii*, *P. canescens*, *P. serotina* and interspecific hybrids with these species show a high level of resistance to *B. jaapii* (Stegmeir et al., 2014a; Budan et al., 2005b; Schuster, 2004; Wharton et al., 2003).

Because of the early blossom time of the cherry flowers, spring frost can damage buds, flowers and young fruits. Especially in areas with cold climates, such as Russia and Canada, spring frost resistance is one of the most important breeding goals. Selection of late blooming genotypes, with higher chilling requirements and with tolerance to spring frost can reduce this risk. Interspecific hybrids with *P. fruticosa* were used as donors for breeding to increase the frost tolerance in sour cherry in Russia and Canada (Bors, 2005; Zhukov and Charitonova, 1988).

2.2.4 Fruit-related traits

Fruit quality has become one of the main selection criteria in sour cherry breeding in recent years. The main quality characteristics are soluble solids, titratable acidity, fruit and juice colour, firmness and good taste. The requirements on the different fruit parameters vary according to utilization. The majority of fruits are used for processing purposes, for example for juice, canning, jam and wine. Only a small part of the sour cherries is produced for fresh market. The quality characteristics are determined by the colour, acid and sugar content of the fruits and their concentrations of volatile compounds. The ideal fruit for processing are of 21–24 mm, have a dark red coloured juice, a high content on sugar and acidity combined with a good aroma. For juice production and fresh consumption the fruit size could be increased. However, in the United States, the bright red 'Montmorency' colour is preferred by processors compared to a very dark red/purple coloured cherry. In the last years, many studies were undertaken to investigate the anthocyanins and the aroma components in sour cherry during the ripening season (Serradilla et al., 2017; Šimunic et al., 2005; Poll et al., 2003; Schmid and Grosch, 1986). Anthocyanins from sour cherry have been shown to possess strong antioxidant and anti-inflammatory activities (Wang et al., 1999). These polyphenols are characterized by their positive influence on human health (Ou et al., 2012).

Mechanical harvesting is gaining importance in cherry growing with the increase of labour costs and the reduction of labour availability. Specific characteristics of cherry fruits and trees are required for these harvest techniques. Those are firm fruits, high resistance to bruising, a low fruit retention force and a dry stem scar to reduce juice loss, uniformity of ripening, and an upright and stable trunk. Only a few new sour cherry cultivars have shown a good suitability for mechanical harvesting. In most cases these are partial self-compatible cultivars with a low fruit set like 'Ujferhertoi fürtös', 'Morina' and 'Pandy'.

2.2.5 Extension of harvest period

The ripening period of sour cherries is about 4 weeks. The most well-known early ripening cultivars originated in Hungary like 'Erdi Jubileum', 'Favorit' and 'Erdi bötermö'. Late ripening cultivars are for example 'Schattenmorelle', 'Fanal' and 'Stevnsbaer'. However, in the major sour cherry production areas, only one or a few cultivars are grown. Therefore, the harvest period is very short. The extension of the harvest period is important for increasing the utilization of mechanical harvest and reducing the labour cost.

3 Methodologies

Breeding methods were reviewed in Iezzoni et al. (2017) and are generally similar between the two crops. In this section, key operations are reviewed.

3.1 Hybridization

Cherry breeding is based on the generation of segregating F_1 families through hybridization between inter-compatible parents. Using self-compatible cultivars as maternal genitors, flowers must be emasculated (through removal of anthers) before they are pollinated. This activity is not necessary for self-incompatible maternal genitors; in this case, only pollinator exclusion through the use of bags is needed.

There are two main strategies for breeders: controlled crosses or open pollinations. In the first case, both the maternal and paternal parent will be known, whereas in the second only through the use of genetic markers might breeders have an idea of how many parental genitors were successfully involved in the cross. With controlled crosses, breeders use knowledge, if available, on the breeding value of pairwise parental combinations, what is called in classical quantitative genetics 'specific combining ability'. Without such prior knowledge, they are simply hoping to combine the best attributes of the two parents in at least some offspring. Use of open pollination, for which only the maternal parent is known, relies on prior knowledge, a suspicion or at least a hope that such maternal parents will generate superior offspring from numerous pollinizers, called a high 'general combining ability'.

Controlled crosses can be done by manual pollination, the use of bumblebees, or where isolated plots are available. Manual pollination requires labour, is time-consuming and the likelihood of success is generally low, lower than for many other fruit crops such as apple or peach. However, it does not require expensive infrastructure and allows breeders to test a potentially high number of specific combinations. In general, the parent that blooms later is chosen as the maternal parent for two reasons: first, it is easier to prepare the pollen prior to crossing so that breeders can use 'fresh' pollen from the current year; second, early ripening parents, which often bloom early, have a higher likelihood of presenting under-developed embryos. Nevertheless, natural fruit set is an important parameter and, when the early blooming parent has the highest fruit set, pollen from the late blooming parent can be collected one year before the cross and freeze-stored. Controlled crosses can be conducted in open-air conditions if the risk of adverse climatic events, such as rain, wind, hail or freezing temperatures, is not too high. Otherwise, maternal trees can be protected with cages with waterproof roofs (special care has to be taken to limit humidity inside the cage to avoid *Monilinia* blossom blight damage on flowers) or potted trees can be used in confined structures, such as tunnels, warehouses or greenhouses. Pollen must be collected at the proper stage of flower development, its viability must be checked and keeping it dry is critical for good conservation.

Hybridizations using bumblebees are preferred when large quantities of seed are sought for a specific pairwise parental combination. This approach requires the isolation of the maternal parent with structures such as those mentioned above. When climatic conditions are particularly favourable, the cheapest system is to install a bee-proof net over the maternal parent. The maternal parent can be a field-planted or a potted tree, as can the paternal parent but for this latter parent a set of cut branches placed in a bucket with water can also be used. When using both maternal and paternal parents in the field, only inter-compatible cultivars with similar blooming periods can be used. Both when using cut branches or potted trees, breeders can fully synchronize the blooming of two cultivars. Two strategies exist: either the late blooming cultivar can be placed in forcing conditions (growth chamber, greenhouse) to accelerate flowering or the early blooming cultivar can be placed in double-wall refrigerated storage to slow down its transition from endodormancy release to flowering.

Another strategy for obtaining a large number of seeds from the cross between two cherry parents is to collect fruits from trees in relatively well-isolated orchards. However, this approach requires collaboration with cherry growers or experimental stations and the number of possible combinations is limited. An intermediate strategy between cross-pollination and open pollination is the so-called poly-cross. Here, a set of inter-compatible cultivars are placed and isolated together and pollination is carried out by bumblebees. Breeders will not initially know the paternal parent for each offspring but with the use of genetic markers it is simple to establish paternity, especially for small numbers of arising advanced selections. The poly-cross system has not been widely used yet for cherry breeding and one issue might be pollen competition among parents.

3.2 Seed germination

Fruits for seed extraction can be collected just prior to optimal harvest maturity or even several weeks before optimal fruit maturation. It is recommended to extract and sterilize the stones immediately after harvest. Empty stones, that is, with no embryo, can generally be easily discarded as they will normally float whereas viable seeds will sink.

Seed dormancy is generally induced by the tree several days before full fruit maturity. An exogenous dormancy mechanism is imposed by intact endocarps, which need to be broken before chilling exposure can occur. In nature, dispersed moist stones are exposed to late summer temperatures and microorganisms present in the environment will produce enzymes that will degrade the cellulose between the carpels. The stone cracks will open, allowing for water and air exchange with the surroundings. Then in winter, the moist seeds will be exposed to chilling temperatures, which will break dormancy and allow the embryo to germinate and grow. The dormancy period can be avoided by harvesting fruit about one to two weeks before full maturity. For such a summer sowing, the stones must be cracked immediately after harvest. Both the endocarp (stone) and testa (seed coat) have to be removed. Thereafter the embryo can be placed in soil for germination.

In breeding programmes that succeed in producing many thousands of seeds per year, a non-sterile warm stratification method is used to overcome the endocarp dormancy. It is based on the mixture of seeds with moist sand, vermiculite or peat moss and a warm stratification period, with good air exchange, of 2–6 weeks at 20°C. This will allow all seeds to split the endocarp and thus become receptive for chilling. Chilling of seeds is most efficient at temperatures between 3°C and 5°C, with high moisture content (above 30% of fresh weight). It will normally take 12–16 weeks before seeds start to germinate in the cold. For smaller quantities of seeds it is recommended to remove the endocarp manually and this can be done on moist seeds just after harvest but also on dry seeds that have been stored. After sterilization, and a chilling exposure period of 3–4 months, seeds are periodically checked for germination and seedlings are transplanted to the greenhouse when the radicle is 1 cm in length.

Embryo *in vitro* culture is generally used when seeds are abnormal or would abort due to insufficient embryo growth during seed development. Early maturing cultivars, such as 'Burlat', are more likely to have seeds that exhibit insufficient embryo growth resulting in little or very poor germination. In this case, the embryos are harvested very immature, before they abort. Embryos should have a length of at least 3–4 mm but even 1-mm sized embryos can be cultured with success (Ivanicka and Pretova, 1986). In general, the success rate of embryo culture is extremely variable (Balla and Brozik, 1996).

3.3 Field selection

Phenotyping protocols for selection of superior seedlings within segregating populations are generally breeding programme-specific. However, published protocols can be used as points of reference (Chavoshi et al., 2014; Stegmeir et al., 2014b).

In sweet cherry, most breeders will base their first level of field selection on desired blooming and fruit maturity times, productivity, fruit size, firmness, appearance and eating quality. At this stage, observations are mostly conducted in the field without any lab-based measurement. For sour cherry, along with yield and fruit quality, disease susceptibility is highly important. Three approaches are common: evaluation of disease resistance over several years in natural orchard conditions; spraying young seedlings with inocula of fungi and bacteria in tunnels with high humidity or in the orchard; and laboratory methods using artificial inocula on detached leaves (Schuster, 2013). Similarly, selection for frost hardiness can be done in the field over several years or with laboratory methods, which can provide a comparison of relative winter hardiness in autumn, midwinter and spring (Jensen and Kristiansen, 2014; Liu et al., 2012). The best-performing selections are subsequently propagated on one or multiple rootstocks to make trees available for replicated field testing at one or multiple locations. In this second level of evaluation, both for sweet and sour cherry, a higher number of traits will be evaluated, including tolerance to biotic and abiotic stresses, and lab measurements for important traits such as fruit size or firmness will be conducted. Traits of focus at this level are those with low heritability for which replication over multiple trees and locations and large fruit numbers are needed.

4 Advances and key cultivars

Tremendous advances have been achieved in cherry breeding since the 1950s, both for sweet and sour cherry, as well as their corresponding rootstocks (the latter are not treated in this chapter, for a comprehensive review, see Hrotko and Rozpara, 2017). In sweet cherry, according to Sansavini and Lugli (2008), during the period 1991–2004, 230 new cultivars were released, and only peach was more active within the *Prunus* genus. Quero-García et al. (2017a) described 22 sweet cherry breeding programmes from 15 countries, mostly totally or partially publicly funded, as well as 119 cultivars of commercial importance. Many private programmes are currently releasing high-quality cultivars. A major advance has been the release of self-compatible cultivars covering a relatively large range of maturity timing. The breeding programme of 'Agriculture and Agri-Food' at Summerland, Canada, has been fundamental in the deployment of self-compatible cultivars derived originally from 'Stella', including 'Lapins', 'Santina', 'Skeena', 'Sumtare' (Sweetheart™), 'Sunburst' and '13S2009' (Staccato™). Other programmes have also released superior self-compatible cultivars such as the University of Bologna ('Early Star' and 'Grace Star') and Zaiger Genetics ('Royal Edie' and 'Royal Helen'). Another major advance has been the breeding of cultivars characterized by high levels of fruit quality (in terms of size, firmness and taste) and productivity. Some of these cultivars (such as 'Lapins', 'Rainier', 'Kordia' and 'Regina') are additionally extremely regular in their annual production and show a very good adaptability potential, which explains their success worldwide. Other programmes have released alternatives to very early traditional cultivars such as 'Burlat' (e.g. 'Rivedel' (syn. 'Early Lory', Earlise™), 'Narana' and 'Kossara') or filled the gap between 'Burlat' and the mid-season slot traditionally represented by 'Summit' ('Early Red' (syn. 'Maraly', Early Garnet™), 'Bedel' (Bellise™), 'Folfer', 'Carmen', 'Giorgia', 'Santina' and 'Tieton'). Recently, important efforts have been invested into the search for low-chilling cultivars; Zaiger Genetics has been very active in this field with the commercialization of several very early blooming cultivars, such as 'Royal Bailey' and 'Royal Marie', and other public and private breeding programmes serving production regions devote considerable resources towards this goal.

In the last century, many new sour cherry cultivars were released from breeding programmes of Europe and North America. The main breeding activities have been conducted by a few breeding programmes in Germany, Hungary, Poland, Serbia and United States in the last few decades (see Schuster et al., 2017 for a detailed description). In the major breeding programmes in Germany and Hungary, a Hungarian local sour cherry cultivar, 'Köröser Weichsel' ('Pandy üvegmeggy'), has been used as donor for high fruit quality and tolerance to diseases. On the other hand, in Hungary, Romania and Serbia many new cultivars resulted from regional clone selections of the landraces

'Pandy', 'Mocanesti' and 'Oblačinska', respectively, or are hybrids between landrace cultivars. One of the main successes in sour cherry breeding has been the use of interspecific hybridization with the objective of searching for tolerance sources to biotic and abiotic stresses. In Russia, crosses between *P. cerasus* and *P. fruticosa* have been aimed at increasing frost tolerance (Bors, 2005; Zhukov and Charitonova, 1988), whereas both in Russia and in Germany interspecific hybridizations of *P. fruticosa* × *P. maackii*, *P. maackii* × *P. cerasus* and *P. cerasus* × *P. serotina* were intended to introgress tolerance to diseases (Schuster et al., 2013; Zhukov, 1979; Mitschurin, 1951).

5 New approaches

5.1 Marker-assisted breeding

5.1.1 Genetic studies

To develop trait-predictive tools for implementation of marker-assisted selection (MAS) strategies in sweet or sour cherry, researchers must determine the linkage relations between marker loci and important agronomic traits. This research is based on the use of genetic maps to identify chromosome regions containing genes influencing qualitative and quantitative traits. The first genetic maps were constructed by working either with intra-specific or interspecific F_1 families, both on sweet and sour cherry (Clarke et al., 2009; Olmstead et al., 2008; Dirlewanger et al., 2004; Bošković and Tobutt, 1998; Wang et al., 1998). These first maps were based on isozyme, RFLP and SSR markers, which were quite laborious to generate (Dirlewanger et al., 2007). With the advent of next-generation sequencing technologies, a new type of marker, called SNP ('single-nucleotide polymorphism'), became available. There are thousands of SNPs within the genomes and they became relatively cheap to detect and incorporate into efficient assays (Peace et al., 2012) and use for constructing high-resolution linkage maps (Balas et al., 2017; Calle et al., 2017; Isuzugawa et al., 2017; Guajardo et al., 2015; Wang et al., 2015; Skipper et al., 2014; Klagges et al., 2013; Cabrera et al., 2012). Contrary to other *Prunus* crops such as peach, most traits of agronomic interest in cherry are polygenic, that is, controlled by a large number of loci. Hence, for most of these traits, QTLs ('quantitative trait loci') individually explaining some of their phenotypic variation have been detected.

Regarding fruit-related traits, a Mendelian trait locus was identified for fruit and skin colour on linkage group (LG) 3, within a family derived from the cross between cultivar 'Emperor Francis' and a mazzard accession, NY 54. Allelic variation in a candidate gene, *PavMYB10*, homologous to apple *MdMYB10* and *Arabidopsis AtPAP1*, was suggested to underlie these traits (Sooriyapathirana et al., 2010). By working on the same cross, Zhang et al. (2010) investigated QTLs for fruit size and its components and found two significant QTLs on LG2 and

LG6. On LG2, a QTL associated with cell number co-located with a fruit size QTL, suggesting that increased fruit size (weight, length and diameter) is associated with an increase in mesocarp cell number. On LG6, pit length and diameter QTLs clustered with those for fruit size, suggesting that the underlying morphological bases of this QTL are differences in pit size. Based on these results, De Franceschi et al. (2013) reported that two cell number regulator genes, which had been previously reported to be involved in fruit size control in tomato, were located within the confidence intervals of the QTLs detected on LGs 2 and 6 in sweet cherry. These two promising candidate genes were named *PavCNR*12 and *PavCNR*20, respectively, and appeared to also be involved in fruit weight control in sour cherry. Nevertheless, Rosyara et al. (2013) detected additional regions of the genome associated with fruit weight in sweet cherry using a five-generation pedigree consisting of 23 founders and parents and 424 progeny individuals. The situation among such a wide set of breeding germplasm is, as expected, more complex than initially reported for a single segregating family, because six QTLs were identified: three on LG2 and one each on LGs 1, 3 and 6. Campoy et al. (2015) conducted QTL analyses on fruit weight and fruit firmness, with two F_1 families derived from the crosses 'Regina' × 'Lapins' and 'Regina' × 'Garnet'. An additional fruit weight QTL was detected on LG5 and having an interesting co-localization with a fruit firmness QTL. Within this study, fruit weight and fruit firmness appeared as highly polygenic but negatively correlated. More recently, Quero-García et al. (2017b) reported a large firmness QTL on LG4 using an F_1 family derived from the cross 'Fercer' × 'X' (the paternal parent is unknown). Interestingly, a large QTL was also found in a nearby region for the trait maturity date, which had already been reported in that genomic region in other *Prunus* crops such as peach and apricot (Salazar et al., 2016; Dirlewanger et al., 2012; Eduardo et al., 2011; Quilot et al., 2004). This maturity date QTL was also found on LG4 in a sweet cherry F_1 family derived from the cross 'Beniyutaka' × 'Benikirari' (Isuzugawa et al., 2017), and two candidate genes of the NAC transcription factor, which had already been identified in peach (Pirona et al., 2013), mapped within the QTL confidence interval. Whether the maturity date and firmness QTLs on LG4 are the result of two closely linked genes or a pleiotropic gene stills needs to be elucidated. Finally, a study conducted by working with an F_1 family derived from the cross 'Ambrunés' × 'Sweetheart' detected a new large QTL for firmness on LG1 and confirmed a previously identified QTL on LG6 (Balas et al., 2017). QTL detection studies have also been conducted for rain-induced cracking tolerance by evaluating the proportion of cracked fruits at maturity and differentiating between three types of cracks: pistillar end, stem cavity and fruit side (Balbontin et al., 2013). This trait proved to be highly complex; most QTLs were not repeatable among years, explained a relatively low proportion of phenotypic variance and were genomically positioned with large confidence intervals. Nevertheless, at least three QTLs appeared to be relatively stable and could explain up to 25% of phenotypic variance in favourable years: one on LG5 for pistillar end cracking, one on LG2 for fruit side cracking and one on LG4 for stem cavity cracking (Quero-García et al., 2017b,c; Quero-García et al., 2014). Finally, two stable QTLs have been detected for fruit acidity on LGs 1 and 6, the latter explaining up to 30% of the phenotypic variance (Quero-García et al., 2017b,c).

Concerning other agronomic traits, several studies have attempted to unravel the genetic determinism of bloom date, productivity and disease resistance in sweet and sour cherry. In sour cherry, two bloom time QTLs, *blm1* and *blm2*, were detected in an F_1 family derived from the cross 'Rheinische Schattenmorelle' × 'Érdi bőtermő'; unfortunately, the genetic effects of the QTL alleles from 'Érdi bőtermő' were to induce early bloom compared to average (Wang et al., 2000). More recently, a QTL was reported on LG4 using an F_1 family derived from 'Újfehértói fürtös' × 'Surefire', which mapped near the *Lb* ('Late Blooming') locus previously identified for almond (Quero-García et al., 2017b). Although the QTL on LG4 had the largest effect on bloom date, QTLs were also identified on LGs 1, 2 and 5 (Cai et al., 2018). In sweet cherry, bloom date QTLs have been found in numerous LGs but the most significant was also detected in the same genomic region of LG4 as for sour cherry, being heterozygous in the late blooming cultivar, 'Regina' (Castède et al., 2014; Dirlewanger et al., 2012). Bloom date was dissected into chilling (CR) and heat (HR) requirements and several co-localizations between CR, HR and bloom date QTLs were reported. Candidate genes underlying the QTL on LG4 were investigated (Dirlewanger et al., 2017; Castède et al., 2015) and are currently being studied through fine-mapping and functional approaches. For productivity, measured qualitatively (score on a scale from 1 to 9), two relatively stable QTLs were recently reported on LGs 1 and 6, each explaining between 10% and 25% of the phenotypic variance depending on the year of study (Quero-García et al., 2017b). To date, the only QTL for disease resistance was reported for cherry leaf spot. By using an F_1 interspecific family of *P. canescens* × *P. avium* (Schuster, 2005), Stegmeir et al. (2014a) identified a significant QTL on LG4, named CLSR_G4. This was validated in *P. canescens* × *P. cerasus* progenies; however, presence of this allele alone did not confer resistance (Stegmeir et al., 2014a).

Overall, numerous QTLs are now available for conversion into trait-predictive DNA tests (steps described by Vanderzande et al., 2018) and others will soon be reported. For example, the recently constructed genetic maps derived from the landrace 'Cristobalina' (Calle et al., 2018) should allow identification of the genomic region(s) responsible for the extra-early bloom date and, hence, the low-chilling requirements, conferred by this cultivar, and the *Pmr1* locus for powdery mildew resistance reported by Olmstead et al. (2001) has been genetically mapped (Y. Zhao, *pers. comm.*, 2018).

5.1.2 DNA tests available for MAS

The first genetic markers made available to cherry breeders were those associated with the *S*-locus on LG6 controlling the GSI system. DNA markers for cross-(in)compatibility and self-compatibility (SC) are based on known base pair and insertion/deletion differences in the DNA sequence of the two *S*-locus genes, the stylar *S-RNase* gene and the pollen-part *F-box* gene (SFB) (reviewed by Herrero et al., 2017). The DNA sequence differences can be targeted by DNA tests to determine if a cross is compatible, to test seedling parentage in cases where contamination may have occurred during crossing and to select for SC at the seedling or later breeding stages. The *S*-genotypes of many sweet and sour cherry cultivars have been published and cultivars are assigned to incompatibility groups (Schuster, 2012, 2018; Sebolt et al., 2017). The predominant SC mutation used in sweet cherry breeding and production is the self-compatible pollen-part mutant S_4' allele that was originally obtained by radiation of 'Napoleon' pollen and led to the cultivar 'Stella' (Lapins, 1975; Lewis, 1949). DNA markers have been developed for detecting this SC mutation, distinguishing it from the non-mutated, common S_4 allele (Haldar et al., 2010; Sonneveld et al., 2005; Ikeda et al., 2004; Zhu et al., 2004). Other sources of SC in sweet cherry are the Italian landrace 'Kronio' (Calabrese et al., 1984) and the Spanish landraces 'Cristobalina' and 'Talegal Ahin' (Cachi and Wünsch, 2014b; Wünsch and Hormaza, 2004b) for which DNA markers have also been developed (Cachi and Wünsch, 2011, 2014b; Marchese et al., 2007). As sour cherry is a tetraploid crop and pollen grains contain two *S*-alleles, the genetics of SI/SC is more complex than for sweet cherry. Indeed, SC of sour cherry is due to the presence of mutations in the *S*-locus genes, resulting in non-functional *S*-haplotypes (Hauck et al., 2006). DNA-based *S*-genotyping of sour cherry cultivars is carried out as in sweet cherry but is more complicated because it involves distinguishing among as many as three variants for a single ancestral *S*-allele haplotype. For example, along with a wild-type functional S_{13} allele, there exists a non-functional pollen-part mutant of this haplotype, S_{13}' and a non-functional stylar-part mutant of this haplotype, S_{13m} (Sebolt et al., 2017). Therefore, the discrimination of certain *S*-haplotypes requires using allele-specific markers and, once the *S*-haplotypes present in each germplasm individual are identified, additional tests may be required for detection of non-functional mutants for that haplotype. dCAP markers have been specifically designed for each mutation (Sebolt et al., 2017; Tsukamoto et al., 2008).

The first QTL for which diagnostic markers were developed and implemented in cherry breeding is the fruit size QTL identified on LG2 (Ru et al., 2015; Zhang et al., 2010). This QTL is flanked by two polymorphic SSR markers, CPSCT038 and BPPCT034, which can be used to define QTL haplotypes (Rosyara et al., 2013). Seven and four alleles have been reported for CPSCT038 and BPPCT034, respectively, providing various possible QTL haplotypes. Several of the LG2 haplotypes are significantly associated with small fruit size, therefore if parents are used that have a small-fruited haplotype (such as CPSCT038-192 and BPPCT034-225), marker-assisted seedling selection can be used to identify these seedlings and eliminate them prior to field planting (Peace, 2011). As the SSR markers that are used to define the LG2 haplotype are approximately 16 cM apart (Klagges et al., 2013), when genotyping for this region, approximately one-sixth of the seedlings will have inherited a recombinant haplotype. Probably because of the co-location of QTLs for several valuable breeding traits, recombination in the LG2 region contributed to several important modern cultivars (Cai et al., 2017). Hence, fine-mapping studies are needed to more precisely locate the loci influencing fruit size and other traits in this 16 cM region, and the development of additional DNA markers that span this region at smaller cM intervals to track recombination. To maximize the probability of planting offspring that will produce large fruit, diagnostic tests are also required for other significant fruit size QTLs stable over several genetic backgrounds, such as those on LGs 3, 5 and 6.

For the sweet cherry skin and flesh colour, Mendelian trait locus was positioned on LG3, mahogany fruit colour (such as that of 'Bing' fruit) being dominant to blush colour fruit (such as that of 'Rainier' fruit) (Sooriyapathirana et al., 2010). A DNA test using an SSR marker for the *MYB10* locus (also named *Rf*) has been developed from SNP array haplotypes and implemented for routine prediction in breeding (Sandefur et al., 2016). The genotypic classes that can be observed within offspring are as follows: *rfrf* (blush skin colour and yellow-white flesh), *Rfrf* (red to mahogany skin and flesh colour) and *RfRf* (usually dark mahogany skin and flesh colour). For those programmes that do not seek to develop blush-type cultivars, this test can be used to cull seedling carrying the *rfrf* genotype. For example, in a cross between cultivars 'Regina' and 'Bing', both of which are *Rfrf*, one quarter of the offspring are predicted to have blushed skin and light-coloured flesh. In sour cherry, SNP haplotypes built from SNPs of the 6K Illumina® II SNP array (Peace et al., 2012) were used to define thirteen haplotypes spanning the *MYB10* region, and a SSR marker was subsequently developed that uniquely identifies the *D1* haplotype associated with dark purple flesh colour in sour cherry (Stegmeir et al., 2015). This marker can be used to distinguish germplasm individuals with dark purple flesh colour from those with lighter flesh colour like 'Montmorency'.

A genetic test was developed for QTL associated with *P. canescens*-derived cherry leaf spot resistance (*CLSR_G4*) (Stegmeir et al., 2014a). Four SSR markers were developed to detect the presence of *P. canescens* resistance allele. Any one of four markers can be used for marker-assisted seedling selection in families derived from *P. canescens*; offspring not having the resistance allele can be discarded prior to field planting.

5.1.3 Deployment strategies

In the area of DNA-informed breeding (Peace, 2017), considerable resources, tools, knowledge and skills have been accumulated for sweet cherry, as described above, such that routine MAS, and other genetic marker applications, are now underway in cherry breeding. One example of a breeding programme that has developed a strategy for routinely using DNA tests is that conducted by Washington State University (WSU).

DNA tests are used for parent and seedling evaluation in a multilayered way. The deployment strategy for each genetic assay depends on what it predicts in the context of breeding targets. Major influences on breeding targets are the regional cherry industry's market class segmentation that is primarily by harvest season and fruit colour type (Oraguzie et al., 2017) and the relative value placed on each genetically achievable threshold (Gallardo et al., 2015a,b; Yue et al., 2014). Deployment first occurs before families are created: crosses are planned that efficiently target desired seedling numbers for each market class, based on trait locus genotypes of parents especially for fruit colour (Sandefur et al., 2016) and maturity date as well as ensuring cross-compatibility of pairs. Also, crosses are designed where possible that do not segregate for undesirable trait locus genotypes, eliminating the need for running such tests on the families. Next, soon after seedling germination in the greenhouse, a DNA test for SC (Haldar et al., 2010) is used on all families segregating for the SC allele. For equivalent reasons, this stage uses the SSR-based DNA test for fruit size for which only those seedlings carrying alleles associated with small or tiny fruit are culled. For remaining seedlings, often less than a quarter of the original number, DNA tests are applied for sorting using the market-class-defining DNA tests for maturity date and fruit colour. Seedlings are sorted into predicted categories of early season mahogany, early season blush, mid-to-late-season mahogany and mid-to-late-season blush. Because each market class has certain needs and thresholds for trait performance, the fourth DNA test deployment step involves running certain DNA tests or applying the cut at certain genotypes. For example, a DNA test for powdery mildew resistance is applied rigorously to the mid-to-late-season category, as susceptibility there is not tolerated. The DNA test distinguishes desirable seedlings carrying one or two dominant-effect alleles for a resistance Mendelian trait locus from those homozygous for the locus's susceptible allele. Similarly, the mid- and late seasons have a more stringent requirement for large fruit, so that only seedlings with the best genotypes of the already applied fruit size DNA test are kept. Finally, if remaining seedling numbers for certain categories allow for further reductions, selection for enhancing attributes is performed. In this last step, higher stringencies can be applied for previously run DNA tests at no further cost – for example, keeping only largest-fruit-size genotypes of early season seedlings. Or additional DNA tests are used that can identify seedlings with valuable but non-essential attributes – for example, using a DNA test for rain-induced fruit cracking that distinguishes rare individuals predicted to be cracking resistant from those predicted to have some susceptibility.

DNA tests are also used in the WSU programme for characterization (identity/relatedness applications) as well as for evaluation (trait performance applications) described above (Peace, 2017). As a by-product from use of the above DNA tests, the genotypic information obtained is routinely used on a family-wide basis to confirm parentage, especially in families for which seedlings are worthless if the intended pollination was unsuccessful. Parentage deduction is also common, especially where the intended father or, even both parents, is found to be incorrect in large numbers – knowledge gained can improve subsequent crossing plans. Parentage confirmations and deductions for individual plants are usually saved until they become elite selections, which is the point at which gaining such knowledge is worth the effort.

5.1.4 Future trends

The capability to obtain routine genome scans of sweet cherry, enabled whether by SNP array genotyping (Peace et al., 2012) or by genotype-by-sequencing (GBS), is powerful for both evaluation and characterization applications. Such genome scanning has been put to breeding use in revealing alleles to target in selection (Cai et al., 2017), determining genome-wide breeding values of breeding germplasm across numerous traits (Piaskowski et al., 2018), predicting maturity date into new environments (Hardner et al., 2018) and predicting bloom date within bi-parental crosses (Fouché et al., 2018). Numerous further applications beckon (Peace, 2017). However, array costs as well as data curation needs currently limit applications to elite germplasm.

Although SSR markers have proven to be very efficient to conduct MAS for certain agronomic traits in cherry, SNP markers are becoming more efficient because SNP genotyping platforms, from single SNPs to dozens to thousands are dropping per-sample costs faster than is occurring for SSR and most other marker types. Low-cost genotyping platforms such as Kompetitive Allele Specific PCR (KASP) are becoming suitable for MAS for many breeding programmes. KASP assays offer high-throughput genotyping that is suited to screening thousands of breeding germplasm individuals for just the few SNPs required to reveal trait locus genotypes with quick turnaround times. This technique has been recently tested in sweet cherry (Fouché et al., 2018). SNP markers located in the QTL regions influencing fruit weight (LG2) and firmness (LG5) (Campoy et al., 2015) were identified and used to design allele-specific primers for the KASP platform. To validate the trait predictiveness of these KASP markers, a genotyping assay was performed on a subset of 'Regina' × 'Lapins' and 'Regina' × 'Garnet' offspring (32 individuals for each cross), as well as 31 accessions of the INRA sweet cherry core collection which had already been genotyped with the 6K array (Campoy et al., 2016). The KASP marker developed for the LG2 fruit weight QTL was able to discriminate three genotypic groups that were significantly associated with fruit weight among these individuals and individuals of the core collection. The KASP marker developed for the LG5 fruit firmness QTL could distinguish firm or soft fruit offspring in the segregating families but not among accessions of the core collection, the latter likely also segregating for the large-effect LG4 firmness QTL. This preliminary analysis indicates that KASP markers could be used in MAS for sweet cherry breeding programmes to help develop cultivars with large and firm fruit. Soon it should also be possible to characterize several KASP markers at a trait locus with more than two effective alleles to define haplotypes that will be more informative for breeding purposes than single bi-allelic SNPs (Peace, 2017), as demonstrated by Tan et al. (2017) for a greenbug resistance in wheat.

6 Phenotyping protocols

Undoubtedly, many more efforts have been devoted to genomics rather than phenomics research. Nevertheless, some progress has been achieved in particular in the field of biotic and abiotic stresses. Concerning bacterial canker, caused by *Pseudomonas* spp., two inoculation methods have been proposed: a leaf node method with inoculation of young seedlings with a mixture of bacteria (Krzesinska and Azarenko, 1992) and a bark inoculation on older trees (Kappel et al., 2012). More recently, rapid screening methods based on immature fruitlet tests have been proposed (Ozaktan, 2015; Kałużna and Sobiczewski, 2014). Mgbechi-Ezeri et al. (2017) reported the assessment of sweet cherry genotypes for response to bacterial canker by using three types of inoculation under controlled conditions: leaf wounding with carborundum, cut wounds in mid-rib and shoot tip. The last method provided the clearest differentiation of susceptibility among the commercial cultivars and advanced selections that were evaluated. Artificial inoculation tests of twigs with conidia suspension of *Monilinia* spp., to evaluate blossom blight resistance, have been described by pioneering work of Schmidt (1937). Concerning brown rot provoked by *Monilinia* spp. on fruits, artificial tests with *Monilinia fructicola* conidia suspensions have been reported on sweet cherry by Kappel and Sholberg (2008) and on both sweet and sour cherry by Brown and Wilcox (1989). Unfortunately, no sources of true resistance were identified within the cultivars analysed. Finally, Wharton et al. (2003) and Schuster (2004) established artificial inoculation test methods with leaf disk in the laboratory to assess cherry leaf spot resistance, which have been incorporated for routine use in different sour cherry breeding programmes, both in the United States and in Europe. In these studies, different levels of resistance were found among sour cherry cultivars, related species (such as *P. canescens*, *P. maackii*, *P. incisa* etc.) but all tested sweet cherry cultivars appeared as susceptible.

Agronomic traits related to fruit quality, such as firmness, soluble sugar content (SSC), titratable acidity (TA) and dry matter content (DMC) are highly important for cherry growers and consumers but are difficult to evaluate on large numbers of samples. Hence, cherry breeders tend to make these time-consuming and destructive measurements only for genitors or advanced selections. Nevertheless, cost-effective and non-destructive analytical systems might facilitate this task and even allow breeders to characterize their materials at earlier stages. SSC can be easily evaluated with a digital refractometer in a very fast and cheap manner. Nevertheless, the evaluation of TA is much more tedious. Near-infrared spectroscopy (NIRS) has been recently gaining much attention as an alternative to classical measurement methods. Carlini et al. (2000) used benchtop NIR spectrophotometers to accurately predict cherry SSC and Lu (2001), by using similar technology obtained very strong models both for SSC and firmness, by working with two sweet cherry cultivars. More recently, Escribano et al. (2017) developed accurate models for a handheld instrument to predict SSC and DMC, by working with 'Chelan' and 'Bing' sweet cherry cultivars. As had already been reported by Crisosto et al. (2003), their consumer preference evaluation study demonstrated that DMC could be equal or superior to SSC for predicting flavour intensity and balance of sweet to sour taste. These studies have however concerned a limited number of cultivars and INRA Bordeaux started recently to test NIRS to predict SSC, TA and firmness by working with a subset of commercial cultivars and several genetic families (C. Lepoittevin, Bordeaux, France, 2018, *pers. comm.*). On sour cherry, NIRS was tested to predict TA but no clear correlation was found.

Other traits for which high-throughput phenotyping protocols are currently being investigated concern the characterization of the different phenological phases of development. For example, one trait which is clearly difficult to evaluate is the amount of chilling and heat requirements (CR and HR, respectively) needed for flowering. Although a certain level of correlation exists between bloom date and CR, it is not significant enough and a precise determination of CR is needed on advanced selections if breeders want to anticipate these selections' behaviours in areas with high frequency of mild winters. Although numerous studies have been conducted on cherries and other tree species in order to correlate different levels of metabolites concentration with the CR, no precise and effective marker for dormancy release has yet been found. Recently, NIRS has also been tested at INRA in order to look for correlation between NIRS spectra and a classic parameter used in dormancy studies, such as the average time to budbreak. The final aim will be to estimate a date of endodormancy release, without the need to conduct fastidious and destructive forcing tests, such as Tabuenca or single-node cutting tests (Lalanne-Tisné et al., 2017). Preliminary results conducted on a small number of cultivars from several forest and fruit species appeared as highly promising (Bonhomme et al., 2015).

7 Future trends and conclusion

One the of main challenges of sweet and sour cherry breeding is, as for many other agricultural crops, the need to enlarge the genetic base for the introgression of new favourable alleles for traits related to biotic and abiotic stresses and for traits such as SC. Breeders can search for favourable alleles within the cultivated elite germplasm pool, wild individuals or related species. The first difficulty for the latter germplasm types is access to them. Exchanges of genetic resources are not always straightforward (reviewed in Iezzoni et al., 2017), expeditions to collect new resources are expensive and might concern regions within the Caucasus area where political instability or military conflicts prevail. Nevertheless, several breeders have attempted in recent years to explore the wide diversity of sweet and sour cherry genetic resources in countries such as Turkey, Iran and Azerbaijan (López-Ortega, 2015). However, additional coordinated efforts at the European or wider level are needed to characterize, preserve and exchange cherry genetic resources, and this becomes urgent in a context of genetic erosion. Two problems are faced by breeders when using germplasm which is genetically distant from commercial cultivars: often the quality of fruit obtained after the first round of crosses is particularly low and laborious backcrossing is then needed; and successes in interspecific hybridization are rare. Obtaining a small number of seed in a crossing season is common even when conducting intra-specific crosses that have previously been successful, and the reasons for these failures are still unknown.

In the case of cherries, given the long juvenility period of these crops, fast cycling breeding approaches developed in other *Rosaceae* tree species (van Nocker and Gardiner, 2014) would be useful. One hurdle specific to *Prunus* crops is the low success of transformation, in most cases due to very low stability of transformed plants. Although several regeneration protocols have been tested, the only reported success concerns the sour cherry cultivar 'Montmorency' and the cherry rootstock 'GiSelA6' (reviewed in Song, 2004).

Modelling and statistical approaches should be more intensively deployed to design cherry 'ideotypes', evaluate genotype × environment interactions and predict the performance of advanced selections and cultivars in a wide range of pedo-climatic conditions, varying as well in terms of cultural practices. It would be interesting to decipher, from a genetics/ecophysiology point of view, why certain cultivars are so 'plastic' and productive in many different situations while others seem only adapted to certain conditions. In the medium to long term, breeders will need to incorporate alleles responsible for tolerance to biotic and abiotic stresses, but also for regularity of production, adaptation to changing climatic conditions, and suitability to reduced labour production systems.

Overall, considering all these aspects, one necessary goal will surely be to increase and strengthen collaboration among cherry breeders worldwide.

8 References

Avramidou, E., Ganopoulos, I. V. and Aravanopoulos, F. A. (2010). DNA fingerprinting of elite Greek wild cherry (*Prunus avium* L.) genotypes using microsatellite markers. *Forestry* 83, 527–33.

Balas, F., López-Corrales, M., Serradilla, M., Cai, L., Iezzoni, A. and Wünsch, A. (2017). Firmness QTL mapping using an 'Ambrunés' × 'Sweetheart' sweet cherry population. *8th International Cherry Symposium*, 5th - 9th June 2017, Yamagata, Japan.

Balbontin, C., Ayala, H., Bastias, R. M., Tapia, G., Ellena, M., Torres, C., Yuri, J. A., Quero-García, J., Rios J. C. and Silva, H. (2013). Cracking in sweet cherries: A comprehensive review from a physiological, molecular and genomic perspective. *Chilean Journal of Agricultural Research* 73, 66–72.

Balla, I. and Brozik, S. (1996). Embryo culture of sweet cherry hybrids. *Acta Horticulturae* 410, 385–6.

Bargioni, G. (1996). Sweet cherry scions: Characteristics of the principal commercial cultivars, breeding objectives and methods. In: Webster, A. D. and Looney, N. E. (Eds), *Cherries: Crop Physiology, Production and Uses.* CAB International, Wallingford, UK, pp. 73–112.

Beaver, J. A. and Iezzoni, A. F. (1993). Allozyme inheritance in tetraploid sour cherry (*Prunus cerasus* L.). *Journal of the American Society for Horticultural Science* 118, 873–7.

Bonhomme, M., Charpentier, J. P., Segura, V., Pâques, L., Audergon, J. M., Quero-García, J., Dirlewanger, E., Beauvieux, R., Wenden, B., Legave, J. M., Farrera, I., Davi, H., Jean, F. and Chuine, I. (2015). A la recherche de méthodes alternatives aux tests biologiques classiques pour la détermination de la date de sortie d'endodormance, la quête du haut débit par le Groupe de Travail Dormance de Perpheclim et d'ODS. Colloque francophone PHENOLOGIE, November 2015, Clermont-Ferrand, France.

Bors, R. H. (2005). Dwarf sour cherry breeding at the University of Saskatchewan. *Acta Horticulturae* 667, 135–40.

Børve, J., Ippolito, A., Tanović, B., Michalecka, M., Sanzani, S. M., Poniatowska, A., Mari, M. and Hrustić, J. (2017). Fungal diseases. In: Quero-García, J., Iezzoni, A., Pulawska, J. and Lang, G. (Eds), *Cherries: Botany, Production and Uses.* CABI, Wallingford, UK, pp. 338–64.

Bošković, R. and Tobutt, K. R. (1998). Inheritance and linkage relationships of isoenzymes in two interspecific cherry progenies. *Euphytica* 103, 273–86.

Bošković, R., Wolfram B., Tobutt K. R., Cerovic R. and Sonneveld, T. (2006). Inheritance and interactions of incompatibility alleles in tetraploid sour cherry. *Theoretical and Applied Genetics* 112, 315–26.

Brettin, T. S., Karle, R., Crowe, E. L. and Iezzoni, A. F. (2000). Chloroplast inheritance and DNA variation in sweet, sour, and ground cherry. *Journal of Heredity* 91, 75–9.

Brown, S. K. and Wilcox, W. F. (1989). Evaluation of cherry genotypes for resistance to fruit infection by *Monilinia fructicola* (Wint.) Honey. *HortScience* 24, 1013–15.

Brown, S. K., Iezzoni, A. F. and Fogle, H. W. (1996). Cherries. In: Janick, J. and Moore, J. N. (Eds), *Fruit Breeding, Vol. I. Tree and Tropical Fruits*. John Wiley & Sons, New York, NY, pp. 213–55.

Budan, S., Mutafa, I., Stoian, I. and Popescu, I. (2005a). Screening of 100 sour cherry genotypes for *Monilia laxa* field resistance. *Acta Horticulturae* 667, 145–51.

Budan, S., Mutafa, I., Stoian, I. and Popescu, I. (2005b). Field evaluation of cultivars susceptibility to leaf spot at Romania's sour cherry genebank. *Acta Horticulturae* 667, 153–7.

Cabrera, A., Rosyara, U. R., De Franceschi, P., Sebolt, A., Sooriyapathirana, S. S., Dirlewanger, E., Quero-García, J., Schuster, M., Iezzoni, A. F. and van der Knaap, E. (2012). Rosaceae conserved orthologous sequences marker polymorphism in sweet cherry germplasm and construction of a SNP-based map. *Tree Genetics & Genomes* 8, 237–47.

Cachi, A. M. and Wünsch, A. (2011). Characterization and mapping of non-S gametophytic self-compatibility in sweet cherry (*Prunus avium* L.). *Journal of Experimental Botany* 62, 1847–56.

Cachi, A. M. and Wünsch, A. (2014a). *S*-genotyping of sweet cherry cultivars from Spain and *S*-locus diversity in Europe. *Euphytica* 197, 229–36.

Cachi, A. M. and Wünsch, A. (2014b). Characterization of self-compatibility in sweet cherry cultivars by crossing experiments and molecular genetic analysis. *Tree Genetics and Genomes* 10, 1205–12.

Cai, L., Voorrips, R. E., van de Weg, E., Peace, C. and Iezzoni, A. (2017). Genetic structure of a QTL hotspot on chromosome 2 in sweet cherry indicates positive selection for favorable haplotypes. *Molecular Breeding* 37, 85.

Cai, L., Stegmeir, T., Sebolt, A., Zheng, C., Bink, M. C. A. M. and Iezzoni, A. (2018). Identification of bloom date QTLs and haplotype analysis in tetraploid sour cherry (*Prunus cerasus*). *Tree Genetics and Genomes* 14, 22.

Calabrese, F., Fenech, L. and Raimondo, A. (1984). Kronio: una cultivar di ciliegio molto precoce e autocompatibile. *Frutticoltura* 46, 27–30.

Calle, A., Cai, L., Iezzoni, A. and Wünsch, A. (2018). High-density linkage maps constructed in sweet cherry (*Prunus avium* L.) using cross- and self-pollination populations reveal chromosomal homozygosity in inbred families and non-syntenic regions with the peach genome. *Tree Genetics and Genomes* 14, 37.

Campoy, J. A., Le Dantec, L., Barreneche, T., Dirlewanger, E. and Quero-García, J. (2015). New insights into fruit firmness and weight control in sweet cherry. *Plant Molecular Biology Reporter* 22, 783–96.

Campoy, J. A., Lerigoleur-Balsemin, E., Christmann, H., Beauvieux, R., Girollet, N., Quero-García, J., Dirlewanger, E. and Barreneche, T. (2016). Genetic diversity, linkage disequilibrium, population structure and construction of a core collection of *Prunus avium* L., landraces and bred cultivars. *BMC Plant Biology*. doi:10.1186/s12870-016-0712-9.

Carlini, P., Massantini, R. and Mencarelli, F. (2000). Vis-NIR measurement of soluble solids in cherry and apricot by PLS regression and wavelength selection. *Journal of Agricultural and Food Chemistry* 48, 5236–42.

Castède, S., Campoy, J. A., Quero-García, J., Le Dantec, L., Lafargue, M., Barreneche, T., Wenden, B. and Dirlewanger, E. (2014). Genetic determinism of phenological traits highly affected by climate change in *Prunus avium*: Flowering date dissected into chilling and heat requirements. *New Phytologist* 202, 703–15.

Castède, S., Campoy, J. A., Le Dantec, L., Quero-García, J., Barreneche, T., Wenden, B. and Dirlewanger, E. (2015). Mapping of candidate genes involved in bud dormancy and flowering time in sweet cherry (*Prunus avium*). *PLoS ONE*. doi:10.1371/journal.pone.0143250.

Chavoshi, M., Watkins, C., Oraguzie, B., Zhao, Y., Iezzoni, A. and Oraguzie, N. (2014). Phenotyping protocol for sweet cherry (*Prunus avium* L.) to facilitate an understanding of trait inheritance. *Journal of the American Pomological Society* 68, 125–34.

Choi, C. and Kappel, F. (2004). Inbreeding, coancestry, and founding clones of sweet cherries from North America. *Journal of the American Society for Horticultural Science* 129, 535–43.

Clarke, J. B., Sargent, D. J., Bošković, R. I., Belaj, A. and Tobutt, K. R. (2009). A cherry map from the inter-specific cross *Prunus avium* 'Napoleon' × *P. nipponica* based on microsatellite, gene-specific and isoenzyme markers. *Tree Genetics & Genomes* 5, 41–51.

Crisosto, C. H., Crisosto, G. M. and Metheney, P. (2003). Consumer acceptance of 'Brooks' and 'Bing' cherries is mainly dependent on fruit SSC and visual skin color. *Postharvest Biology and Technology* 28, 159–67.

de Franceschi, P., Stegmeir, T., Cabrera, A., van der Knapp, E., Rosyara, U., Sebolt, A., Dondini, L., Dirlewanger, E., Quero-García, J., Campoy, J. and Iezzoni, A. (2013). Cell number regulator genes in *Prunus* provide candidate genes for the control of fruit size in sweet and sour cherry. *Molecular Breeding* 32, 311–26.

Demir, T., Demirsoy, L., Demirsoy, H., Kaçar, Y. A., Yilmaz, M. and Macit, I. (2009). Molecular characterization of sweet cherry genetic resources in Giresun, Turkey. *Fruits* 66, 53–62.

Dirlewanger, E., Graziano, E., Joobeur, T., Garriga-Calderé, F., Cosson, P., Howad, W. and Arús, P. (2004). Comparative mapping and marker-assisted selection in Rosaceae fruit crops. *Proceedings of the National Academy of Sciences* 101, 9891–6.

Dirlewanger, E., Claverie, J., Wünsch, A. and Iezzoni, A. F. (2007). Cherry. In: Kole, C. (Ed.), *Genome Mapping & Molecular Breeding, Volume 4: Fruits and Nuts*. Springer, Berlin, pp. 103–18.

Dirlewanger, E., Quero-García, J., Le Dantec, L., Lambert, P., Ruiz, D., Dondini, L., Illa, E., Quilot-Turion, B., Audergon, J-M., Tartarini, S., Letourmy, P. and Arùs, P. (2012). Comparison of the genetic determinism of two key phenological traits, flowering and maturity dates, in three *Prunus* species: Peach, apricot and sweet cherry. *Heredity* 109, 280–92.

Dirlewanger, E., Campoy, J. A., Wenden, B., Castède, S., Le Dantec, L., Barreneche, T. and Quero-García, J. (2017). Genetic analyses of chilling requirement and flowering date in sweet cherry, two key traits for breeding programs. *Acta Horticulturae* 1172:299–306. doi:10.17660/ActaHortic.2017.1172.57.

Eduardo, I., Pacheco, I., Chietera, G., Bassi, D., Pozzi, C., Vecchietti, A. and Rossini, L. (2011). QTL analysis of fruit quality traits in two peach intraspecific populations and importance of maturity date pleiotropic effect. *Tree Genetics & Genomes* 7(2), 323–35.

Ercisli, S., Agar, G., Yildirim, N., Duralija, B., Vokurka, A. and Karlidag, H. (2011). Genetic diversity in wild sweet cherries (*Prunus avium*) in Turkey revealed by SSR markers. *Genetics and Molecular Research Journal* 10(2), 1211–19.

Escribano, S., Biasi, W. V., Lerud, R., Slaughter, D. C. and Mitcham, E. J. (2017). Non-destructive prediction of soluble solids and dry matter content using NIR spectroscopy and its relationship with sensory quality in sweet cherries. *Postharvest Biology and Technology* 128, 112–20.

Faust, M. and Surányi, D. (1997). Origin and dissemination of cherry. In: Janick, J. (Ed.), *Horticultural Reviews*, Volume 19. John Wiley & Sons, Inc., New York, NY.

Fouché, M., Zaracho Echagüe, N. H., Flutre, T., Vimont, N., Le Dantec, L., Richard, L., Wenden, B., Quero-García, J, Barreneche, T and Dirlewanger, E. (2018). Last advances on sweet cherry genomic tools for the pre-breeding. *Plant and Animal Genome Conference*, 13–17 January, San Diego, USA.

Gallardo, R. K., Li, H., McCracken, V., Yue, C., Luby, J. and McFerson, J. R. (2015a). Market intermediaries' willingness to pay for apple, peach, cherry, and strawberry quality attributes. *Agribusiness* 31, 259–80.

Gallardo, R. K., Li, H., Yue, C., Luby, J., McFerson, J. R. and McCracken, V. (2015b). Market intermediaries' ratings of importance for rosaceous fruits' quality attributes. *International Food and Agribusiness Management Review* 18, 121–54.

Ganopoulos, I. V., Kazantzis, K., Chatzicharisis, I., Karayiannis, I. and Tsaftaris, A. (2011). Genetic diversity, structure and fruit trait associations in Greek sweet cherry cultivars using microsatellite based (SSR/ISSR) and morpho-physiological markers. *Euphytica* 181(2), 237–51.

Guajardo, V., Simon, S., Sagredo, B., Gainza, F., Munoz, C., Gasic, K. and Hinrichsen, P. (2015). Construction of high density sweet cherry (*Prunus avium* L.) linkage maps using microsatellite markers and SNPs detected by genotyping-by-sequencing (GBS). *PLoS ONE* 10(5), e0127750.

Guarino, C., Santoro, S., De Simone, L. and Cipriani, G. (2009). *Prunus avium*: Nuclear DNA study in wild populations and sweet cherry cultivars. *Genome* 52, 320–37.

Haldar, H., Haendiges, S., Edge-Garza, D., Oraguzie, N., Olmstead, J., Iezzoni, A. and Peace, C. (2010). Applying genetic markers for self-compatibility in the WSU Sweet Cherry Breeding Program. *Acta Horticulturae* 859, 375–80.

Hardner, C., Hayes, B., Kumar, S., Vanderzande, S., Cai, L., Piaskowski, J., Quero-García, J., Barreneche, T., Giovannini, D., Liverani, A., Charlot, G., Villamil-Castro, M., Campoy, J., Oraguzie, N. and Peace, C. (2018). Prediction of genetic value for sweet cherry fruit maturity among environments using a 6K SNP array. *Horticulture Research* (accepted).

Hauck, N. R., Yamane, H., Tao, R. and Iezzoni, A. F. (2006). Accumulation of non-functional *S*-haplotypes results in the breakdown of gametophytic self-incompatibility in tetraploid *Prunus*. *Genetics* 172, 1191–8.

Hedrick, U. P. (1915). The cherries of New York. Report of the New York Agricultural Experiment Station for the Year 1914 II. J.B. Lyon, Albany, NY .

Herrero, M., Rodrigo, J. and Wünsch, A. (2017). Flowering, fruit set and development. In: Quero-García, J., Iezzoni, A., Pulawska, J. and Lang, G. (Eds), *Cherries: Botany, Production and Uses*. CABI, Wallingford, UK, pp. 14–35.

Hrotko, K. and Rozpara, E. (2017). Rootstocks and improvement. In: Quero-García, J., Iezzoni, A., Pulawska, J. and Lang, G. (Eds), *Cherries: Botany, Production and Uses*. CABI, Wallingford, UK, pp. 117–39.

Iezzoni, A., Schmidt, H. and Albertini, A. (1990). Cherries (*Prunus*). In: Moore, J. N. and Ballington Jr., J. R (Eds), *Genetic Resources of Temperate Fruit and Nut Crops*, Volume 1. International Society for Horticultural Science, Wageningen, The Netherlands, pp. 111–73.

Iezzoni, A., Wünsch, A., Höfer, M., Giovannini, D., Jensen, M., Quero-García, J., Campoy, J. A., Vokurka, A. and Barreneche, T. (2017). Biodiversity, germplasm resources and breeding methods. In: Quero-García, J., Iezzoni, A., Pulawska, J. and Lang, G. (Eds), *Cherries: Botany, Production and Uses*. CABI, Wallingford, UK, pp. 36–59.

Ikeda, K., Watari, A., Ushijima, K., Yamane, H., Hauck, N. R., Iezzoni, A. F. and Tao, R. (2004). Molecular markers for the self-compatible $S_{4'}$-haplotype, a pollen-part mutant in sweet cherry (*Prunus avium* L.). *Journal of American Society of Horticultural Science* 129, 724–8.

Isuzugawa, K., Shirasawa, K., Kurosaka, S., Takahashi, Y., Saito, Y., Adachi, E., Mitsunobu, I. and Yamamoto, T. (2017). QTL analysis and candidate gene mapping for harvest day in sweet cherry (*Prunus avium* L.). *8th International Cherry Symposium*, 5th - 9th June, Yamagata, Japan.

Ivanicka, J. and Pretova, A. (1986). Cherry (*Prunus avium* L.). In: Bajai, Y. P. S. (Ed), *Biotechnology in Agriculture and Forestry, Volume 1: Trees 1*. Springer Verlag, Berlin, Germany, pp. 154–69.

Jensen, M. and Kristiansen, K. (2014). Aspects of freezing tolerance in sour cherry. *Proceedings of the EU COST action FA1104 Sustainable Production of High-quality Cherries for the European Market*. COST meeting Plovdiv, Bulgaria, May 2014. https://www.bordeaux.inra.fr/cherry/docs/dossiers/Activities/Meetings/2014%2005%2026-27%20WG1-WG3%20Meeting_Plovdiv/Presentations/Jensen_Plovdiv2014.pdf.

Joublan, J. P., Serri, H. and Ocompo, J. (2005). Evaluation of sweet cherry germplasm in southern Chile. *Acta Horticulturae* 667, 69–74.

Kaçar, Y. A., Çetiner, M. S., Cantini, C. and Iezzoni, A. F. (2006). Simple sequence repeat (SSR) markers differentiate Turkish sour cherry germplasm. *Journal of American Society of Horticultural Science* 60, 136–43.

Kałużna, M. and Sobiczewski, P. (2014). Bacterial canker of cherry - methods of susceptibility testing. *Proceedings of the EU COST Action FA1104 Sustainable Production of High-quality Cherries for the European market*. COST meeting Plovdiv, Bulgaria, May 2014. https://www.bordeaux.inra.fr/cherry/docs/dossiers/Activities/Meetings/2014%2005%2026-27%20WG1-WG3%20Meeting_Plovdiv/Presentations/Kaluzna_Plovdiv2014.pdf.

Kappel, F. and Sholberg, P. L. (2008). Screening sweet cherry cultivars from the Pacific Agri-Food Research Centre Summerland breeding program for resistance to brown rot (*Monilinia fructicola*). *Canadian Journal of Plant Science* 88, 747–52.

Kappel, F., Granger, A., Hrotkó, K. and Schuster, M. (2012). Cherry. In: Badenes, M. L. and Byrne, D. H. (Eds), *Fruit Breeding, Handbook of Plant & Breeding 8*. Springer Science + Business Media, LLC, New York, NY, pp. 459–504.

Klagges, C., Campoy, J. A., Quero-García, J., Guzman, A., Mansur, L., Gratacos, E., Silva, H., Rosyara, U. R., Iezzoni, A., Meisel, L. A. and Dirlewanger, E. (2013). Construction and comparative analyses of highly dense linkage maps of two sweet cherry intra-specific progenies of commercial cultivars. *PLoS ONE* 8, e54743.

Knoche, M. and Winkler, A. (2017). Rain-induced cracking of cherries. In: Quero-García, J., Iezzoni, A., Pulawska, J. and Lang, G. (Eds), *Cherries: Botany, Production and Uses*. CABI, Wallingford, UK, pp. 140–65.

Krzesinska, E. Z. and Azarenko, A. N. M. (1992). Excised twig assay to evaluate cherry rootstocks for tolerance to *Pseudomonas syringae* pv. *syringae*. *Hortscience* 27, 153–5.

Küssmann, G. (1962). *Handbuch der Laubgehölze*. Paul Parey Verlag, Berlin/Hamburg.

Lacis, G., Rashal, I., Ruisa, S., Trajkovski, V. and Iezzoni, A. F. (2009). Assessment of genetic diversity of Latvian and Swedish sweet cherry (*Prunus avium* L.) genetic resources collections by using SSR (microsatellite) markers. *Scientia Horticulturae* 121, 451–7.

Lalanne-Tisné, G., Quero-García, J., Lafargue, M., Joly, J., Fouilhaux, L., Dirlewanger, E., Costes, E. and Legave, J. M. (2017). Comparison of phenotypic methodologies to characterize chilling and heat requirements of peach and apple cultivars. *Acta Horticulturae*. doi:10.17660/ActaHortic.2017.1172.65.

Lapins, K. O. (1975). 'Compact Stella' cherry. *Fruit Cultivars Journal* 29, 20.

Lewis, D. (1949). Structure of the incompatibility gene. II. Induced mutation rate. *Heredity* 3, 339–55.

Liu, L., Pagter, M. and Andersen, L. (2012). Preliminary results on seasonal changes in flower bud cold hardiness of sour cherry. *European Journal of Horticultural Science* 77, 109–14.

López-Ortega, G. (2015). Field explorations of natural populations of sweet and sour cherries in Azerbaijan. *Proceedings of the EU COST Action FA1104 Sustainable Production of High-quality Cherries for the European Market*. COST meeting Dresden, Germany, July 2015. https://www.bordeaux.inra.fr/cherry/docs/dossiers/Activities/Short%20Term%20Scientific%20Missions/STSM%20Scientific%20Report_Lopez-Ortega%202.pdf.

Lu, R. (2001). Predicting firmness and sugar content of sweet cherries using near-infrared diffuse reflectance spectroscopy. *Transactions of the ASAE* 44, 1265–74.

Marchese, A., Bošković, R. I., Caruso, T., Raimondo, A., Cutuli, M. and Tobutt, K. R. (2007). A new self-compatibility haplotype in the sweet cherry 'Kronio', S5', attributable to a pollen-part mutation in the SFB gene. *Journal of Experimental Botany* 58(15/16), 4347–56.

Mariette, S., Tavaud, M., Arunyawat, U., Capdeville, G., Millan, M. and Salin, F. (2010). Population structure and genetic bottleneck in sweet cherry estimated with SSRs and the gametophytic self-incompatibility locus. *BioMed Central Genetics* 11, 77.

Mgbechi-Ezeri, J., Porter, K. L., Johnson, K. B. and Oraguzie, N. C. (2017). Assessment of sweet cherry (*Prunus avium* L.) genotypes for response to bacterial canker disease. *Euphytica*. doi:10.1007/s10681-017-1930-4.

Mitschurin, I. W. (1951). *Ausgewählte Schriften*. Verlag Kultur und Fortschritt, Berlin.

Olden, E. J. and Nybom, N. (1968). On the origin of *Prunus cerasus* L. *Hereditas* 59, 327–45.

Olmstead, J. W., Lang, G. A. and Grove, G. G. (2001). Inheritance of powdery mildew resistance in sweet cherries. *HortScience* 36, 337–40.

Olmstead, J., Sebolt, A., Cabrera, A., Sooriyapathirana, S., Hammar, S., Iriart, G., Wang, D., Chen, C. Y., van der Knapp, E. and Iezzoni, A. F. (2008). Construction of an intra-specific sweet cherry (*Prunus avium* L.) genetic linkage map and synteny analysis with the *Prunus* reference map. *Tree Genetics & Genomes* 4, 897–910.

Oraguzie, N. C., Watkins, C. S., Chavoshi, M. S. and Peace, C. (2017). Emergence of the Pacific Northwest sweet cherry breeding program. *Acta Horticulturae* 1161, 73–7.

Ou, B., Bosak, K. N., Brickner, P. R., Iezzoni, D. G. and Seymour, E. M. (2012). Processed tart cherry products – comparative phytochemical content, *in vitro* antioxidant capacity and *in vitro* anti-inflammatory activity. *Journal of Food Science* 77, 105–12.

Ozaktan, H. (2015). Screening the susceptibility of some sweet cherry cultivars to *Pseudomonas syringae* pv. *syringae* isolates by immature fruitlet test. *Proceedings of the EU COST Action FA1104 Sustainable Production of High-quality Cherries for the European Market*. COST meeting Plovdiv, Bulgaria, May 2014. https://www.bordeaux.inra.fr/cherry/docs/dossiers/Activities/Meetings/2014%2005%2026-27%20WG1-WG3%20Meeting_Plovdiv/Presentations/Ozaktan_Plovdiv2014.pdf.

Papadopoulos, N. T., Lux, S. A., Köpler, K. and Bëlien, T. (2017). Invertebrate and vertebrate pests: Biology and management. In: Quero-García, J., Iezzoni, A., Pulawska, J. and Lang, G. (Eds), *Cherries: Botany, Production and Uses*. CABI, Wallingford, UK, pp. 305–37.

Peace, C. (2011). A success story in Rosaceae marker-assisted breeding: Larger fruit for sweet cherry. What can we learn? Community Breeders' Page, *RosBREED Quarterly Newsletter* 2(4), 7.

Peace, C. (2017). DNA-informed breeding of rosaceous crops: Promises, progress, and prospects. *Horticulture Research* 4, 17006.

Peace, C., Bassil, N., Main, D., Ficklin, S., Rosyara, U. R., Stegmeir, T., Sebolt, A., Gilmore, B., Lawley, C., Mockler, T. C., Bryant, D. W., Wilhelm, L. and Iezzoni, A. (2012). Development and evaluation of a genome-wide 6K SNP array for diploid sweet cherry and tetraploid sour cherry. *PLoS ONE* 7(12), e48305.

Pérez-Sánchez, R., Gómez-Sánchez, M. A. and Morales-Corts, R. (2008). Agromorphological characterization of traditional Spanish sweet cherry (*Prunus avium* L.), sour cherry (*Prunus cerasus* L.) and duke cherry (*Prunus* × *gondouinii* Rehd.) cultivars. *Spanish Journal of Agricultural Research* 6, 42–55.

Piaskowski, J., Hardner, C., Cai, L., Iezzoni, A., Zhao, Y. and Peace, C. (2018). Genomic heritability estimates in sweet cherry indicate non-additive genetic variance is relevant for industry-prioritized traits. *BMC Genomics* 19, 23.

Pirona, R., Eduardo, I., Pacheco, I., Da Silva Ling, C., Miculan, M., Verde, I., Tartarini, S., Dondini, L., Pea, G., Bassi, D. and Rossini, L. (2013). Fine mapping and identification of a candidate gene for a major locus controlling maturity date in peach. *BMC Plant Biology*. doi:10.1186/1471-2229-13-166.

Poll, L., Peterson, M. B. and Nielson, G. S. (2003). Influence of harvest year and harvest time on soluble solids, titrateable acid, anthocyanins content and aroma components in sour cherry (*Prunus cerasus* L. cv. 'Stevnsbaer'). *European Food Research and Technology* 216, 212–16.

Puławska, J., Getaz, M., Kałużna, M., Kuzmanović, N., Obradović, A., Pothier, J. F., Ruinelli, M., Boscia, D., Saponari, M., Vegh, A. and Palkovics, L. (2017). Bacterial diseases. In: Quero-García, J., Iezzoni, A., Pulawska, J. and Lang, G. (Eds), *Cherries: Botany, Production and Uses*. CABI, Wallingford, UK, pp. 365-85.

Quero-García, J., Fodor, A., Reignier, A., Capdeville, G., Joly, J., Tauzin, Y., Fouilhaux, L. and Dirlewanger, E. (2014). QTL detection of important agronomic traits for sweet cherry breeding. *Acta Horticulturae* 1020, 57–64.

Quero-García, J., Schuster, M., López-Ortega, G. and Charlot, G. (2017a). Sweet cherry cultivars and improvement. In: Quero-García, J., Iezzoni, A., Pulawska, J. and Lang, G. (Eds), *Cherries: Botany, Production and Uses*. CABI, Wallingford, UK, pp. 60–94.

Quero-García, J., Campoy, J. A., Barreneche, T., Le Dantec, L., Wenden, B., Fouché, M., Dirlewanger, E., Silva, H., Cai L and Iezzoni, A. (2017b). Present and future of marker-assisted breeding in sweet and sour cherry. *VIII International Cherry Symposium*, 5–9 June 2017, Yamagata, Japan.

Quero-García, J., Campoy, J. A., Castède, S., Pitiot, C., Barreneche, T., Lerigoleur-Balsemin, E., Wenden, B., Le Dantec, L. and Dirlewanger, E. (2017c). Breeding sweet cherries at INRA-Bordeaux: From conventional techniques to marker-assisted selection. *Acta Horticulturae* 1161, 1–14.

Quilot, B., Wu, B. H., Kervella, J., Genard, M., Foulongne, M. and Moreau, K. (2004). QTL analysis of quality traits in an advanced backcross between *Prunus persica* cultivars and the wild relative species *P. davidiana*. *Theoretical and Applied Genetics* 109(4), 884-97.

Redalen, G. (1984). Fertility in sour cherries. *Gartenbauwissenschaft* 49, 212–17.

Rodrigues, L. C., Morales, M. R., Fernandes, A. J. B. and Ortiz, J. M. (2008). Morphological characterization of sweet and sour cherry cultivars in a germplasm bank at Portugal. *Genetic Resources and Crop Evolution* 55, 593-601.

Rosyara, U., Bink, C. A. M., van de Weg, E., Zhang, G., Wang, D., Sebolt, A., Dirlewanger, E., Quero-García, J., Schuster, M. and Iezzoni, A. (2013). Fruit size QTL identification and the prediction of parental QTL genotypes and breeding values in multiple pedigreed populations of sweet cherry. *Molecular Breeding* 32, 875–87.

Ru, S., Main, D., Evans, K. and Peace, C. (2015). Current applications, challenges, and perspectives of marker-assisted seedling selection in Rosaceae tree fruit breeding. *Tree Genetics and Genomes* 11, 8.

Salazar, J. A., Ruiz, D., Campoy, J. A., Tartarini, S., Dondini, L. and Martínez-Gómez, P. (2016). Inheritance of reproductive phenology traits and related QTL identification in apricot. *Tree Genetics and Genomes* 12, 71.

Sandefur, P., Oraguzie, N. and Peace, C. (2016). A DNA test for routine prediction in breeding of sweet cherry fruit color, Pav-Rf-SSR. *Molecular Breeding* 36, 33. doi:10.1007/s11032-016-9458-y.

Sansavini, S. and Lugli, S. (2008). Sweet cherry breeding programs in Europe and Asia. *Acta Horticulturae* 795, 41–58.

Schmid, W. and Grosch, W. (1986). Identifizierung flüchtiger Aromastoffe mit hohen Aromawerten in Sauerkirschen (*Prunus cerasus* L.). *Zeitschrift Lebensmittel Untersuchung Forschung* 182, 407–12.

Schmidt, M. (1937). Infektionversuche mit *Sclerotina cinerea* an Süß- und Sauerkirschen. *Gartenbauwisswissenschaft* 11, 167–82.

Schuster, M. (2004). Investigation on resistance to leaf spot disease, *Blumeriella jaapii* in cherries. *Journal of Fruit and Ornamental Plant Research* 12, 275–9.

Schuster, M. (2005). Meiotic investigations in a *Prunus avium* × *P. canescens* hybrid. *Acta Horticulturae* 667, 101–2.

Schuster, M. (2011). Sauerkirsche 'Achat'. *JKI Datenblätter – Obstsorten*. Heft 2. doi:10.5073/jkidos.2012.00.

Schuster, M. (2012). Incompatible (*S*-) genotypes of sweet cherry cultivars (*Prunus avium* L.). *Scientia Horticulturae* 148, 59–73.

Schuster, M. (2013). Resistance breeding in cherries – goals and results. *Proceedings of the EU COST Action FA1104 Sustainable Production of High-quality Cherries for the European Market*. COST meeting October 2013, Pitesti, Romania. https://www.bordeaux.inra.fr/cherry/docs/dossiers/Activities/Meetings/15-17%2010%202013_3rd%20MC%20and%20all%20WG%20Meeting_Pitesti/Presentations/16_Schuster_Mirko.pdf.

Schuster, M. (2018). Self-incompatibility (*S*) genotypes of cultivated sweet cherries – An overview 2017 https://www.openagrar.de/receive/openagrar_mods_00034381. doi:10.5073/20171213-111734.

Schuster, M. and Schreiber, H. (2000). Genome investigations in sour cherry, *P. cerasus* L. *Acta Horticulturae* 538, 375–9.

Schuster, M., Flachowsky, H. and Köhler, D. (2007). Determination of self-incompatible genotypes in sweet cherry (*Prunus avium* L.) accessions and cultivars of the German Fruit Gene Bank and from private collections. *Plant Breeding* 126, 533–40.

Schuster, M., Grafe, C., Hoberg, E. and Schütze, W. (2013). Interspecific hybridization in sweet and sour cherry breeding. *Acta Horticulturae* 976, 79–86.

Schuster, M., Apostol, J., Iezzoni, A., Jensen, M. and Milatović, D. (2017). Sour cherry cultivars and improvement. In: Quero-García, J., Iezzoni, A., Pulawska, J. and Lang, G. (Eds), *Cherries: Botany, Production and Uses*. CABI International, Wallingford, UK, pp. 95–116.

Sebolt, A. M., Iezzoni, A. F. and Tsukamoto, T. (2017). *S*-genotyping of cultivars and breeding selections of sour cherry (*Prunus cerasus* L.) in the Michigan State University Sour Cherry breeding program. *Acta Horticulturae* 1161, 31–40.

Serradilla, M. J., Fotirić Akšić, M., Manganaris, G. A., Ercisli, S., González-Gómez, D. and Valero, D. (2017). Fruit chemistry, nutritional benefits and social aspects of cherries. In: Quero-García, J., Iezzoni, A., Pulawska, J. and Lang, G. (Eds), *Cherries: Botany, Production and Uses*. CABI International, Wallingford, UK, pp. 420–41.

Šimunic, V., Kovač, S., Gašo-Sokač, D., Pfannhauser, W. and Murkovic, M. (2005). Determination of anthocyanins in four Croatian cultivars of sour cherries (*Prunus cerasus*). *European Food Research and Technology* 220, 575–8.

Skipper, E., Sargent, D. J. and Fernández-Fernández, F. (2014). Linkage map development using the cherry 6k whole genome genotyping array and the identification of a novel locus controlling flesh colour. *Proceedings of the EU COST Action FA1104 Sustainable Production of High-quality Cherries for the European Market*. COST meeting Plovdiv, Budapest, March 2014. https://www.bordeaux.inra.fr/cherry/docs/dossiers/Activities/Meetings/03-05%2003%202014%20WG1%20Use%20of%20molecular%20markers%20for%20diversity%20studies/Skipper.pdf.

Song, G. Q. (2014). Recent advances and opportunities in cherry biotechnology. *Acta Horticulturae* 1020, 89–98.

Sonneveld, T., Tobutt, K. R., Vaughan, S. P. and Robbins, T. P. (2005). Loss of pollen-*S* function in two self-compatible selections of *Prunus avium* is associated with deletion/mutation of an *S* haplotype-specific F-box gene. *Plant Cell* 17, 37–51.

Sooriyapathirana, S. S., Khan, A., Sebolt, A. M., Wang, D., Bushakra, J. M., Lin-Wang, K., Allan, A. C., Gardiner, S. E., Chagne, D. and Iezzoni, A. F. (2010). QTL analysis and candidate gene mapping for skin and flesh color in sweet cherry fruit (*Prunus avium* L.). *Tree Genetics and Genomes* 6, 821–32.

Stainer, R. (1975). 'Stevnsbaer' - eine interessante Sauerkirsche für die Safterzeugung. *Obstbau/Weinbau* 5, 142–5.

Stanys, V., Stanytė, R., Stanienė, G. and Vinskienė, J. (2008). *S*-allele identification by PCR analysis in Lithuanian sweet cherries. *Biologija* 54, 22–6.

Stegmeir, T., Schuster, M., Sebolt, A., Rosyara, U., Sundin, G. S. and Iezzoni, A. (2014a). Cherry leaf spot resistance in cherry (*Prunus*) is associated with a quantitative trait locus on linkage group 4 inherited from *P. canescens*. *Molecular Breeding* 34(3), 927–35.

Stegmeir, T., Sebolt, A. and Iezzoni, A. (2014b). Phenotyping protocol for sour cherry (*Prunus cerasus* L.) to enable a better understanding of trait inheritance. *Journal of the American Pomological Society* 68, 40–7.

Stegmeir, T., Cai, L., Basundari, R. A., Sebolt, A. M. and Iezzoni, A. F. (2015). A DNA test for fruit flesh color in tetraploid sour cherry (*Prunus cerasus* L.). *Molecular Breeding* 35, 149.

Tan, C. T., Yu, H., Yang, Y., Xu, X., Chen, M., Rudd, J. C., Xue, Q., Ibrahim, A. M. H., Wang, S., Sorrels, M. E. and Liu, S. (2017). Development and validation of KASP markers for the greenbug resistance gene *Gb7* and the Hessian fly resistance gene *H32* in wheat. *Theoretical and Applied Genetics* 130(9), 1867–84.

Tavaud, M., Zanetto, A., Santi, F. and Dirlewanger, E. (2001). Structuration of genetic diversity in cultivated and wild cherry trees using AFLP markers. *Acta Horticulturae* 546, 263–9.

Tavaud, M., Zanetto, A., David, J. I., Laigret, F. and Dirlewanger, E. (2004). Genetic relationships between diploid and allotetraploid cherry species (*Prunus avium, Prunus × gondouinii* and *Prunus cerasus*). *Heredity* 93, 631–8.

Tobutt, K. R., Bošković, R., Cerović, R., Sonneveld, T. and Ružić, D. (2004). Identification of incompatibility alleles in the tetraploid species sour cherry. *Theoretical and Applied Genetics* 108, 775–85.

Truchseß, C. (1819). *Systematische Classification und Beschreibung der Kirschensorten*. Cotta'sche Buchhandlung, Stuttgart.

Tsukamoto, T., Tao, R. and Iezzoni, A. F. (2008). PCR markers for mutated *S*-haplotypes enable discrimination between self-incompatible and self-compatible sour cherry selections. *Molecular Breeding* 21, 67–80.

van Nocker, S. and Gardiner, S. E. (2014). Breeding better cultivars, faster: Applications of new technologies for the rapid deployment of superior horticultural tree crops. *Horticulture Research* 1, 14022.

Vanderzande, S., Piaskowski, J. L., Luo, F., Edge-Garza, D. A., Klipfel, J., Schaller, A., Martin, S. and Peace, C. (2018). Crossing the finish line: How to develop diagnostic DNA tests as breeding tools after QTL discovery. *Journal of Horticulture* 5, 228.

Wang, D., Karle, R., Brettin, S. and Iezzoni, A. F. (1998). Genetic linkage map in sour cherry using RFLP markers. *Theoretical and Applied Genetics* 97, 1217–24.

Wang, H., Nair, M. G., Strasburg, G. M., Chang, Y., Booren, A. M., Gray, J. I. and DeWitt, D. L. (1999). Antioxidant and antiinflammatory activities of anthocyanins and aglycon, cyanidin, from tart cherries. *Journal of Natural Products* 62, 294–6.

Wang, D., Karle, R. and Iezzoni, A. F. (2000). QTL analysis of flower and fruit traits in sour cherry. *Tree Genetics & Genomes* 100, 535–44.

Wang, J., Zhang, K., Zhang, X., Yan, G., Zhou, Y., Feng, L., Ni, Y. and Duan, X. (2015). Construction of commercial sweet cherry linkage maps and QTL analysis for trunk diameter. *PLoS ONE* 10, e0141261.

Watkins, R. (1976). Cherry, plum, peach, apricot and almond. In: Simmonds, N. W. (Ed.), *Evolution of Crop Plants*. Longman, New York, NY, pp. 242–7.

Webster, A. D. (1996). The taxonomic classification of sweet and sour cherries and a brief history of their cultivation. In: Webster, A. D. and Looney, N. E. (Eds), *Cherries: Crop Physiology, Production and Uses*. CAB International, Wallingford, UK, pp. 3–24.

Wenden, B., Campoy, J. A., Jensen, M. and López-Ortega, G. (2017). Climatic limiting factors: Temperature. In: Quero-García, J., Iezzoni, A., Pulawska, J. and Lang, G. (Eds), *Cherries: Botany, Production and Uses*. CABI, Wallingford, UK, pp. 166–88.

Wharton, P. S., Iezzoni, A. and Jones, A. L. (2003). Screening cherry germ plasm for resistance to leaf spot. *Plant Disease* 87, 471–7.

Whiting, M. and Perry, R. L. (2017). Fruit harvest methods and technologies. In: Quero-García, J., Iezzoni, A., Pulawska, J. and Lang, G. (Eds), *Cherries: Botany, Production and Uses*. CABI, Wallingford, UK, pp. 442–59.

Wünsch, A. and Hormaza, J. I. (2004a). Molecular evaluation of genetic diversity and *S*-allele composition of local Spanish sweet cherry (*Prunus avium* L.) cultivars. *Genetic Resources and Crop Evolution* 51, 635–41.

Wünsch, A. and Hormaza, J. I. (2004b). Genetic and molecular analysis in Cristobalina sweet cherry, a spontaneous self-compatible mutant. *Sexual Plant Reproduction* 17, 203–10.

Yamane, H., Tao R., Sugiura A., Hauck N. R. and Iezzoni, A. F. (2001). Identification and characterization of *S*-RNases in tetraploid sour cherry (*Prunus cerasus*). *Journal of the American Society for Horticultural Science* 126, 661–7.

Yue, C., Gallardo, R. K., Luby, J. J., Rihn, A. L., McFerson, J. R., McCracken, V., Oraguzie, N., Weebadde, C., Sebolt, A. and Iezzoni, A. (2014). An evaluation of U.S. tart and sweet cherry producers trait prioritization: Evidence from audience surveys. *HortScience* 49, 931–7.

Zhang, G., Sebolt, A. M., Sooriyapathirana, S. S., Wang, D., Bink, M. C. A. M., Olmstead, J. W. and Iezzoni, A. F. (2010). Fruit size QTL analysis in an F_1 population derived from a cross between a domesticated sweet cherry cultivar and a wild forest sweet cherry. *Tree Genetics & Genomes* 6, 25–36.

Zhu, M., Zhang, X., Zhang, K., Jiang, L. and Zhang, L. (2004). Development of a simple molecular marker specific for detecting the self-compatible S_4' haplotype in sweet cherry (*Prunus avium* L.). *Plant Molecular Biology Reporter* 22, 387–98.

Zhukov, O. S. (1979). Anwendung neuer genetischer Methoden in der Sauerkirschenzüchtung. *Tag.-Ber., Akad. Landwirtsch.-Wiss. DDR, Berlin* 174:203–6.

Zhukov, O. S. and Charitonova, E. N. (1988). *Selektzia vishi*. Agroproizdat, Moscow.

Zhukovsky, P. M. (1965). Main gene centers of cultivated plants and their wild relations within territory of the USSR. *Euphytica* 14, 177–88.

Zwitzscher, M. (1961). Kirschen. In: Kappert, H. and Rudolf, W. (Eds), *Handbuch der Pflanzenzüchtung. Volume V. Züchtung der Sonderkulturpflanzen*. Paul Parey Verlag, Berlin, pp. 573-602.

Chapter 4

Sustainable sweet cherry cultivation: a case study for designing optimized orchard production systems

Gregory A. Lang, Michigan State University, USA

1 Introduction – opportunities and challenges

1.1 Production sustainability

Sweet cherries (*Prunus avium*) are produced in temperate zones worldwide, though among tree fruits they are often associated with a wide range of risks to sustainable production. These include susceptibility to low temperature damage during winter in more northern climates, poor bloom and fruit set due to insufficient chilling in warm Mediterranean-type climates, frost damage to blossoms in most climates due to bloom in early spring, fruit cracking due to rain during ripening, decreased fruit quality and/or abnormal flower formation due to high temperatures during summer, and susceptibility to a host of serious diseases and insect pests. Additionally, harvest of the fruit is very labor-intensive due to its relatively small size (in comparison with other stone fruits and pome fruits) and delicate nature.

Sweet cherry production worldwide has increased dramatically over the past two decades, facilitated by strong market demand but especially driven by four major advances in plant materials and orchard technologies. The sweet cherry production revolution has been facilitated by (1) self-fertile cultivars, which have increased yields and consistency in production due to improved

http://dx.doi.org/10.19103/AS.2018.0040.18

pollination in spring year-in and year-out; (2) precocious and vigor-controlling rootstocks, which have yielded earlier returns-on-investment and easier tree and harvest management; (3) orchard covering systems to protect against rain-induced fruit cracking and other climatic risks, improving yield consistency for market supply; and (4) intensive orchard training systems, which have improved fruit quality, yields, and orchard labor efficiencies.

1.2 Market opportunities

The production of sweet cherries has increased more than 50% around the world since the beginning of the twenty-first century, an indication of strong consumer demand and favorable economic returns to growers, in spite of the many challenges for sustainable production. The leading countries for domestic plus export production during this time have been Turkey and the United States, followed by Spain and Italy. In the past 10 years, Chile has moved past both the United States and Turkey to become the top exporting country, based largely on its exports to the rapidly growing middle-class population of China. Consequently, Chinese market demand also has driven a rapid expansion of China's domestic sweet cherry production, which has now possibly surpassed the production levels of Turkey and the United States, although exact numbers are hard to confirm at this time. Iran is also thought to be among the top five countries worldwide for cherry production, though like China, production is mostly for the domestic market and therefore reliable figures are difficult to obtain. Current Iranian cherry cultivation is unlikely to be comparable to that in modern orchards of other leading producers. Production also has increased significantly in Kazakhstan, which is a major supplier to the Russian market that has, in recent years, been unavailable to exports from the United States and many European nations due to political barriers.

From 1984 to 2000, sweet cherry production in the United States only increased by 14%, from 164 926 to 188 604 mt, whereas from 2000 to 2016, production increased by 68%, to 317 732 mt. This is reflective of not only strong consumer demand but also an increasing number of middle-class consumers for exports, and more stable annual production due to improved genetic materials (including self-fertile cultivars) and production technologies (including orchard rain covers and improved frost protection measures). Sweet cherry market growth is anticipated to continue in China, and the potential exists for middle-class consumer market demand in India to follow a similar trajectory. Growth in market demand from other Asian countries has followed rising middle classes as well (e.g., Korea, Taiwan, Vietnam). Many of these countries have very few areas well-suited climatically to sweet cherry production.

The strength of consumer demand can be implied from the price per ton paid to US growers over the past three decades. The average 6-year price per ton for fresh market sweet cherries increased from $992 in 1980–85 to $1865 in 1998–2003, and then to $3100 in 2011–16. This contrasts with the average price paid for processing sweet cherries, which only increased from $472 to $741 over the same period. The price per ton for processing sour cherries actually fell from $628 in 1980–85 to lower values throughout 1986–2010, although the period 2011–16 saw a recovery to $810. Still, if annual inflation was estimated at 2% during this time frame, neither sweet nor sour processed cherry prices kept up with inflation, while fresh sweet cherry prices exhibited an annual average increase of 2.2% in addition to the 2% annual inflation factor.

1.3 Climate change

All tree fruits are subject to production limitations associated with adverse climatic conditions, and sweet cherries are at or near the top of the list of species particularly susceptible to such risks. Consequently, changes in climate also may affect the sustainability of cherry production. In Michigan, since World War II, average annual precipitation has increased by ~10%, the average date for bloom has advanced to occur ~10 days earlier, the average incidence of damaging frosts that occur after buds begin to swell in late winter/early spring has more than doubled, and the lowest temperatures experienced during the spring frost season have become more extreme (see Blanke et al., 2017). Sweet cherries are very susceptible to flower bud, shoot, and even tree trunk damage from unseasonably warm periods in winter that are followed by sudden low temperatures, as well as moderate fall temperatures followed by an early sudden low temperature event before leaf senescence.

In some regions with typically mild winter temperatures, such as California and Spain, the changing climate has increased the frequency of insufficient chilling temperatures for breaking endodormancy, leading to protracted and weak flowering and canopy foliation. Often, the average orchard yields in such locations are half of those for orchards in more temperate regions with longer winters. In some growing regions, the changing climate also may increase the frequency or intensity of droughts, making irrigation a necessity to achieve consistent, sustainable yields and fruit quality. The occurrence of unusual climatic events that can damage the crop or trees, such as hail, intensive rainfall, or unusually strong winds, seems to be increasing in many regions around the world, driving the development and adoption of orchard technologies to help mitigate such risks.

1.4 Precision fruit growing

Finally, as sweet cherry orchards have become more intensive, with plantings at higher densities to achieve earlier and higher yields, greater attention must be paid to precision in developing and maintaining the foliar canopy to achieve adequate photosynthesis for the desired growth of higher fruit loads on smaller trees. Therefore, canopy structural development and crop load management can be critical to achieve consistently high fruit quality. The relatively small size and delicate nature of sweet cherry fruits require significant and timely harvest labor, as does annual canopy training and pruning. Therefore, sustainable production is improved when strategies for significant labor efficiencies are considered, beginning with the design of the new orchard to be planted. Sustainable orchard planning includes consideration of the potential use of trellises to guide precise canopy structural development, simplification of canopy structure to facilitate easier crop load and leaf area management, and tree training system designs that reduce the use of ladders (which are labor-inefficient and dangerous to workers) or even facilitate the mechanization of some tasks. Increasing the precision of each aspect of fruit growing can save time, supplies, and labor, while increasing the fruit yields and quality necessary to meet targeted market opportunities.

2 Morphology, growth, and fruiting

2.1 Vegetative morphology and canopy structure

In the past several decades, tree fruit orchards have evolved to more intensive production systems, with the objectives of earlier and higher yields, more consistent fruit quality, and often simpler, or at least more precise, procedures for developing the tree canopy. The foundation for these systems is a focus on optimization of light interception and photosynthesis. At its most foundational level, this leads to an examination of the pattern of node production, or phyllotaxy, for sweet cherry canopy development, growth, and leaf-to-fruit relationships (which reflect crop load carrying capacity).

Both vegetative and reproductive growth of sweet cherry is comprised of simple buds, with each bud containing only preformed vegetative leaves or preformed reproductive flowers. Initial growth is comprised of an elongating terminal meristem with periodic formation of nodes with a single lateral meristem. At each node, the lateral meristem will form a leaf that retains the meristematic point of potential growth within the leaf axil. This axillary meristem usually remains vegetative during the year of formation, but in some cases, it may be induced to become reproductive and develop flowers the following spring. This occurs most commonly at the base of current-season shoots, on the nodes that were preformed in the bud before the vegetative meristem elongated in the spring.

The phyllotaxy of these nodes evolved in a repeating pattern of spirals, with the placement of each node described mathematically by a Fibonacci number, yielding the divergence angle of 137.5° that generally forms between two successive leaves relative to the 360° circumference of the growing shoot (this is also known as the 'golden angle'). Consequently, for a vertically oriented shoot, as nodes and leaves form concomitant with shoot extension, each successive leaf contributes to the maximum interception of light with the minimum overlap and shading of lower leaves (Fig. 1). This is the result of optimized efficiency in the capture of light in the forest setting where sweet cherry evolved to compete with other tall forest trees (compared to understory bushes or rosetted herbaceous annuals). This is also known as Hofmeister's rule: each successive primordium forms at the least crowded spot along the meristematic shoot, resulting in the most efficient packing of organs on the tree structure as they develop. Each successive leaf develops at a position that is furthest from the previous two leaves (137.5°). Vegetative shoot meristems that develop in the axils of these leaves that form may, in subsequent years, become active lateral shoots that are thus positioned for the most efficient interception of light; however, because of their less than vertical orientation, while the phyllotactic development of their nodes and leaves follows the same Fibonacci-based pattern, the orientation of their leaf angles change, also being more horizontal to optimize light interception.

Consequently, evolutionary efficiency is maximized with the leaf orientation of vertical growth, and as vegetative growth deviates from vertical, leaf angles become reoriented to better intercept light but this also has the result of increasing shade below. This shading is of competitive benefit to reduce the fitness and survival of competing plants, but it also can be of significant energetic cost to the sweet cherry tree itself since it reduces the efficiency of older, lower parts of its own canopy, leading initially to their reduced productivity and eventually to their likely dieback and abscission as the productive portion of the canopy moves ever higher and outward.

Figure 1 Sweet cherry vegetative shoot formation and leaf phyllotaxy, with leaves forming at 137.5° angles in a Fibonacci spiral arrangement to maximize light interception while minimizing shading of lower leaves.

The second morphological trait of importance for understanding canopy morphology and growth habit is that sweet cherry is acrotonic in its branch development and exhibits strong apical dominance. That is, the terminal meristem tends to elongate vigorously while inhibiting growth of the subtending lateral meristems that develop that same season. At bud break of the next season, the terminal meristem again elongates vigorously, and the next several upper lateral meristems may also elongate at a slightly lesser rate, forming acute angles and providing potential replacements for the terminal meristem if it becomes damaged. Most other lower lateral meristems, though, remain inhibited from elongating. In nature, this results in a strong central leader with distinct tiers of lateral branches and branch-less gaps associated with each previous season of growth. The strongest growth occurs at the top of the central leader and follows a decreasing gradient of vigor from the top to the bottom due, in part, to progressive shading as well as apical dominance and positional gradients in hormonal growth promotion. This is the most efficient use of growth resources for competing for light ever higher in a diverse forest canopy. Understanding this natural sweet cherry growth habit is vital for developing horticultural training and pruning strategies to adapt such a forest tree to a more manageable, labor-efficient orchard production system, as will be discussed later in this chapter.

2.2 Reproductive morphology, flowering, and fruit growth

As noted above, a small proportion of nodes near the base of current-season shoot growth may be induced to begin reproductive development, resulting in a population of solitary (one per node) basal buds that will flower the following spring. On a typical sweet cherry tree, these comprise a minority of the flowering sites in the canopy. These solitary basal flower buds are more rounded than the solitary vegetative buds that form at the majority of the nodes on the previous season's growth. There may be zero to six or more non-spur flower buds at the base of the shoots, depending on variety, rootstock, and overall tree age. Precocity-inducing rootstocks, such as the Gisela series, tend to develop these basal flower buds on shoot growth of very young trees, whereas on vigorous non-precocious rootstocks like Mazzard or Colt, trees may be 5 years old or more before basal flowers begin to form.

The remaining nodes on current-season shoot growth remain primordially vegetative through the completion of their season of formation, resulting in a single vegetative bud per node that becomes prominent and slightly pointed as it develops primordially. The following spring, a few of the more terminal of these buds may elongate into new lateral shoots, as described above. The majority of the buds open to form a whorl of five to nine leaves, each constituting a new secondary node comprised of the leaf, a primordial subtending meristem, and a minimal internode, plus the continuing primordial terminal meristem. These non-elongated vegetative structures comprised of multiple secondary nodes are known as 'spurs.' Since there are no flower buds associated with these spurs, these are termed 'non-fruiting spurs' (Fig. 2).

For shoots going into their third season of growth, the original primary nodes (from base to apex) are comprised of (1) basal nodes that are now 'blind,' that is, having no meristem remaining if they were the site of previous basal non-spur flower buds since those flowers would have abscised or led to fruits which subsequently were harvested or abscised, with no remaining vegetative meristem; (2) 'fruiting spur' nodes, which were induced to form a solitary flower bud in the axillary node of one or more previous leaves in the whorl, resulting in generally one to nine flower buds per fruiting spur, all subtending the ongoing vegetative terminal meristem; (3) development and extension of one or more secondary lateral shoots, particularly if pruning has been imposed somewhere higher on the shoot; and (4) continued elongation of any secondary lateral shoots that first elongated the previous year, often with a few basal non-spur fruiting secondary nodes.

Consequently, the fundamental complete fruiting unit of a sweet cherry tree, comprising all of the potential leaf and fruit populations, is a 3-year-old branch that has (1) new shoot leaves (large, single leaves at each node on elongating new shoot growth), (2) non-fruiting spur leaves (five to nine per primary node on the portion of the shoot that grew the previous season), (3) non-spur solitary flower buds, one per primary node near the base of previous season shoot growth, (4) fruiting spur leaves (five to nine per primary node on the portion of the shoot that grew two seasons previous), and (5) fruiting spur flower buds,

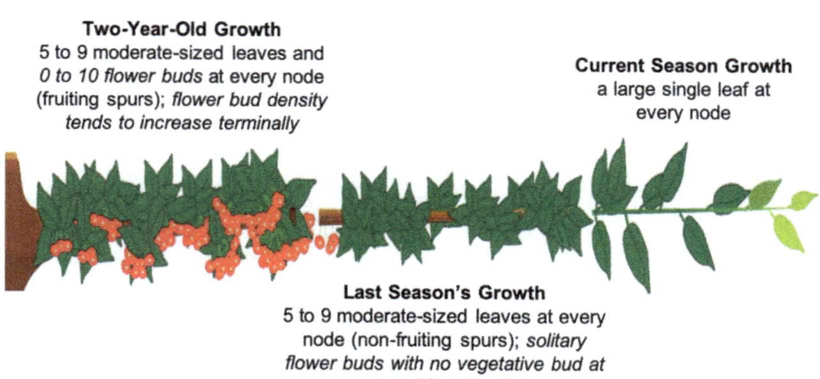

Figure 2 Deconstruction of sweet cherry vegetative and reproductive growth into three leaf populations and two populations of fruiting sites: 2-year-old shoot segments with fruiting spurs and fruiting spur leaves, previous season shoot segments with non-fruiting spur leaves and non-spur basal fruit at the segment base, and new extension shoot growth leaves (Lang, 2005).

zero to ten per primary node on the portion of the shoot that grew two seasons previous (Fig. 2; Lang et al., 2004).

In tree canopies that are old, overcropped, or otherwise of extremely low vigor with negligible extension shoot growth (including weak, thin pendent shoots), shoots may terminate in a flowering spur. This is usually an indication of an orchard in need of significant pruning, fertilization, and renovation management, since the fruit quality on such low vigor shoots is generally unsatisfactory.

Sweet cherry flower buds are induced during late spring (e.g., May–June in the northern hemisphere, November–December in the southern hemisphere) of the season prior to fruiting. That is, hormonal signals received at the undifferentiated meristems in the axils of leaves initiate the expression of genes that begin directing those meristematic cells toward the formation of primordial floral tissues rather than vegetative tissues. In early summer (in most cases just after fruit ripening), the induced meristems begin differentiating into floral organs (Guimond et al., 1998a,b). High temperatures during mid-summer, thought to be most damaging during the transition from sepal to petal formation (Beppu and Kataoka, 2011), can cause abnormal primordial pistil development, resulting in fruits the next season that may range from fused doubles to single fruits with deep ventral sutures. The susceptibility to such abnormal development under high temperatures varies by genotype; 'Bing' and 'Tieton' are particularly susceptible.

Differentiation continues through summer into early fall, but is arrested with the onset of endodormancy. Once adequate chilling has occurred for the transition from endodormancy to ecodormancy (usually in late winter, but can be quite variable depending on location and year-to-year climatic variation), differentiation continues until the flowers open (Gibeaut et al., 2017), culminating with dehiscence of the anthers and release of pollen for transfer to receptive stigmas on the pistils. The literature of cherry flower bud induction, formation, and fruit development has been reviewed extensively by Herrero et al. (2017).

Insufficient chilling for the complete transition from endodormancy to ecodormancy can result in abnormal flowers with less viable pollen or less receptive stigmas. The use of dormancy-breaking chemicals, such as hydrogen cyanamide (Dormex, see Raffo et al., 2009), Waiken® (a vegetable oil compound, see Bound and Miller, 2006), or other nitrogen-based compounds (such as Erger or calcium ammonium nitrate, for example, CAN-17; and potassium nitrate), can help complete the endodormancy to ecodormancy transition and improve bloom quality in low-chilling environments or where it may be desirable to advance bloom date.

Most sweet cherries are self-infertile, requiring cross-pollination and fertilization with compatible pollen, although there is an increasing number of improved cultivars that are self-fruitful (see Section 3.1). Whether a cultivar is self-fruitful or self-infertile, pollinators are required to move pollen from the dehiscing anthers of a suitable pollinizer to the receptive stigma of the pistil, since sweet cherry pollen is relatively heavy and sticky, not light and wind-borne. Sweet cherry pollination is usually ensured by placing four to six honeybee (e.g., *Apis mellifera*) hives (or colonies) per ha (two to three hives per acre) in the orchard just as flowers begin to open. Pollination in higher density orchards is improved by increasing the number of pollinators. In recent years, bumblebee (*Bombus* spp.) hives also have become commercially available for orchard pollination. There are far fewer bumblebees per hive, but bumblebees can be more efficient pollinators since they work longer under adverse conditions like lower temperatures or higher winds. They also tend to work better under plastic covers; depending on the light transmission properties of the plastic, honeybees may have difficulty navigating under plastic, reducing their efficiency. There also are many native pollinators of sweet cherry, including solitary bees (e.g., *Osmia* spp.) that may be present, particularly for orchards near woodlands or meadows. The application of any pesticides, especially insecticides, during bloom should be eliminated or minimized by relegating applications to night sprays when pollinators are not active.

Sweet cherry fruit growth typically follows the classic double sigmoid growth curve of stone fruits: Stage I (bloom, fruit set, and initial fruit growth comprised of cell division and elongation), Stage II (embryo growth and endocarp lignification/pit hardening, during which the fleshy mesocarp grows very little), and Stage III (mesocarp cell elongation, rapid fruit growth, embryo maturation, flesh and skin color change, and ripening events such as sugar and organic acid accumulation, volatile development, and flesh softening). Typically, for a given cultivar, the earliest-opening flowers in the blossom population result in the largest fruit at harvest, similar to the situation with king bloom in apple (*Malus domestica*). In a study of early, mid-, and late ripening cultivars, Gibeaut et al. (2017) found that genotypic differences in ripening time are manifested exclusively during Stage III; the developmental periods for pre-anthesis, Stage I, and Stage II were similar across cultivars.

The use of gibberellin applications during early Stage III has become a standard practice in North America and other production areas around the world to increase sweet cherry fruit firmness and size, greatly improving shipping and postharvest quality. Some commercial formulations of gibberellin are even certifiable for organic production. Targeted applications of cytokinins during the early stages of growth and cell division, and of various gibberellins during the later growth stages and cell elongation, continue to be studied to promote increased fruit size and crop value (Zhang and Whiting, 2011), as does the ever-increasing number of biostimulants and foliar nutritional supplements available to growers. To date, only the use of gibberellins has become widespread commercially (Einhorn et al., 2013).

2.3 Endodormancy, ecodormancy, and bud break

Typically, sweet cherry flowers open before leaves begin expanding; vegetative budbreak and leaf emergence tend to begin during fruit set. In colder temperate growing regions, if winter chilling greatly exceeds the amount needed for transitioning from endodormancy to ecodormancy (as during extended winters and cold early springs), when temperatures eventually do become warmer, bud break occurs rapidly and extensively, often with leaf expansion quickly following the opening of flowers while pollination is still ongoing. Conversely, insufficient chilling for completing the endo- to ecodormancy transition can result in delayed and incomplete foliation. In this case, as spring temperatures increase when chilling has been incomplete, vegetative and reproductive growth may eventually occur, but it will be weak, with fewer buds breaking over a longer period of time and poor growth of the resulting shoots, resulting in smaller leaves and often incompletely formed flowers in the fruiting spurs. Insufficient chilling tends to result in a reduced and delayed flow of storage reserves to the activated meristems, resulting in poor fruit set and sections of shoots with unopened, and often eventually dead, buds. Thus, promotion of the transition from endodormancy to ecodormancy, whether from sufficient winter chilling or from manipulation strategies such as the use of dormancy-breaking chemicals, or mid-winter shading or evaporative cooling of buds in areas with sunny, mild winters (see below), is a critical process for achieving adequate and consistent annual yields and therefore sustainable cherry production.

Air temperatures considered to be most conducive to alleviating endodormancy in sweet cherry range from around 1°C to 9°C (34°F to 48°F); temperatures colder than this have little effect in promoting endodormancy alleviation, while temperatures above this range have a limited effect of negating earlier promotive chilling. Air temperatures do not equate directly to actual bud temperatures, since at a specific air temperature that might be promotive, inhibitory, or neutral for endodormancy alleviation, radiant solar heating can alter bud temperatures in variable ways due to clouds, fog, or clear sky. For example, air temperatures may be slightly below the endodormancy alleviation range, but if the day is sunny and clear, the buds may be warmed just enough to be in the 1–9°C range. Conversely, air temperatures may be in the endodormancy alleviation range, but if the day is sunny and clear, the buds may be so warm as to negate previous chilling. Thus, manipulating the exposure of buds to solar radiation, via shade or evaporative cooling with over-tree microsprinklers (Rijal et al., 2015a,b), can be used to help improve endodormancy alleviation in low-chilling areas. The same solar radiation–bud temperature manipulation principles can be used during ecodormancy, to reduce bud temperatures after endodormancy has been alleviated and thereby prolong ecodormancy. The strategy of prolonging ecodormancy may be useful where delayed budbreak would reduce the risk of damaging spring frosts, or to shift bloom, fruit development, and ripening to a later (more valuable) market window.

2.4 Photosynthesis and the leaf area-to-fruit relationship

Regardless of the species, sustainable tree fruit production begins with light interception by the leaves, driving photosynthesis for fruit development. Taking a close look at sweet cherry leaves and canopy structure is the first step for optimizing orchard development and management. As noted above, the fundamental sweet cherry vegetative growth units are comprised of two types of leaves (large shoot leaves and typically smaller spur leaves) and three leaf populations (shoot leaves, non-fruiting spur leaves, and fruiting spur leaves).

The shoot leaves function largely to provide photosynthates for shoot extension growth early in the season, shifting to export of photosynthates for fruit development during the final rapid stage of fruit growth and ripening (Ayala and Lang, 2015, 2018), and finally a postharvest shift to export of photosynthates for trunk and root growth and storage reserves for the next season. The first shoot leaves to appear in the spring are preformed in the terminal bud of dormant shoot tips, and these tend to be smaller than subsequent leaves that are neo-formed as the new shoot extends, developing new nodes. Shoot leaf area continues increasing as long as shoot elongation occurs, with terminal budset usually occurring around harvest or shortly afterward. The neo-formed leaves generally provide a good indication of the adequacy of current-season nitrogen availability, as shoot leaf size can be modest when nitrogen is limited and very large when nitrogen is sufficient to excessive.

Non-fruiting spur leaves develop from single vegetative buds at primary nodes on previous season shoot growth, appearing in the spring as a whorl of modest-sized leaves on an extremely short shoot (the 'spur') that has no associated flowers or fruits. On the sections of branches that developed two seasons or more earlier, spurs develop a similar whorl of leaves but there may be flowers or fruits that subtend the leaves. The subtending fruit alter the export of photosynthates compared to the non-fruiting spurs, as will be discussed below.

All spur leaves are preformed, and depending on spring temperatures, the expansion of the entire spur leaf area on the tree is usually completed within 2-3 weeks of bud break, becoming the primary initial source of photosynthesis for subsequent shoot and fruit growth. The development of these leaves is wholly dependent on storage reserves of carbohydrates and nitrogenous compounds, so leaf health and orchard fertility (and pest) management of the previous season can significantly influence spur leaf size in the subsequent spring (Ouzounis and Lang, 2011).

Initial shoot extension in spring also is dependent on storage reserves. As ecodormant buds begin to respond to warming temperatures (quantified as growing degree days at a particular base temperature, often 7.2°C [45°F]) in the late winter or early spring, storage carbohydrates and amino acids flow into the buds for utilization in final differentiation of flower and leaf primordia, culminating with bud break, flowering, leaf expansion, and initial shoot elongation. Only after (1) bloom and fruit set have occurred (about 14 days after full bloom), (2) the majority of the preformed spur and shoot leaves have expanded, and (3) the shoots have begun elongating, does the supply of growth resources begin shifting from storage reserves to current-season carbohydrates from photosynthesis and nitrogenous compounds from root uptake (Ayala and Lang, 2015). The maturation of the first leaves leads to their becoming photosynthetically competent and establishing an evapotranspirational flow from the roots to the canopy growing points. Young green fruits are photosynthetically competent and can provide up to about 15% of their carbohydrate needs during Stages I and II of fruit growth (Ayala and Lang, 2017).

Using $^{13}CO_2$ and isolated (girdled) 3-year-old sweet cherry fruiting branches, Ayala and Lang (2018) compared the proportion of current photosynthates from leaves on fruiting spurs, non-fruiting spurs, and current-season shoot extension growth that was translocated from each leaf population to developing spur fruit during Stage I, Stage II, and early, mid-, and late Stage III. Of the ^{13}C-labeled carbon from all leaf sources that was recovered in the spur fruit, the majority was provided by the fruiting spur leaves, ranging from 44% to 54% (Table 1). The proportion provided by the non-fruiting spur leaves ranged from 28% to 41%, and that provided by the shoot leaves ranged from 10% (during Stage I when shoot leaf area was lowest) to 25% (during late Stage

Table 1 The relative contribution of photosynthetic carbon from sweet cherry fruiting spur, non-fruiting spur, and extension shoot leaves to developing spur fruit

Fruit growth sampling	Days after full bloom	Leaf source of ^{13}C recovered in spur fruit (%)		
		Fruiting spur	Non-fruiting spur	Extension shoot
Stage I	25	49	41	10
Stage II	40	51	30	19
Early Stage III	44	54	31	15
Mid-Stage III	56	44	38	18
Late Stage III	75	47	28	25

Source: adapted from Ayala and Lang (2018).

III when shoot elongation and consequently leaf area were at their seasonal maximum). Clearly, spur leaf area (and therefore leaf size) is critical for spur fruit development, but shoot leaf area also can be a significant source of carbon for sweet cherry fruit growth.

When the relative partitioning of carbon from each leaf source to the spur fruit was evaluated 2 days after pulsing the leaves with $^{13}CO_2$, the amount of fruiting spur leaf fixed carbon recovered from fruit ranged from 57% at late Stage III to 79% at mid-Stage III, and the majority of the remainder was recovered still in the spur leaves or wood. The amount of non-fruiting spur leaf fixed carbon recovered from fruit ranged from 31% at early Stage III to 71% at mid-Stage III, and the majority of the remainder was recovered still in the non-fruiting spur leaves or wood, although a small portion was also detected as extension shoot growth. The amount of extension shoot leaf fixed carbon recovered from fruit ranged from 18% at early Stage III to 59% at mid-Stage III, and the majority of the remainder was recovered still in the extension shoot leaves or wood. These data indicate that during mid-Stage III, when the fruit are growing at their greatest rate, the majority of the carbon being fixed by all three leaf populations was being used for fruit growth. Fruiting spur leaves provided minimal carbon to extension shoot growth regardless of growth stage, and the carbon proportion being partitioned by non-fruiting spur and extension shoot leaves to extension shoot growth was greatest during Stage II and early Stage III, followed by late Stage III.

Before the advent of dwarfing rootstocks in the 1990s, the primary fresh market sweet cherry growing regions of the United States (Washington, Oregon, California) grew trees on vigorous (usually Mazzard or *Prunus mahaleb* seedling) rootstocks, which tended to have relatively high leaf-to-fruit ratios and very good fruit quality. This was an almost inadvertent result of the annual pruning required to keep such large trees manageable in the orchard, which stimulated new growth and removed fruiting sites, creating high leaf area-to-fruit (LA:F) ratios. Several growers in Washington who were early adopters of the Gisela dwarfing rootstocks in the late 1990s began reporting that trees on these rootstocks also grew smaller fruit. A preliminary crop load adjustment study with 'Rainier' on Gisela 7 showed clearly that large fruit could be grown on dwarf trees when the crop load was reduced to improve the LA:F ratio (G.A. Lang and M. D. Whiting, unpublished). Whiting's doctoral research pursued this further, demonstrating that below a LA:F ratio of ~200 cm^2, 'Bing' fruit quality (mass and soluble solids content) on Gisela 5 declined rapidly (Whiting and Lang, 2004a), presumably due to limited availability of photosynthates (i.e., a photosynthetic source limitation). Similar results have been reported with 'Skeena' on a range of Gisela rootstocks and across various training systems: fruit mass decreased rapidly on trees with LA:F ratios below ~200 cm^2 (Neilsen et al., 2016). These quantifications of the component and overall leaf and fruit relationships in sweet cherry canopies have been used to model tree growth, yields, fruit quality, and responses to pruning (Lang et al., 2004; Lang, 2008a; Lang and Lang, 2009). Consequently, for orchards on dwarfing rootstocks, canopy management strategies have been developed to promote balance in the LA:F ratio by reducing fruiting sites and stimulating new leaf area, to assure production of high-quality fruit (Lang, 2000, 2005, 2008a). This means that annual pruning is a vital tool for sustainable production of premium quality fruit.

3 Plant materials for sustainable production

3.1 Cultivars

Cultivars play a critical role in the sustainability of sweet cherry production. First and foremost, a cultivar must be reasonably productive, of reasonable quality in terms of appearance, flavor, and texture, and in demand by the targeted consumer market. For example, in the United States, the blush cherry 'Rainier' is considered to be the highest value cultivar (on average across the entire marketing season) in grocery stores, often retailing for $2/kg ($1/lb) or more than dark red cultivars at the same time; it also commands a high consumer demand and price when exported to many Asian markets. However, it has minimal to no value as an export to most European markets, where consumers prefer dark red cherries almost exclusively. Similarly, the relatively small-sized, stemless cherries produced in Spain and marketed as Picota cherries command good prices in specific European markets (such as Spain, the United Kingdom, and Germany) that value their renowned flavor, but they would be unlikely to sell well in most US markets that put a premium on size and expect fruit with stems.

A major advance in cultivar development occurred with the 1970 release of 'Stella,' the first self-compatible (also known as self-fertile) cultivar with commercially acceptable fruit quality (Lapins, 1975). 'Stella' fruits are relatively large, with medium-firm dark red flesh, a shiny attractive appearance, and the tree is very productive. Prior to 'Stella,' all sweet cherry cultivars required one or more alternative sources of pollen, from other genotypes, in order to set fruit through successful fertilization due to gametophytic incompatibility (Herrero et al., 2017). Although incompatible pollen can germinate on the stigma of the target flower, if the S-allele associated with the pollen tube sperm cells is the same as either of the two S-alleles in the pistil, growth of the pollen tube is arrested before fertilization of the ovule egg and central cells can occur. Each cultivar genotype has two S-alleles, and each pollen grain carries one or the other of the S-alleles. Unrelated cultivars also may be incompatible with each other, if they both carry the same two S-alleles. If the pollenizer cultivar shares one S-allele with the target fruiting cultivar, half of its pollen will be incompatible;

if no S-alleles are shared, all of the pollen will be compatible. In addition to pollen compatibility of each cultivar, their bloom times must overlap to assure adequate pollen availability when the fruiting cultivar stigma is receptive.

Consequently, one of the two S-alleles carried by self-compatible sweet cherries is a mutant S-allele that is universally compatible. Thus, self-compatible cultivars do not need a pollenizer for ovule fertilization nor is bloom overlap a problem since the pollen and stigma are open and viable at the same time. However, a pollinator (e.g., honeybees, etc.) is still needed to move the pollen from the anthers to the stigma; this is sometimes misunderstood by some growers. 'Stella' gave rise to the next generation of self-compatible cultivars, which include 'Lapins' and 'Sunburst' (both released in 1983), and the third generation, which includes 'Sweetheart' and 'Celeste' (both released in 1993). Both 'Lapins' and 'Sweetheart' have become internationally important cultivars and have been used (along with others from the earlier generations) for the additional development of many self-compatible cultivars from breeding programs around the world. Although the list of commercial-quality self-compatible cultivars does not yet encompass the full 8+ week ripening period of traditional cultivars, the gaps are narrowing considerably. For extensive lists and descriptions of current cultivars, one is referred to Quero-Garcia et al. (2017) and numerous other sources such as annual commercial nursery catalogues and the Brooks and Olmo lists of new cultivar releases published periodically by the American Society for Horticultural Science (e.g., Lang, 2002a, 2006b, 2008b, 2016b).

Some sweet cherry breeding programs are increasingly focused on the development of early ripening cultivars for areas with mild winters, such as southern California, Spain, and north central Chile. As the availability of such commercial-quality cultivars increases, it will become feasible to expand production into new areas with low winter chilling and earlier (higher value) market opportunities, filling the current gaps in production (March–April in the northern hemisphere, September–October in the southern hemisphere). Such cultivars also may become important for some of the warmer current growing areas where climate change trends may reduce winter chilling below the levels needed for consistent production of standard cultivars.

3.2 Rootstocks

Until the turn of the twenty-first century, sweet cherries were propagated primarily on vigorous seedling rootstocks like Mazzard (*Prunus avium*) or mahaleb (*P. mahaleb*) or the vigorous clonal rootstock Colt (*P. avium* x *P. pseudocerasus*). These are still the primary rootstocks used in many parts of the world. Other variations on these rootstocks have been used to a significant but lesser extent, such as clonal or in-bred seed selections of *P. mahaleb* (e.g.,

Sante Lucie 64 and 405, respectively), *P. avium* (e.g., Charger), *P. cerasus* (e.g., Stockton Morello), and *P. avium* x *P. mahaleb* hybrids (the 'MaxMa' series) (Wertheim, 1998).

A major advance in cherry rootstock development occurred with the commercial release in the early 1990s of several precocious, highly productive, and vigor-controlling interspecific hybrid clonal rootstocks from Justus Liebig University in Giessen, Germany (Gruppe, 1985; Lang, 2000). Since that time, semi-dwarfing Gisela 5 and semi-vigorous Gisela 6 have become major rootstocks used around the world, with more recent additions of dwarfing Gisela 3 and semi-vigorous Gisela 12. Other rootstock breeding and selection programs have generated somewhat similar rootstocks (e.g., the Gran Manier series from Gembloux, Belgium; Edabriz from France; the Weiroot series from Weihenstephan, Germany; the PiKu series from Pillnitz, Germany; the PHL series from Holovousy, Czech Republic; the CAB series from Bologna, Italy; the Krymsk series from the Vavilov Institute in Russia; the Corette series from Michigan State University in the United States), but none have yet achieved the current success of Gisela 5 and 6.

The current Gisela rootstocks exhibit good graft compatibility and minimal suckering and impart a high productive potential and a range of scion vigor compared to full-vigor Mazzard seedling rootstocks. When grown in most temperate regions with adequate annual pruning to balance crop loads, Gisela 3 is dwarfing with vigor ranging from 30% to 45%, Gisela 5 is semi-dwarfing ranging from 45% to 65%, and Gisela 6 and 12 are semi-vigorous ranging from 70% to 90% of the vigor typical of seedling rootstocks (Kappel et al., 2013). These ranges are all derived from irrigated trees, as they all tend to have relatively shallow, limited root systems that absolutely require irrigation to establish well in the orchard and maintain adequate growth rates to achieve high fruit quality. This limited root system trait, along with a reduced hydraulic conductivity at the graft union (Olmstead et al., 2006, 2010), are factors that appear to limit successful utilization in hot, stressful environments (e.g., California, north central Chile, Spain, Turkey). Even with daily drip irrigation, when evapotranspiration rates are extremely high, stomatal closure may occur from midday through late afternoon, depressing photosynthesis and growth incrementally. On the other hand, in moderate temperate climates (e.g., the midwestern United States and northern Europe), excellent growth and production can be maintained with irrigation provided throughout the season or at least during drought conditions, plus tree management and pruning to adequately balance crop loads with canopy leaf area (Lang et al., 2019). Therefore, sustainable sweet cherry production on Gisela rootstocks requires planting in the appropriate environment, with orchard and canopy management matched to site vigor and the inherent productivity of the scion cultivar. Extremely productive cultivars like 'Lapins' and 'Sweetheart' may require more aggressive pruning and crop load management on extremely productive rootstocks like Gisela 5, or the use of somewhat less precocious, productive rootstocks like CAB6P or Krymsk 6.

Small tree stature for higher density orchards can be achieved most consistently by the use of vigor-controlling precocious rootstocks, though other methods are possible with plant growth regulators such as prohexadione-Ca (e.g., Elfving et al., 2003) or deficit irrigation (naturally or artificially) in arid, hot climates where current precocious rootstocks do not grow well. In growing regions with high pH calcareous soils, adaptation and performance under such conditions tend to be more important for orchard sustainability than vigor control for intensive production (Hrotkó et al., 2014). In such locations, mahaleb rootstocks have performed particularly well. Conversely, in growing regions with heavy soils that may be prone to *Phytophthora* root rot infections, mahaleb rootstocks are usually more sensitive than rootstocks with *P. avium* or *P. cerasus* in their genetic backgrounds. For extensive lists and descriptions of the various traits for current rootstocks worldwide, one is referred to Hrotkó and Rozpara (2017) and other sources such as the Brooks and Olmo lists of rootstock releases published periodically by the American Society for Horticultural Science (e.g., Lang, 2002b, 2006a, 2016a).

4 Tools for optimizing orchard tree development

4.1 Nursery tree establishment

Contemporary sweet cherry orchards have become increasingly intensive, with a focus on filling the orchard space more quickly with higher planting densities and beginning production earlier with precocious rootstocks and scion cultivars. Quality nursery trees with traits matched to the planned training system(s) are a critical component of achieving these goals. Sweet cherry nursery tree quality parameters include caliper of the leader, height, branching (or 'feathers'), and extensiveness of the root system. In recent years, specialized nursery tree products for cherry have expanded to include 'well-feathered' 1- or 2-year-old trees developed through application of growth regulators (such as Promalin®) or knip-boom trees, respectively, or even bi-leader trees developed through opposite side double budding. Knip-boom nursery trees advance structural development and early fruiting for training systems that develop spindle-like canopy architectures, but are most expensive and relatively rare in the United States, though more common in Europe. Growth regulator-induced feathered trees also advance early canopy structural development and fruiting, though to a lesser extent. Of critical importance to added value in a feathered nursery tree is whether the branches have desirable wide crotch angles, are of moderate vigor, are located where the first branches of the orchard tree are desired, and are numerous enough to create balanced, rather than dominant, lateral growth. Feathered nursery tree value is decreased if a significant number of the feathers must be removed due to poor placement or must be pruned back to initiate more desirable numbers or positions.

In the United States, the most common type of nursery tree for sweet cherries is the 1-year-old single leader or 'whip' tree, though these may have a couple of side branches. Usually when nursery trees have only a few side branches, the branches vary in caliper, length, and position, and most often are cut back at planting to basal vegetative buds to be regrown during the first year in the orchard concomitant with other newly formed lateral branches. Such nursery trees are also more suitable for multiple leader training systems such as Spanish Bush (SB), Steep Leader (SL), Kym Green Bush (KGB), or Upright Fruiting Offshoots (UFO). Specific training systems will be addressed in the next section.

After planting of the nursery tree, lateral branches can be induced with heading of the leader, or leaving the leader intact with activation of specific lateral buds by scoring, bud selection and removal, or growth regulator applications (Lang, 2005; Long et al., 2015). The best results with these techniques tend to occur when imposed at the beginning of the second year in the orchard, after the root systems have become established, rather than at planting when the nursery tree is subject to transplant shock and reestablishment of the root system damaged during nursery tree digging and storage. Potted nursery trees are now becoming available, with the advantage of minimal transplant shock and an ability to plant later in spring, but such plants are also usually smaller than typical field-grown bare-root nursery trees. If the potted trees are dormant, the bud activation techniques (described below) can be imposed as with bare-root trees. If the potted trees are already foliated, techniques have yet to be developed for inducing branches at the time of planting.

Scoring just above the bud where a new lateral branch is desired interrupts the basipetal flow of auxin from the more terminal meristems, releasing that lateral bud from growth inhibition. Treatment of the bud with a high dose of cytokinin and gibberellin (e.g., Promalin, often mixed with paint and applied directly to the target buds) at the green tip stage of bud swell also can overcome this inhibition, but response is temperature-dependent and may give poor results in cold springs. Spray applications of lower doses of cytokinin and gibberellin as leaves are emerging (by which time spring temperatures are usually a little warmer) have also shown promise for initiating new branches. Bud selection involves the manual removal of several buds above and below the target buds selected for branch development, from the terminal down to the lowest selected bud, such that about 60–80% of the buds are removed, which increases the potential for the remaining buds to elongate. This has a wide response window for imposing, but is most easily applied during bud swell since buds snap off readily at this stage. The resulting branches from bud selection/removal tend to be more vigorous than those induced by scoring or growth regulators.

4.2 Pruning and growth manipulation

Pruning strategies for developing and maintaining the sweet cherry canopy structure can be partitioned by timing, typically ranging from dormant to late spring, to pre- and postharvest summer pruning as described below.

Dormant pruning (mid-winter through spring bud break) removes some storage carbon and nitrogen reserves, but more importantly removes potential vegetative and reproductive buds, thereby increasing the relative amount of carbon and nitrogen reserves that can be partitioned to the remaining, pruning-prioritized growing points. This is particularly valuable for (1) enhancing and directing growth during the initial training of the canopy architecture or structure of the tree and (2) removing areas of potentially excessive future crop loads, to improve the LA:F ratios for premium fruit quality. The absence of leaves during dormant pruning makes it much easier to evaluate the canopy structure for identifying undesirably weak or strong growth to be removed, or sections of branches with an excessive number of fruiting spurs. While dormant pruning reduces the overall size of the tree, it also invigorates the remaining parts of the tree structure.

Late spring pruning (3–4 weeks after bud break, after the transition from storage reserves to current-season photosynthates to support new growth) removes active leaf area and therefore is used primarily to remove or reduce excessive tree vigor. The regrowth that may result is almost entirely dependent upon current photosynthesis and nitrogen uptake, in competition with already established growth of fruit and elongating shoots.

Preharvest summer pruning (10 days to 2 weeks before anticipated harvest, late Stage III of fruit development) removes active leaf area and, if done during the period of greatest carbohydrate demand by fruits, may negatively affect fruit size, depending on the severity of the summer pruning. However, if it is imposed by the modest hedging of still-elongating extension shoots, it may remove vegetative competition for carbohydrates as well as reduce shading and provide more light to the spurs for optimizing spur leaf photosynthesis during the final period of carbon demand by the fruit.

Postharvest summer pruning (generally, in July–August in the northern hemisphere or January–February in the southern hemisphere) removes leaf area that is primarily contributing photosynthates to the storage tissues for the following season. This is increasingly being accomplished by mechanical hedging, particularly as tree row canopies become more hedge-like, and by mechanical topping (removal of terminal growth above the maximum desired tree height, as with KGB tree training). Earlier timing of postharvest summer pruning improves light to spur leaves and differentiating spur floral buds, but it may stimulate a minor amount of late summer shoot regrowth that could be more susceptible to low temperature damage where winters are very cold. Later timing of postharvest summer pruning also can improve light to spur buds, but by this time, some interior leaves may already have been lost to shading; however, this timing is unlikely to stimulate late summer shoot regrowth. Consequently, later summer pruning can help reduce vigor by removing leaf area, with a concomitant potential reduction in storage carbohydrates and amino acids for subsequent spring growth, and minimal regrowth that would be susceptible to winter cold damage.

As noted above, the development of improved rootstocks that impart precocious flowering for earlier cropping in the life of the orchard has been a major advance in sustainable sweet cherry production systems. However, even for trees on vigorous, non-precocious rootstocks, precocity can be improved by bending or tying down or twisting/cracking branches, which slows growth and usually induces earlier flower induction. The Solaxe training system is predicated on the tying of lateral branches such that their growing tip is below the point of their insertion into the central leader, which slows growth and promotes spur formation (Lauri and Claverie, 2005). Of course, this requires additional orchard labor, but is a useful tool where precocious rootstocks have not been used. Vigor also can be reduced for such trees with other horticultural management techniques, such as root pruning in the spring (the cutting of roots with a tractor-pulled shank or blade through the soil about 0.5 m from the trunk) or deficit irrigating during summer in climates where regular rainfall is scarce. The greatest impact on vegetative growth from root pruning is generally from bud break through Stage I of fruit growth, which reduces storage reserves for allocation to elongating shoots. Depending on the severity of vigor control required, root pruning can be imposed on one or both sides of each tree row, or alternated from one side to the other every year. Foliar applications of growth inhibitors, such as prohexadione-calcium (Apogee™, Regalis™; Elfving et al., 2003) or paclobutrazol (Cultar™), may be useful where regulations allow. Additionally, some cultivars are inherently more precocious (e.g., 'Sweetheart') than others (e.g., 'Early Robin'). Varietal growth habits also differ, from somewhat pendent and weeping ('Anderson') to spreading ('Sweetheart') to strongly upright ('Lapins').

5 Designing optimized orchard production systems

5.1 The evolution of intensive cherry training systems

Sweet cherries have followed apples into the realm of high-density orchards, driven by the same relatively universal production advantages and goals. All sweet cherry orchard production systems utilize tree training and pruning primarily to (1) promote specific structural canopy elements for light interception, (2) limit excessive vigor to minimize canopy shade, (3) control the development of fruiting sites in balance with supportive leaf area (e.g., crop load management), and (4) carry out most orchard tasks as labor-efficiently as possible. Optimized light interception and shade minimization promote higher yields and more uniform fruit development and ripening, and crop load management promotes premium fruit quality. The smaller tree statures gained by using vigor-controlling rootstocks facilitate more efficient pruning and harvest labor. Precision fruit growing brings a greater level of intensity, yet simplification where possible, to each of these facets, increasingly with an eye toward partial mechanization and ideally with some level of quantification for achieving specific outcomes.

Following the lead of modern apple orchards, the majority of high-density sweet cherry orchards are planted on dwarfing (e.g., Gisela 3) to semi-dwarfing (e.g., Gisela 5) rootstocks, trained as variations of spindle trees following the lead of Fritz Zahn in Germany in the 1990s. These have evolved to include the Vogel spindle, the Tall Spindle Axe (TSA), and the Solaxe (Laurie and Claverie, 2005; Long et al., 2015), as well as other variants. However, the majority of current high-density orchards in the United States that utilize precocious rootstocks are planted on semi-vigorous Gisela 6, Gisela 12, or Krymsk 6, and most orchards in hot, dry climates around the world still use vigorous rootstocks like Mahaleb, Mazzard, Colt, CAB6P, and others. This is likely an indication that the evolutionary timeline for most cherry growers to adopt new orchard systems is still progressing from vigorous to truly dwarfing rootstocks. A similar progression took many decades for apple orchards, which evolved from vigorous seedling to semi-vigorous M7-type rootstocks to semi-dwarfing M26-type rootstocks to M9-type dwarfing rootstocks.

However, sweet cherry training systems have subsequently deviated somewhat from apple (and in fact, led the way) in the exploration of simplified multiple leader training systems that can utilize precocious rootstocks ranging from dwarfing to semi-vigorous. Some of the concepts learned through this exploration (Lang, 2001) provide guidelines for new ways to utilize moderate-vigor rootstocks while still achieving high density, intensive orchard goals, as will be discussed below. In fact, some of the new cherry training concepts are increasingly being explored and adopted for other tree fruits as well, including apples, pears (*Pyrus communis*), peaches (*Prunus persica*), apricots (*Prunus armenica*), and plums (*Prunus salicina* and *P. domestica*).

Since the growth habit of sweet cherry trees evolved to compete in forest ecosystems, vigor management tends to be more of a problem, in the long run, for trees trained to a single central leader where low vigor is not achieved through dwarfing rootstocks or otherwise growth-limiting conditions such as poor soils or cool, short-season climates. As noted earlier in this chapter, natural sweet cherry growth is apically dominant and acrotonic, with the greatest annual vigor and branching at the top of the tree and with lower branches often becoming progressively shaded and weaker. In orchard settings, of course, this evolutionary growth habit simply creates greater challenges in redirecting growth and fruiting to the lower portions of the canopy for the goal of being more labor efficient for training and harvest. The vigor of the central leader can be reduced somewhat by training it on a trellis at an oblique angle, such as the Marchand Drapeau canopy (Moreno et al., 1998), or alternating the angle of every other tree tilted into the tractor alley to create a V-canopy, but these are primarily of benefit only until the tree reaches its target maximum height. Without additional interventions such as root pruning or plant growth regulators, the repeated pruning (for the duration of the life of the orchard) of the central leader at the point of maximum orchard height often invigorates that portion of the canopy, particularly if done during dormant pruning. Early and heavy cropping can compete significantly with vegetative vigor; however, the opposite is also true, that light cropping (chronic or periodic, as due to spring frost damage) can lead to periods of dramatically increased vigor. Highly vigorous vegetative growth tends to initiate fewer flower buds, leading to a negative cycle of excessive vigor that forms fewer flowers, which leads to less of a crop to compete for resources, which leads to more excessive vigor. Consequently, battling excessive vigor in the top of the tree and trying to increase vigor in the bottom of the tree is an annual challenge once central leader trees reach maturity. This tends to be true for most temperate tree fruits.

Training trees to multiple leaders has been a relatively effective way to reduce the height of apically dominant species like sweet cherry, since the tree's natural vigor and apically dominant upright growth habit are diffused into multiple leaders, each being a fraction of the vigor and dominance that a single leader would have. Although these mini-leaders still maintain individual apical dominance, each is proportionally less vigorous than a single leader. Once they have reached the target height of the orchard, the regrowth that results from repeated topping is also proportionally less vigorous. This concept has long been utilized for cherry, peach, and other stone fruits on vigorous rootstocks trained to open vase or Goblet canopy architectures, with variations and improvements resulting in modified multiple leader canopies such as SL in Washington state (Long et al., 2015) and the SB and Cataluña Vase in Spain and Italy.

5.2 Innovations in cherry canopy architectures

Various sweet cherry training strategies have been investigated to bridge the gap between single leader and traditional multiple leader systems, including dual-leader and tri-leader trees that reduce the vigor of each leader to about one-half to one-third that of a single leader, and multi-scaffold planar palmette canopies (e.g., Moreno et al., 1998). In the 1990s, the exploration of sweet cherry canopy architectural concepts to utilize an even greater number of multiple leaders to manage tree vigor and crop loads was undertaken from two different viewpoints. In 1994 in Australia, cherry grower Kym Green began developing what has become known as the KGB training system which utilizes vigorous rootstocks and numerous heading cuts during the first 2–3 years of orchard establishment to produce 25–30 upright, uniform leaders of moderate (diffused) vigor that can be picked almost entirely without ladders (Green, 2005). The largest leaders, about 10–15% of the canopy, are cut back to stubs every year for renewal of new fruiting wood to maintain high fruit quality. Consequently, the majority of the tree remains physiologically young, even as the base trunk and sub-trunks age chronologically. This system achieves the goals of vigor control, simplified and relatively uniform canopy development and maintenance, utilization of the tree's natural growth habit, high yields, minimal use of ladders (each mini-leader is young and flexible enough to be bent down for harvest), and consequently high labor efficiency. Tree height can be maintained by mechanical hedging. The KGB training system has subsequently been adopted in many cherry production areas around the world, particularly Australia, the United States, and Chile.

In 1999 in Washington state, tree fruit physiologist Greg Lang began developing a cordon-like cherry tree canopy to produce 12–15 upright, uniform leaders of moderate (diffused) vigor for precise training to a trellis to create a very narrow, two-dimensional canopy that would be suitable for simplified training and crop load management, mechanical hedging and high harvest efficiency. Whip nursery trees were headed and two uniform leaders were developed, which were tied down horizontally to a bottom trellis wire at the end of the first growing season to form two cordons similar to the Vertical Shoot Position (VSP) training system for winegrapes (*Vitis vinifera*). At the beginning of the second season, the bottom and lower side buds on each cordon were removed, promoting uniform upright growth of new shoots to create many vertical leaders about 20 cm apart (Fig. 3a), based somewhat on the natural phyllotaxy and repeating pattern of sweet cherry growth. This concept was established with the very upright-growing 'Lapins' on both vigorous Mazzard rootstock at wide tree spacing and semi-dwarfing Gisela 5 rootstock at closer spacing. This provided a demonstration that the same fundamental, simplified multi-leader canopy architecture could be adopted for rootstocks of varying vigor by proportionally increasing the length of the cordons and hence the number of upright leaders per tree to diffuse vigor. This also provides flexibility to manage anticipated orchard vigor due to other components beyond rootstock, such site fertility, growing season length, cultivar productivity, and potential orchard management tools like deficit irrigation or plant growth inhibitors. Such precision and flexibility in tree development has consequently moved modern orchard planning and management beyond thinking about trees per ha and toward managing a target number of simplified leaders per ha, much like a super spindle orchard but with the option to plant far fewer

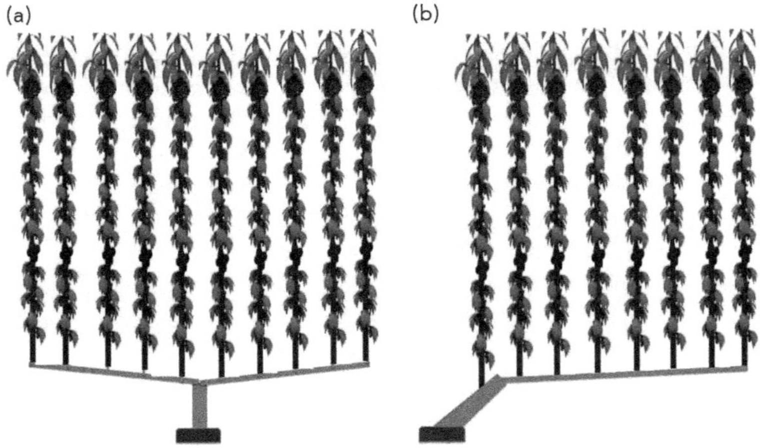

Figure 3 (a) Dual- and (b) single-cordon UFO sweet cherry planar canopy architectures.

trees and utilize a greater number of suitable rootstocks for adaptation to a particular site. Similar to KGB, about 10–15% of the leaders in the canopy are renewed annually to maintain physiologically young fruiting wood and high fruit quality across the entire tree canopy.

Like the precise quantification and management of spurs, buds, and fruit clusters in the VSP grape canopy, the development of relatively uniform simplified leaders on the cherry canopy cordons facilitates easier estimation of leaf area and fruiting sites to achieve balanced annual crop loads. The narrow upright leaders optimize light distribution uniformity and minimize internal canopy shade, improving synchronization of all stages of crop development from bloom through ripening (and also optimization and uniformity of foliar applications of pesticides, nutrients, and growth regulators). Furthermore, the vertically oriented, simplified (non-branching) leaders utilize the natural phyllotaxy and efficiency of the Fibonacci spiral arrangement of leaf area and fruiting spurs for light interception and carbohydrate supply to fruit clusters. As if a VSP grape-like cherry canopy was not strange enough, this canopy architecture evolved further in collaboration with tree fruit physiologist Matthew Whiting who subsequently utilized the whip nursery tree as a single cordon, planted at an oblique angle and tied to the bottom trellis wire to begin development of the upright leaders 1 year earlier (Fig. 3b). At this point, such cordon-based planar cherry canopies became known as the UFO training system, and the optimum planting angle was identified as 45° (Law and Lang, 2016). Both single- and dual-cordon UFO sweet cherry canopies have subsequently been adopted in various cherry production areas around the world, particularly New Zealand, the United States, and Chile (Fig. 4). Similar cordon canopy architectures also are increasingly being tested and refined

Figure 4 Dual-cordon vertical plane (a) and single-cordon dual-angled plane (b) UFO sweet cherry canopy architectures in Washington, USA.

for adoption to other tree fruits, including apples, apricots, plums, pears, and peaches as interest increases in the various advantages of narrow, two-dimensional fruiting wall orchard production systems.

5.3 Cherry training systems for sustainable production

Guidelines for implementing these (KGB and UFO) and five other types of canopy training systems have been described recently in year-by-year developmental detail by Long et al. (2015): spindle trees such as TSA, Super Slender Axe (SSA), and Vogel Central Leader (VCL); multiple leader three-dimensional canopy trees such as SB and SL; and also dual inclined planar trellised trees such as UFO-Y or UFO-V canopies. Some key aspects to these contemporary training systems are noted in the remainder of this section.

Three-dimensional central leader canopies (Vogel and Zahn spindles, TSA, and other central leader permutations such as Solaxe) utilize developmental training techniques (such as activation of lateral buds by scoring, bud selection, or promotive growth regulator applications) to increase the number of lateral branches formed during the establishment years. The level of lateral branch induction success can vary by scion cultivar and is critical for moderating overall vigor (both tree height and strength of the lateral branches) unless dwarfing or semi-dwarfing rootstocks are used. Heading of the leader also can be used to promote the formation of a whorl of lateral branches (plus the replacement for the central leader), though the resulting branches tend to be vigorous and require further manipulation during growth to assure wide crotch angles, such as the use of clothespins or toothpicks to set good angles as the lateral shoots form. The Solaxe system relies on the tying down of most branches to reduce their vigor and increase their fruitfulness. However, if branches become too pendent, fruit size and quality may decline concomitant with their vigor. Likewise, pruning decisions must focus on both renewal of fruiting sites and good light distribution to insure that lower tier branches don't decline in productivity due to inner canopy shade.

The two-dimensional SSA canopy training system incorporates several innovations for sweet cherry production: (1) very high density planting (less than 1 m apart, Fig. 5a), which introduces significant root competition that reduces vigor (though not enough for vigorous rootstocks to be successful without additional vigor control measures like root pruning, deficit irrigation, or plant growth inhibitor applications); (2) short pruning (Fig. 5b), which minimizes the formation of fruiting spurs and promotes the majority of the crop to form on non-fruiting spur basal buds on previous season shoot growth, resulting in favorable LA:F ratios and high fruit quality; and (3) annual renewal of the fruiting units on the entire tree (Fig. 5c), rather than only a minor proportion of the canopy. The sustainability of the SSA system is somewhat cultivar-dependent, since cultivars can vary in their propensity for non-spur basal flowering, and given the extensive detailed (but simple) pruning, it is relatively labor-intensive.

The SL canopy training system integrates aspects of both spindle and multiple leader canopy architectures. Vigor is diffused into three to four leaders that are spaced very close together, upon each of which lateral branches are developed on the outward-facing 25–35%. This creates an

Figure 5 Very high density Super Slender Axe (SSA) sweet cherry trees in Italy (a), with short pruning (b) to promote cropping on non-spur basal flowers and renewal of one to two vegetative extension shoots per fruiting unit (c).

overall tree canopy shape somewhat comparable to an expanded conical (spindle) central leader (narrow at the top, broad at the bottom). However, each leader is proportionally less vigorous than a single leader tree, and the narrowness of the leader group creates an open 'chimney' effect for at least some light penetration into the middle of the canopy if interior-growing shoots are removed annually.

Both KGB and UFO training systems eliminate the formation of a single dominant central leader, produce primarily on spurs with minimal development of lateral branches, and are comprised of multiple leaders of moderate vigor that arise from a low, multibranched stump in the case of the KGB or a low, nearly horizontal single or double cordon in the case of the UFO. This allows the growth of each leader to follow the inherent apically dominant growth habit of most sweet cherry cultivars, yet at a usually eightfold or more lower vigor than a single central leader. Furthermore, these training systems utilize a cyclical, patterned renewal of each of the multiple leaders over time, such that they may only have to be topped (which invigorates the top of the leader) a couple of times before renewal after 6 or 7 years, rather than throughout the remaining life of the orchard. Additionally, their renewal at the bottom of the canopy (at the UFO cordon or the KGB branched stump) localizes the invigoration at the base, rather than the top, of the canopy fruiting unit. The KGB is pruned and harvested without ladders by bending each leader down to the level of a worker on the ground; leaders are renewed such that no fruiting wood becomes too old to bear high-quality fruit or to be bent down for picking from the ground. The UFO can be summer-pruned mechanically by hedging and can be harvested and detail-pruned during dormancy without ladders by laborers on motorized orchard platforms. However, because these canopy architectures produce fruit primarily on spurs, cultivar spur and branching traits can impact system management and yields. Furthermore, with the free-standing nature of the KGB, rootstocks that do not promote strong vertical leader growth (such as Gisela 3, 5, and 6) may be unsuitable due to inordinate bending of the leaders under crop loads.

Other canopy training systems exist and continue to be developed, and good yields of good quality fruit are possible on most systems when the fundamental concepts of light interception, minimization of shade, and periodic renewal of fruiting wood are incorporated into basic orchard management priorities. Where training systems begin to differentiate from each other are in other operational facets that may be of greater or lesser priority to the individual grower, such as initial tree numbers and cost, capital costs for trellising, optimization for yields, optimization for fruit size, optimization for labor or mechanization efficiency, worker skill level, and/or the need to train new workers every year.

6 Mitigating abiotic and biotic risks to sustainable production

6.1 Managing soil water and fertility

The provision of water and nutrients for sustainable sweet cherry production follows the same fundamental principles as for most other temperate-zone tree fruits. Any periods of low soil water content may cause stomates to close, thereby not only reducing water loss from leaves but also limiting carbon uptake through photosynthesis and reducing plant turgor pressure for fruit growth. The adoption of vigor-limiting rootstocks can magnify these detrimental effects through at least two mechanisms. First, root systems tend to be proportional to canopy size, so smaller trees explore less soil volume for water and nutrient uptake and therefore may experience root zone depletion more quickly than traditional orchards on vigorous rootstocks with extensive root zones. When planted at higher densities, there is also greater tree-to-tree root competition for available water and nutrients. Second, as noted earlier, water conductance across the graft union also may be decreased for dwarfing rootstocks (Olmstead et al., 2006), leading to more frequent periods of diurnal water stress that can limit not only tree growth but also fruit expansion. Consequently, high-density sweet cherry growth and fruiting has been shown to increase significantly with high frequency, short duration irrigation and fertigation regimes (Neilsen et al., 2010, 2014). Daily drip irrigation to replace the previous day's evapotranspiration, divided into four equal applications over the course of the day, resulted in higher leaf and fruit nutrient levels, higher yields, and greater trunk circumference, compared to trees provided with the same amount of water once every 2 days.

Various studies have examined cherry rootstocks for differential nutrient uptake (e.g., Hrotkó et al., 2014; Jiménez et al., 2007; Sitarek et al., 1998). Soil characteristics, orchard fertility management, and climate can affect nutrient uptake and partitioning within the tree, so it can be difficult to draw generalizations regarding cherry rootstock genotype versus phenotype effects on plant nutrition (Lang et al., 2011). However, foliar fertilization also can provide a valuable tool for managing cherry nutrition and performance. Of particular significance for sustainable production of cherries under high density, vigor-limiting rootstock conditions is the use of fall foliar applications of nitrogen to increase storage levels of amino acids for availability during spring bud swell, bloom, and initial fruit and spur leaf growth (Ouzounis and Lang, 2011). The small trunks and root systems of trees in such orchards provide less nutrient storage capacity compared to standard trees, so supplemental foliar applications are largely partitioned directly to spurs and can result in larger spur leaves in the spring, which improve early photosynthetic capacity and transpirational flow of water and nutrients during fruit set and Stage I growth of fruits as well as new shoots. This increase in reserves due to translocated foliar nitrogen in early fall also resulted in better cold acclimation. However, such foliar applications have less impact when nitrogen levels are already luxuriant.

Soil health, abiotic and biotic soil conditions, tree nutrient requirements, and management strategies for cherry orchard delivery of water and nutrients have been discussed recently in great detail by Neilsen et al. (2017).

6.2 Microclimate modification

In many parts of the world, the greatest challenge for sustainable sweet cherry production is the potential incidence of a range of adverse climatic events, including insufficient winter chilling; dormant tissue damage from low winter temperatures; damage to emerging tissues from spring frosts prior to, during or after bloom; rain-induced fruit cracking during ripening; summer heat-induced double ovary formation in flower buds (which leads to double fruit the following season); and hail damage to fruit and leaves, or rain dissemination of diseases, throughout the growing season. Wenden et al. (2017) have recently reviewed the temperature-related limitations to cherry production, including recent information regarding the molecular genetics of cherry responses to the environment. To date, however, the main success in genetic advances for climatic adaptation has been the development of cherry cultivars with increasingly lower chilling requirements (e.g., genotypes developed by the private breeding programs of International Fruit Genetics and Zaiger Genetics in California; Quero-Garcia et al., 2017). The ability to mitigate most of the other climate-induced risks continues to rely on horticultural solutions.

Certainly, the horticultural strategies for mitigation of climatic risks begin with selecting an orchard site with a history of minimal production risk factors. Ideally, orchard establishment should be in a region with adequate winter chilling, rare incidence of spring frosts, a warm moderate to long growing season, and a relatively dry and sunny spring and summer with adequate availability of good quality water for irrigation. Where spring frosts may be common, sloping orchard sites help the coldest (dense) air to drain out of the orchard; for this reason, sites at the bottom of slopes are especially prone to frost damage. Orchard sites on flat ground should utilize frost protection measures such as wind machines (purpose-built fans on towers above the orchard canopy) to mix warmer (inversion) layers of air above the orchard with the colder air in the orchard during radiative frost events. The burning of fuels such as with propane gas or diesel heaters, or solid fuel sources such as wood piles or straw bales, can raise air temperatures within the orchard on cold nights, but often this is energy inefficient since the added heat quickly rises above the tree canopy and is effectively lost. For this reason, the use of added heat can be particularly effective in combination with wind machines to pull the heated air back down into the orchard. However, propane can be expensive, and diesel and solid fuel sources can create unsatisfactory air quality and may be banned or restricted in many locations. While these measures are most commonly used to mitigate spring frost damage, they also may be of some use to mitigate mid-winter low temperature damage to tissues, depending on the nature and extent of the low temperature event (e.g., whether it is a wind-driven cold air mass or whether an inversion layer of warmer air from orchard radiative heat loss is present).

Some frost protection also can be achieved with irrigation sprinklers, though these strategies can create other problems. Under-tree sprinklers will release a few degrees of heat to the air in the orchard, depending on the temperature differential between the water and the air. As with burned fuels, this can be particularly effective in combination with wind machines. Over-tree sprinklers can be used to keep bud temperatures above freezing during a frost event through the continuous release of heat as applied water freezes on the buds (the latent heat of fusion), but such applications also may promote increased incidence of diseases such as bacterial canker (*Pseudomonas syringae* pv. *syringae* and *P. syringae* pv. *morsprunorum*), which thrive in cool, wet conditions. Both types of sprinklers can saturate the orchard soil, temporarily reducing machinery and labor mobility within the orchard and prolonging cold orchard soils. Alternatively, over-tree sprinklers can be utilized for very brief, periodic pulsed applications of water during ecodormancy (after the chilling requirement has been met and endodormancy has been broken) to evaporatively cool the buds on sunny, warm days, thereby delaying bud break by a week or more (Rijal et al., 2015a,b) since the risk of spring frost events decreases with time.

Where winters are mild and chilling may be suboptimal in some years, the use of low-chilling requirement cultivars is the most sustainable solution. Effective dormancy-breaking chemicals (such as hydrogen cyanamide) are available, but do not necessarily guarantee a strong bud break and bloom year-in and year-out. Recently, microclimate modification has been explored to enhance the accumulation of chilling temperatures at the bud (rather than air) level to promote the breaking of endodormancy in low-chilling regions. Similar to the pulsed use of over-tree sprinklers described above to evaporatively cool buds during ecodormancy to delay bloom, weather station sensors to detect air temperature and solar radiation can facilitate controlled applications of water during endodormancy to increase the de facto physiological chilling on days when sun-warmed buds would otherwise experience no or even negative chilling (Ampatzidis et al., 2018). Such pulsed over-tree sprinkler systems also may be useful during summer to evaporatively cool differentiating sweet cherry flower buds in hot climates and to reduce the risk of double ovary/double fruit formation.

Rain-induced fruit cracking during Stage III of cherry ripening is one of the most common climatic problems around the world and is a physiologically and biophysically complex phenomenon that is not yet fully understood (Knoche and Winkler, 2017). Typically, the major sweet cherry production regions tend to be arid or have relatively dry periods from late spring through early summer, since few cherry cultivars are reliably resistant to rain-induced cracking and therefore the most sustainable production is in areas with a low incidence of rainfall. However, even in the most favorable climates like the rain shadows east of the Cascade mountains in the Pacific Northwest United States and Canada, damaging rain events can severely impact yields, fruit quality, market supply and prices in some years. In more humid production regions, rain events can reduce yields to varying extents more than 60–75% of the time. Spray-applied protective fruit coatings, ranging from hydrophobic biofilms (e.g., RainGuard™, Parka™) to osmoticums (e.g., calcium chloride) that slow the uptake of water through the fruit cuticle, can provide some reduction in damage from rain events that maintain a wet fruit surface for an extended period of time. However, uptake of water through the root system is an additional factor that can drive rain-induced fruit cracking even when uptake is reduced through the cuticle. Consequently, some growers have increasingly turned to microclimate modification through the use of orchard covering techniques to minimize exposure of both the fruit and the roots to potentially damaging levels of rainfall.

Sweet cherry orchard covering strategies include various types of plastic sheeting that is supported by poles and cables over individual rows, by steel hoops (high tunnels) over multiple rows, and by steel and cable (automated retractable roof structures) or entirely steel (greenhouse) structures over entire orchards (Lang, 2009, 2013, 2014; Lang et al., 2016). When entire orchards are covered, rainwater infiltration of orchard soils is usually prevented, but when high tunnels or individual row covers are used, rainwater shed by the covers must be collected in gutters or soil-level drainage systems to prevent potential root system saturation. Rows under such covers also can be planted on raised beds, which can be particularly effective in orchards on dwarfing rootstocks to maintain the majority of the root zone above the level where rainwater might saturate the soil. Individual row covers can be seasonally fixed or movable, periodically closed and retracted as conditions dictate. Covering systems that are closed during bloom can provide an extra degree or two of frost protection during some frost events. Honeybee navigation may be impaired by fully closed covers due to the blockage of polarized and ultraviolet light by most plastic formulations, but coverage gaps between individual row covers or partially open retractable roof or high tunnel covers can facilitate normal honeybee navigation during pollination. Bumblebee navigation and pollination is not affected negatively by covers. Furthermore, most covering systems reduce incident wind speed as well, which can facilitate improved pollinator flights.

Consequently, orchard covering systems provide many options for microclimate modification, most importantly mitigating the risks of rain-cracking and secondarily, spring frosts, but there are numerous other potential impacts on cherry production as well (Blanke et al., 2017). Covers and sidewalls on the perimeter of the orchard covering system can be closed prior to bloom to accumulate earlier heat units to advance bloom, as well as for a period after bloom to further promote growing degree days to advance ripening. Care must be taken with covers during pollination, however, to prevent the occurrence of supraoptimal temperatures that could decrease flower (ovule) longevity. Covers can mitigate the risk of hail and wind damage to fruit and trees, as well as provide a sustainable approach to reducing the incidence of rain-disseminated diseases such as cherry leaf spot (*Blumeriella jaapii*) and bacterial canker. Some orchard covers also can be used to optimize the microclimate in various ways, such as reducing excessive light intensity and temperature to maintain optimal photosynthesis over a longer period during the day; preventing excessive evapotranspiration to better maintain plant and fruit turgor (leading to larger fruit size); providing some partial shade during sunny winter days to improve bud chilling unit accumulation; or providing some partial shade during sunny summer days to reduce bud temperatures that would lead to double ovary development. However, since covering systems can trap heat, operational care must be taken to prevent excessive temperatures at key stages of the tree life cycle that could reduce fruit quality, reduce photosynthesis, or cause de-acclimation of cold hardiness if in use during winter.

6.3 Diseases and insects

Sustainable production for any tree fruit must include reliable prevention and prophylactic measures for diseases and pests, including insects and mammalian predators such as birds and deer. The availability and regulations for use of synthetic and organic chemicals for insect and disease control vary from growing region to region, and from year to year. Consequently, no specific recommendations will be made in this chapter, but rather local university/extension and private advisers, agrichemical suppliers, and regulatory authorities should be consulted for up-to-date management materials and best application practices. Similarly, cherry packing and marketing entities should be consulted regarding the use and residue tolerances for allowable management materials in the destinations where the fruit will be marketed. As with control materials, important diseases and insect pests of sweet cherry also vary from growing region to region. For example, in North America, a major factor for the predominance of sweet cherry production in the arid regions of the Pacific Northwest United States and Canada is the relatively minor incidence of major diseases and insect pests (as well as lower rain-cracking). Conversely, in the rainy and humid eastern United States, disease and insect pressures are magnitudes higher, requiring more costly control programs and risking greater crop losses due to a greater incidence of sublethal infections and infestations.

Some of the major worldwide sweet cherry pathogens include bacterial diseases such as bacterial canker, fungal diseases such as cherry leaf spot and brown rot (*Monilinia fructicola*, *M. fructigena*, and *M. laxa*), and viral diseases such as prune dwarf virus (PDV) and Prunus necrotic ringspot virus (PNRSV). Bacterial canker may be the most serious pathogen found in essentially every growing region, as it tends to be ubiquitous in the environment and opportunistic, becoming pathogenic when populations increase to infectious levels under favorable climatic conditions. Infections are favored by rain for dissemination (hence, orchard covers can reduce infection incidence), frost for tissue injury, and cool-to-mild wet conditions for population growth. However, infections can occur not only with frost injury (such as 'blossom blast' death of frost-injured flowers or vegetative meristems) but also from pruning cuts, bark inclusions where crotch angles are acute, and even leaf scars from leaf abscission in the fall. Brown rot can attack flowers and fruits, ultimately killing spurs in addition to rotting or mummifying the fruit. Cherry leaf spot can cause significant tree defoliation, which reduces storage reserves and can lead to increased low temperature damage during winter and spring. There are no cures for viruses, other than tree removal when infections become apparent. Many strains of PDV and PNRSV, which can spread from tree to tree through infected pollen, are relatively symptomless, allowing spread without serious negative impacts, but some strains can reduce growth, yields, or even lead to tree mortality, particularly in combination with other viruses or if certain hybrid *Prunus* rootstocks are used that are sensitive to these viruses (e.g., Lang and Howell, 2001).

Some other serious diseases vary in prevalence from region to region, including the bacterial pathogens crown gall (*Agrobacterium* and *Rhizobium* spp.) and bacterial leaf spot (*Xanthomonas arboricola*); the fungal pathogens Armillaria root rot (*Armillaria* spp.), Phytophthora root rot (*Phytophthora* spp.), gray mold (*Botrytis cinerea*), powdery mildew (*Podosphaera clandestina*), silver leaf (*Chondrostereum purpureum*), Leucostoma canker (*Leucostoma* spp.), Verticillium wilt (*Verticillium dahliae*), anthracnose or bitter rot (*Colletotrichum acutatum*), and various other fruit rots (*Rhizopus* spp., *Mucor* spp., *Cladosporium* spp., *Alternaria* spp., *Sclerotinia sclerotiorum*, *Penicillium expansum*); the viruses plum pox (PPV or sharka), little cherry, cherry mottle leaf (ChMLV), cherry leaf roll (CLRV), tomato ringspot (ToRSV), and the phytoplasma Western X-disease. For extensive descriptions of, and management strategies for, sweet cherry diseases, one is referred to Pulawska et al. (2017) for bacterial diseases, Børve et al. (2017) for fungal diseases, and James et al. (2017) for viral and phytoplasma diseases and genetic disorders.

Some of the major worldwide sweet cherry insect pests include fruit flies (*Rhagoletis cerasi*, *R. cingulata*, *R. indifferens*, *R. fausta*, and *Drosophila suzukii*), black cherry aphid (*Myzus cerasi*), cherry slug (*Caliroa cerasi*), spider mites (various genera and species), and scales (e.g., San Jose, *Quadraspidiotus perniciosus*; Lecanium, *Lecanium corni*). However, there are many other insects that vary in importance from region to region, such as plum curculio (*Conotrachelus nenuphar*) and Japanese beetle (*Popillia japonica*) in eastern North America, various borers (e.g., peach tree borer, *Synanthedon exitiosa*; shothole borer, *Scolytus rugulosus*; American plum borer, *Euzophera semifuneralis*; and Pacific flatheaded borer, *Chrysobothris mali*), leafrollers (e.g., oblique-banded leafroller, *Choristoneura rosaceana*; pandemis leafroller, *Pandemis pyrusana*), fruitworms (green, *Lithophane antennata*; cherry, *Grapholita packardi*), and other pests (eastern tent caterpillar, *Malacosoma americanum*) across North America. For extensive descriptions of, and management strategies for, sweet cherry pests, one is referred to Papadopoulos et al. (2017).

7 Conclusions and future trends

The revolution in cherry production technologies over the past two decades has advanced the production sustainability of sweet cherries perhaps more than any other tree fruit. These technologies include self-compatible cultivars that achieve more consistent yields year-in and year-out; precocious and vigor-controlling rootstocks that facilitate earlier returns-on-investment and smaller trees that can be managed and protected more precisely; orchard covering systems that protect against rain-induced fruit cracking, as well as providing a number of other significant benefits such as frost protection, manipulation of ripening, and reduction of some diseases; intensive innovative orchard training systems that improve fruit quality, yields, and orchard labor efficiencies; and improvements (not discussed in this chapter) in postharvest optical sorting technologies for packing houses to increase efficiency as well as optimize returns for exceptional fruit.

Some of the remaining major technological advances to be achieved include mechanized or robotic harvesters, which present a significant challenge due to the relatively small size and delicate nature of sweet cherry fruit, and greater genetic resistance to rain-cracking and critical diseases, such as bacterial canker, brown rot, and cherry leaf spot, and insect pests, such as plum curculio, spotted wing drosophila, and other cherry fruit flies. Increasing interest and research in biological control strategies (biocontrols) for some persistent diseases and insects portend well for expanding future sustainable production strategies that may be less problematic than historic pesticides, with regard to both potential resistance/efficacy and secondary impacts in the environment.

8 Where to look for further information

Foundational and contemporary compilations of sweet cherry research and knowledge include the seminal work of Webster and Looney (1996), *Cherries: Crop Physiology, Production, and Uses* (CAB International) and its 20-year update by Quero-Garcia et al. (2017), *Cherries: Botany, Production, and Uses* (CAB International).

Continuously updated and archived reports on sweet cherry rootstock research in North America can be found on the website of the North Central Regional Project #140 (NC-140) at http://www.nc140.org. The site includes annual reports for on going and completed coordinated rootstock evaluation research across North America, as well as contact information for participating scientists.

Coordinated sweet cherry research efforts and project compilations in Europe can be accessed at the website of the COST Action FA1104 ('Sustainable Production of High-Quality Cherries for the European Market') at http://www.bordeaux.inra.fr/cherry.

Furthermore, under the auspices of the International Society for Horticultural Science (ISHS), the world's cherry scientists meet every four years at the Symposium on Cherry Production to present state-of-the-art research, the proceedings of which are archived in volumes of *Acta Horticulturae*. To date, eight such Symposia have been held, the most recent being convened in Yamagata, Japan, in 2017 and published as Volume 1235.

9 References

Ampatzidis, Y., Kiner, J., Abdolee, R., and Ferguson, L. 2018. Voice-controlled and wireless solid set canopy delivery (VCW-SSCD) systems for mist-cooling. *Sustainability* 10(2), 421. doi:10.3390/su10020421.

Ayala, M. and Lang, G. A. 2015. ^{13}C-photoassimilate partitioning in sweet cherry (*Prunus avium* L.) during early spring. *Cienc. Investig. Agr.* 42, 191–203. Available at: http://www.rcia.uc.cl/index.php/rcia/article/view/1508.

Ayala, M. and Lang, G. A. 2017. Morphology, cropping physiology and canopy training. In: Quero-Garcia, J., Iezzoni, A., Pulawska, J., and Lang, G. (Eds), *Cherries: Botany, Production, and Uses*. CAB International, Wallingford, UK, pp. 269–304.

Ayala, M. and Lang, G. A. 2018. Current season photoassimilate distribution in sweet cherry. *J. Amer. Soc. Hort. Sci.* 143(2), 110–7. doi:10.21273/JASHS04200-17.

Beppu, K. and Kataoka, I. 2011. Studies on pistil doubling and fruit set of sweet cherry in warm climate. *J. Japan. Soc. Hort. Sci.* 80(1), 1–13. doi:10.2503/jjshs1.80.1.

Blanke, M. M., Lang, G. A., and Meland, M. 2017. Orchard microclimate modification. In: Quero-Garcia, J., Iezzoni, A., Pulawska, J. and Lang, G. (Eds), *Cherries: Botany, Production, and Uses*. CAB International, Wallingford, UK, pp. 244–68.

Børve, J., Ippolito, A., Tanovic, B., Michalecka, M., Sanzani, S. M., Poniotowska, A., Mari, M., and Hrustic, J. 2017. Fungal diseases. In: Quero-Garcia, J., Iezzoni, A., Pulawska, J., and Lang, G. (Eds), *Cherries: Botany, Production, and Uses*. CAB International, Wallingford, UK, pp. 338–64.

Bound, S. A. and Miller, P. 2006. Effects of Waiken® on flowering and spring growth in apple. *Acta Hortic.* 727(727), 167–76. doi:10.17660/ActaHortic.2006.727.19.

Einhorn, T. C., Wang, Y., and Turner, J. 2013. Pre-harvest applications of gibberellic acid (GA_3) improve fruit firmness and postharvest fruit quality of late-season sweet cherry culitivars. *HortScience* 48(8), 1010–7. doi:10.21273/HORTSCI.48.8.1010.

Elfving, D. C., Lang, G. A., and Visser, D. B. 2003. Prohexadione-Ca and ethephon reduce shoot growth and increase flowering in young, vigorous sweet cherry trees. *HortScience* 38(2), 293–8. doi:10.21273/HORTSCI.38.2.293.

Gibeaut, D. M. D., Whiting, M. D., and Einhorn, T. 2017. Time indices of multiphasic development in genotypes of sweet cherry are similar from dormancy to cessation of pit growth. *Ann. Bot.* 119(3), 465–75. doi:10.1093/aob/mcw232.

Green, K. 2005. High density cherry systems in Australia. *Acta Hortic.* 667(667), 319–24. doi:10.17660/ActaHortic.2005.667.46.

Gruppe, W. 1985. An overview of the cherry rootstock breeding program at Giessen. *Acta Hortic.* 169, 189–98.

Guimond, C. M., Lang, G. A., and Andrews, P. K. 1998a. Timing and severity of summer pruning affects flower initiation and shoot regrowth in sweet cherry. *HortScience* 33, 647–9.

Guimond, C. M., Andrews, P. K., and Lang, G. A. 1998b. Scanning electron microscopy of floral initiation in sweet cherry. *J. Amer. Soc. Hort. Sci.* 123(4), 509–12. doi:10.21273/JASHS.123.4.509.

Herrero, M., Rodrigo, J., and Wusch, A. 2017. Flowering, fruit set, and development. In: Quero-Garcia, J., Iezzoni, A., Pulawska, J., and Lang, G. (Eds), *Cherries: Botany, Production, and Uses*. CAB International, Wallingford, UK, pp. 14–35.

Hrotkó, K. and Rozpara, E. 2017. Rootstocks and improvement. In: Quero-Garcia, J., Iezzoni, A., Pulawska, J. and Lang, G. (Eds), *Cherries: Botany, Production, and Uses*. CAB International, Wallingford, UK, pp. 117–39.

Hrotkó, K., Magyar, L., Borsos, G., and Gyeviki, M. 2014. Rootstock effect on nutrient concentration of sweet cherry leaves. *J. Plant Nutr.* 37(9), 1395–409. doi:10.1080/01904167.2014.911317.

James, D., Cieslinska, M., Pallás, V., Flores, R., Candresse, T., and Jelkmann, W. 2017. Viruses, viroids, phytoplasmas and genetic disorders of cherry. In: Quero-Garcia, J., Iezzoni, A., Pulawska, J., and Lang, G. (Eds), *Cherries: Botany, Production, and Uses*. CAB International, Wallingford, UK, pp. 386–419.

Jiménez, S., Pinochet, J., Gogorcena, Y., Beltrán, J. A., and Moreno, M. A. 2007. Influence of different vigor cherry rootstocks on leaves and shoots mineral composition. *Sci. Hortic.* 112(1), 73–9. doi:10.1016/j.scienta.2006.12.010.

Kappel, F., Lang, G., Azarenko, A., Facteau, T., Gaus, A., Godin, R., Lindstrom, T., Nuñez-Elisea, R., Pokharel, R., Whiting, M., et al. 2013. Performance of sweet cherry rootstocks in the 1998 NC-140 regional trial in western North America. *J. Amer. Pomol. Soc.* 67, 186–95.

Knoche, M. and Winkler, A. 2017. Rain-induced cracking of sweet cherries. In: Quero-Garcia, J., Iezzoni, A., Pulawska, J., and Lang, G. (Eds), *Cherries: Botany, Production, and Uses*. CAB International, Wallingford, UK, pp. 140–65.

Lang, G. A. 2000. Precocious, dwarfing, and productive – how will new cherry rootstocks impact the sweet cherry industry? *HortTechnology* 10, 719–25. doi:10.21273/HORTTECH.10.4.719.

Lang, G. A. 2001. Critical concepts for sweet cherry training systems. *Compact Fruit Tree* 34(3), 70-3.
Lang, G. A. 2002a. Cherry - sweet. In: Okie, W. R. (Ed.), *Register of New Fruit and Nut Cultivars: List 41*, pp. 253-5. *HortScience* 37, 620-52.
Lang, G. A. 2002b. Cherry rootstock. In: Okie, W. R. (Ed.), *Register of New Fruit and Nut Cultivars: List 41*, p. 255. HortScience 37, 251-72.
Lang, G. A. 2005. Underlying principles of high density sweet cherry production. *Acta Hortic.* 667(667), 325-36. doi:10.17660/ActaHortic.2005.667.47.
Lang, G. A. 2006a. Cherry rootstock. In: Clark, J. R. and Finn, C. E. (Eds), *Register of New Fruit and Nut Cultivars: List 43*, pp. 1109-10. *HortScience* 41, 1101-33.
Lang, G. A. 2006b. Cherry - sweet. In: Clark, J. R. and Finn, C. E. (Eds), *Register of New Fruit and Nut Cultivars: List 43*, pp. 1110-11. *HortScience* 41, 1101-33.
Lang, G. A. 2008a. Sweet cherry orchard management: from shifting paradigms to computer modeling. *Acta Hortic.* 795(795), 597-604. doi:10.17660/ActaHortic.2008.795.94.
Lang, G. A. 2008b. Cherry - sweet. In: Finn, C. E. and Clark, J. R. (Eds), *Register of New Fruit and Nut Cultivars: List 44*, pp. 1324-5. *HortScience* 43, 1321-43.
Lang, G. A. 2009. High tunnel tree fruit production - the final frontier? *HortTechnology* 19, 50-5. doi:10.21273/HORTSCI.19.1.50.
Lang, G. A. 2013. Tree fruit production in high tunnels: current status and case study of sweet cherries. *Acta Hortic.* 987(987), 73-81. doi:10.17660/ActaHortic.2013.987.10.
Lang, G. A. 2014. Growing sweet cherries under plastic covers and tunnels: physiological aspects and practical considerations. *Acta Hortic.* 1020(1020), 303-12. doi:10.17660/ActaHortic.2014.1020.43.
Lang, G. A. 2016a. Cherry rootstock. In: Gasic, K. and Preece, J. E. (Eds), *Register of New Fruit and Nut Cultivars: List 48*, p. 631. *HortScience* 51, 620-52.
Lang, G. A. 2016b. Cherry - sweet. In: Gasic, K. and Preece, J. E. (Eds), *Register of New Fruit and Nut Cultivars: List 48*, pp. 631-4. *HortScience* 51, 620-52.
Lang, G. A. and Howell, W. 2001. Lethal sensitivity of some new cherry rootstocks to pollen borne viruses. *Acta Hortic.* 557(557), 151-4. doi:10.17660/ActaHortic.2001.557.19.
Lang, G. A. and Lang, R. J. 2009. VCHERRY - an interactive growth, training, and fruiting model to simulate sweet cherry tree development, yield and fruit size. *Acta Hortic.* 803, 235-42.
Lang, G. A., Olmstead, J. W., and Whiting, M. D. 2004. Sweet cherry fruit distribution and leaf populations: modeling canopy dynamics and management strategies. *Acta Hortic.* 636(636), 591-9. doi:10.17660/ActaHortic.2004.636.74.
Lang, G., Valentino, T., Robinson, T., Freer, J., Larsen, H., and Pokharel, R. 2011. Differences in mineral nutrient contents of dormant cherry spurs as affected by rootstock, scion, and orchard site. *Acta Hortic.* 903, 963-71.
Lang, G. A., Sage, L., and Wilkinson, T. 2016. Ten years of studies on systems to modify sweet cherry production environments: retractable roofs, high tunnels, and rain-shelters. *Acta Hortic.* 1130(1130), 83-90. doi:10.17660/ActaHortic.2016.1130.12.
Lang, G. A., Wilkinson, T., and Larson, J. E. 2019. Insights for orchard design and management using intensive sweet cherry canopy architectures on dwarfing to semi-vigorous rootstocks. *Acta Hortic.* 1235, 161-8.
Lapins, K. O. 1975. 'Compact Stella' cherry. *Fruit Var. J.* 29, 20.

Laurie, P. E. and Claverie, J. 2005. Sweet cherry training to improve fruit size and quality - an overview of some recent concepts and practical aspects. *Acta Hortic.* 667(667), 361-6. doi:10.17660/ActaHortic.2005.667.51.

Law, T. L. and Lang, G. A. 2016. Planting angle and meristem management influence sweet cherry canopy development in the 'Upright Fruiting Offshoots' training system. *HortScience* 51(8), 1010-5. Available at: http://hortsci.ashspublications.org/content/51/8/1010.full.

Long, L., Lang, G., Musacchi, S., and Whiting, M. 2015. Cherry training systems. Pacific Northwest Ext. Publ. 667, 63pp.

Moreno, J., Toribio, F., and Manzano, M. A. 1998. Evaluation of palmette, marchand and vase training systems in cherry varieties. *Acta Hortic.* 468(468), 485-90. doi:10.17660/ActaHortic.1998.468.61.

Neilsen, G. H., Neilsen, D., Kappel, F., Toivonen, P., and Herbert, L. 2010. Factors affecting establishment of sweet cherry on Gisela 6 rootstock. *HortScience* 45(6), 939-45. doi:10.21273/HORTSCI.45.6.939.

Neilsen, G. H., Neilsen, D., Kappel, F., and Forge, T. 2014. Interaction of irrigation and soil management on sweet cherry productivity and fruit quality at different crop loads that simulate those occurring by environmental extremes. *HortScience* 49(2), 215-20. doi:10.21273/HORTSCI.49.2.215.

Neilsen, D., Neilsen, G. H., Forge, T., and Lang, G. A. 2016. Dwarfing rootstocks and training systems affect initial growth, cropping and nutrition in 'Skeena' sweet cherry. *Acta Hortic.* 1130, 199-206. doi:10.17660/actahortic.2016.1130.29.

Neilsen, G. H., Neilsen, D., and Forge, T. 2017. Environmental limiting factors for cherry production. In: Quero-Garcia, J., Iezzoni, A., Pulawska, J., and Lang, G. (Eds), *Cherries: Botany, Production, and Uses*. CAB International, Wallingford, UK, pp. 189-222.

Olmstead, M. A., Lang, N. S., Lang, G. A., Ewers, F. W., and Owens, S. A. 2006. Examining the vascular pathway of sweet cherries grafted onto dwarfing rootstocks. *HortScience* 41(3), 674-9. doi:10.21273/HORTSCI.41.3.674.

Olmstead, M. A., Lang, N. S., and Lang, G. A. 2010. Carbohydrate profiles in the graft union of young sweet cherry trees grown on dwarfing and vigorous rootstocks. *Sci. Hortic.* 124(1), 78-82. doi:10.1016/j.scienta.2009.12.022.

Ouzounis, T. and Lang, G. A. 2011. Foliar applications of urea affect nitrogen reserves and cold acclimation of sweet cherries (*Prunus avium* L.) on dwarfing rootstocks. *HortScience* 46(7), 1015-21. doi:10.21273/HORTSCI.46.7.1015.

Papadopoulos, N. T., Lux, S. A., Köppler, K., and Beliën, T. 2017. Invertebrate and vertebrate pests: biology and management. In: Quero-Garcia, J., Iezzoni, A., Pulawska, J., and Lang, G. (Eds), *Cherries: Botany, Production, and Uses*. CAB International, Wallingford, UK, pp. 305-37.

Pulawska, J., Gétaz, M., Kaluzna, M., Kuzmanovic, N., Obradovic, A., Pothier, J. F., Ruinelli, M., Boscia, D., Saponari, M., Végh, A., et al. 2017. Bacterial diseases. In: Quero-Garcia, J., Iezzoni, A., Pulawska, J., and Lang, G. (Eds), *Cherries: Botany, Production, and Uses*. CAB International, Wallingford, UK, pp. 365-85.

Quero-Garcia, J., Schuster, M., López-Ortega, G., and Charlot, G. 2017. Sweet cherry varieties and improvement. In: Quero-Garcia, J., Iezzoni, A., Pulawska, J., and Lang, G. (Eds), *Cherries: Botany, Production, and Uses*. CAB International, Wallingford, UK, pp. 60-94.

Raffo, M. D., Mañueco, L., Candan, A. P., Santagni, A., and Menni, F. 2009. Dormancy breaking and advancement of maturity induced by hydrogen cyanamide:

a strategy to improve profits in sweet cherry production. *Acta Hortic.* 1020, 497–502.

Rijal, I., Flore, J., and Andresen, J. 2015a. Final report for GNC13-18: use of water mist to protect tree fruit from spring frost damage. USDA-SARE program. Available at: https://projects.sare.org/project-reports/gnc13-181/

Rijal, I., Flore, J., and Andresen, J. 2015b. Mist-cooling to delay bloom and prevent frost damage - old idea, new technology. In: *Proceedings of the ASABE 1st Climate Change Symposium: Adaptation and Mitigation Conference*, Chicago, IL, 3–5 May 2015. American Society of Agricultural and Biological Engineers, St. Joseph, MI.

Sitarek, M., Gryzyb, Z. S., and Olszewski, T. 1998. The mineral elements concentration in leaves of two sweet cherry cultivars grafted on different rootstocks. *Acta Hortic.* 468(468), 373–6. doi:10.17660/ActaHortic.1998.468.46.

Webster, A. D. and Looney, N. E. 1996. *Cherries: Crop Physiology, Production, and Uses.* CAB International, Wallingford, UK.

Wenden, B., Campoy, J. A., Jensen, M., and López-Ortega, G. 2017. Climatic limiting factors: temperature. In: Quero-Garcia, J., Iezzoni, A., Pulawska, J., and Lang, G. (Eds), *Cherries: Botany, Production, and Uses.* CAB International, Wallingford, UK, pp. 166–88.

Wertheim, S. J. 1998. Rootstock guide. Publication nr. 25. Fruit Research Station, Wilhelminadorp, the Netherlands, 144pp.

Whiting, M. D. and Lang, G. A. 2004. 'Bing' sweet cherry on the dwarfing rootstock Gisela 5: thinning affects fruit quality and vegetative growth, but not net CO_2 exchange. *J. Amer. Soc. Hort. Sci.* 129, 407–15.

Zhang, C. and Whiting, M. D. 2011. Improving 'Bing' sweet cherry fruit quality with plant growth regulators. *Sci. Hortic.* 127(3), 341–6. doi:10.1016/j.scienta.2010.11.006.

Chapter 5

Challenges and opportunities in pear breeding

Danielle Guzman and Amit Dhingra, Washington State University, USA

1 Introduction

Pear (*Pyrus* spp.) is an economically important fruit worldwide. It is a member of the subfamily Maloideae in the Rosaceae family. There are two distinct types of pear: European (*Pyrus communis*) and Asian (*Pyrus pyrifolia*). The European pear is the most widely consumed pear outside of Asia and it is highly desired for its unique organoleptic profile and health benefits (Morgan, 2015). Pears are amongst the oldest cultivated tree fruit species in the world. The first reference of pears being planted in tended plots dates to the second millennium BC in ancient Assyria (Morgan, 2015). During this period, pears were considered a luxury commodity and were coveted as an elite fruit tree because of their wide range of uses, including consumption of the fresh fruit, use in baking, and for making perry (pear cider), as well as use as an ornamental, as a rootstock material, and for timber.

In 2015, China was the largest producer of pears with 9 000 000 metric tons, the bulk of which was comprised of Asian Pears, while the European Union was second, producing 2 450 000 metric tons. The United States ranked as the third largest (fresh) pear-producing country at 665 000 metric tons (USDA, 2016). In the United States, the Pacific Northwest (PNW) states produce approximately 80% of the pear crop (USA Pears, 2018). This region has more than 1600 pear

http://dx.doi.org/10.19103/AS.2018.0040.19

growers recognized by the Pear Bureau Northwest and they produce over 84% of fresh pears in the United States (USA Pears, 2018). Pears have been an important agricultural crop in the United States since the eighteenth century. By the early 1900s, the US pear industry was the largest in the world, worth approximately $19 million, with California leading in production (Morgan, 2015). The pear industry in the United States flourished by taking advantage of both fresh and processed fruit markets.

Pear cultivation in typical production regions faces two main challenges. The first is competition from other ready-to-eat temperate zone fruits such as apples (*Malus domestica*), stone fruits (*Prunus* spp.), and berries (e.g., *Fragaria* spp. and hybrids, *Rubus* spp. and hybrids, *Vaccinium* spp. and hybrids), given undesirable characteristics in pears such as inconsistent ripening. As a result, total US pear consumption has remained relatively constant over the past several decades despite population growth. The consumer market for pears could be improved by generation and introduction of superior varieties.

The second challenge relates to the role of rootstocks in cultivation. European and Asian pears, such as other temperate fruit species, are highly heterozygous and obligate outcrossers (Zuccherelli et al., 2002). Therefore, seed-based methods of propagation are rarely utilized because they don't reproduce true-to-type. Commercial pear cultivars are clonally propagated and grafted to seedling or clonal rootstocks to maintain the selected genotype. Rootstocks play a vital role in tree fruit production, as the choice of rootstock can dictate overall orchard efficiency (Elkins et al., 2012b). For commercial production, pears are grafted on rootstocks that are well-adapted to local soils and confer desirable characteristics to the fruit-bearing portion of the tree or the scion. Dwarfing pear rootstocks have been used for several decades in Europe. However, in the United States, 97% of the pear orchards still represent low-density plantings with large three-dimensional trees that can reach up to 4.6 m in height (USDA, 2016; Elkins et al., 2008). Large vigorous trees require difficult and labor-intensive management, and are inefficient in terms of application of other inputs such as water, pesticides, bio-regulators, and so forth. Pear growers in the United States are reluctant to adopt dwarfing quince (*Cydonia oblonga*) rootstocks, widely utilized in Europe, due to their limited cold hardiness and graft incompatibility with major pear cultivars. The development of a dwarfing pear rootstock acclimated to the US conditions is thus vital for the future growth and competitiveness of the US pear industry.

2 Pear cultivars

Genetically, pears belong to the Rosaceae subfamily *Maloideae* and are an allopolyploid with a haploid chromosome number of n = 17. There are approximately 24 main species of *Pyrus* recognized worldwide, most of which are found scattered from Europe to China, though a few species are found in the Atlas Mountains. Cultivated *Pyrus* species can be categorized into three groups–large fruited species, small fruited species, and their hybrids (Itai, 2007). The main large fruiting species utilized as scion varieties are *P. communis, P. bretschneideri, P. pyrifolia,* and *P. ussuriensis,* all of which are cultivated in temperate regions (Bell et al., 1996).

The most important *Pyrus* species in the fresh and processed market is *P. communis.* In addition, hybrids of European and Asian pear are also used throughout North America for processing. For pear cider or perry, the most desired cultivar is the snow pear (*P. nivalis*). The most popular pear cultivated in parts of China and Japan is the sand pear (*P. pyrifolia*). Additionally, pear germplasm used throughout China and Japan includes *P. ussuriensis, P. bretschneideri, and P. pashia (*Hancock and Lobos 2008).

The most important European pear cultivars grown in the United States include "Bartlett," "D'Anjou," and "Comice." The chance seedling "Bartlett," also known as "Williams' Bon Chretien," originated in England during the 1700s (Dondini and Sansavini, 2012). "Comice" (Doyenné Du Comice) is considered to have originated in France during the 1800s and "d'Anjou" (Beurré d' Anjou) in Belgium during the 1900s (Dondini and Sansavini, 2012). Of these varieties, "Bartlett" has been the leading cultivar in the fresh and processed market, and it is mainly produced in Washington and California. "Bartlett" has a short growing period compared to other pear varieties and produces a consistent heavy crop load; however, this cultivar is self-sterile. The most important winter pear is "d'Anjou," which is produced mainly in Washington and Oregon. "D' Anjou" tends to produce vigorous trees that require cross-pollination, with major pollinizers being "Bartlett" or "Bosc."

3 Pear rootstocks

Some of the most desirable characteristics imparted by a rootstock are vigor control (dwarfing), precocity, disease resistance, and environmental adaptation (Webster, 1995; Koepke and Dhingra, 2013). Rootstocks are known to influence crop production by manipulating canopy architecture, flowering, nutrient uptake, fruit quality, and overall yield (Nimbolkar et al., 2016; Koepke and Dhingra, 2013). The development of dwarfing rootstocks in combination with horticultural practices has decreased the overall height and canopy size of the composite tree, producing a more compact and manageable tree (Elkins et al., 2008). The utilization of dwarfing rootstocks has revolutionized the production of apple and sweet cherry (*Prunus avium*) by enabling the planting of high-density orchard systems (Cummins and Aldwinckle, 1983; Meland, 1998; Lang, 2005; Fazio, 2015). This type of orchard system produces sustainable fruit yields earlier, generates higher quality fruit, provides a safer work environment, and economizes on labor inputs (Elkins et al., 2008; Heinicke, 1975; Webster, 1995). Additionally, dwarfing rootstocks can increase the amount of floral buds formed annually and reduce the inherent vigor of the clonal scion, allowing for a more compact and easily managed tree (Jackson, 2003; Foster et al., 2014).

Apple growers have a lot of options when it comes to selecting dwarfing rootstocks. In contrast, the pear industry is still searching for the ideal dwarfing rootstock to facilitate the adoption of modern high-density orchard systems, especially for the United States. Today, most pear cultivars are grafted onto European pear or quince rootstocks (Brewer and Palmer, 2011). The range of scion vigor control is far more limited when it comes to pear rootstocks, and few are equivalent to the M9 apple rootstock (Brewer and Palmer, 2011). Furthermore, the limited range of current pear rootstocks sometimes exhibit graft incompatibility – a biochemical mismatch between the two genotypes used as rootstock and scion. In such cases, an interstock may be used in combination with the traditional rootstock to overcome graft incompatibility reactions (Samad et al., 1999), though this approach is cumbersome and expensive.

The standard pear rootstock utilized worldwide is a *Pyrus communis* seedling, usually a seedling of "Bartlett." "Bartlett" seedling rootstocks tend to induce a sufficient fruit set and display satisfactory orchard uniformity. However, these seedling rootstocks tend to produce large dense tree canopies and lack fire blight resistance. In some parts of the world, dwarfing quince rootstocks are utilized for high-density pear orchards. Most conventional orchards in Europe use quince rootstocks such as "Sydo" and Provence "BA29-C." For high-density orchards (3000–4000 trees/ha), rootstocks such "Adams" and "Quince C" are growers' first choice (Dondini and Sansavini, 2012). In general, quince rootstocks are precocious and less vigorous, and have shown good resistance to wooly pear aphid and root lesion nematodes (Webster, 2002). However, quince rootstocks tend to be less cold hardy, have limited fire blight resistance, and are more sensitive to calcareous soils, when compared to *P. communis* rootstocks.

In the United States, harsher winters and incompatibility with several major US scion cultivars, particularly "Bartlett" (Elkins et al., 2012b), have precluded the use of quince rootstocks. Some newer quince selections from areas with cold winters should not be ruled out as parental material in the PNW, as they may possess improved hardiness (Einhorn et al., 2011; Webster and Wertheim, 2003). Recent evaluation of quince cold-hardy taxa indicate that 25 quince accessions have achieved similar or greater cold hardiness levels than cold-hardy pear clones (Einhorn et al., 2011). However, graft incompatibility may still limit the use of these hardier quince rootstocks, making the development of suitable rootstocks derived from *Pyrus* spp. a more pragmatic alternative.

In 1984, Oregon nurseryman Lyle Brooks developed clonal pear rootstocks from the progeny population of the cross "Old Home" (OH) × "Farmingdale"

(F), which were evaluated based on ease of propagation. From a population of 516 OH×F individuals, 13 were selected and extensively evaluated for disease resistance, dwarfing, fruit production and quality, and environmental adaptation. The most common "Bartlett"-type rootstocks utilized in the United States are still the OH×F rootstocks, the most popular being OH×F 87 and 97. OH×F 87 is considered a semi-dwarfing rootstock, and can reduce scion growth by up to 39% (Brooks, 1988). This rootstock is productive, tolerant to fire blight and pear decline, and is precocious compared to OH×F 97, which is resistant to pear decline and fire blight, and is cold hardy, but produces larger trees than OH×F 87. Interestingly, recent genetic analysis has shown that the male parent of OH×F 87 and OH×F 97 is not "Farmingdale," but "Bartlett" (Postman et al., 2013).

4 Germplasm resources

Germplasm commonly used for pear rootstock breeding includes the small-fruited species *P. betulifolia* and *P. calleryana* and the large-fruited species *P. ussuriensis, P. pyrifolia,* and *P. communis.* Most mainstream rootstocks that breeders use for European pear scion cultivars are derived from germplasm of *P. communis* and quince. The East Malling Research Station in England collected various quince rootstocks from surrounding nurseries during the 1900s to eliminate confusion in their classification (Hummer et al., 2012). These East Malling quince rootstocks were designated with an alphabetical letter for identification, which are now known as "Quince A"–"Quince G" (Hummer et al., 2012). Quince rootstock selections that have been widely utilized in Europe for many years are "Provence Quince" (same as BA 29-C), "Quince A (QA)," and "Quince C (QC)" (Postman, 2009). In addition to these selections, quince rootstocks derived from hybridization include "Sydo," "Adams332," "Ct.S. 212," "Ct.S.214," and "QR 193-16." However, of these selections, only a few are available in the United States such as "Quince A," "Quince C," and "Provence Quince," all of which are mainly paired with "Comice" since they demonstrate graft incompatibility with other cultivars (Postman, 2009).

To utilize desirable germplasm in breeding projects, which often represent a wide range of phenotypic diversity, it is essential to understand the genetic interrelatedness of selected parental material. This can be accomplished by the use of DNA-based molecular markers to identify polymorphic genomic regions, which facilitate evaluation of the degree of diversity and interrelatedness using statistical methods. Plant breeders utilize this information to develop appropriate crossing plans. Over the years, several studies analyzing genetic diversity in *Pyrus* subspecies (ssp.) germplasm and interrelatedness of pear selections have been reported. These studies used DNA-based molecular markers, such as Simple Sequence Repeats (SSRs), Random Amplified Polymorphic DNA (RAPD), and microsatellites (Liu et al., 2015; Teng et al., 2002;

Song et al., 2014; Volk et al., 2006; Miranda et al., 2010; Kimura et al., 2002; Erfani-Moghadam et al., 2012).

Of these, analysis of the wild *Pyrus communis* and ancient Spanish cultivars was very informative, which aided in establishing a core collection of highly diverse *Pyrus* germplasm. The differentiation and diversity among individuals of *P. communis* ssp. *pyraster* and *P. communis* ssp. *caucasica* collected from the National Plant Germplasm System National Clonal Germplasm Repository in Corvallis, Oregon (USA) were evaluated using microsatellite markers (Volk et al., 2006; Miranda et al., 2010). A total of 13 microsatellite loci were utilized to identify the broad diversity present in the population comprising 145 wild and cultivated individuals. A hierarchical clustering method grouped the individuals into 12 clusters and two main populations. Population structure revealed that there was a distinct genetic differentiation between *P. communis* ssp. *pyraster* from *P. communis* ssp. *caucasica*. A similar study evaluated the genetic diversity of 154 Spanish pear accessions, all belonging to *Pyrus communis* using eight microsatellite loci (Miranda et al., 2010). Population structure analysis revealed that most of the germplasm collection displayed genetic distinctness when compared to pear varieties cultivated in European orchards. Based on the genetic variability found in this germplasm collection, a core collection of the Spanish pear cultivars was created to be used as a breeding source for desirable traits and to optimize breeding process.

With the advancements in accessing DNA-based polymorphism information through next-generation sequencing, a recent report employed a genotyping by sequencing approach to assess the genetic diversity and interrelatedness across 214 *Pyrus* accessions using approximately 15 000 single nucleotide polymorphisms (SNPs) (Kumar et al., 2017). The *Pyrus* collection was comprised of Asian and European pears and interspecific hybrids. The SNP data revealed 11% and 25% of individuals held alleles specific for Asian and European species, respectively, while the interspecific hybrids revealed an average of 45% and 55% introgression from Asian and European ancestors, respectively. Furthermore, phenotypic information was gathered for fruit quality traits from all *Pyrus* accessions. From the phenotypic analysis, a new quantitative trait loci (QTL) for fruit firmness on LG16, and a SNP responsible for 5% of variation in fruit firmness located on LG15 was identified, respectively.

While the genetic diversity analysis enables the selection of the most diverse parents and streamlines breeding experiments, the next task is to establish segregating populations where the desirable traits are expected to combine in a few individuals. The populations can then be used to establish gene-, or genomic region, trait associations through the development of linkage maps. In the case of *Pyrus communis*, several segregating populations have been developed, which subsequently have been utilized to develop linkage maps using DNA-based molecular markers (Yamamoto et al., 2007,

2013; Yamamoto and Terakami, 2016; Wu et al., 2014; Montanari et al., 2015; Chen et al., 2015).

As highlighted in Table 1, several genetic maps have been constructed using populations that segregate for important traits. Resistance to pear psylla, *Psylla pyri*, was mapped in a population derived from a cross between a resistant hybrid PEAR3 ("*Pyrus* × bretschneideri" × "*P. Communis*") and susceptible "Moonglow" (*P. communis*) where the individuals were genotyped using SNPs and SSR markers (Montanari et al., 2015). The research revealed that parents "PEAR3" and "Moonglow" respond similarly to Psylla infestation. However, they were able to detect QTLs on LG5 and LG8 of PEAR3 and on LG15 of "Moonglow" (Montanari et al., 2015).

Over the last few years, several genomes of European pears have been released (Chagné et al., 2014; Wu et al., 2013; Dhingra, 2013). With these resources now available, it is expected that highly desirable traits such as cold tolerance, disease resistance, and vigor control will be mapped successfully, and molecular markers can be utilized to introgress or pyramid these traits in future rootstock selections.

For identifying QTLs associated with important fruit traits, a linkage map was developed using SSR and SNP markers from a population derived from a cross between "Bayuehong" and "Dangshansuli." A total of 3143 SNP and 93 SSR markers were mapped to 17 linkage groups. Based on the high-density linkage map and phenotypic data, 31 potential QTLs were identified for a variety of fruit traits, including pedicel length, single fruit weight, soluble solids content, transverse diameter, vertical diameter, calyx status, flesh color, juice content, number of seeds, skin color, and skin smoothness (Wu et al., 2014). Li et al. (2017) developed a high-density consensus map based on SSR and SNP

Table 1 A representative list of *Pyrus* spp.-derived segregating populations and the traits for which they segregate

Traits	Population	Publication
Pest resistance	Genetic mapping of *Cacopsylla Pyri* resistance in an interspecific pear (*Pyrus* spp.) population	Montanari et al. (2015)
Cold tolerance	Genetic maps for two pear cultivars "Red Bartlett" (*Pyrus communis*) and "Nanguo pear" (*Pyrus ussuriensis*) were constructed using SRAP molecular markers	Zhao et al. (2013)
Leaf traits	A genetic linkage map was constructed using a population developed from a cross between two pear cultivars, "Yali" and "Jingbaili," using AFLP and SSR markers	Sun et al. (2009)
Fruit traits	High-density genetic linkage map construction and identification of fruit-related QTLs in pear using SNP and SSR markers	Wu et al. (2014)
Dwarfing	Genetic mapping of PcDw determining pear dwarf trait	Wang et al. (2011)

markers from nine different pear-based maps. This integrated genetic map was developed with the integration of 1232 SSR, and 3853 SNP markers spanning 3266 cM in genetic distance, making it the longest linkage map constructed for pear. The development of this consensus map will help in developing molecular markers for desirable traits, and facilitate pyramiding of desirable traits through breeding.

5 Pear breeding techniques

There are approximately 15 pear breeding programs in Europe, ten in North America, and numerous programs in South Africa, New Zealand, and Australia. Internationally, pear cultivar and rootstock breeding and evaluation programs are focused on mitigating barriers hindering industry advancement and profitability (Dondini and Sansavini, 2012). Common breeding target traits include productivity, disease resistance, precocity, and dwarfing (Wertheim, 2002; Elkins et al., 2012b). However, traits of interest vary depending on the region. In general, European rootstock breeding programs focus on refining the limitations of dwarfing quince rootstocks by improving cold hardiness, fire blight resistance, tolerance to pyslla, graft compatibility, and tolerance to iron chlorosis (Bell and Zwet, 1993; Fischer, 2009; Brewer and Palmer, 2011; Elkins et al., 2012a).

There is a shortage of pear breeding programs in the United States. Currently, breeding and genomics programs led by Kate Evans (http://dialogue.tfrec.wsu.edu/breed/kate-evans/) and Amit Dhingra (https://genomics.wsu.edu/), respectively, at Washington State University represent the only active pear breeding program dedicated to pear rootstock improvement and development. Another US pear breeding program is at the USDA-ARS Appalachian Fruit Research Station in Kearneysville, West Virginia. However, the focus of the USDA-ARS program has been fire blight resistance.

The challenge faced by US pear breeding programs is finding suitable parental material for developing dwarfing, precocious, and cold-tolerant rootstocks. Utilization of genetically diverse germplasm is essential in the development or improvement of cultivars because it maximizes the possibility of incorporating beneficial traits which may be lacking in commonly cultivated germplasm (Hawkes, 1991). There is a great amount of genetic diversity in the *Pyrus* germplasm; however, incorporating traits from wild species into breeding programs is time-intensive and may require extensive backcrossing to eliminate graft incompatibility issues and undesirable traits. An alternative approach, which is suitable for highly heterozygous crops such as pear, is to cross locally adapted cultivars and screen for individuals which may manifest not only the desirable traits of the parents, but also other desirable traits such as dwarfing. Such individuals have the added benefit of requiring fewer or no backcross generations to become suitable for widespread use. When parental material has insufficient genetic diversity, chemical or radiation mutagenesis can be used to generate additional genetic diversity.

Breeding of self-incompatible perennial crops requires carefully planned strategies to develop desirable cultivars. The overall aim is to produce new cultivars, which harbor novel traits that provide growers a market advantage through improved horticultural characteristics or through higher consumer preference leading to a greater market share (Dondini and Sansavini, 2012). Plant breeding approaches can be categorized into biparental crosses and biotechnology approaches. These approaches often are used complementarily to produce desirable cultivars. In general, both approaches follow three principal steps (Hawkes, 1991; Bell and Itai, 2011) which include:

1 creation of phenotypic and genetic variation;
2 phenotypic evaluation of individuals; and
3 selection of elite parental material and extended evaluation of elite selections or segregants.

Plant breeding utilizes hybridization with unadulterated pollen and involves crossing of two compatible individuals to generate genetically variable offspring or segregants. The biotechnology approaches involve the use of recombinant DNA techniques to generate genetic variability, which can include the use of mutagenesis or transgenic approaches, including the recent use of genome editing (Acquaah, 2012; Yin et al., 2017; Karkute et al., 2017).

Evaluation of the recombinant segregants or genetically altered plants requires subsequent phenotyping for desirable traits. With the increased availability of genomic resources, and high-throughput genotyping, DNA-based molecular markers are being used increasingly in conjunction with phenotyping to track traits at a much earlier developmental stage than possible by typical phenotyping methods, allowing for earlier selection, and focus of time and resources on elite lines (Huang et al., 1997; Evans and James, 2003; Sharma, 2003; Miklas et al., 2006; Serdani et al., 2006; Bell, 2013; Govindaraj et al., 2015). Due to the advances in molecular marker technology, breeding objectives can be obtained faster and more efficiently (Acquaah, 2012; Brewer and Palmer, 2011).

All commercially available pear cultivars to date have been developed through the hybridization process. Breeding objectives for pear cultivar improvement include, but are not limited to, fruit quality, storage ability, and disease, and pest resistance (Cummins and Aldwinckle, 1983; Fischer, 2009; Dondini and Sansavini, 2012). The major breeding objectives for rootstock improvement include vigor control (dwarfing), precocity, environmental adaptation, disease and pest resistance, and graft compatibility (Wertheim,

2002; Webster and Wertheim, 2003; Brewer and Palmer, 2011; Elkins et al., 2012b). New pear cultivars continue to be released and successfully adopted internationally; however, in the United States, replacing old cultivars remains challenging.

While desirable traits can be introgressed from sexually compatible species and genera, there is a greater risk of introducing linked deleterious traits and of graft incompatibility. An attractive alternative is to generate diversity in *Pyrus* using mutagenesis. This approach has been utilized since the early 1900s by plant breeders to accelerate the process of identifying plants with desirable agronomic traits such as disease resistance, adaptation to harsh environments, traits that affect reproductive abilities, and more (Acquaah, 2012). According to the FAO-IAEA, more than 3000 mutant plant varieties have been released for commercial use worldwide (FAO/IAEA 2018).

Mutagenesis is a process that alters the genetic information of an organism in a stable manner. Genetic mutations are produced naturally due to errors in repair of damaged DNA (Shu et al., 2012). DNA damage can be induced in an organism by means of physical or chemical mutagens such as UV irradiation, ionizing radiation, and alkylating agents (Mba et al., 2010). These mutagenic agents simply imitate the process of natural (spontaneous) mutations on an accelerated timescale (Acquaah, 2012). There are several types of alterations to DNA, but they can be classified into two broad categories: gene mutations and genomic mutations (Maluszynski et al., 1995; Ahloowalia and Maluszynski, 2001). Gene mutations are small-scale alterations to a DNA sequence, specifically the alteration, addition, or deletion of nucleotides (Acquaah, 2012). Genomic mutations are large-scale alterations to chromosomes due to errors in cell division or chromosomal breakage (Acquaah, 2012). Such errors cause improper distribution of chromosomes in daughter cells. If there is any change in the inherited distribution of chromosomes, the new state is termed aneuploidy, defined as the duplication or deletion of one or more chromosomes or chromosomal fragments (Huettel et al., 2008). The resulting gene imbalance from aneuploidy results in a noticeable effect on an organisms' phenotype (Oladosu et al., 2016). In plant breeding, the development of aneuploid plants is one of the major ways of generating genetic diversity (Henry et al., 2005). Mutagenesis is of great interest to tree fruit breeders for addressing the challenges related to self-incompatibility (SI), long juvenile phase, graft compatibility, vigor control, and other traits for which natural variation is insufficient (Pandey, 1974; Zhang and Lespinasse, 1991; Simard et al., 2011).

While mutagenesis generates random mutations, genome editing allows for introducing targeted mutations, insertions or deletions by use of a transgenic intermediate, or by transient expression of the editing mechanism (Yin et al., 2017; Karkute et al., 2017). This is only useful when a gene responsible for the trait has already been described (e.g. *Dwf1* in apple). The advantage of utilizing genome editing is that it doesn't require the presence of exogenous DNA in the final cultivar, thus not requiring undergoing the process of deregulation (Bortesi and Fischer, 2015). There are various types of genome-editing approaches available, the most commonly used being the CRISPR-Cas9 as it is user friendly and has a relatively high efficiency. As an example of application of genome editing in the case of perennial tree crops, CRISPR-Cas9 was successfully utilized to target the endogenous apple phytoene desaturase gene, which is a gene required for carotenoid biosynthesis (Nishitani et al., 2016). In this study, leaflets of the semi-dwarfing rootstock "JM2" were used for transformation. The transgenic lines produced clear and partially albino phenotypes in apple. Other than this report, to date there seems to be limited application of genome editing in tree fruit (Karkute et al., 2017). However, as gene-trait associations are established and as transformation techniques improve, genome editing may offer a rapid way to improve perennial crops.

6 Improving particular traits

6.1 Dwarfing

The precise mechanism(s) of how dwarfing rootstocks exert their effects on the scion has been under investigation for many years. The underlying process of dwarfing and how tree development is affected is poorly understood, but there are several proposed mechanisms including anatomical, hormonal, nutritional, and molecular (Clearwater et al., 2007; Tombesi et al., 2014; Koepke and Dhingra, 2013; Foster et al., 2015; Knäbel et al., 2015).

Several recent studies in apple have led to the identification of at least three loci associated with dwarfing: *Dw1*, located on Linkage Group (LG) 5 (Celton et al., 2009a,b; Pilcher et al., 2008); *Dw2*, located at the top of LG11 (Fazio et al., 2014); and, most recently, *Rb3*, or *Dw3*, located on LG13 (Harrison et al., 2016). Collectively, the data suggests that multiple loci confer a dwarfing phenotype and that the interactions between these loci are additive, which combined confer the more extreme dwarfing properties of commercial apple rootstocks.

A study by Wang et al. (2011) was done using a bulked segregant analysis approach with SSR and RAPD markers derived from apple and pear to identify markers associated with the dwarf trait. Four markers co-segregated with the dwarf trait, two being SSR (KA14 and TsuENH022) and two being RAPD (S1212 and S1172) markers. Mapping of the SSR markers unraveled the location of the dwarfing PcDw locus, which is located on LG 16. Unraveling the genetic basis of the dwarfing trait will have great implications for the future of the pear industry.

The recent characterization of the genetic traits present in apple rootstocks and the microsynteny between apple and pear offers the opportunity to utilize the information from apple to evaluate analogous genetic elements in pear germplasm or, alternatively, transfer the apple dwarfing/precocity genes directly into pear rootstocks through transgenic or gene-editing strategies (Celton et al., 2009b; Harrison et al., 2016; Fazio et al., 2014; Knäbel et al., 2015; Pilcher et al., 2008). Such a strategy is supported by a recent report (Knäbel et al., 2015) that mapped rootstock architectural and flowering traits in pear using a biparental cross "Old Home" × "Louise Bonne de Jersey." The report found that most pear architectural traits, as well as flowering, co-localized with the apple *Dw1* on LG5 (Knäbel et al., 2015). However, the overall effect was small. These findings suggest that rootstock–scion interactions in pear may be controlled by at least some of the same loci as apple, although strong effect alleles are yet to be identified (Knäbel et al., 2015). Therefore, it is critical to develop genetic resources specific for *Pyrus* species that segregate for desirable traits.

Despite this overlap with *Dw1* between apple and pear, it is likely that other dwarfing mechanisms could be present in the *Pyrus* genus that are not present or have not yet been described in *Malus*. Dwarfing can be manifested by several different mechanisms, and a rootstock can have a variety of influences on the vigor and tree architecture of a grafted scion depending on the mechanism (Webster, 1995; Suzuki et al., 1988; Clearwater et al., 2007; Tombesi et al., 2014; Koepke and Dhingra, 2013; Foster et al., 2015; Knäbel et al., 2015). Proposed potential mechanisms include physical influence of the graft union on solute transportation, size differences between the vasculature of rootstock and scion, and complex interactions involving phytohormones such as cytokinins, auxins, and abscisic acid (Webster and Wertheim, 2003).

Unlike in other crop species, the role of gibberellins on dwarfing genotypes has not been adequately investigated in tree fruits, and contradictory evidence exists. Although a brachytic (reduced length of internodes) dwarf peach (*Prunus persica*) used as a rootstock did not dwarf a standard peach scion (Suzuki et al., 1988), the dwarf plum (*Prunus salicina*) mutant DGO24, containing a mutant *gibberellin-2 oxidase* gene, resulted in reduced growth of the standard scion "Early Golden" (El-Sharkawy et al., 2012). However, smaller fruit and delayed fruit development also were seen attesting to potential negative effects (El-Sharkawy et al., 2012). In addition, transgenic overexpression of endogenous gibberellin metabolism genes in poplar (a *Populus tremula* × *P. alba* clone) resulted in both semi-dwarfing and precocious flowering in some lines, albeit with altered catkin morphology (Zawaski et al., 2011). Gibberellin may play a role in dwarfing; however, little is known if this phytohormone-based mechanism is transmitted from the rootstocks to the scion in orchard settings (Suzuki et al., 1988; Webster, 2002; El-Sharkawy et al., 2012). The development of aneuploidy in plants is another way to generate genetic diversity, resulting in desirable phenotypes, such as dwarfing (Shu et al., 2012; Bradshaw, 2013). Cytogenetic analysis of interspecific crosses between the female *P. communis* rootstock "Pyriam" and different male *P.* spp. in the INRA program produced aneuploid individuals which had very low vigor (Simard et al., 2011). However, these individuals need to be evaluated to determine if the dwarfing phenotype of these potential rootstocks will be transmitted to the grafted scions.

There are a variety of dwarfing *Pyrus* rootstocks that have been developed around the world. The *P. communis* rootstock, "BP1," was bred in South Africa, and is classified as a semi-dwarfing rootstock (Chevreau and Bell, 2004). However, adoption has been limited due to extreme difficulties in propagation. In addition, French breeding programs have developed a series of clones called "Brossier" and "Retuzière," which were selected from seedlings of perry pear and seedlings of "Beurrè Hardy," "Old Home," and "Kirstensaller," respectively (Hancock and Lobos, 2008). Similarly, Germany produced a *P. communis* dwarfing rootstock "Pyrodwarf," which is easy to propagate, but produces suckers and reduces fruit set. In addition, "Pyrodwarf" and seedlings of "Bartlett" are also utilized as rootstocks in the United States (Reimer, 1925; Jacob, 1998; Reil et al., 2007). However, as noted above, the most widely utilized *Pyrus* rootstock series in the United States is the OH×F series (Webster, 1998). A summary of rootstocks used for Pyrus production along with their impact on tree size is depicted in Fig. 1.

6.2 Precocity

Pears, like most woody plants, have a long genetically predetermined juvenile period that can last up to ten years or more. The length of this juvenile phase strongly affects breeding timelines (Acquaah, 2012). Therefore, the goal of many tree fruit breeding programs is to shorten the time between each generation (Webster and Wertheim, 2003). Horticultural practices such as root pruning, application of growth retardants, artificial growth cycling, and grafting to precocious rootstocks are utilized to address this issue (Petracek et al., 2003; Hirst and Ferree, 1995; Janick and Paull, 2008; Mitre et al., 2012). However, these approaches vary in the degree of effectiveness depending upon the species and the environment they inhabit. A transgenic approach increased the apple precocity via the expression of the *BpMADS4 gene,* a member of the *APETALA1/ FRUITFULL* group of MADS genes in silver birch (*Betula pendula*) (Flachowsky et al., 2007; Elo et al., 2007). The early flowering genotypes can be used as transgenic intermediates for the development of desirable selections (van Nocker and Gardiner, 2014), but this approach requires maintaining the transgene throughout intervening generations and segregating the transgene out of the final generation to comply with regulations, making it somewhat cumbersome to use. Discovering a source of precocity from natural or artificially induced genetic variation within *Pyrus* would enable such work to be done without the prolonged regulations, and would be immensely helpful to all future pear breeding projects.

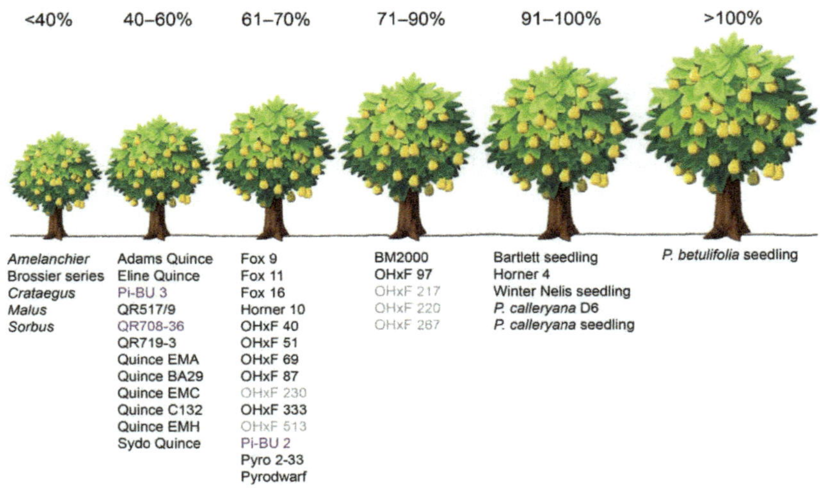

Figure 1 Illustration of overall influence on tree size* by various rootstocks compared to a Pyrus seedling. Abbreviations and names: BM = *P. communis* series from Australia; Brossier = *P. nivalis* series from Angers, France; Fox = *P. communis* series from the University of Bologna, Italy; Horner = OH×F clonal series from D. Horner (Oregon nurseryman) and selections by OSU-MCAREC; OH×F = "Old Home × Farmingdale" series; Pi-BU = Pyrus series from Pillnitz, Germany; Pyro and Pyrodwarf = *P. communis* selections from Germany; QR = *P. communis* selections; "Adams," "BA29 (= BA29-C = Provence Quince)," "EMA (= Quiince A)," "EMC (= Quiince C)," "EMH (= Quince H)," "Sydo" = Quince dwarfing rootstocks (they all require interstem grafts for most pear cultivars). Selections shown in gray text indicate antiquated selections no longer in commercial production. Selections shown in purple text indicate possible susceptibility to pear decline. *This classification may vary for different cultivars due to cultivar/rootstock interactions. Illustration adapted from the article by Elkins et al. (2012b), WSU_TFREC.

6.3 Cold hardiness

Environmental adaptation is an important target trait for US pear breeders. In the temperate fruit production regions of the PNW, cold hardiness is one of the most important traits for new cultivars and rootstocks. Lack of cold hardiness is a limiting factor affecting tree fruit production in many regions of the world. Injuries caused by cold damage usually can be associated with low fall and winter temperatures and late spring frosts, resulting in tree or bud damage, and flower damage, respectively. Cold hardiness is a complex phenomenon with multiple factors, making this trait difficult to evaluate. Factors impacting cold hardiness include temperature, photoperiod, the plants' genetic background, and physiological status (Palonen and Buszard, 1997). Cold hardiness is a dynamic trait and is subject to temporal variation. These temporal variables include the age and rate at which trees develop hardiness, the trees' ability to retain hardiness, the rate and time of dehardening, and the trees' ability to reharden. Due to the complexity of this trait, development of cold-hardy cultivars is difficult and largely dependent upon factors that are difficult to control and phenotype in breeding projects.

Plants have developed physiological and molecular defense mechanisms to adapt to cold stress. Deep supercooling of water is a freeze avoidance mechanism caused by controlling mineral and solute content of water in the vascular tissue, which can allow water to exist as a liquid below 0°C, and thus avoid damage to certain stem xylem tissues and flower buds in many deciduous tree fruit species. However, if temperature decreases below a low nucleation point, deep supercooled water in these tissues will freeze intracellularly, causing lethal injury to the plant (Lindstrom et al., 1995). The nucleation temperature of supercooled water specific to pear has restricted its cultivation to mid-latitude regions of North America (Palonen and Buszard, 1997).

Cold hardiness has been mapped using SRAP markers. The population used for mapping consisted of individuals derived from a cross between "Red Bartlett" (*Pyrus communis*) and "Nanguopear" (*Pyrus ussuriensis*). "Bartlett" was selected based on having good fruit quality while "Nanguopear" was selected for cold tolerance. From this research, a linkage map was constructed for both parents, in which 103 markers were mapped to 20 linkage groups (Zhao et al., 2013). This information is expected to aid in the identification of QTL for cold hardiness.

6.4 Fire blight resistance

The most serious disease affecting pome fruit production is fire blight caused by a gram-negative bacterium, *Erwinia amylovora*, with the major affected areas being North America and Europe (Bell and Itai, 2011). Fire blight is a highly infectious and deadly disease that enters the plant through natural openings such as flowers or wounds (Nuclo et al., 1998; Eastgate, 2000). Once present in the plant, the disease spreads systemically, causing rapid black necrosis and bacterial ooze in infected tissue. The disease can overwinter in cankers and spread rapidly to neighboring trees by rainfall, wind, and insects. Damage from infection can cause death to major tree scaffolds or entire trees, resulting in reduced orchard efficiency and yields (Zwet and Keil, 1979; Özaktan and Bora, 2004; van der Zwet et al., 2016). In addition, the presence of *E. amylovora* on fruit and plant material may restrict export due to domestic and international regulations intended to restrict the spread of the disease.

Managing this pathogen can be difficult because no single strategy is completely effective. Common management strategies can include a combination of antibiotic sprays, pruning, and biological control agents (Rademacher, 2004; Johnson and Temple, 2013). The threat of fire blight increases with increasing orchard density, making the use of fire blight-resistant rootstocks and orchard management practices mandatory for sustaining a productive orchard. Consequently, breeding for fire blight resistance has become a highly desirable and promising disease management strategy. Much research has been done to gain a better understanding of the inheritance of resistance and susceptibility of pears to fire blight.

In an early attempt to address the fire blight issue, fire blight-resistant *Pyrus* germplasm was collected from around the world (Reimer, 1925). From the germplasm collection, diverse seedling populations were generated and evaluated for disease resistance and fruit quality. Out of the entire *Pyrus* collection, Reimer's most valuable selections came from a nursery in Paris, France, and a fruit grower's orchard in Farmingdale, Illinois (USA). The selected individuals were known as "Old Home" and "Farmingdale," respectively. In 1925, Reimer made crosses between highly fire blight-resistant cultivars "Old Home" and "Farmingdale" (OH×F) to produce a seedling population that produced a high percentage of individuals that exhibited fire blight resistance and displayed uniform growth (Reimer, 1950).

Resistance to fire blight in pear is polygenic, and several QTLs associated with resistance have now been described (Bokszczanin et al., 2009; Montanari et al., 2016; Dondini et al., 2005). True immunity to this disease has not been observed, but varying degrees of resistance exists within the *Pyrus* genus. Resistance to fire blight in *P. communis* is moderate compared to the highly resistant Asian species, *P. ussuriensis* and *P. calleryana* (Fischer, 2009). In general, *Pyrus* germplasm has a wide range of resistance to pathogens and a large portion of its germplasm has yet to be characterized.

6.5 Tree architecture

Trees have a high degree of architectural variation, indicating that branch angles and leaf arrangements are genetically regulated, as it is their ability to respond to environmental cues such as overcrowding and light, nutrient, and water availability (Nagashima and Hikosaka, 2011). In tree fruit production, there are several parameters that contribute to tree architecture, including overall tree height, pattern and periodicity of branching, and the angle and orientation of branching (Hollender and Dardick, 2015). The ideal architecture for fruit trees in a commercial setting is to be compact, contain fewer branches, and exhibit a balance between vegetative and reproductive growth, which has a major impact on the economics of production (Jackson, 2003). Manipulating tree architecture facilitates pruning, fruit thinning, harvesting, maximizes light interception and distribution, and improves overall fruiting habit (Sansavini and Musacchi, 1994; Costes et al., 2006). These variables can increase orchard productivity by increasing the density of plantings per acre and aiding the efficient use of inputs. In the interest of reducing the costs of manual tree manipulation, cultivars with improved tree architectural traits are preferred.

6.6 Self-incompatibility

Self-incompatibility (SI) prevents self-fertilization and is common among tree fruits in the Rosaceae family. Self-compatible cultivars have been established in a variety of tree fruits by natural selection or induced mutations. Successful examples include almond (*Prunus dulcis*), apricot (*P. armeniaca*), sweet cherry, peach, and Japanese pear (Lapins, 1970; Socias i Company, 1998; Wünsch and Hormaza, 2004; Mehlenbacher et al., 1991; Vilanova et al., 2006; Tao et al., 2007; Hirata, 1989; Predieri et al., 1997; Lansari and Iezzoni, 1990). Self-fertility has been shown to increase yield, as in sweet cherry (Tehrani and Brown, 2010). Almost all *Pyrus* spp. are self-incompatible, and therefore it is essential to interplant pollinizer cultivars to obtain acceptable and consistent yields (Stephene, 1958). The benefit of having self-compatible cultivars is that crop yields are more predictable, and the need for interplanting pollinizer species is reduced (Stephene, 1958). While self-compatibility is a desirable scion trait, having self-compatible rootstock material can facilitate genetics research to identify genes linked to important traits by enabling development of recombinant inbred lines.

The utilization of *Pyrus* germplasm in a breeding program is impacted by SI, which is a genetic mechanism that prevents self-crossing or inbreeding, and necessitates outcrossing (McClure and Franklin-Tong, 2006). SI renders floral styles to reject self-pollen, which is the reason there is a high degree of heterozygosity in *Pyrus*. This type of SI is known as gametophytic self-incompatibility (GSI), and is common in the Rosaceae family (Takayama and Isogai, 2005). Molecular studies have identified that GSI is controlled by two tightly linked genes at the S-locus: one is a pollen-specific gene that encodes an F-box protein motif, and the other encodes an extracellular ribonuclease S-RNAse glycoprotein that is expressed in the pistil (Sanzol and Robbins, 2008; Sassa, 2016). F-box proteins lend specificity to the SCF E3 ubiquitin ligase complex, which targets proteins for degradation by the 26S proteasome, while RNAses trigger pollen tube arrest due to interruption of transcription and translation. Both genes are multiallelic; however, they are so tightly linked that specific alleles appear as set pairs in a given individual. The rejection of pollen in the pistil happens on a like-matches-like basis, meaning when the S-haplotype of the pollen matches either of the S-haplotypes in the pistil, pollen will be rejected due to the S-RNAses, resulting in incompatibility (Sanzol and Robbins, 2008). Thus, incompatibility occurs when cultivars share the same S-allele.

As pears are functionally diploid, each individual has two S-alleles, allowing varieties to have different classes of (in)compatibility: incompatible, half-incompatible, and compatible. In tree fruit breeding, S-genotyping has proven to be valuable for determining cross-fertility. To perform S-allele genotyping, polymerase chain reaction is used to amplify S-RNAse (S-alleles), which are ordered into five highly conserved regions that include C1, C2, C3, RC4, and C5, and a hypervariable region. These conserved regions define the function of the protein, while the hypervariable region plays an important role in distinguishing the difference between self- from non-self-pollen (Yamane and Tao, 2009). To overcome SI issues, self-fertile cultivars have been developed by natural selection or by inducing mutations, with examples including "Stella" sweet cherry and "Osa-Nijisseiki" Japanese pear (Lapins, 1970; Furuta and Imai, 1987).

7 Future trends

A generalized definition of sustainable agriculture is that it aims to sustain the resources and communities (including farmers) that comprise profitable farming practices and environmental stewardship (Velten et al., 2015). Genetic improvement of staple crops led to a massive increase in yields in the mid-1900s to usher in the first green revolution (The Green Revolution, 1969). Now, there is a call for ushering in the next green revolution using a sustainable approach (Martin-Guay et al., 2018). Integration of genomics, genetics, and high-throughput phenotyping along with transgenic and gene-editing approaches will enable the achievement of targeted crop improvements in a timely manner critical for feeding the planet's human population. In terms of genetic improvement of pears, this translates to higher yields with a reduced environmental footprint.

As has been discussed, pear production in the United States urgently needs a suite of locally acclimatized and dwarfing rootstocks. Other than this fundamental trait, improvement of traits falls under two major categories. One set of traits can be categorized as production or agronomic traits, which enable efficient production in terms of biotic and abiotic compatibility with specific production environments. In the PNW United States, cold hardiness is a necessity and in California, resistance to root and crown rot (caused by *Armillaria mellea*) is essential. Other key traits include fire blight resistance, tolerance to iron chlorosis, and resistance to pear decline.

One of the major pear industry issues is low per capita consumption. As genetic and genomic knowledge increases, pear research is expected to target development of varieties that ripen consistently and have other favorable appearance and flavor traits that are attractive to consumers and obviate the need for extensive post-harvest manipulations. This information also can be harnessed in combination with other approaches, such as phenomics to enable the identification of genes and pathways that may be targeted by chemical compounds to manipulate physiological processes. Such an approach, termed as chemogenomics, has been utilized well in drug discovery (Bredel and Jacoby, 2004). In plants, application of a similar approach can help in circumventing the need for transgenic plants enabling direct application for improvement of food crops (Stokes and McCourt, 2014). As an illustration of utilizing this approach in pear, a ripening delay was achieved by targeting a non-ethylene pathway in which pre-harvest application of glycine betaine delayed harvest by a few days and also improved storage of fruit in controlled atmosphere conditions (Dhingra and Schaeffer, 2014). In another example, the cold conditioning physiological model for ripening pears and gene expression approach were combined to identify metabolic pathways that are involved in conditioning and ripening. This enabled the development of a chemical intervention strategy to overcome the ripening limitation imposed by 1-methycyclopropene (Dhingra and Hendrickson, 2017). Pears as a crop have been somewhat frozen in time when it comes to advances in genetics and production. However, all the progress and technologies developed for other crops can now be rapidly adapted for improving the pear industry, with the caveat that genetic improvement will need to be accomplished first. From that standpoint, pear research stands at an interesting threshold (Dhingra, 2016).

8 Where to look for further information

Further information regarding ongoing and future genomics and breeding research in pears can be accessed from the WSU Genomics and Biotechnology Research Program (www.genomics.wsu.edu), Pear Genomics Research Network (https://ucanr.edu/sites/peargenomics/), and WSU Tree Fruit website (http://treefruit.wsu.edu/). The annual Plant and Animal Genome Conference (www.intlpag.org) is expected to continue to provide the latest information as genomics and breeding research generates information that will contribute to sustainable production of pear in the future.

9 References

Acquaah, G. 2012. *Principles of Plant Genetics and Breeding* (2nd edn.). Wiley-Blackwell, Hoboken, NJ. doi:10.1002/9781118313718.

Ahloowalia, B. S., and Maluszynski, M. 2001. Induced mutations - a new paradigm in plant breeding. *Euphytica* 118(2), 167–73. doi:10.1023/A:1004162323428.

Bell, R. L. 2013. Host resistance to pear *Psylla* of breeding program selections and cultivars. *HortScience* 48(2), 143–5. doi:10.21273/HORTSCI.48.2.143.

Bell, R. L. and Itai, A. 2011. *Pyrus*. In: Kole, C. (Ed.), *Wild Crop Relatives: Genomic and Breeding Resources*. Springer, Berlin Heidelberg, pp. 147–77.

Bell, R. L. and Zwet, T. 1993. New fire blight resistant advanced selections from the USDA pear breeding program. *Acta Horticulturae* 338, 415–20. doi:10.17660/ActaHortic.1993.338.69.

Bell, R. L., Quamme, H. A., Layne, R. E. C., and Skirvin, R. M. 1996. Pears. In: Janick, J. and Moore, J. N. (Eds), *Fruit Breeding. Tree and Tropical Fruits, vol. 1*. John Wiley & Sons, London, UK, pp. 441–514.

Bokszczanin, K., Dondini, L., and Przybyla, A. A. 2009. First report on the presence of fire blight resistance in linkage group 11 of *Pyrus ussuriensis* Maxim. *Journal of Applied Genetics* 50(2), 99–103. doi:10.1007/BF03195660.

Bortesi, L. and Fischer, R. 2015. The CRISPR/Cas9 system for plant genome editing and beyond. *Biotechnology Advances* 33(1), 41-52. doi:10.1016/j.biotechadv.2014.12.006.

Bradshaw, J. E. 2013. *Plant Mutation Breeding and Biotechnology*. Edited by Q. Y. Shu, B. P. Forster, and H. Nakagawa. CABI, Wallingford, UK, 608pp. ISBN 978-178064-085-3.

Bredel, M. and Jacoby, E. 2004. Chemogenomics: an emerging strategy for rapid target and drug discovery. *Nature Reviews. Genetics* 5(4), 262-75. doi:10.1038/nrg1317.

Brewer, L. R. and Palmer, J. W. 2011. Global pear breeding programmes: goals, trends and progress for new cultivars and new rootstocks. *Acta Horticulturae* 909, 105-19. doi:10.17660/ActaHortic.2011.909.10

Brooks, L. A. 1988. *Pear Tree Old Home X Farmingdale Variety No. 87*. U.S. Patent No. PP6362.

Celton, J. M., Chagné, D., Tustin, S. D., Terakami, S., Nishitani, C., Yamamoto, T., and Gardiner, S. E. 2009a. Update on comparative genome mapping between *Malus* and *Pyrus*. *BMC Research Notes* 2(1), 182. doi:10.1186/1756-0500-2-182.

Celton, J. M., Tustin, D. S., Chagné, D., and Gardiner, S. E. 2009b. Construction of a dense genetic linkage map for apple rootstocks using SSRs developed from *Malus* ESTs and *Pyrus* genomic sequences. *Tree Genetics and Genomes* 5(1), 93-107. doi:10.1007/s11295-008-0171-z.

Chagné, D., Crowhurst, R. N., Pindo, M., Thrimawithana, A., Deng, C., Ireland, H., Fiers, M., Dzierzon, H., Cestaro, A., Fontana, P., et al. 2014. The draft genome sequence of European pear (*Pyrus communis* L. "Bartlett"). *PLOS ONE* 9(4), e92644. doi:10.1371/journal.pone.0092644.

Chen, H., Song, Y., Li, L.-T., Khan, M. A., Li, X.-G., Korban, S. S., Wu, J., and Zhang, S.-L. 2015. Construction of a high-density simple sequence repeat consensus genetic map for pear (*Pyrus* spp.). *Plant Molecular Biology Reporter* 33(2), 316-25. doi:10.1007/s11105-014-0745-x.

Chevreau, E. and Bell, R. 2004. Pears (*Pyrus* spp.) and quince (*Cydonia* spp.). In: Litz, R. (Ed.), *Biotechnology of Fruit and Nut Crops*. CABI Publishing, Wallingford, UK, pp. 543-65.

Clearwater, M. J., Blattmann, P., Luo, Z., and Lowe, R. G. 2007. Control of scion vigour by kiwifruit rootstocks is correlated with spring root pressure phenology. *Journal of Experimental Botany* 58(7), 1741-51. doi:10.1093/jxb/erm029.

Costes, E., Lauri, P.-E., and Regnard, J. L. 2006. Analysing fruit tree architecture - Implications for tree management and fruit production. *Horticultural Reviews* 32, 1-61. doi:10.1002/9780470767986.ch1.

Cummins, J. N. and Aldwinckle, H. S. 1983. Breeding apple rootstocks. In: Janick, J. (Ed.), *Plant Breeding Reviews* (vol. 1). Springer US, Boston, MA, pp. 294-394. doi:10.1007/978-1-4684-8896-8_10.

Dhingra, A. 2013. Pre-publication release of Rosaceae genome information. Washington State University. Available at: https://genomics.wsu.edu/research/ (accessed on December 10, 2016).

Dhingra, A. 2016. The age of the pear. In: *Good Fruit Grower* (vol. 67). Washington State Fruit Commission, Wenatchee, WA. Available at: https://www.goodfruit.com/the-age-of-the-pear/.

Dhingra, A. and Hendrickson, C. 2017. Control of ripening and senescence in pre-harvest and post-harvest plants and plant materials by manipulating alternative oxidase activity. USA Patent 9,591,847, March 14, 2017.

Dhingra, A. and Schaeffer, S. 2014. Use of photosynthetic pigment stabilizing agents to regulate ripening and quality in fruits and vegetables. Available at: https://www.google.com/patents/US8901039.

Dondini, L. and Sansavini, S. 2012. European pear. In: Badenes, M. L., and Byrne, D. H. (Eds), *Fruit Breeding*. Springer US, Boston, MA, pp. 369–413. doi:10.1007/978-1-4419-0763-9_11.

Dondini, L., Pierantoni, L., Gaiotti, F., Chiodini, R., Tartarini, S., Bazzi, C., and Sansavini, S. 2005. Identifying QTLs for fire-blight resistance via a European pear (*Pyrus communis* L.) genetic linkage map. *Molecular Breeding* 14(4), 407–18. doi:10.1007/s11032-005-0505-6.

Eastgate, J. A. 2000. *Erwinia amylovora*: the molecular basis of fireblight disease. *Molecular Plant Pathology* 1(6), 325–9. doi:10.1046/j.1364-3703.2000.00044.x.

Einhorn, T. C., Turner, J., Gibeaut, D., and Postman, J. D. 2011. Characterization of cold hardiness in quince: potential pear rootstock candidates for northern pear production regions. *Acta Horticulturae* 909(909), 137–43. doi:10.17660/ActaHortic.2011.909.13.

Elkins, R. B., Turner, J., Seavert, C. F., Castagnoli, S., Mitcham, E. J., Biasi, W. V., Colonna, A., and Bell, R. 2008. Evaluation of potential alternative European pear cultivars for U.S. West Coast growers. *Acta Horticulturae* 800, 483–90. doi:10.17660/ActaHortic.2008.800.61.

Elkins, R., Bell, R., and Einhorn, T. 2012a. Needs assessment for future US pear rootstock research directions based on the current state of pear production and rootstock research. *Journal of the American Pomological Society* 66(3), 153–63.

Elkins, R., Bell, R., and Einhorn, T. C. 2012b. Needs assessment for future US pear rootstock research directions based on the current state of pear production and rootstock research. *Journal of the American Pomological Society* 66(3), 153–63.

Elo, A., Lemmetyinen, J., Novak, A., Keinonen, K., Porali, I., Hassinen, M., and Sopanen, T. 2007. BpMADS4 has a central role in inflorescence initiation in silver birch (*Betula pendula*). *Physiologia Plantarum* 131(1), 149-58. doi:10.1111/j.1399-3054.2007.00947.x.

El-Sharkawy, I., El Kayal, W., Prasath, D., Fernández, H., Bouzayen, M., Svircev, A. M., and Jayasankar, S. 2012. Identification and genetic characterization of a gibberellin 2-oxidase gene that controls tree stature and reproductive growth in plum. *Journal of Experimental Botany* 63(3), 1225–39. doi:10.1093/jxb/err345.

Erfani-Moghadam, J., Ebadi, A., Abdollahi, H., and Fatahi, R. 2012. Genetic diversity of some pear cultivars and genotypes using simple sequence repeat (SSR) markers. *Plant Molecular Biology Reporter* 30(5), 1065-72. doi:10.1007/s11105-012-0421-y.

Evans, K. M. and James, C. M. 2003. Identification of SCAR markers linked to *Pl-w* mildew resistance in apple. *Theoretical and Applied Genetics* 106(7), 1178–83. doi:10.1007/s00122-002-1147-2 Pl-w.

FAO/IAEA. 2018. *Manual on Mutation Breeding* (3rd edn.). Food and Agriculture Organization of the United Nations, Rome, Italy.

Fazio, G. 2015. The Geneva apple rootstock breeding program. In: Janick, J. (Ed.), *Plant Breeding Reviews* (vol. 39). Wiley-Blackwell, pp. 379-424. doi:10.1002/9781119107743.ch8.

Fazio, G., Wan, Y., Kviklys, D., Romero, L., Adams, R., Strickland, D., and Robinson, T. 2014. *Dw2*, a new dwarfing locus in apple rootstocks and its relationship to induction of early bearing in apple scions. *Journal of the American Society for Horticultural Science* 139(2), 87-98. doi:10.21273/JASHS.139.2.87.

Fischer, M. 2009. Pear breeding. In: Priyadarshan, P. M., and Jain, S. M. (Eds), *Breeding Plantation Tree Crops: Temperate Species*. Springer, New York, NY, pp. 135–60.

Flachowsky, H., Peil, A., Sopanen, T., Elo, A., and Hanke, V. 2007. Overexpression of *BpMADS4* from silver birch (*Betula pendula* Roth.) induces early-flowering in apple (*Malus x domestica* Borkh.). *Plant Breeding* 126(2), 137–45. doi:10.1111/j.1439-0523.2007.01344.x.

Foster, T. M., Watson, A. E., van Hooijdonk, B. M., and Schaffer, R. J. 2014. Key flowering genes including FT-like genes are upregulated in the vasculature of apple dwarfing rootstocks. *Tree Genetics and Genomes* 10(1), 189–202. doi:10.1007/s11295-013-0675-z.

Foster, T. M., Celton, J. M., Chagné, D., Tustin, D. S., and Gardiner, S. E. 2015. Two quantitative trait loci, *Dw1* and *Dw2*, are primarily responsible for rootstock-induced dwarfing in apple. *Horticulture Research* 2, 15001. doi:10.1038/hortres.2015.1.

Furuta, O. and Imai, T. 1987. Pomological characters of a new, self-fertilizing Japanese pear cultivar "Osa-Nijisseiki" and the optimal fruit thinning method. *Tottori Fruit Tree Experiment Station* 10, 1–19.

Govindaraj, M., Vetriventhan, M., and Srinivasan, M. 2015. Importance of genetic diversity assessment in crop plants and its recent advances: an overview of its analytical perspectives. *Genetics Research International* 2015, 431487. doi:10.1155/2015/431487.

Hancock, J. and Lobos, G. 2008. Pears. In: Hancock, J. (Ed.), *Temperate Fruit Crop Breeding*. Springer, Kluwer Academic Publishers, Dordrecht, the Netherlands, pp. 299–355. doi:10.1007/978-1-4020-6907-9-10.

Harrison, N., Harrison, R. J., Barber-Perez, N., Cascant-Lopez, E., Cobo-Medina, M., Lipska, M., Conde-Ruíz, R., Brain, P., Gregory, P. J., and Fernández-Fernández, F. 2016. A new three-locus model for rootstock-induced dwarfing in apple revealed by genetic mapping of root bark percentage. *Journal of Experimental Botany* 67(6), 1871–81. doi:10.1093/jxb/erw001.

Hawkes, J. G. 1991. The importance of genetic resources in plant breeding. *Biological Journal of the Linnean Society* 43(1), 3–10. doi:10.1111/j.1095-8312.1991.tb00578.x.

Heinicke, D. R. 1975. *Agriculture Handbook - High-Density Apple Orchards--Planning, Training, and Pruning*. Report number 58. USDA.

Henry, I. M., Dilkes, B. P., Young, K., Watson, B., Wu, H., and Comai, L. 2005. Aneuploidy and genetic variation in the *Arabidopsis thaliana* triploid response. *Genetics* 170(4), 1979–88. doi:10.1534/genetics.104.037788.

Hirata, N. 1989. Self-compatible mutant in Japanese pear. *Gamma Field Symposium* 28, 71–80.

Hirst, P. M. and Ferree, D. C. 1995. Rootstock effects on the flowering of "delicious" apple. I. Bud development. *Journal of the American Society for Horticultural Science* 120(6), 1010–7. doi:10.21273/JASHS.120.6.1010.

Hollender, C. A. and Dardick, C. 2015. Molecular basis of angiosperm tree architecture. *The New Phytologist* 206(2), 541–56. doi:10.1111/nph.13204.

Huang, N., Angeles, E. R., Domingo, J., Magpantay, G., Singh, S., Zhang, G., Kumaravadivel, N., Bennett, J., and Khush, G. S. 1997. Pyramiding of bacterial blight resistance genes in rice: marker-aided selection using RFLP and PCR. *Theoretical and Applied Genetics* 95(3), 313–20. doi:10.1007/s001220050565.

Huettel, B., Kreil, D. P., Matzke, M., and Matzke, A. J. M. 2008. Effects of aneuploidy on genome structure, expression, and interphase organization in *Arabidopsis thaliana*. *PLOS Genetics* 4(10), e1000226. doi:10.1371/journal.pgen.1000226.

Hummer, K. E., Pomper, K., Postman, J., Graham, J. C., Stover, E., Mercure, E., Aradhya, M., Crisosto, C. H., et al. 2012. Emerging fruit crops. In: Badenes, M. L. and Byrne, D. H. (Eds), *Fruit Breeding*. Springer, Boston, MA, pp. 97–147. doi:10.1007/978-1-4419-0763-9_4.

Itai, A. 2007. Genome mapping and molecular breeding in plants. In: Kole, C. (Ed.), *Fruit and Nuts*. Springer, Berlin. doi:10.1007/978-3-540-34533-6_6.

Jackson, J. E. 2003. *The Biology of Apples and Pears*. Cambridge University Press, Cambridge. doi:10.1017/CBO9780511542657.

Jacob, H. B. 1998. Pyrodwarf, a new clonal rootstock for high density pear orchards. *Acta Horticulturae* 38(475), 169–78. doi:10.17660/ActaHortic.1998.475.20.

Janick, J. and Paull, R. E. 2008. *The Encyclopedia of Fruit and Nuts*. CABI Publishing, Wallingford, UK. doi:10.1079/9780851996387.0000.

Johnson, K. B. and Temple, T. N. 2013. Evaluation of strategies for fire blight control in organic pome fruit without antibiotics. *Plant Disease* 97(3), 402–9. doi:10.1094/PDIS-07-12-0638-RE.

Karkute, S. G., Singh, A. K., Gupta, O. P., Singh, P. M., and Singh, B. 2017. CRISPR/Cas9 mediated genome engineering for improvement of horticultural crops. *Frontiers in Plant Science* 8, 1635. doi:10.3389/fpls.2017.01635.

Kimura, T., Zhong Shi, Y. Z., Shoda, M., Kotobuki, K., Matsuta, N., Hayashi, T., Ban, Y., and Yamamoto, T. 2002. Identification of Asian pear varieties by SSR analysis. *Breeding Science* 52(2), 115–21. doi:10.1270/jsbbs.52.115.

Knäbel, M., Friend, A. P., Palmer, J. W., Diack, R., Wiedow, C., Alspach, P., Deng, C., Gardiner, S. E., Tustin, D. S., Schaffer, R., et al. 2015. Genetic control of pear rootstock-induced dwarfing and precocity is linked to a chromosomal region syntenic to the apple *Dw1* loci. *BMC Plant Biology* 15(1), 230. doi:10.1186/s12870-015-0620-4.

Koepke, T. and Dhingra, A. 2013. Rootstock scion somatogenetic interactions in perennial composite plants. *Plant Cell Reports* 32(9), 1321–37. doi:10.1007/s00299-013-1471-9.

Kumar, S., Kirk, C., Deng, C., Wiedow, C., Knaebel, M., and Brewer, L. 2017. Genotyping-by-sequencing of pear (*Pyrus* spp.) accessions unravels novel patterns of genetic diversity and selection footprints. *Horticulture Research* 4, 17015. doi:10.1038/hortres.2017.15.

Lang, G. A. 2005. Underlying principles of high-density sweet cherry production. *Acta Horticulturae* 667, 325–36. doi:10.17660/ActaHortic.2005.667.47.

Lansari, A. and Iezzoni, A. 1990. A preliminary analysis of self-incompatibility in sour cherry. *HortScience* 25(12), 1636–8.

Lapins, K. O. 1970. The. Stella cherry. *Fruit Varieties and Horticultural Digest* 24(1), 19–20.

Li, L., Deng, C. H., Knäbel, M., Chagné, D., Kumar, S., Sun, J., Zhang, S., and Wu, J. 2017. Integrated high-density consensus genetic map of *Pyrus* and anchoring of the "Bartlett" v1.0 (*Pyrus communis*) genome. *DNA Research* 24(3) 1, 289–301. doi:10.1093/dnares/dsw063.

Lindstrom, O. M., Anisko, T., and Dirr, M. A. 1995. Low-temperature exotherms and cold hardiness in three taxa of deciduous trees. *Journal of the American Society for Horticultural Science* 120(5), 830–4. doi:10.21273/JASHS.120.5.830.

Liu, Q., Song, Y., Liu, L., Zhang, M., Sun, J., Zhang, S., and Wu, J. 2015. Genetic diversity and population structure of pear (*Pyrus* spp.) collections revealed by a set of core genome-wide SSR markers. *Tree Genetics and Genomes* 11(6), 128. doi:10.1007/s11295-015-0953-z.

Maluszynski, M., Ahloowalia, B. S., and Sigurbjörnsson, B. 1995. Application of *in vivo* and *in vitro* mutation techniques for crop improvement. In: Cassells, A. C. and Jones, P. W. (Eds), *The Methodology of Plant Genetic Manipulation: Criteria for Decision Making: Proceedings of the Eucarpia Plant Genetic Manipulation Section Meeting held at Cork*, Ireland from September 11–14, 1994. Springer Netherlands, Dordrecht, pp. 303–15. doi:10.1007/978-94-011-0357-2_36.

Martin-Guay, M. O., Paquette, A., Dupras, J., and Rivest, D. 2018. The new Green Revolution: sustainable intensification of agriculture by intercropping. *The Science of the Total Environment* 615, 767–72. doi:10.1016/j.scitotenv.2017.10.024.

Mba, C., Afza, R., Bado, S., and Jain, S. 2010. Induced mutagenesis in plants using physical and chemical agents. In: Davey, M. R. and Anthony, P. (Eds), *Plant Cell Culture: Essential Methods*. Wiley & Sons Ltd., Chichester, UK, pp. 111–30. doi:10.1002/9780470686522.ch7.

McClure, B. A. and Franklin-Tong, V. 2006. Gametophytic self-incompatibility: understanding the cellular mechanisms involved in "self" pollen tube inhibition. *Planta* 224(2), 233–45. doi:10.1007/s00425-006-0284-2.

Mehlenbacher, S. A., Cociu, V., and Hough, F. L. 1991. Apricots (*Prunus*). *Acta Horticulturae* 290, 65–110. doi:10.17660/ActaHortic.1991.290.3.

Meland, M. 1998. Yield and fruit quality of "van" sweet cherry in four high density production systems over seven years. *Acta Horticulturae* 468, 425–32. doi:10.17660/ActaHortic.1998.468.52.

Miklas, P. N., Hu, J., Grünwald, N. J., and Larsen, K. M. 2006. Potential application of TRAP (targeted region amplified polymorphism) markers for mapping and tagging disease resistance traits in common bean. *Crop Science* 46(2), 910–6. doi:10.2135/cropsci2005.08-0242.

Miranda, C., Urrestarazu, J., Santesteban, L. G., Royo, J. B., and Urbina, V. 2010. Genetic diversity and structure in a collection of ancient Spanish pear cultivars assessed by microsatellite markers. *Journal of the American Society for Horticultural Science* 135(5), 428–37. doi:10.21273/JASHS.135.5.428.

Mitre, V., Mitre, I., Sestras, A. F., and Sestras, R. E. 2012. Effect of roots pruning upon the growth and fruiting of apple trees in high density orchards. *Bulletin UASVM Horticulture* 69(1). doi:10.15835/buasvmcn-hort:8527.

Montanari, S., Guérif, P., Ravon, E., Denancé, C., Muranty, H., Velasco, R., Chagné, D., Bus, V. G. M., Robert, P., Perchepied, L., et al. 2015. Genetic mapping of *Cacopsylla pyri* resistance in an interspecific pear (*Pyrus* spp.) population. *Tree Genetics and Genomes* 11(4), 74. doi:10.1007/s11295-015-0901-y.

Montanari, S., Perchepied, L., Renault, D., Frijters, L., Velasco, R., Horner, M., Gardiner, S. E., Chagné, D., Bus, V. G. M., Durel, C.-E., et al. 2016. A QTL detected in an interspecific pear population confers stable fire blight resistance across different environments and genetic backgrounds. *Molecular Breeding* 36(4). doi:10.1007/s11032-016-0473-z.

Morgan, J. 2015. *The Book of Pears: the Definitive History and Guide to Over 500 Varieties*. Chelsea Green Publishing, Hartford, VT.

Nagashima, H. and Hikosaka, K. 2011. Plants in a crowded stand regulate their height growth so as to maintain similar heights to neighbours even when they have potential advantages in height growth. *Annals of Botany* 108(1), 207–14. doi:10.1093/aob/mcr109.

Nimbolkar, P. K., Awachare, C., Reddy, Y. T. N., Chander, S., and Hussain, F. 2016. Role of rootstocks in fruit production–a review. *Journal of Agricultural Engineering and Food Technology* 3, 183–8.

Nishitani, C., Hirai, N., Komori, S., Wada, M., Okada, K., Osakabe, K., Yamamoto, T., and Osakabe, Y. 2016. Efficient genome editing in apple using a CRISPR/Cas9 system. *Scientific Reports* 6, 31481. doi:10.1038/srep31481. Available at: https://www.nature.com/articles/srep31481#supplementary-information.

Nuclo, R. L., Johnson, K. B., Stockwell, V. O., and Sugar, D. 1998. Secondary colonization of pear blossoms by two bacterial antagonists of the fire blight pathogen. *Plant Disease* 82(6), 661–8. doi:10.1094/PDIS.1998.82.6.661.

Oladosu, Y., Rafii, M. Y., Abdullah, N., Hussin, G., Ramli, A., Rahim, H. A., Miah, G., and Usman, M. G. 2016. Principle and application of plant mutagenesis in crop improvement: a review. *Biotechnology and Biotechnological Equipment* 30(1), 1–16. doi:10.1080/13102818.2015.1087333.

Özaktan, H. and Bora, T. 2004. Biological control of fire blight in pear orchards with a formulation of *Pantoea agglomerans* strain Eh 24. *Brazilian Journal of Microbiology* 35(3), 224–9. doi:10.1590/S1517-83822004000200010.

Palonen, P. and Buszard, D. 1997. Current state of cold hardiness research on fruit crops. *Canadian Journal of Plant Science* 77(3), 399–420. doi:10.4141/P96-013.

Pandey, K. K. 1974. Overcoming interspecific pollen incompatibility through the use of ionising radiation. *Heredity* 33(2), 279–84. doi:10.1038/hdy.1974.93.

Petracek, P. D., Silverman, F. P., and Greene, D. W. 2003. A history of commercial plant growth regulators in apple production. *HortScience* 38(5), 937–42. doi:10.21273/HORTSCI.38.5.937.

Pilcher, R. L. R., Celton, J.-M., Gardiner, S. E., and Tustin, D. S. 2008. Genetic markers linked to the dwarfing trait of apple rootstock "malling 9." *Journal of the American Society for Horticultural Science* 133(1), 100–6. doi:10.21273/JASHS.133.1.100.

Postman, J. 2009. *Cydonia oblonga*: the unappreciated quince. *Arnoldia* 67(1), 2–9.

Postman, J., Kim, D., and Bassil, N. 2013. OH × F paternity perplexes pear producers. *Journal of American Pomological Society* 67, 157–67.

Predieri, S., Magli, M., and Zimmerman, R. H. 1997. Pear mutagenesis: *in vitro* treatment with gamma-rays and field selection for vegetative form traits. *Euphytica* 93(2), 227–37. doi:10.1023/A:1002984516192.

Rademacher, W. 2004. Prohexadione-Ca induces resistance to fireblight and other diseases. *EPPO Bulletin* 34(3), 383–8. doi::doi:10.1111/j.1365-2338.2004.00765.x.

Reil, W., Ireland, J., and Elkins, R. 2007. Propagation and rootstock selection. In: *Pear Production and Handling Manual*. Publication 3483. University of California Agriculture and Natural Resources, Davis, CA.

Reimer, F. C. 1925. *Blight Resistance in Pears and Characteristics of Pear Species and Stocks*. Oregon Agricultural College, Corvallis, OR.

Reimer, F. C. 1950. *Development of Blight Resistant French Pear Rootstocks*. Oregon Agricultural College, Corvallis, OR.

Samad, A., McNeil, D. L., and Khan, Z. U. 1999. Effect of interstock bridge grafting (M9 dwarfing rootstock and same cultivar cutting) on vegetative growth, reproductive

growth and carbohydrate composition of mature apple trees. *Scientia Horticulturae* 79(1-2), 23-38. doi:10.1016/S0304-4238(98)00179-4.
Sansavini, S. and Musacchi, S. 1994. Canopy architecture, training and pruning in the modern European pear orchards: an overview. *Acta Horticulturae* 367, 152-72. doi:10.17660/ActaHortic.1994.367.20.
Sanzol, J. and Robbins, T. P. 2008. Combined analysis of S-Alleles in European pear by pollinations and PCR-based S-Genotyping; correlation between S-Phenotypes and S-RNase genotypes. *Journal of the American Society for Horticultural Science* 133(2), 213-24. doi:10.21273/JASHS.133.2.213.
Sassa, H. 2016. Molecular mechanism of the S-RNase-based gametophytic self-incompatibility in fruit trees of *Rosaceae*. *Breeding Science* 66(1), 116-21. doi:10.1270/jsbbs.66.116.
Serdani, M., Spotts, R. A., Calabro, J. M., Postman, J. D., and Qu, A. P. 2006. Evaluation of the USDA national clonal *Pyrus* germplasm collection for resistance to *Podosphaera leucotricha*. *HortScience* 41(3), 717-20. doi:10.21273/HORTSCI.41.3.717.
Sharma, T. 2003. Molecular diagnosis and application of DNA markers in the management of fungal and bacterial plant diseases. *Indian Journal of Biotechnology* 2, 99-109.
Shu, Q., Forster, B., and Nakagawa, H. 2012. *Plant Mutation Breeding and Biotechnology*. International Atomic Energy Agency; Food and Agriculture Organization of the United Nations (FAO).
Simard, M. H., Guisnel, R., Daguin, F., Demilly, D., Billy, B., and Honoré, D. 2011. Is dwarfing in pear rootstocks due to aneuploid genetic structures? In: *XI International Pear Symposium 909*, 2010, pp. 59-66.
Socias i Company, R. 1998. Fruit tree genetics at a turning point: the almond example. *Theoretical and Applied Genetics* 96(5), 588-601. doi:10.1007/s001220050777.
Song, Y., Fan, L., Chen, H., Zhang, M., Ma, Q., Zhang, S., and Wu, J. 2014. Identifying genetic diversity and a preliminary core collection of *Pyrus pyrifolia* cultivars by a genome-wide set of SSR markers. *Scientia Horticulturae* 167, 5-16. doi:10.1016/j.scienta.2013.12.005.
Stephene, W. P. 1958. Pear pollination studies in Oregon. *Technical Bulletin of the Oregon Agricultural Experiment Station* 43, 1-43.
Stokes, M. E. and McCourt, P. 2014. Towards personalized agriculture: what chemical genomics can bring to plant biotechnology. *Frontiers in Plant Science* 5, 344. doi:10.3389/fpls.2014.00344.
Sun, W., Zhang, Y., Le, W., and Zhang, H. 2009. Construction of a genetic linkage map and QTL analysis for some leaf traits in pear (*Pyrus* L.). *Frontiers of Agriculture in China* 3(1), 67-74. doi:10.1007/s11703-009-0013-2.
Suzuki, K., Murase, S., and Yamazaki, T. 1988. *Preliminary Research for Dwarfing Rootstock of Peach*. The Peach, Gainesville, FL.
Takayama, S. and Isogai, A. 2005. Self-incompatibility in plants. *Annual Review of Plant Biology* 56(1), 467-89. doi:10.1146/annurev.arplant.56.032604.144249.
Tao, R., Watari, A., Hanada, T., Habu, T., Yaegaki, H., Yamaguchi, M., and Yamane, H. 2007. Self-compatible peach (*Prunus persica*) has mutant versions of the S haplotypes found in self-incompatible *Prunus* species. *Plant Molecular Biology* 63(1), 109-23. doi:10.1007/s11103-006-9076-0.
Tehrani, G. and Brown, S. 2010. Pollen-incompatibility and self-fertility in sweet cherry. In: Janick, J. (Ed.), *Plant Breeding Reviews* (vol. 9). Wiley, pp. 367-88. doi:10.1002/9780470650363.ch10.

Teng, Y., Tanabe, K., Tamura, F., and Itai, A. 2002. Genetic relationships of *Pyrus* species and cultivars native to East Asia revealed by randomly amplified polymorphic DNA markers. *Journal of the American Society for Horticultural Science* 127(2), 262–70. doi:10.21273/JASHS.127.2.262.

The Green Revolution. 1969. *Nutrition Reviews* 27(5), 133–6. doi:10.1111/j.1753-4887.1969.tb04993.x. Available at: https://onlinelibrary.wiley.com/doi/10.1111/j.1753-4887.1969.tb04993.x.

Tombesi, S., Nardini, A., Farinelli, D., and Palliotti, A. 2014. Relationships between stomatal behavior, xylem vulnerability to cavitation and leaf water relations in two cultivars of *Vitis vinifera*. *Physiologia Plantarum* 152(3), 453–64. doi:10.1111/ppl.12180.

USA Pears. 2018. Pear Bureau Northwest. Available at: https://usapears.org/.

USDA. 2016. *Agricultural Statistics 2016*. USDA NASS, Washington DC.

van der Zwet, T. and Keil, H. L. 1979. *Fire Blight, a Bacterial Disease of Rosaceous Plants*. United States Department of Agriculture, Beltsville.

van der Zwet, T., Orolaza-Halbrendt, N., and Zeller, W. 2016. Front matter. In: *Fire Blight: History, Biology, and Management*. Bacteriology. The American Phytopathological Society, St. Paul, MN, pp. i–ix. doi:10.1094/9780890544839.fm.

van Nocker, S. and Gardiner, S. E. 2014. Breeding better cultivars, faster: applications of new technologies for the rapid deployment of superior horticultural tree crops. *Horticulture Research* 1, 14022. doi:10.1038/hortres.2014.22.

Velten, S., Leventon, J., Jager, N., and Newig, J. 2015. What is sustainable agriculture? A systematic review. *Sustainability* 7(6), 7833–65. doi:10.3390/su7067833.

Vilanova, S., Badenes, M. L., Burgos, L., Martínez-Calvo, J., Llácer, G., and Romero, C. 2006. Self-compatibility of two apricot selections is associated with two pollen-part mutations of different nature. *Plant Physiology* 142(2), 629–41. doi:10.1104/pp.106.083865.

Volk, G. M., Richards, C. M., Henk, A. D., Reilley, A. A., Bassil, N. V., and Postman, J. D. 2006. Diversity of wild *Pyrus communis* based on microsatellite analyses. *Journal of the American Society for Horticultural Science* 131(3), 408–17. doi:10.21273/JASHS.131.3.408.

Wang, C., Tian, Y., Buck, E. J., Gardiner, S. E., Dai, H., and Jia, Y. 2011. Genetic mapping of PcDw determining pear dwarf trait. *Journal of the American Society for Horticultural Science* 136(1), 48–53. doi:10.21273/JASHS.136.1.48.

Webster, A. D. 1995. Rootstock and interstock effects on deciduous fruit tree vigour, precocity, and yield productivity. *New Zealand Journal of Crop and Horticultural Science* 23(4), 373–82. doi:10.1080/01140671.1995.9513913.

Webster, A. D. 1998. A brief review of pear rootstock development. *Acta Horticulturae* 475, 135–42. doi:10.17660/ActaHortic.1998.475.16.

Webster, T. 2002. Dwarfing rootstocks: past, present and future. *Compact Fruit Tree* 35, 67–72.

Webster, A. D. and Wertheim, S. J. 2003. Apple rootstocks. In: Ferree, D. C. and Warrington, I. J. (Eds), *Apples: Botany, Production and Uses*. CABI Publishing, Wallingford, UK, pp. 91–124.

Wertheim, S. J. 2002. Rootstocks for European pear: a review. *Acta Horticulturae* 596, 299–309. doi:10.17660/ActaHortic.2002.596.47.

WSU_TFREC. Rootstocks for pear. WSU Tree Fruit Research and Extension Center. Available at: http://treefruit.wsu.edu/web-article/pear-rootstocks/ (accessed on December 4, 2018).

Wu, J., Wang, Z., Shi, Z., Zhang, S., Ming, R., Zhu, S., Khan, M. A., Tao, S., Korban, S. S., Wang, H., et al. 2013. The genome of the pear (*Pyrus bretschneideri* Rehd.). *Genome Research* 23(2), 396–408. doi:10.1101/gr.144311.112.

Wu, J., Li, L. T., Li, M., Khan, M. A., Li, X. G., Chen, H., Yin, H., and Zhang, S. L. 2014. High-density genetic linkage map construction and identification of fruit-related QTLs in pear using SNP and SSR markers. *Journal of Experimental Botany* 65(20), 5771–81. doi:10.1093/jxb/eru311.

Wünsch, A. and Hormaza, J. I. 2004. Molecular evaluation of genetic diversity and S-allele composition of local Spanish sweet cherry (*Prunus avium* L.) cultivars. *Genetic Resources and Crop Evolution* 51(6), 635–41. doi:10.1023/B:GRES.0000024649.06681.43.

Yamamoto, T. and Terakami, S. 2016. Genomics of pear and other Rosaceae fruit trees. *Breeding Science* 66(1), 148–59. doi:10.1270/jsbbs.66.148.

Yamamoto, T., Kimura, T., Terakami, S., Nishitani, C., Sawamura, Y., Saito, T., Kotobuki, K., and Hayashi, T. 2007. Integrated genetic linkage maps for pear based on SSR and AFLP markers. *Breeding Science* 57(4), 321–9. doi:10.1270/jsbbs.57.321.

Yamamoto, T., Terakami, S., Moriya, S., Hosaka, F., Kurita, K., Kanamori, H., Katayose, Y., Saito, T., and Nishitani, C. 2013. DNA markers developed from genome sequencing analysis in Japanese pear (*Pyrus pyrifolia*). *Acta Horticulturae* 976, 477–83. doi:10.17660/ActaHortic.2013.976.67.

Yamane, H. and Tao, R. 2009. Molecular basis of self-(in)compatibility and current status of S-genotyping in Rosaceous fruit trees. *Journal of the Japanese Society for Horticultural Science* 78(2), 137–57. doi:10.2503/jjshs1.78.137.

Yin, K., Gao, C., and Qiu, J. L. 2017. Progress and prospects in plant genome editing. *Nature Plants* 3, 17107. doi:10.1038/nplants.2017.107.

Zawaski, C., Kadmiel, M., Pickens, J., Ma, C., Strauss, S., and Busov, V. 2011. Repression of gibberellin biosynthesis or signaling produces striking alterations in poplar growth, morphology, and flowering. *Planta* 234(6), 1285–98. doi:10.1007/s00425-011-1485-x.

Zhang, Y. X. and Lespinasse, Y. 1991. Pollination with gamma-irradiated pollen and development of fruits, seeds and parthenogenetic plants in apple. *Euphytica* 54(1), 101–9. doi:10.1007/BF00145636.

Zhao, Y., Lin, H., Guo, Y., Liu, Z., Guo, X., and Li, K. 2013. Genetic linkage maps of pear based on srap markers. *Pakistan Journal of Botany* 45, 1265–71.

Zuccherelli, S., Broothaerts, W., Tassinari, P., Tartarini, S., Dondini, L., Bester, A., and Sansavini, S. 2002. S-allele characterization in self-incompatible pear (*Pyrus communis*): biochemical, molecular and field analyses. *Acta Horticulturae* 596, 147–52. doi:10.17660/ActaHortic.2002.596.18.

Chapter 6

Challenges and opportunities in pear cultivation

Todd Einhorn, Michigan State University, USA

1 Introduction

Fresh market European pears (*Pyrus communis*) are a high-value commodity. Global pear consumption, however, has declined over the past few decades. Concomitantly, increasing market demand for large size-classes of pear, shifting consumer preference toward sustainable production practices, and greater governmental oversight (to ensure food safety and traceability, immigration and labor regulation, or chemical use restrictions) increase pressure on producers to remain profitable. Moreover, pear is an inherently vigorous species with few rootstock options for dwarfing (outside of Europe), requiring intensive horticultural management to reduce canopy size and promote early cropping. Thus, to remain profitable, producers' decisions to capitalize land to European pear over alternative, high-value tree fruit crops depend heavily on innovation, demonstrable early returns on investment, and high productivity.

Presently, high-density plantings offer the greatest potential to achieve these objectives. The innovative high-density orchard systems developed over decades in Europe were driven by high production/labor costs and availability

http://dx.doi.org/10.19103/AS.2018.0040.20

of dwarfing rootstocks. High-density pear production seems novel in many regions of the world, but consider that during the 1930s, 2300 to 4000 trees/ha were being trained to the vertical cordon 'Ferraguti system in the Po Valley of Italy (Sansavini, 1982). Globally, many regions have adopted high-density plantings and intensive training systems from Europe or extended familiar apple (*Malus domestica*) production models to pear. These expensive pear plantings, however, require significant horticultural intervention to manage vigor and encourage early cropping given the dearth of dwarfing rootstocks in many locations. Such challenges are, comparatively, practically non-existent for modern apple production worldwide. Further, several notable physiological differences between apple and pear preclude simple transfer of horticultural techniques from one crop to the other, as will be discussed herein. Unlike apple, pear varieties have not changed much over the past century of production. Thus, market expansion or displacement of antiquated varieties with new, exciting and profitable varieties lags behind apple. Despite these challenges, European pears are unique among temperate-zone tree fruit for their distinctive soft, buttery, melting flesh and wide-ranging aromatic flavor profile. For consumers who do not prefer soft, ripened pears, the recent release of several interesting, crisp-fleshed European pear cultivars may generate interest. These delectable and diverse fruit are worthy of far greater consumer attention. Marketing, new cultivar development, improved fruit quality, and higher production efficiency are all integral to the evolution and sustainable future of pear.

This review will focus on the main production challenge for pear trees, low yield potential, which is inevitably linked to excessive vigor and poor precocity. A summary of the underlying biological, physiological, and biophysical mechanisms that regulate these processes will be advanced, as will practical strategies to reverse or mitigate production challenges they pose. There have been significant advances in plant growth regulator (PGR) use in pear production. New chemistries and novel uses of existing chemistries provide compelling opportunities to manage high-density plantings. One caveat regarding utilization of the PGR chemistries discussed herein is the lack of universality of labeling laws among countries. Prioritization was given to PGRs with broad global use for pear or closely related tree fruit crops, but depending on the region, some may not be labeled for pear use. Several interesting, once-useful chemistries no longer labeled for pear were omitted from this review.

2 Mechanisms of flowering and pollination

2.1 Flower inflorescence and fertility status

The flowering structure of pear is an indeterminate raceme with ~6 to 9 flowers, depending on the cultivar. Unlike the cyme inflorescence of apple with a developmentally advanced central flower (i.e., 'king'), the basal flowers of a pear inflorescence (umbel-like raceme) are more developmentally advanced than distal flowers and have higher fruit set potential (T. Einhorn, personal observation). Pear flowers produce abundant pollen, but their nectar has low sugar content compared with other tree fruits; thus, pollen-foraging bees will collect the protein-rich pollen of pear, but nectar-foraging bees prefer other floral sources (Vansell, 1946). Pollination depends primarily upon pollinators, but controlled pollination studies indicate a limited degree of fertilization and fruit set occurs from wind disseminated pollen if compatible sources are near (Lombard et al., 1971).

Pear flowers are self-infertile and exhibit gametophytic self-incompatibility (GSI) from a single multi-allelic locus (S-locus), consistent with genera of Rosaceae and other flowering plant families (see review by De Franceschi et al., 2012). Compatibility between pear genotypes was historically determined from controlled test-crosses (e.g., Olenz and Zielinski, 1965). Zuccherelli et al. (2002) first characterized the polymorphic S-determinant stylar ribonucleases (S-RNase) of European pear. Further discrimination of S-RNases and identification of pollen S-locus F-box genes (SLF) have resulted in S-allele genotyping of well over 100 pear cultivars (Goldway et al., 2009; Moriya et al., 2007; Mota et al., 2007; Quinet et al., 2014; Sanzol and Robbins, 2008; Sassa et al., 2007; Zisovich et al., 2009). This information broadly benefits breeding programs and production systems alike. Compatible pollen is of little consequence if anthesis dates for pollinizers and main cultivars do not overlap to ensure potential for fertilization. An inverse relationship between fruit set and pollinizer distance is well established (Lombard et al., 1971; Westwood and Grim, 1962). Modern orchards typically establish compatible cultivars in alternating groups of single or multiple rows to facilitate fertilization and harvest management. These designs are an improvement from individual, interspersed pollinizer trees (typically at ratios of 1:8 with the 'main' cultivar), but do not obviate orchard blocks of multiple cultivars. Surprisingly, exploitation of the *Pyrus* germplasm for compatible pollen sources for *P. communis* has been limited (Ketchie et al., 1996). Phenotyping of vegetative and reproductive characteristics of *P. betulafolia* and *P. calleryana* clones identified several genotypes with compact growth and profuse, annual flowering that could potentially serve as pollen donors for European pear orchards (Castagnoli, 2008; Ketchie et al., 1996; LeLezec et al., 2005). Although high rates of pollen sterility were observed with *P. nivalis* (Zielinski and Thompson, 1966), several *P. nivalis* genotypes successfully pollinated 'Conference' pear (Kemp et al., 2008). Incorporation of alternative species of *Pyrus* into high-density orchards, analogous to crabapples (e.g., *Malus* spp.) for pollination of apple, could further improve management efficiency.

Despite the GSI system of pear, several cultivars (i.e., 'Bosc', 'Bartlett', 'Flemish Beauty', 'Beurre Hardy', 'Conference', 'Rocha') have the capacity to set seedless fruit when self-pollinated (e.g., stimulative parthenocarpy; Westwood, 1993). Quinet and Jacquemart (2015) did not detect significant differences between fruit set for bagged, self-pollinated 'Conference' flowers, open pollinated 'Conference' flowers, or cross-pollinated 'Conference' flowers using compatible pollen in two of three years. Interestingly, the presence of seeds in 'Bartlett', 'd'Anjou', and 'Bosc' pears following self-pollination experiments implies a limited degree of self-fertilization (Stephen, 1958). The mechanisms controlling set of parthenocarpic fruit have not been fully elucidated. Westwood et al. (1968) documented a three-fold increase in indole-3-acetic acid (IAA) of 'd'Anjou' fruiting spurs bearing seeded fruit compared to spurs with seedless fruit. The subsequent year, caged trees (to exclude pollinators) set a full crop of seedless fruit, but over the next two years, fruit set declined markedly. IAA content was not measured during these years, but exogenously applied 2,4,5 T-P (an auxin or auxin-like compound) restored seedless set of caged trees, implicating auxin as a translocatable fruit set 'promoter' (Westwood et al., 1968). Whatever the mechanism(s), parthenocarpic fruit or fruit with low seed content appear to have decreased sink strength compared with fruit of higher seed content (Callan and Lombard, 1978; Lombard et al., 1971; Quinet and Jacquemart, 2015; Stephen, 1958; Weinbaum et al., 2001), though this was not evident for all cultivars evaluated (i.e., 'd'Anjou', Stephen, 1958). In apple, cell division of fruit cortical tissue was fertilization dependent (Malladi and Johnson, 2011).

Fruit set of facultative parthenocarpic cultivars is aided by temperature (Griggs, 1953). For example, in California, warm temperatures in the Sacramento river delta induce full commercial crops of parthenocarpic 'Bartlett' pears. This is the basis for solid-block orchards in this region, although Griggs (1953) summarized research showing distinct advantages of cross-pollination for fruit set and size in this environment. In northern Pacific US regions (Hood River and Wenatchee Valleys of Oregon and Washington), however, cross-pollination of 'Bartlett' is recommended for commercial fruit production (Stebbins et al., 1979). 'Bartlett' orchards located between California and northern Oregon and Washington, i.e., in the Rogue River district and Willamette Valley, are regarded as partially self-fertile with crop variation among years presumably due to temperature. Quinet and Jacquemart (2015) also documented temperature effects on parthenocarpic set of 'Conference'.

The influence of seeds on return bloom of pear is not clear. The flowering and bearing habits of most *P. communis* cultivars are annual, with relatively few strongly biennial-bearing cultivars (e.g., 'Bosc' and 'Comice'). Diffusates from seeded and seedless 'Bartlett' pears varied considerably over the season and did not provide compelling support for seed-induced hormonal regulation of floral bud production (Griggs et al., 1970). While seed content of 'Spencer Seedless' apples induced biennial bearing, Weinbaum et al. (2001) disproved the broad applicability of the findings to pear by showing that seedless 'Bartlett' pears also strongly inhibited flowering of spurs the subsequent season.

2.2 Effective pollination period (EPP)

Several important pear cultivars ('Comice', 'Abate Fetel', 'Packham's Triumph', 'd'Anjou', etc.) lack precocity. The situation for 'Packham's Triumph' and 'd'Anjou', however, is compounded by the non-use of size-controlling rootstocks, which are not readily available in the regions where production of these two cultivars is greatest. Cultivars with low productivity are typically classified as 'shy' bearers, but the mechanisms limiting fruit set vary considerably. Non-precocious, highly vigorous 'd'Anjou' trees, for example, transition abruptly at some point during the 5th–12th leaf and produce prolific bloom, yet fruit set is characteristically poor until trees settle into regular and often heavy bearing. The productivity of 'Comice', on the other hand, is universally underwhelming throughout its life. A simple, conceptual model has been useful for integrating the variables that limit fertility, termed the effective pollination period (EPP; Williams, 1965; see excellent review by Sanzol and Herrero, 2001). The EPP is the duration of time the ovule remains viable minus the time required for pollen grain germination and growth to the base of the ovary. Stigmatic receptivity, pollen genotype, pollen tube growth kinetics, pollen–pistil interactions, proportion of non-functional ovules, etc., all affect the EPP and fertilization potential, which are affected by environmental factors (mainly temperature), nutrition, hormones, and carbohydrate status.

To complicate the picture, varying methodologies to determine EPP can produce disparate results when evaluating a specific cultivar. For instance, 'Comice' is widely recognized to have a short EPP, though reports have ranged from 1 day (Lombard et al., 1972), 2–5 days (Crisosto et al., 1992; Williams, 1965), 6 days (Tromp and Borsboom, 1994) and, after chemical treatment, up to 9 days (Crisosto et al., 1986). Low ovule viability is often acknowledged to limit EPP and fruit set of 'Comice', but Callan and Lombard (1978) were unable to link early ovule degeneration to low female fertility (i.e., <5% of all embryo sacs evaluated 4 days after anthesis were malformed or degenerative). Temperature strongly affects EPP (Lombard et al., 1972). Optimal temperatures for pear pollen germination appear to be lower than those for pollen tube growth (Vasilakakis and Porlingis, 1985). Pollen germination among several cultivars and species of *Pyrus* varied considerably, whether provided with 12°C or 20.5°C with little consistent difference between temperatures (Kemp et al., 2008). 'Bartlett' pollen tubes required 24 h or 120 h at 21.1°C and 10°C, respectively, to reach the base of 'd'Anjou' pistils (Mellenthin et al., 1972). Under field conditions using caged limbs, fruit set was reduced and seed content was significantly lower at 13°C than at higher temperatures, though an inadequate experimental design and lack of replication limit the interpretation of these data. These authors further attempted to correlate a decade of 'd'Anjou' yield records with the maximum temperature recorded during 0–10 days after full bloom (dafb). A linear function suggested that the maximum temperature explained a fair portion of the yield variability (r = 0.668), but variation of 100% in the middle of the curve clearly indicates involvement of additional factors. There is wide acceptance that high temperatures speed up the rate of pollen tube growth, but also advance ovule degeneration (Sanzol and Herrero, 2001). Temperature chamber experiments often do not report humidity, yet Lombard et al. (1972) ascribed poor fertility of 'Comice' to high humidity and suggested an adverse role of the stigma on fruit set. Sanzol et al. (2003) elucidated stigmatic receptivity as the limiting factor for 'Agua de Aranjuez' fruit set, whereby a 2-day EPP was not associated with ovule degeneration or pollen attributes. Clearly, further work in this interesting area would be helpful to clarify which component(s) limit EPP for a given cultivar and climate.

2.3 Floral initiation

Pear floral initiation has been proposed to occur ~ 60 dafb (Westwood, 1993) after completion of shoot extension (i.e., June–July in the northern hemisphere), as similarly summarized for apple (Bubán and Faust, 1982). Apex broadening is the first visible indication of apple floral initiation (Foster et al., 2003), preceding doming by several weeks. The general timeline for pear initiation varies widely among cultivars and seasons. For a given year, flower initiation of 'Agua de Aranjuez', 'Beurre Hardy', 'Conference', 'Comice', and 'Passe Crassane' differed by as many as 40 days (Carrera, 1982). Within-cultivar variation in the timing of flower initiation was markedly lower, but seasonal anomalies over a 4-year period produced a range of ±7 days (Carrera, 1982). Presumably, anthesis dates were similarly shifted, but only calendar dates of floral initiation events were provided. Similar accounts of cultivar differences were reported in South Africa, where 'Forelle' initiated flowers by 56 dafb, but 'Rosemarie' required 77 dafb (Reynolds et al., 2005a). These authors noted that 'Forelle' terminated extension growth a few weeks before 'Rosemarie', which is consistent with observations that floral initiation depends on completion of annual shoot growth. Flower initiation was asynchronous, requiring ~5 weeks from the first to last meristems to initiate floral primordia, irrespective of the cultivar (Reynolds et al., 2005a). The number of vegetative primordia developing on the bud axis and the time between each successive primordium (i.e., the plastochron) have been established for some cultivars of apple and can range from 8 to 18 days (Buban and Faust, 1982). Similar data for pear either do not exist or are difficult to locate. The high variations associated with genotype and environment limit extrapolation of data from outside regions of interest. The timing of floral initiation of apple differed depending on the branch type (Buban and Faust, 1982). These data would be quite useful for pear, especially in combination with previously published pear cropping models (Sansavini and Musacchi, 1994).

3 Chemical manipulation of flowering

3.1 Chemical promotion of flower initiation

With apples, Harley et al. (1958) demonstrated a positive effect of postbloom naphthaleneacetic acid (NAA) applications on return bloom, even though NAA did not reduce fruit set in the year of application. Delaying NAA application to 30 dafb also markedly increased return bloom, leading to the conclusion that NAA was florigenic. Multiple applications of NAA (5 to 10 ppm) between 36 and 140 dafb increased return bloom of apple cultivars in over half of the trials reported by McArtney et al. (2007, 2013). These authors also documented a significant increase in return bloom from late-summer NAA applications (30 days from harvest), which challenge the dogma that such treatments need to preempt natural floral initiation events, and instead suggest that quiescent, vegetative meristems have a higher degree of plasticity than previously thought (McArtney et al., 2013). Ethylene evolution of apple spurs was increased by postbloom NAA applications (McArtney, 2002; Walsh et al., 1979). Interestingly, the ethylene inhibitor, aminoethoxyvinylglycine (AVG), applied in combination with NAA, reduced, but did not negate, the florigenic properties of NAA (McArtney et al., 2013), suggesting that apple floral induction by auxin may be mediated at least partially by auxin-induced ethylene (Curry, 1991). Williams (1972) increased return bloom of apple using ethephon at 5 weeks after bloom or, in the fall, 1 month prior to leaf abscission.

With pear, however, mid-season NAA applications over a range of similar timings were generally non-florigenic for 'd'Anjou', 'Bosc', and 'Packham's Triumph' trees in Oregon, New York, and Brazil, respectively, though significant improvement in 'Bosc' return bloom was noted in one of two years (Einhorn, Francescatto and Pasa, data not published). Compared to NAA, greater persistence and significantly higher ethylene evolution were measured following 2-chloroethylphosphonic acid (ethephon) sprays on apple (McArtney, 2002). Thus, a more successful approach to promoting flowering in pear may be through direct application of ethephon. Ethephon improved return bloom and return yield of 'd'Anjou' pear when applied at 57 dafb (Einhorn et al., 2014). More recent work showed marked increases in return bloom of 'Bartlett', 'Bosc', and 'd'Anjou' when applications were made between 40 and 60 dafb (Einhorn, manuscript in preparation). Efficacy was optimized at 300 ppm, which was only one-third to one-fifth the concentrations required to improve return bloom of 'Northern Spy' apple (Duyvelshoff and Cline, 2013), although dose responses depend on apple cultivar (Bukovac et al., 2006; McArtney et al., 2007; Schmidt et al., 2009). The positive effect of ethephon on pear flowering was consistent over multiple years and even reversed floral inhibition from the gibberellic acid (GA) inhibitor, prohexadione-calcium (Table 1; section 4.1). Fruit ripening or harvest quality was not affected by ethephon in the year of application or following 3 months of postharvest storage (Einhorn et al., manuscript in preparation), but ethephon's effects on fruit quality following prolonged storage requires further elucidation. The potential of ethephon to improve pear return bloom is deserving of additional research.

3.2 Chemical inhibition of flower initiation

There is substantial evidence for GA-inhibition of pome fruit flowering (Chan and Cain, 1967; Cristoferi et al., 1981; Hoad, 1978; Tromp, 1982). Floral initiation of apple is ostensibly sensitive to combinations of GA_3, GA_4, and GA_7 (Hoad, 1984); purportedly, GA_7 exerts disproportionate control over flowering (Tromp, 1982). Application of GA_3 significantly reduced floral induction of 'Bartlett' and 'Winter Nelis' trees (Griggs and Iwakiri, 1961; Gil et al., 1972). However, due to the generally poor fruit set and/or limited biennial bearing of major commercial pear cultivars, the need to reduce floral initiation is uncommon. A fair amount of evidence for GA-induced floral inhibition of pear is from use of high rates of gibberellin (GA) for 'rescuing' frost-injured flowers (see Section 4.1). Practical use of GA to inhibit floral initiation in the nursery or first orchard leaf to promote vegetative growth would be unnecessary for pear scions on vigorous rootstocks, but may benefit trees on highly precocious and dwarfing rootstocks.

Table 1 The effect of whole-tree application of prohexadione-calcium (P-Ca) and/or ethephon on return bloom (percentage of spurs and 1-year-old shoot populations with bloom) and return yield of 16-year-old 'd'Anjou'/OH × F 87 pear trees

	Return bloom		Return yield and fruit weight		
Treatments	Spurs (%)	1-year shoots (%)	Tree yield (kg)	Fruit no.	Fruit wt. (g)
Control	43 b	42 b	113.8 b	534 b	212
P-Ca 250 ppm	17c	13 c	83.4 c	365 c	227
Ethephon 300 ppm	64 a	60 a	149.7 a	634 a	235
Ethephon 450 ppm	48 b	49 b	138.3 a	659 a	209
P-Ca 250 ppm + Ethephon 300 ppm	57 ab	28 abc	124.3 ab	610 a	203
P-Ca 250 ppm + Ethephon 450 ppm	43 b	34 b	111.6 b	504 b	220

P-Ca was applied when new shoot emergence reached 5 cm and again ~70 days later. Ethephon was applied at 45 days after full bloom.
Different letters within columns signify significant difference at $P < 0.05$ by Fisher's protected LSD test.
Data are means of six single-tree replicates.
Source: data from T. Einhorn (unpublished).

4 Chemical manipulation of fruit set

4.1 Gibberellins

Promotion of fruit set by GA has been achieved when factors favoring pollination and fertilization are limited, most commonly by episodic spring frosts. The prevailing theory is that unfertilized ovules or non-functional embryos do not produce GA, which, in turn, retards growth and/or reduces the sink strength of the fruit. Thus, exogenous 'replacement' of seed-derived GA prevents abscission. Frost rescue of pear requires low GA concentrations (5–10 ppm); higher concentrations reduce return bloom (Deckers and Schoofs, 2002; Sansavini et al., 1981). The efficacy of different GA types varies considerably, but induction of parthenocarpic fruit set of pear was not consistently related to GA_{4+7} or GA_3 (Deckers and Schoofs, 2002; Gil et al., 1972; Quinet and Jacquemart, 2015). Other studies show no positive effects of GA on fruit set (Lafer, 2008; Silva et al., 2008; Vercammen and Gomand, 2008). Vercammen and Gomand (2008) documented few positive yield effects from GA_{4+7} or GA_3 applied to open-pollinated 'Conference' trees and indicated that initially higher fruit set from GA treatment was lost during June drop. Unpollinated 'Winter Nelis' flowers treated with GA_3 or GA_{4+7} set roughly 40% of a full crop of normal fruit, but only if GA_3 was applied multiple times after bloom; otherwise, all fruit abscised prematurely (Gil et al., 1972). Chitu et al. (2008) also demonstrated significantly high fruit set for 'Bosc' with 20 and 50 ppm GA_3 applied at 80% full bloom and again 5 days after petal fall. These data may help explain the inconsistencies of single GA applications intended to increase fruit set. Rapid catabolism of the physiologically active forms of GA *in-planta* may necessitate frequent applications to maintain sufficient levels over a wider time expanse.

Despite an apparent regulatory role of GA in parthenocarpic fruit set, interaction with other phytohormones, such as IAA, may be required for action (Bangerth, 2005; Sastry and Muir, 1963). Exogenous applications of IAA, 6-BA, and GA to bagged, unpollinated pear flowers produced parthenocarpic fruit set equivalent to that of self-pollinated flowers when weather conditions were warm; however, only GA increased fruit set when conditions were unfavorable (Quinet and Jacquemart, 2015). In cases where GA did improve fruit set, fruit size was not improved compared to unpollinated, non-GA treated flowers. Combinations of GA_{4+7} and BA increased fruit set of 'Abate Fetel' in half of the trials, but fruit size and seed content were not reported (Vilardell et al., 2008). Although exogenous GA increased mobilization of soluble carbohydrate from leaves to fruit (Sansavini et al., 1981), rarely does GA actually promote fruit size. Failure of low-dose GA applications to consistently set fruit following frosts and the possibility that seedless fruit will be small or misshapen may still be insufficient to discourage utilization of this relatively low-cost option, since the potential benefit of vigor control is of great value.

4.2 Nutrients

Crisosto et al. (1992) increased the EPP of 'Comice' flowers and subsequent fruit set via bloom sprays of the polyamine, putrescene. Putrescene delayed ovule senescence, resulted in faster rates of pollen tube growth, and was associated with higher nitrogen (N) content in treated flowers. These results corroborated with previous reports of the positive relationship between the N status of flowers and pear fruit set (Batjer and Thompson, 1949; Curetti et al., 2013; Williams, 1965). Nitrogen appears to affect flowering 'strength', though N partitioning to pear organs depends on fertilizer timing (see reviews by Sánchez, 2002, 2015; Sanchez et al., 1995). The experiments of Sánchez and Righetti (1990) and Sánchez et al. (1990a,b, 1991, 1992) demonstrated N allocation to current season fruit depended on application timing. Spring applications ~1 month after bloom resulted in high N recovery in extension shoot leaves, excessive vigor, and variable fruit N content depending on crop load. Applications made later in the season (3–6 weeks before harvest), however, increased N reserves in aerial portions of the tree. The relatively weak fruit sink for summer-applied N had little effect on fruit quality. Postharvest delivery of N markedly increased storage in roots compared to other timings and was related to improved pear flowering, fruit set, and early fruit cell division compared to earlier applications. Indeed, remobilization of stored N was the primary source for rapidly growing 'Abbe Fetel' organs during spring development (Tagliavini et al., 1997). However, current-season N uptake corresponded with the onset of shoot growth and contributed markedly to N content of all organs, including spur leaves, which was not observed for apple (Nielsen et al., 1997). These differences were attributable to the late leaf emergence of pear compared to apple, resulting in relatively higher percentages of current-season N supply in pear organs (Sánchez et al., 1991; Tagliavini et al., 1997). Thus, pear production in low-fertility soils would benefit from spring-applied N, whereas fertile soils may only require a postharvest application of N.

Boron (B) is also positively related to tree fruit flowering, and foliar application of soluble forms of B is the most efficient means to correct transient B deficiencies near flowering (Brown, 2001). The mechanism by which B improves fruit set remains unclear, though pollen germination has been implicated (Stanley and Lichtenerg, 1963). Postharvest B sprays increased 'Hosui' Asian pear pollen weight, germination rate, and pollen tube growth (Lee et al., 2009). In European plum (or prune) (*Prunus domestica*), B from autumn foliar sprays was remobilized to fruiting spurs where an increased B content in spring was associated with higher fruit set (Hanson and Breen, 1985). Export of B from apple, pear, prune, and sour cherry (*Prunus cerasus*) spur leaves occurred within a week of application (Hanson, 1991) and ~75% of B was symplastically translocated in spring (Hanson and Breen, 1985). Picchioni and Weinbaum (1995) determined that 88–96% of foliar uptake from labeled B sprays occurred within 24 h of application to apple, pear, prune, and sweet cherry (*Prunus avium*). Further, B was highly mobile once absorbed. Leaf surface characteristics were the primary determinants for penetration; pear leaves were only half as efficient as apple, though genotypic differences among cultivars are expected. Postharvest B uptake by 'Comice' leaves led to higher concentration in flower buds in spring (Sanchez et al., 1998). The addition of urea aided uptake. Wojcik and Wojcik (2003) measured significantly higher B in 'Conference' flowers, leaves, and fruitlets for sprays applied postharvest or just before full bloom. These changes were associated with increased fruit set and yield.

4.3 Aminoethoxyvinylglycine (AVG)

A large body of literature implicates ethylene in fruit abscission (see reviews by Morrison Baird and Webster, 1979; Bangerth, 2000). Direct attempts to reduce ethylene using the ethylene inhibitor, AVG, significantly increased 'Comice' fruit set when applied between bloom and 4 weeks after bloom (Lombard and Richardson, 1982). The authors did not measure ethylene production, but hypothesized that AVG inhibited high internal ethylene 4 to 6 weeks after bloom, which promotes 'June drop'. AVG reduced fruit mass, which was not entirely explained by the increased crop load. The high AVG concentration (i.e., 1000 ppm) applied in these experiments was likely a key factor limiting growth. More recent work using 100–250 ppm AVG increased 'Packham's Triumph' and 'Abate Fetel' fruit set by 50–100% when applied 2 week after bloom (Dussi, 2011; Sanchez et al., 2011). AVG applied 7 dafb at rates between 60 and 125 ppm significantly improved fruit set of 'Rocha' and 'Packhams Triumph' in Brazil and 'Williams' in Austria (Carra et al., 2018; Lafer, 2008).

Recently, the ethylene production rates of developing fruitlets of 'Comice' and 'd'Anjou' were characterized between anthesis and 30 dafb (T. Einhorn, unpublished data). Season and cultivar effects contributed to variation in the amplitude of ethylene peaks, but temporal ethylene production patterns were quite similar: low ethylene (<1 ppm) at anthesis steadily increased to maximum values by 2 weeks after bloom (which were 5- to 10-fold those at bloom), before rapidly declining and stabilizing at prebloom values by 20 dafb (Fig. 1a and b). AVG had no appreciable effect on fruit respiration (Fig. 1c) and practically eliminated ethylene within 24 h from application - an effect that persisted for 15 days, albeit with diminishing strength (Fig. 1d). It is notable that AVG efficacy and persistence varied among experiments with equivalent rates. Foliar penetration of AVG depends on temperature, relative humidity (RH), and rewetting (Knoche and Petrachek, 2014), and these factors require consideration to optimize efficacy. AVG increased fruit set depending on rate (between 30 and 120 ppm, typically saturating around 80 ppm), but only when applied postbloom. Dussi and Sugar (2011) identified 100–150 ppm AVG as

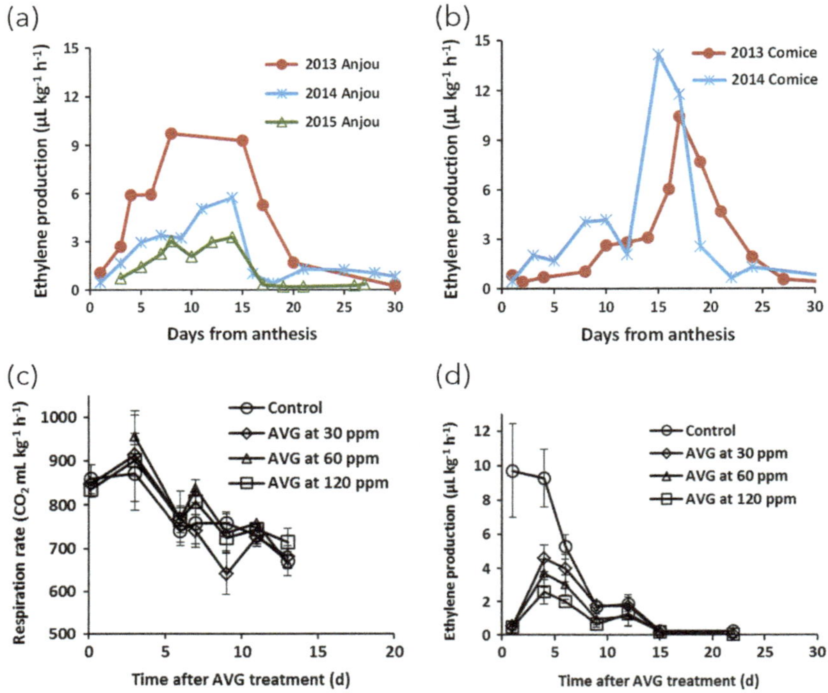

Figure 1 Ethylene production rate of flowers and developing fruits of 'd'Anjou' and 'Comice' pear between 0 and 30 days after full bloom (top panels) and effect of aminoethoxyvinylglycine (AVG) on respiration rate and ethylene production of 'd'Anjou' fruitlets. AVG applications were made ~10 days after bloom. Source: T. Einhorn and Y. Wang (unpublished data).

the most economical rate when factoring product and application costs, yield improvement, and fruit size. In these studies, fruit size was negatively affected only when AVG induced excessive fruit set. Within this concentration range, fruit growth was unaffected (T. Einhorn, unpublished data). Seed content was not influenced by AVG, which agrees with earlier work on apple (Williams, 1981). Collectively, these data indicate that (1) ethylene is associated with fruitlet abscission, possibly at rates exceeding 1 $\mu l \cdot kg^{-1} \cdot h^{-1}$; (2) pollination and fertilization were non-limiting factors to fruit set in these studies; (3) AVG is not a remedy for retention of frost-injured or otherwise seedless fruit; and (4) multiple pear cultivars appear to have similar ethylene biology.

4.4 Compounds that promote fruitlet abscission (i.e., thinning)

Pear cultivars differ widely in fruit set efficiency. In the US, where fully dwarfing rootstocks are not yet widely available, most cultivars do not require thinning, the exceptions being 'Bartlett' and to a lesser extent, 'Bosc'. In other parts of the world, productive scion-rootstock combinations, i.e., 'Conference' on dwarfing quince (*Cydonia oblonga*), are prone to oversetting. Williams et al. (1978) suggested a ratio of 30–40 leaves per fruit was necessary to attain commercial fruit size for a range of pear cultivars. Modern apple trees require roughly 10 leaves per fruit (Warmund, 2008). Based on similar maximum leaf photosynthetic potential and dry matter content of mature apples and pears, this author would expect far lower leaf-to-fruit ratios for canopies of high-density pear trees on dwarfing rootstocks, but data are not readily accessible. Thinning of pear is primarily achieved using hormone classes of chemical thinners. Caustic compounds are used infrequently because of erratic performance, potential for phytotoxicity, and their narrow window of efficacy (Wertheim, 2000), which many tree fruit producers in northern regions contend is too risky given the potential for frost events after bloom. Similar concerns exist for apple, but Musacchi and Greene (2017) show data indicating a relatively low risk of over-thinning during bloom. Bloom thinning of apple is aided by the predictive pollen tube growth model (Yoder et al., 2009, 2013). To this author's knowledge, a similar model has not been adopted by pear producers. Time and retrospect are major advantages when delaying thinning until petal fall, since conditions at bloom (bloom density, pollen availability, pollinator activity, and weather conditions) can be integrated into a thinning strategy. Delaying thinning poses little risk for pear, because unlike apple, return bloom is not nearly as negatively affected by the presence of developing fruitlets. Reasons for this could be associated with a greater lag between early fruitlet (and seed) development and flower initiation of pear than apple, or the relatively low fruit set of pear (between 30 and 80 fruit per 100 clusters; Lombard et al., 1988). Indeed, few pear cultivars are prone to biennial bearing (e.g., 'Comice' and 'Bosc'), and it is uncommon for thinning to increase return bloom of pear (Greene, 2012).

Over the past several decades, the thinning efficacy of many compounds has been assessed for pear. Ethephon, NAA, and 6-benzyladenine (BA) tend to have the strongest and most consistent effects on pear abscission (see comprehensive reviews by Bangerth, 2000; Dennis, 2000; Dussi, 2011; Wertheim, 2000). Presently, BA is the most widely utilized pear thinner with fairly consistent response across diverse cultivars and geographic regions (Curetti et al., 2010; Dussi and Sugar, 2011; Gimenez et al., 2009; Greene, 2012; Maas et al., 2010; Theron et al., 2011; Wertheim, 2000). Generally, rates between 50 and 150 ppm are efficacious. A major limitation of BA in northern climates is the relatively high temperature requirement (>18°C) for effective foliar penetration and activity (Bubán, 2000; Wertheim, 2000). Anecdotal accounts that BA increases fruit size independent of thinning are corroborated by enhanced cell division activity of BA-treated apples (Wismer et al., 1995). Another cytokinin, *N*-(2-chloro-4-pyridyl)-*N*'-phenylurea (CPPU), was shown to promote fruit size of 'Spadona' pear by increasing cell division (Flaishman et al., 2001, 2005). An assessment of CPPU on 'd'Anjou' fruit growth (applied at 16 mm) resulted in significant fruitlet abscission and significantly larger fruit, likely a consequence of reduced cropload. Reports of misshapen fruit from CPPU treatments (Wertheim, 2000) were not observed (Table 2; T. Einhorn, unpublished data). Flaishman et al. (2005) did not provide fruit set data or the total number of harvested fruit. New commercial formulations of CPPU implore efficacy studies on pear thinning. Since the most recent pear thinning reviews (Dussi, 2011; Wertheim, 2000), a fair amount of activity has been conducted on two novel compounds that warrant discussion herein: metamitron and abscisic acid (ABA). Both of these compounds affect carbohydrate status, but act through different mechanisms.

Metamitron is a triazinone herbicide that blocks electron transfer in Photosystem II (PSII). At low concentrations (100–500 ppm), metamitron causes a transitory reduction in net photosynthesis (Pn) that presumably induces carbohydrate deficit and fruit abscission. Metamitron reduced the maximum potential quantum efficiency of PSII (Fv/Fm) of peach (*Prunus persica*) and apple leaves in a rate-dependent manner (McArtney et al., 2012). At the highest concentration (300 ppm), metamitron reduced rates of chlorophyll fluorescence from 7 to 12 days for peach and apple, respectively, but only fruit set of apple was reduced in a rate-responsive manner (McArtney et al., 2012). Maas et al. (2010) reported effective thinning of 'Conference' pear with 175–300 ppm metamitron. Metamitron reduced 'Bartlett' pear leaf photosynthesis by 50–80% and persisted for 15 days at rates between 150 and 600 ppm (Fig. 2); thinning was commensurate with these photosynthetic reductions (Elsysy et al., manuscript in review). Optimum thinning efficacy was achieved with 150 ppm under Hood River Valley, Oregon, conditions on multiple orchards from 5- to 40-year-old 'Bartlett' when applications were made at 10–12 mm fruit diameter.

Table 2 The effect of CPPU applications on fruit set, yield components, and fruit quality at harvest of 'd'Anjou'/OH × F 87 pear trees

Treatments	Fruit set (%	Yield/tree		Fruit	Fruit shape		
CPPU (ppm)	fruit/cluster)	fruit no.	weight (kg)	wt. (g)	Height:Diameter	FF (kg)	SSC (%)
0	62.0 a	499.0 a	97.8 a	206.6 c	1.23	6.9	13.0
10	46.0 ab	191.4 b	54.7 b	276.9 ab	1.21	7.1	12.9
15	41.5 ab	199.8 b	54.3 b	269.3 b	1.17	6.8	13.0
20	35.0 b	125.8 b	33.8 c	265.5 b	1.18	7.1	13.5
25	35.3 b	115 b	34.4 c	296.0 a	1.17	7.1	13.1
Significance	*0.05*	*<0.0001*	*<0.0001*	*<0.0001*	*0.06*	*0.07*	*0.55*

Data are based on five single-tree replicates.
Applications were made to whole canopies when fruit diameter was 14.4 mm.
Source: data from T. Einhorn (unpublished).

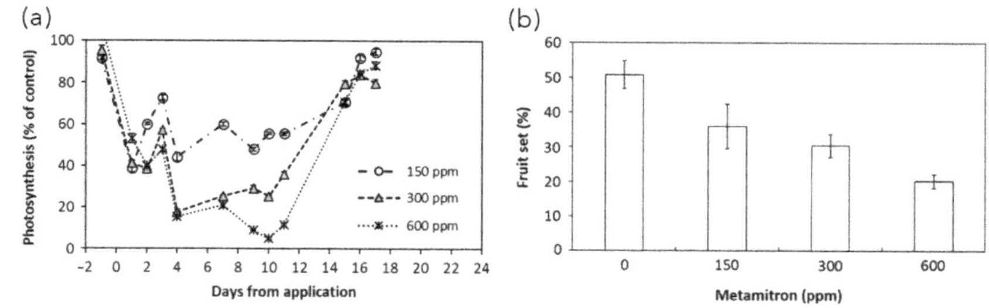

Figure 2 Effect of metamitron rate on single-leaf photosynthesis (a) and fruit set (b) of 'Bartlett' pear trees. Metamitron was applied to whole canopies at 10.9 mm fruitlet diameter. Source: T. Einhorn and A. Sarret (unpublished data).

Severe thinning was noted in two trials, though in both cases a combination of young trees and high temperatures likely contributed to the response. High temperatures also were associated with similar observations following commercial applications of metamitron in the Po Valley, Italy (G. Costa, personal communication). Rates exceeding 300 ppm induced mild chlorosis, but symptoms were generally reversible; however, marginal leaf burn at 600 ppm can reduce functional leaf area (T. Einhorn, unpublished data). Translocation of metamitron within the plant appears negligible (S. Eskelsen, personal communication), requiring direct foliar contact; however, the importance of healthy spur leaves on early fruit development of apple (Wünsche et al., 1996; Hansen, 1971), sweet cherry (Ayala and Lang, 2018), and Japanese pear (*Pyrus pyrifolia*) (Zhang et al., 2005a,b) would suggest that appropriate applications should be at concentrations below the threshold for leaf injury.

Water stress-inducible ABA production elicits stomatal closure and subsequent water conservation (Salisbury and Ross, 1992; Davies and Zhang, 1989). ABA is also an abscission agent (Addicott and Lyon, 1969). Both of these mechanisms implicate ABA as a potential fruit thinner. Exogenous application of ABA caused abscission of mature apples (Edgerton, 1971); however, the mode of action of ABA on abscission of young 'Bartlett' fruit was Pn inhibition as an indirect consequence of partial stomatal closure (Einhorn and Arrington, 2018). These authors tested single and additive effects of shade (44% and 77%) and ABA (125 ppm) on 'Bartlett' leaf physiology and fruit set. Individually, these factors increased fruitlet abscission but, when combined, they were not additive. The abscission response was closely associated with photosynthesis (carbohydrate deficit), controlled by whichever factor exerted stronger inhibition of Pn when the two factors were applied together. Had ABA caused fruitlet abscission by an alternative mode of action (i.e., via ethylene regulation), one would have expected increased thinning when combined with shading. Interestingly, stomatal response to ABA persisted for 7–10 days, albeit with diminishing strength (Arrington et al., 2017; Einhorn and Arrington, 2018) as similarly demonstrated for apple (McArtney et al., 2014). When applied at 10–12 mm fruit diameter, ABA consistently thinned 'Bartlett' (Table 3) in markedly different environments (Arrington et al., 2017; Greene, 2012), 'Cold Snap', and 'Bosc' (Cline et al., 2018). Petal fall sprays had a weaker effect, but significantly reduced crop load, increased fruit size, and promoted return bloom (Greene, 2012). The thinning efficacy of ABA for pear cultivars in the Pacific Northwestern and Eastern United States and Eastern Canada seems optimum between 125 and 300 ppm (Arrington et al., 2017; Cline et al., 2018; Greene, 2012). Higher concentrations, while increasing stomatal effects and abscission, resulted in leaf yellowing or abscission (Arrington et al., 2017; Greene, 2012). The addition of BA to counteract ABA-induced leaf yellowing was ineffective (Greene, 2012). In the United States, the commercial formulation of ABA (ProTone®, Valent Biosciences Corp.) meets organic materials standards and is presently labeled in Washington for organic pear production.

Given that both of these compounds induce carbohydrate deficit, the carbon balance model used to predict and optimize apple thinning (Lakso, 2011; Lakso et al., 2006) may be adaptable to pear. Basic physiological data describing temperature optima for pear fruit growth, carbohydrate allocation among organs, respiratory costs, etc., are lacking; however, similar C3 leaf

Table 3 Effect of abscisic acid (S-ABA, ProTone®, Valent BioSciences) applications to whole 'Bartlett' pear trees on fruit set (expressed as percent set of untreated control trees) for five separate trials

	2012	2013	2013	2014	2014
ABA (ppm)	8-year-old	18-year-old	10-year-old	19-year-old	10-year-old
0	100	100	100	100	100 a
50		100	71	95	
100		99	57	86	
125	59				65 b
150		96			
200		78	18	79	
250	12				
400			7	58	
500	2				
Significance					
Linear	0.0113	N.S.	0.0476	<0.001	
Quadratic	0.0459	N.S.	N.S.	<0.001	
Fisher's protected LSD					0.0077

The lack of significance in 2013 (18-year-old trees) was attributed to 10 days of high light conditions following applications. N.S. = not significant at $P < 0.05$.

physiology (Section 6.1), thinning response, and fruit growth and development of pear and apple suggest sufficient overlap to warrant testing.

5 Chemical manipulation of vegetative growth

5.1 GA inhibition

GA is actively produced in young, rapidly growing leaves and shoots and regulates internodal shoot elongation (Srivastava, 2002). Inhibition of GA synthesis, therefore, has been an effective strategy to control vegetative growth (Rademacher, 1991). There are several chemical classes of GA inhibitors that affect metabolism at different steps of the GA biosynthesis pathway (Rademacher, 2000). Daminozide, paclobutrazol, and prohexadione-calcium (P-Ca) have all been widely used to regulate cropping and vegetative growth of tree fruit, but due to use restrictions, the latter is presently the most widely used and commercially available GA inhibitor. P-Ca disrupts the conversion of GA from the inactive to active forms by interfering with dioxygenases late in the GA pathway (Evans et al., 1999; Rademacher, 2000). P-Ca consistently reduced shoot growth of diverse pear cultivars over a wide range of climates (Asín et al., 2007; Costa et al., 2001, 2004; Einhorn et al., 2014; Pasa and Einhorn, 2014, 2017; Rademacher et al., 2004; Smit et al., 2005; Southwick et al., 2004; Sugar et al., 2004). Most of these studies report ~50% reduced shoot length. P-Ca concentrations were typically between 100 and 250 ppm. Shorter shoots had higher node densities, potentially increasing future fruiting sites (Pasa and Einhorn, 2014). P-Ca degradation can result in sudden and rapid shoot growth that ultimately exceeds non-treated shoots if left unchecked. Because P-Ca inhibits GA biosynthesis late in the pathway, Elfving (D. Elfving, personal communication) proposed that a buildup of GA precursors may facilitate high rates of GA synthesis after P-Ca effects subside. While several experimenters used fixed application intervals to sustain growth regulation, weekly growth rate analyses characterized P-Ca degradation and optimized chemical-use efficiency (Einhorn et al., 2014; Pasa and Einhorn, 2014, 2017). These data showed that, unlike apple, growth inhibition of pear shoots persists for several months after application, depending on cultivar and climate (Einhorn et al., 2014). For example, in the cool and short-growing season of the upper Hood River Valley, a single P-Ca application at 5 cm shoot length controlled 'd'Anjou' shoot growth for the entire season (Einhorn et al., 2014). In contrast, P-Ca typically persisted for 60 to 80 days in the lower Hood River Valley, an effect that was consistently observed over multiple years.

While vegetative response to P-Ca has been consistent, P-Ca effects on pear yield and fruit size have been variable. A yield increase in the year of application was proposed to reduce the sink strength of shoots (Costa et al., 2004; Rademacher et al., 2004; Vilardell et al., 2008). Alternatively, P-Ca interference with dioxygenases may also inhibit ACC production (Rademacher, 2000). Reduced ethylene synthesis (as described earlier) can enhance fruit retention. However, the majority of pear studies did not report on an increased fruit set or yield by P-Ca (Costa et al., 2004; Einhorn et al., 2014; Pasa and Einhorn, 2017; Smit et al., 2005; Rademacher et al., 2004; Sugar et al., 2004). Fruit growth of 'Bartlett', 'Rosemarie', 'd'Anjou', and 'Bosc' was reduced, though not commensurate with shoots (Smit et al., 2005; Sugar et al., 2004). In the case of 'd'Anjou' and 'Bosc', these deleterious effects were not consistent among experiments (Sugar et al., 2004) and were lacking altogether from other reports (Einhorn et al., 2014; Smit et al., 2005). P-Ca seemed to directly affect 'Bartlett' fruit growth (i.e., not crop load related). Cell division in the cortex might plausibly be limited by GA inhibition given the overlapping timing of P-Ca applications. In this case, the extended duration of cell division of pear fruit (Westwood, 1993) is insufficient to offset the extensive persistence of P-Ca *in-planta*. Still, fruit size of 'Bartlett', 'Red Clapp', 'Blanquilla', 'Abbé Fétel', 'Packham's Triumph', 'Forelle', 'Flamingo', and 'Early Bon Chretien' was unaffected by P-Ca (Asín et al., 2007; Pasa and Einhorn, 2017; Smit et al., 2005; Southwick et al., 2004).

The most surprising, and perplexing, response of pear to P-Ca is floral inhibition (Einhorn et al., 2014; Maas, 2005; Rademacher et al., 2004; Smit et al., 2005; Sugar et al., 2004). Suppressing the biosynthesis of a hormone negatively associated with floral induction would be expected to promote, not limit, flowering. The phenomenon is both cultivar dependent (Smit et al., 2005; Sugar et al., 2004) and rate responsive (Rademacher et al., 2004). Asín et al. (2007) attributed a lack of P-Ca effect on return bloom of 'Blanquilla' to low P-Ca rates (i.e., 120 ppm). Rademacher et al. (2004) showed that reduced return bloom of 'Conference' pear did not translate to lower yields, though this could be related to the prolific flowering and fruitfulness of 'Conference'. As will be detailed later, fruit set efficiency of pear is inversely related to the number of flowers. Unfortunately, applying low P-Ca concentrations is not a universal solution, given the variable dose response of cultivars. For example, 125 ppm P-Ca induced less floral inhibition of 'd'Anjou' than that caused by 250 ppm P-Ca, but provided insufficient vigor control to warrant use at this rate (Einhorn et al., 2014).

Flower inhibition of different fruiting structures was differentially affected by P-Ca. Terminal bloom on short, one-year-old 'd'Anjou' shoots was inhibited far greater than spurs (Einhorn et al., 2014). Interestingly, the natural development of floral initiation of apple occurred later on short shoots than spurs (Buban and Faust, 1982). Unfortunately, these observations do not explain why P-Ca may have inhibited return bloom of spurs; they do, however, reinforce the widely accepted inhibitory role of GA on floral initiation since the delayed degradation of P-Ca in pear would have coincided with the potentially later onset of floral induction of shoots. Irrespective, the deleterious effects of P-Ca on bloom of sensitive cultivars requires alternative strategies for managing vigor via P-Ca. Elfving et al. (2003a) limited P-Ca applications to the tops of large 'Bartlett' trees where the need for vigor management was greatest (Elfving et al., 2003a). This strategy is supported by acropetal translocation of P-Ca *in-planta* (Pasa and Einhorn, 2014; Rademacher, 2000), thereby precluding undesirable effects in subtending organs. Recently, the addition of ethephon prior to floral initiation was shown to completely restore 'd'Anjou', 'Bosc', and 'Bartlett' flower inhibition by P-Ca (Einhorn et al., manuscript in preparation; Table 1). These results were consistently observed over several years and were not diminished by multiple P-Ca applications (125 or 250 ppm). This program requires validation for different genotypes and diverse regions.

Finally, P-Ca appears to suppress fire blight (Costa et al., 2001; Yoder et al., 1999). Multiple applications of P-Ca (100 ppm) reduced 'Abbé Fétel' shoot growth and *Erwinia amylovora* (fire blight) infection (Costa et al., 2001). A single application of 250 ppm P-Ca 2 weeks prior to inoculation with *E. amylovora* significantly reduced lesion length, but not disease severity or incidence. Interestingly, low rates of P-Ca (50 ppm) also reduced the number of infected organs on 'Abbé Fétel' trees despite having no influence on growth. More recently, P-Ca was shown to drastically reduce disease incidence and severity of inoculated potted apple trees despite high bacterial populations within shoots (McGrath et al., 2009). These authors reported pronounced thickening of mid-vein cell walls of P-Ca-treated apple leaves and hypothesized that such anatomical changes may create a barrier to xylem entry and spread of *E. amylovora*. These results have spawned experimentation with low concentration sprays of P-Ca to newly established apple orchards with the objective of suppressing fire blight, but not canopy growth (P. Schwallier, personal communication). This author is unaware of similar uses in new pear orchards.

5.2 Ethephon

The reduced shoot extension following ethephon treatment to apple (Cline and Bakker, 2017) and sweet cherry (Elfving et al., 2003b) trees was not observed for pear at concentrations between 150 and 450 ppm (Einhorn et al., 2014; Einhorn, data not published). However, Elfving et al. (2003b) demonstrated additive growth inhibition from combined P-Ca and ethephon sprays to sweet cherry shoots. Similar combinations produced only minor effects on the growth of pear shoots compared with P-Ca alone (Einhorn et al., 2014).

5.3 GA and BA

While promotion of branching does not directly affect flowering, larger, well-branched pear nursery trees tend to be more precocious and productive than smaller trees (Deckers and Daemen, 1998; Lawes et al., 1997). Thus, use of BA or proprietary combinations of GA_{4+7} and BA to induce branches (i.e., feathering) of nursery trees has been widely investigated (Miller, 1988). Pear has received less attention than apple, but branching of 'Bartlett', 'Honey Sweet', 'Seckel', and 'Comice' was increased significantly by GA_{4+7} and BA (Cody et al., 1985; Jacyna, 1996; Palmer et al., 2011). Cultivar selection had a pronounced effect on branching with GA_{4+7} and BA (Jacyna, 1996) as similarly reported for apple (Lordan et al., 2017). Palmer et al. (2011) demonstrated that GA, alone, induced branching, but thorniness and poor branch position too high on the central axis were undesirable effects of this treatment. In all cases, GA improved 'Comice' leader height when combined with BA (Palmer et al., 2011) as observed similarly for apple trees (Lordan et al., 2017), a desirable attribute for promoting earlier canopy development after transplanting. Four weekly applications proved advantageous at concentrations of 500–2,000 ppm once trees reached the desirable height for feathering, which tended to be around 60–70 cm (Palmer et al., 2011).

6 Physical manipulation of flowering and fruit set

6.1 Girdling, notching, and scoring

Interruption of phloem transport can affect vegetative (mainly branching) and reproductive (floral initiation and/or fruit set) processes. The role of auxin in apical dominance and suppression of lateral bud break is well established (Thimann and Skoog, 1933). Techniques that temporarily interrupt the basipetal transport of auxin in phloem, such as notching or scoring above lateral buds, help release buds from paradormancy. Vegetative growth of pear trees is notoriously acrotonic, though cultivars vary in the degree of apical control (du Plooy et al., 2002). Concentrations of IAA, zeatin, and zeatin riboside in vertically positioned apple shoots were ~2-fold higher than in horizontal shoots (Sanyal and Bangerth, 1998). Branch induction from notching or girdling is associated with reduced IAA and may also be due to a lower ratio of auxin to GA and cytokinin (Bangerth, 2005; Müller and Leyser, 2011). Branch induction from exogenous BA + GA, in lieu of heading, supports this hypothesis (Section 4.3). As noted earlier, branching increases potential fruiting sites. Precision branch management by notching above individual buds may require the addition of complete circumferential scoring (at set intervals of ~30 cm) on one- and two-year-old wood to improve the response (Musacchi, 2015) (Fig. 3). Notching or girdling is not limited to green tip bud phenology and can be used throughout the growing season, though these techniques will be ineffective once buds reach vegetative maturity in late summer. The use of pruners with opposing knives enables practitioners to easily feel differences in the resistance of phloem and xylem when scoring,

Figure 3 Initiation of branches from two girdling cuts made ~30 cm apart on 2-year-old section of 'Bartlett'/OH × F 87 pear leaders. Photo: T. Einhorn.

but the cambial layer may only be a single to a few cell layers thick (depending on time of year) and, therefore, practically impossible to feel. Thus, formation and differentiation of new parenchyma cells into vascular tissue will largely depend on how severely the cambium is injured. Callusing and rebridging severed tissues depend on plant factors (cultivar, rootstock, nutrition, water, and vigor) and environmental conditions (temperature, light, and humidity). Other factors that reduce branching success include the presence of large limbs, i.e., ~50% of the circumference of the branch or leader being notched, proximal to the girdle or notch.

A more direct effect on flower promotion and fruit set of pear was achieved by girdling (in this case, removing an entire strip of phloem) the base of trunks (Reynolds et al., 2005b). These effects may be due to changes in carbohydrate and/or hormone distribution, but supporting evidence for either mechanism is inconclusive (see reviews by Goren et al., 2004; Theron and Steyn, 2008). Spatial and temporal factors play an important role in tree response to girdling (Reynolds et al., 2005b; Westwood and Bjornstad, 1974). From a practical standpoint, girdling should precede floral induction and initiation events. Girdling close to the timing of floral initiation had significantly less of an effect on return bloom of 'Rosemarie' than earlier timings, even though all treatments occurred prior to the experimentally derived floral initiation date (Reynolds et al., 2005a,b). Like scoring, causes for poor responses to girdling may be attributable to time required for regrowth to bridge severed vascular tissue, which, in turn, is proportional to the width of removed phloem tissue (see reviews by Goren et al., 2004; Theron and Steyn, 2008). Westwood and Bjornstad (1974) described a study (E.S. Degman, unpublished data) in which 'd'Anjou' limb girdling at full bloom reduced fruit set above the girdle and improved set below the girdle, but when girdling was delayed until 7 weeks after bloom, the opposite was observed. Depending on girdling time, a plethora of opposing lines of reasoning can be proposed for any combination of basipetal and acropetal transport of either growth inhibitors or promoters. Webster (2002) reported insignificant effects of girdling on floral initiation of young 'Comice' and 'Concorde' trees, irrespective of timing. Westwood and Bjornstad (1974) evaluated single and interactive effects of thinning and girdling of 'd'Anjou' and 'Comice' limbs and determined that girdling only increased set when combined with reducing the number of fruit per cluster, which led to the conclusion that less competition among fruit within a cluster enhanced fruit perception to a setting factor. In this case, girdling was performed 3 weeks after full bloom and clusters were thinned to either one or two fruit and compared to non-thinned clusters. Clearly, complex interaction among many factors seems necessary to elicit the desired effects of these techniques. Given the relatively few girdling experiments performed on pear compared with other species (Goren et al., 2004; Sousa et al., 2008; Theron and Steyn, 2008), there is a need for reductionist style experimentation to elucidate which mechanisms, if any, elicit flowering.

6.2 Canopy pruning

Pruning is required to manipulate tree growth and development and to manage flowering and fruiting. The emphasis on pear tree pruning has shifted from a heavy reliance on heading cuts to develop structural scaffolding of low-density plantings (Forshey et al., 1992). However, unlike the minimalistic pruning of modern high-density apple systems, in which vigor is controlled by precocious fruiting, dwarfing rootstocks, and limb bending below horizontal, significant pruning is required for pears since cultivars are inherently vigorous, dwarfing rootstocks are limited in use, and shoots respond poorly to below-horizontal positions (see Section 8). Interestingly, work from New Zealand showed a positive effect of heading one-year-old 'Concorde' and 'Comice' trees on sylleptic branch formation, resulting in greater floral development in the second and third years after heading compared to unheaded trees (Seleznyova et al., 2013).

The type of pruning cut (i.e., thinning to remove a branch entirely or heading to reduce branch length) and its placement relative to other organs of the branch should be informed by the vegetative and fruiting habits of the cultivar (du Plooy et al., 2002; Sansavini, 1966; Sansavini and Musacchi,

1994). Readers are directed to Sansavini (1966) and Sansavini and Musacchi (1994) for their description of the five classifications of pear fruiting habits and examples of representative pear cultivars for each class. These classifications were made according to the proportion of fruit borne on different vegetative structures and/or ages of wood that, in some cases, change over time. For the group V cultivar, 'Passe Crassane', pruning severity (e.g., heading into two- and three-year-old wood) was positively related to fruit set of the remaining flowers (Sansavini and Zocca, 1968, cited in Sansavini and Musacchi, 1994). Thus, competitive sinks appeared to limit cropping. De Maeyer and Deckers (1984) also demonstrated improved fruit set of the group II cultivar 'Comice' as pruning intensity increased. Westwood and Bjornstad (1974), through spur thinning, proposed that sink competition among fruit limited set of 'Comice'. Competing sinks can be vegetative and/or reproductive and may involve correlative inhibition by hormones (Bangerth, 2005). To test the inhibitory effect of shoot extension on fruit set of 'Packham's Triumph', Saunders et al. (1991) removed either half or entire sections of the previous season's growth segment at different times relative to bloom. They then counted the number of fruit on the two-year-old wood. Compared to unpruned limbs, both pruning treatments improved fruit set, but removal of the entire section of 1-year-old growth had a greater effect. Delaying heading until bloom or longer negatively affected fruit set. The authors implicated an inhibitory hormonal factor since the demand for carbohydrate and nutrients among organs at bloom would be expectedly low. Einhorn (T. Einhorn, unpublished data) evaluated a similar protocol on 'd'Anjou' trees over 2 years, but failed to observe a positive effect from either the degree of branch removal or timing of the cut (Fig. 4), though, in this experiment, timing only varied by 2 weeks (±1 week from bloom). 'D'Anjou' benefits from long pruning, though shortening 2-year-old wood just above a spur can be effective, especially with respect to fruit set, which will limit a strong vegetative response. In the experiment described previously, branches typically possessed ~four spurs.

The linkage between proximity of new shoots and the health and vigor of subordinate spurs is the basis for 'Click' pruning. This relatively new approach removes one-year-old wood at a basal position above the growth ring between 1- and 2-year-old shoot segments (Musacchi and Greene, 2017; Dallabetta, 2014). Musacchi (2015) demonstrated that Click pruning of 'Bartlett' pear trees reduced blind wood of older nodes and promoted fruit size by forcing the fruiting zone to spurs of 2-year and older wood as opposed to the tips (i.e., brindilla). Click pruning has been used on Type IV apple varieties that exhibit a propensity for developing blind wood (Musacchi and Greene, 2017).

Given the need to reduce labor, mechanical pruning (hedging) of planar canopy systems has become increasingly popular (Sazo et al., 2010; Sansavini, 1974; Sansavini and Musacchi, 1994). Hedging is non-selective. Strong vertical

Treatment d'Anjou' limbs	Fruit set (% of flowers with fruit)		Harvested fruit wt. (g)	
2010	*(-)7 d anthesis*	*anthesis*	*(-)7 d anthesis*	*anthesis*
Control, unheaded	11.1	11.6	194.4	198.5
1-year wood headed	8.9	9.3	191.4	204.5
1-year wood removed	16.9	10.5	191.6	194.8
P < F	*0.3447*	*0.4995*	*0.9405*	*0.6233*
2011	*(-)7 d anthesis*	*anthesis*	*(-)7 d anthesis*	*anthesis*
Control, unheaded	8.8	n.d.	198.1	n.d.
1-year wood headed	12.3	n.d.	201.6	n.d.
1-year wood removed	8.5	n.d.	198.2	n.d.
P < F	*0.2154*		*0.8678*	

Figure 4 Effect of pruning severity of 1-year-old wood on 'd'Anjou' pear fruit set and fruit weight at harvest. Limbs were either unheaded (not shown), headed by removing ~50% of 1-year-old wood (top left) or headed by removing entire 1-year-old section (top right). All limbs possessed four subtending flowering spurs on the 2-year-old section of wood. Pruning was performed 7 days before anthesis (2010) or at full bloom (2010 and 2011). Trees had uniform flower load (n = 30). Photo and data: T. Einhorn and J. Turner (unpublished).

shoots or branches growing parallel to the tree row will not be pruned. Annual machine-hedging of pear canopies significantly increased leaf density and canopy volume compared to hand-pruned counterparts, resulting in severe shading and reduced fruiting (Baldini et al., 1984). The authors suggested that periodic hand-pruning was necessary to open up machine-hedged canopies. The temporal effects of hedging on vegetative regrowth, and pear quality and crop load, require further exploration. Sansavini and Musacchi (1994) suggest limiting summer pruning to postharvest timings, plausibly because buds are vegetatively mature at this time which precludes regrowth. An improvement in the quality of fruiting wood may occur from an enhanced light environment.

The long photosynthetic aging of pear leaves (L. Corelli-Grappadelli, personal communication) and higher intra-canopy light might increase carbohydrate reserves of fruiting wood and buds. Short-season cultivars and/or temperate regions with long postharvest periods may particularly benefit from late summer pruning.

6.3 Root pruning

Pruning the root systems of temperate-zone fruit trees has been extensively investigated (see reviews by Ferree et al., 1992; Geisler and Ferree, 1984; Saure, 2007). The primary objective of root pruning is to reduce current-season vegetative growth. The practice is relatively crude, forcing the plant to re-establish a functional equilibrium between the root system and canopy. The immediate loss of reserve/storage carbohydrate and nutrients, reduced capacity for water and nutrient acquisition, and changes in hormone balance purportedly contribute to the response. Compared to apple, pear has received surprisingly little of this attention despite its inherently high vigor and relatively poor precocity (Arrington, 2015; Asín et al., 2007; Janssens et al., 2011; Maas, 2008; Vercammen et al., 2005; Yehia et al., 2011). Irrespective, root pruning is commercially practiced in pear orchards throughout Europe and to a limited extent elsewhere. Reduced canopy size following root pruning is a consistent response of pear (Arrington, 2015; Asín et al., 2007; Vercammen et al., 2005). The magnitude and duration of growth inhibition are proportional to the volume of roots excised and the time required for the tree to re-establish a balanced root-to-shoot ratio (Arrington, 2015; Poni et al., 1992).

Root pruning implements are generally inserted to a depth of 45 cm, based on general root architectural patterns of clonal apple and pear rootstocks, although *Pyrus* spp. rootstocks may have more sinker roots than apple (Atkinson, 1980). Most studies show comparatively greater benefits from pruning both sides of the tree row than one side. Tree response seems optimal when applications are performed near bloom (Geisler and Ferree, 1984), as storage reserves are being remobilized to meristematic points of growth. Delayed root pruning at or beyond 'June drop' had less of a reductory effect on extension shoot growth than earlier timings, but strongly inhibited root regeneration of apple trees (Schupp and Ferree, 1987). Ferree (1990) concluded that 30–50% of roots needed to be removed to significantly reduce shoot growth. Therefore, the distance between trunks and root pruning implements should vary according to tree age; generally, younger trees are pruned 45 cm from the trunk and older trees up to a maximum of 90 cm (Ferree, 1990). However, tree age can be a poor proxy for root volume. For example, root pruning sixth-leaf 'd'Anjou'/Old Home × Farmingdale (OH × F) 87 trees 45 cm from trunks severely inhibited shoot and fruit growth for three seasons after pruning

(Arrington, 2015). In addition to tree age and/or vigor, annual precipitation, irrigation management, soil fertility, rootstock, ground cover system, and crop load all influence the volume and location of roots and should be considered when establishing pruning distances. The value of soil excavation to assess root volume and inform pruning distance cannot be overemphasized. Annual root pruning at equivalent distance and depth produced growth inhibition similar to that observed during the first year of application (Arrington, 2015; Ferree, 1990). Tree response to root pruning is the primary factor when considering annual reapplication. Vercammen et al. (2005) proposed a simple decision-aid model for repeat pruning based on the quantity and quality of fruit buds in early spring.

Root pruning limits canopy volume primarily by inhibiting the growth rate of shoots, but branch initiation is also affected. Changes in 'd'Anjou' shoot morphology were attributed to shorter internodes, as opposed to the number of nodes or leaf area per node (Arrington, 2015). Internodal growth inhibition may be attributed to reduced GA activity in shoots due to less cytokinin synthesis from the immediate loss of root tips (Saure, 2007). Vercammen et al. (2005) attributed a 44% decrease in the canopy volume of root-pruned 'Conference' trees to fewer shoots and not reduced shoot length. Asín et al. (2007) observed a combination of shorter shoots and fewer shoots for root-pruned 'Blanquilla' trees. Plausibly, roots generated after pruning would restore cytokinin levels and, in turn, shoot growth and branch initiation, but the time dependency of these processes is unclear and multiple endogenous factors and environmental signals are involved in their regulation. Arrington (2015) inferred that new root formation was responsible for the sudden, high shoot growth rates of single-side root pruned, fourth-leaf 'd'Anjou'/OH × F 87 trees 50 days after root pruning. Poni et al. (1992) showed a recovery of shoot-to-root ratios of potted pear trees 50 days after root pruning, due to a combination of high rates of new root production and limited shoot growth.

Enhanced intra-canopy light distribution is a byproduct of reduced vegetative growth (Arrington, 2015; Ferree, 1990; McArtney and Belton, 1992). Asín et al. (2007) documented a 40% cumulative increase in 'Blanquilla' flower clusters over a 3-year root pruning study. Arrington (2015) observed a ~50% increase in return bloom of root-pruned 'd'Anjou' trees. Maas (2008) and Yehia et al. (2011) also reported higher return bloom of 'Conference' and 'Le Conte' pear, respectively, 1 year after root pruning. These data would appear to underscore the importance of light environment on flowering, but it remains unclear if flower induction is promoted by an increase in light, changes in the hormone balance or carbohydrate and nutrient allocation, and/or water stress. Although higher light transmission may help transition vegetative meristems to floral meristems, root pruning is not a corrective practice for poor branch architecture, high proportion of blind wood, and/or low node density. In contrast to pear, root pruning typically did not affect return bloom of apple trees (Ferree et al., 1992). The disparate responses of pear and apple may be attributed to the inherently lower flower density of pear. Greater return bloom is often associated with higher fruit set and yield (Arrington, 2015; Maas, 2008; Yehia et al., 2011) but not always (Asín et al., 2007).

Despite these positive effects, root pruning reduced fruit size of apple and pear in some trials (Arrington, 2015; Ferree, 1989, 1992; Schupp and Ferree, 1988, 1990; Vercammen et al., 2005), but not others (Asín et al., 2007; Khan et al., 1998; Maas, 2008; Sosna, 2002; Yehia et al., 2011). Variation among experiments is expectedly high since the experimental protocols, plant material, soil, and climatic factors differed widely among studies. Fruit growth of root-pruned trees may be limited during cell division by decreased cytokinin supply on IAA and GA activity (Saure, 2007), reduced carbohydrate supply, or greater competition between newly initiated roots and above-ground sinks (Geisler and Feree, 1984; McArtney and Belton, 1992).

Water stress does not appear to play a significant role despite the abrupt change in accessible soil volume. Janssens et al. (2011) documented slightly lower stem water potential of root-pruned 'Conference' pear trees, but without deleterious effects to yield components compared to non-pruned trees over a multi-year study. Stem water potential of field-grown 'd'Anjou' trees was unaffected by root pruning (Arrington, 2015), although measurements of water potential were not conducted the first month after root pruning. Marsal et al. (2008) showed that transitory reductions in stem water potential after root pruning recovered within 30 days. These data corroborate the hypothesis that reduced gas exchange concomitant with less vegetative growth serves to maintain a favorable water status in root-pruned trees (Poni et al., 1992). Schupp and Ferree (1990) and Poni et al. (1992) did document lower leaf gas exchange of potted, root-pruned apple and pear trees, but the percent reductions were relatively minor and short-lived. Though clonal pear rootstocks are void of taproots, they generate numerous vertical sinkers (Atkinson, 1980; Lombard and Westwood, 1987) inaccessible to root pruning that may preclude development of water stress. In unirrigated situations, drought stress was implicated in smaller fruit and higher percentage of premature yellowing of 'Conference' (Vercammen et al., 2005). Nutrient deficiencies in root-pruned canopies have been proposed to limit fruit growth as well (Vercammen et al., 2005), but direct measures of leaf and fruiting spur nutrient content from root-pruned pear and apple trees do not support this assertion (Arrington, 2015; Schupp and Feree, 1990). Additional fruit quality effects from root pruning have received far less attention in pear studies. Skin color of apple is generally improved by root pruning, presumably as a result of an improved light environment. Thus, new blushed pear varieties may similarly benefit from this practice.

Compared to alternative methods of vigor management, root pruning has relatively low investment costs (time, labor, equipment, and application frequency). In this author's opinion, the practice deserves greater research attention. The combination of reduced canopy volume and similar or increased yield results in higher yield efficiency. In the case of 'd'Anjou', yield efficiency of root-pruned trees was increased by 100% (Arrington, 2015). Given the global transition to higher planting densities, especially in combination with *P. communis* rootstocks (Robinson, 2011), root pruning represents a viable strategy to manage vigor and improve pear tree efficiency.

7 Rootstocks for high-density orchards

Barritt (1992) described the dwarfing rootstock as 'the key piece of the orchard system puzzle, the piece to which all the other pieces must fit to complete the puzzle and thus develop a successful new orchard'. No single factor can alter the precocity, bearing potential, and efficiency of pear trees more profoundly than the rootstock. The transition to high-density pear orchards in many regions of the world has been severely hindered by a lack of commercially available, dwarfing rootstocks (Elkins et al., 2012), rendering many of the interventions described herein necessary to suppress vigor and induce cropping. Rootstocks from *Pyrus* germplasm have yet to confer sufficient precocity and/or dwarfing to pear scions. Interesting interspecific hybridization between triploid *P. nivalis* accessions and diploid species of *Pyrus* yielded dwarfing aneuploids that may lead to rootstocks with desirable agronomic traits (Simard et al., 2011).

In the Pacific Northwestern and Eastern US, OH × F 87 induced greater precocity, higher yield efficiency, and improved fruit size compared to other *Pyrus* rootstocks (Einhorn et al., 2013; Robinson and Dominguez, 2015). Pyro 2-33 has performed similarly to OH × F 87 throughout the first 5 years of high-density trials with 'Bartlett', 'Bosc', and 'd'Anjou' (T. Einhorn, R. Elkins and T. Robinson, unpublished data), confirming earlier observations of Robinson and Dominguez (2015). A sister of Pyro 2-33, Pyrodwarf (Jacob, 2002) was purported to be dwarfing and productive, but was proven unacceptable for use in the Pacific Northwestern US due to profuse suckering, low yield, poor fruit size, and insufficient dwarfing (Einhorn et al., 2013). Despite the advantages of OH × F 87, tree size remains 70–80% of that for trees on seedling rootstock (Elkins et al., 2012).

Given the dearth of dwarfing *Pyrus* candidates, new rootstocks for pear are more likely to come from alternative genera, though traditionally these sources are not without shortcomings (e.g., susceptibility to low temperature injury, incompatibility, pear decline and fire blight sensitivity, and lime-induced chlorosis). Space does not permit a comprehensive appraisal of the pear rootstock literature, but readers are referred to Ch. 5 in this book for a more detailed review and are encouraged to peruse other reviews on the subject (Elkins et al., 2012; Lombard and Westwood, 1987; Webster, 1998; Wertheim, 1998, 2002). Since the most recent of these reviews, quince 'Eline' (of Romanian origin, but produced in The Netherlands), quince H (also known as MH or EMH) (UK), and the inter- and intraspecific hybrids of *Amelanchier* (Germany) are exciting new candidates that warrant a brief discussion.

'Eline' conferred dwarfing to 'Conference' that was similar to the rootstock Quince C (also known as MC or EMC), but with notable improvements in fruit size and yield (Maas, 2008). 'Eline' has purported cold hardiness, which was evident following a mid-winter orchard freeze of ~−23°C, after which the rootstock shanks exhibited limited browning compared to other quince (Maas, 2015). A visible assessment of 'Conference' tree growth and leaf color the ensuing spring suggested that trees on 'Eline' were the least affected. These results corroborate unpublished controlled-rate freezing data of above-ground cuttings sampled from 1-year-old, own-rooted 'Eline' trees. Mid-winter xylem browning of 'Eline' was 50% at −25°C, but >50% at −30°C (T. Einhorn, unpublished data). A threshold of 50% xylem browning following exposure to −30°C was the main criterion for the advancement of 20 quince accessions from the quince core collection of the USDA Clonal Germplasm Repository as 'cold hardy' candidate rootstocks (Einhorn et al., 2011). Ten of these rootstocks were micropropagated and established in research orchards in 2017 with the 'Bartlett' and 'd'Anjou' using 'Comice' interstems (T. Einhorn, unpublished data). Additional cold-hardy quince selections that have received limited testing are the S-series from Poland and the K series from Lithuania.

Cumulative yields of 'Conference' on these rootstocks were only one-third those of Quince C or Sydo over the first 4 years, indicating incongruence between dwarfing and precocity (Kviklys and Kvikliene, 2004). 'Eline' appears to confer both traits to the scion (Maas, 2008). Chip buds of 'Conference' on 'Eline' were evidently compatible (Maas, 2015). Direct grafts of 'd'Anjou' on 'Eline' were incompatible (T. Einhorn, unpublished data). Incompatibility is expected to vary with scion. Further, the virus status of these trees was uncertain. Virus-infected material affects grafting compatibility, quite possibly having contributed to historical reports of graft incompatibility between pear and quince taxa (J. Postman, personal communication). Virus issues aside, new insights into the anatomical and biochemical factors causing failure of quince/pear combinations provide direct evidence of incompatibility between species (Ermel et al., 1999; Espen et al., 2005; Irisarri et al., 2015; Nocito et al., 2010).

Quince H was developed at East Malling, UK (Browning and Watkins, 1991) and confers significant dwarfing to pear scions, but results from field performance trials have been variable. Reasons for this could be scion variety, soil conditions, horticultural management, and/or environmental factors. Compared to Quince C, trees on Quince H were slightly smaller (Maas, 2015), similar (Iglesias and Batlle, 2011), or more vigorous, comparable to Quince A (also known as MA or EMA), but smaller than Provence Quince (also known as BA29-C) (Johnson et al., 2005; Sugar and Basile, 2011). Quince H dwarfed 'Conference' greater than Adams, Quince C, or 'Eline', but trees on Quince H were the least precocious or productive and failed to induce large fruit size (Maas, 2015). Quince H and Quince C conferred similar dwarfing to 'Conference' in Spain, but produced lower yields than Quince A, Adams, and Sydo (Iglesias and Batlle, 2011). However, yield efficiency was highest for Quince H and Sydo. Yield efficiency of 'Comice' was also highest on Quince H than Quince C or Provence Quince (Sugar and Basile, 2011). Musacchi et al. (2011) grouped Quince H with rootstocks of similar vigor (e.g., Sydo, Adams and Provence Quince) in multiple high-density orchard system configurations (2000–4000 trees/ha), but not in ultra-high-density plantings (4000–7000 trees/ha), which were limited to Quince C. 'Abbe Fetel' had the highest yield efficiency on Quince C, but fruit size was smallest. Quince H performed similarly to Sydo, Provence Quince, and Adams, although training system influenced production and fruit size. Hardiness data are limited for Quince H, but trees were considered less healthy than 'Eline' following the natural freeze event described earlier, though no mortality was observed (Maas, 2015).

Plants belonging to the genus *Amelanchier* (Family: *Rosaceae*) have extreme cold hardiness with many species surviving temperatures below −40°C (Vertucci and Stushnoff, 1992). Lombard et al. (1984) evaluated the suitability of *Amelanchier* seedlings, along with other related genera (*Crataegus, Sorbus, Cydonia*) as pear rootstocks. Graft compatibility between *Amelanchier alnifolia* and pear was deemed adequate and trees were extremely dwarfing with high yield efficiency; however, trees were non-uniform (Lombard et al., 1984). In 1999, the Bavarian Centre of Pomology and Fruit Breeding (BCPFB) at Hallbergmoos, Germany, began developing intra- and interspecific *Amelanchier* hybrids from a collection of species (primarily *A. alnifolia* and *A. spicata)* (M. Neumüller, unpublished data). Selection pressure was toward dwarfing, cold hardiness, productivity, and sensitivity to iron-chlorosis. Graft compatibility was evaluated with 'Comice', and 'Beurre Hardy' compatible selections were screened for resistance to *Candidatus Phytoplasma pyri* (pear decline phytoplasma). Several Amelanchier selections showed insignificant xylem browning after controlled-rate freezing to −40°C when dormant (T. Einhorn, unpublished). Field performance trials were established at the BCPFB using 'Comice' and 'Beurré Hardy' and in Hood River, Oregon, USA using 'd'Anjou'.

In Germany, *Amelanchier* rootstocks were more dwarfing than Pyrodwarf, Quince A, or Quince C. Yield efficiency was more than two-fold that of 'Pyrodwarf' and similar to 'Quince C' (M. Neumüller, unpublished data). In the

USA, a 'd'Anjou' trial was established at 1 × 4 m (2,989 trees/ha) comparing three *Amelanchier* genotypes to OH × F 87. Third-leaf 'd'Anjou' trees on *Amelanchier* rootstocks produced over 50 flower clusters per tree, which was ten-fold more than OH × F 87 (Fig. 5). At harvest, trees on *Amelanchier* produced 20 fruits per tree with an average weight of 277 g, resulting in 18.5 t/ha. Trees on OH × F 87 produced no fruit. *Amelanchier* rootstocks reduced trunk cross-sectional area and canopy volume by ~50% compared to OH × F 87, yet total leaf area was not affected by rootstock. Trees on *Amelanchier* rootstocks produced markedly more spurs with rosettes of leaves and shorter internodes than OH × F 87. Based on these preliminary data, new trials in Germany and the US are underway using new, superior selections of *Amelanchier* rootstocks. Compatibility is excellent with 'Comice' and 'Beurre Hardy', but not 'd'Anjou'

Rootstock	Flower clusters/ tree (no.)	Yield/ tree (kg)	Fruit/ tree (no.)	Fruit wt. (g)	TCA (cm^2)	Yield efficiency (kg/cm^2 TCA)
Amelanchier 10	64.2 a	4.3 a	20.6 a	209.2 ab	13.3 b	0.33 a
Amelanchier 2	61.3 a	4.4 a	19.7 a	225.6 a	11.4 b	0.39 a
Amelanchier 7	58.8 a	4.1 a	20.5 a	204.6 ab	10.8 b	0.39 a
OH × F 87	11.3 b	0.4 b	1.8 b	187.4 b	20.1 a	0.02 b

Figure 5 Fourth-leaf 'd'Anjou' pear trees trained to a V trellis by tipping trees ~10° opposite neighboring trees. Photo shows a single, 7-tree replicate of 'd'Anjou'/*Amelanchier* 10 on left and 'd'Anjou'/OH × F 87 on right prior to harvest. Summary table provides fourth-leaf data for three *Amelanchier* clones vs. OH × F 87. Data are from seven multi-tree replicates (n = 7 trees). Photo: T. Einhorn; Data: T. Einhorn and M. Neumüller (unpublished).

or 'Bartlett', which require interstems (M. Neumüller, personal communication). *Amelanchier* rootstocks hold quite a bit of promise, but require intensive testing in diverse regions.

8 Canopy training in high-density orchard systems

An increase in intensive, high-density pear orchard systems in several regions of the world can be attributed to the localized availability of dwarfing rootstocks (Musacchi, 2011; Sansavini and Musacchi, 1994; Vercammen, 2011). Many of these systems are derivatives of spindle or V canopy architectures. Perhaps the most compelling innovation in pear tree training is the bi-axis system, Bibaum® (Musacchi, 2006, 2008). The development of Bibaum® trees begins in the nursery (Fig. 6) by inserting two chip buds in opposite positions on the rootstock. The technique encourages uniform leader development from the outset, attributed to co-dominant growth of the scion buds (Fig. 7). Alternatively, bi-axis trees can be developed in the field using a heading cut; however, cropping is delayed for 1 year and there is a greater potential for non-uniform growth between the developing axes that may lead to imbalanced leaders, compared to Bibaum® nursery trees. Competition between the two leaders of a bi-axis tree diffuses vigor, which is advantageous for use with vigorous *Pyrus* rootstocks.

A factorial trial was established in 2013 in Oregon, USA, using the non-precocious cultivar 'd'Anjou' to evaluate the performance of four *Pyrus* rootstocks (OH × F 87, OH × F 69 and Pyro 2-33) at three in-row spacings (1, 1. 5, 2 m each with 4 m alley spacing) and three training systems (V, bi-axis, or spindle) (T. Einhorn, unpublished data). The mean trunk cross-sectional area of individual axes of the bi-axis trees was ~55% of that of the single axis

Figure 6 Nursery production of bi-axis pear trees in Italy using the patented Bibaum® double budding technique. Photo: T. Einhorn.

Figure 7 Third-leaf 'd'Anjou'/OH×F 87 bi-axis pear trees (2017 trees ha^{-1}). Photo: T. Einhorn.

trees. Bi-axis trees flowered earlier than the other training systems (third-leaf), but fruit set was extremely low, which is characteristic of 'd'Anjou'. Fourth and fifth leaf yields were similar among training systems, but the bi-axis system produced calmer, more balanced trees with a higher number of lateral branches of reduced diameter than the single axis spindle or angled V trees. While the bi-axis trees may facilitate lower tree density plantings compared to single axis trees, 1 m in-row spacing facilitated 0.5 m spacing between axis. For type II and III cultivars which fruit predominantly on older wood, this may be the minimum spacing to balance fruiting with canopy volume, and 1.5 m may be optimal. This trial also demonstrated a positive relationship between in-row spacing and tree size, with the highest density plantings producing the smallest trees, irrespective of rootstock or training system effects. These data agree with a previous report of high-density pears on *Pyrus* rootstocks (Robinson, 2011).

Several orchard training systems use branch positioning to induce precocity and productivity. The horticultural practice of bending apple limbs at or below the horizontal reduced terminal shoot extension (Myers and Ferree, 1983), induced spur formation of nodes, promoted flowering, and resulted in a two- to three-fold increase in the number of fruit compared to branches set to 45°or vertical (Robbie et al., 1993). Promotion of flowering in 45°-positioned Japanese pear limbs was associated with increased cytokinin and ABA concentrations and decreased IAA and GA concentrations in lateral buds compared to vertical shoots (Ito et al., 1999). Shoots of mature apple trees bent to horizontal positions had significantly lower IAA and zeatin/zeatin riboside than vertical shoots (Sanyal and Bangerth, 1998). Shoot bending is also likely to promote ethylene production (Sanyal and Bangerth, 1998). Horizontal limb positioning of 'Bartlett'/OH × F 97 trees induced watersprout production at basal nodes that precluded spur production and shifted fruiting sites to distal positions (Fig. 8). Limb treatments were imposed in the fourth-leaf when scaffolds, initially trained to a 9-wire system (wires 45 cm apart), were renewed with bevel cuts and

Limb angle (°)	Primary scaffolds per tree: Avg. length (m)	Primary scaffolds per tree: Total length (m)	Secondary shoots per tree: Number	Secondary shoots per tree: Avg. length (cm)	Secondary shoots per tree: Total length (m)	Leaf area (m^2)
45	1.61 a	24.9 a	93 a	54.6 b	57.8 a	21.6 a
30	1.22 b	19.7 b	64 b	58.1 b	39.9 b	8.1 b
0	1.17 b	17.1 b	89 a	67.4 a	62.7 a	21.1 a

Limb angle (°)	Fruit set (fruit/tree)	Yield: Per tree (kg)	Yield: Per ha (tons)	Avg. fruit wt. (g)
45	141 b	17.2 b	38.4	198 a
30	216 a	22.7 a	50.7	194 a
0	75 c	12.2 c	27.2	183 b

Figure 8 Effect of limb orientation (0°, 30°, or 45° from horizontal) on vegetative and reproductive growth of seventh leaf 'Bartlett'/OH × F 97 pear trees (2270 trees ha^{-1}) trained to an 8-wire trellis. Wires are spaced 45 cm apart. All scaffold limbs were renewed with bevel cuts in the fifth leaf and trained to three orientations. Data are from third leaf limbs. Photo and data: T. Einhorn and J. Turner (unpublished).

newly emerging shoots were oriented 45°, 60°, or 90° from the vertical once 15 cm of growth accrued. Limbs trained to 60° angles were highly florigenic in the spring of the third year, producing significantly more flower clusters that resulted in higher fruit set and more fruit per limb than limbs trained to 45° or 90° (Fig. 8). The latter treatment produced the fewest flower clusters, least number of harvestable fruit, and smallest fruit size, which was opposite to the results of apple limb positioning studies (Robbie et al., 1993). Growth of the terminal was the least for shoots positioned horizontally. These results come as no surprise – the pear palmette hedgerow with oblique branch training evolved from the high-density vertical cordons of the 1930s and 1940s in the Po Valley, Italy (Sansavini, 1982). Branch angles of these plantings were between 45° and 90° from the vertical. The tall spindle pruning system of apple promotes early fruiting by planting well-feathered, unheaded trees; limiting pruning to annual renewal of the largest limb(s); and branch bending below horizontal (Robinson, 2011). Application of tall spindle concepts to pear resulted in extremely high cumulative yields (60 t/ha in the fourth leaf), but with small fruit size (160–180 g 'Bosc' and 'Bartlett') (Robinson, 2011; Robinson and Dominguez, 2015). It is unclear whether the size potential was limited as a physiological response to branch bending, or some other factor such as crop load or climate.

9 Environmental factors that affect flowering, fruit set, and yields

9.1 Light

Pear is a C3 species with a sigmoidal photosynthetic light response that reaches asymptotic values at ~40% of full photosynthetically active radiation (PAR) (DeJong, 2007; Einhorn et al., 2012). Genetic variability among green-skinned European pear cultivars did not have a marked effect on Pn, though incompatibility between rootstocks and scions can markedly alter photosynthesis (Corelli-Grapadelli et al., 2001). Pigmented leaves of red-fruited 'Bartlett', 'Clapp's Favorite', and 'd'Anjou' sports were only ~70% as photosynthetically efficient as their green-skinned progenitors (Martin et al., 1997). Pear leaves that develop in the shade have characteristic shade-leaf physiology with lower light compensation points, dark respiration (Rs), and maximum Pn than leaves developed under moderate or high PAR (Einhorn et al., 2012). 'Bartlett' leaves exposed to 20% illumination for 12 weeks had significantly lower specific leaf mass (SLM, g·cm^{-2} leaf area) than well-illuminated leaves (Garriz et al., 1997). Similar data were published for 'Forelle' in South Africa (Dussi and Huysamer, 1995) and 'Bartlett' and 'd'Anjou' in Canada (Kappel and Neilsen (1994).

Light management of pear canopies is critical to optimizing potential yield. The relationship between dry matter production of apple trees and orchard light interception is linear to maximum values of ~84% (summarized by Lakso, 1994). Similar data are not available for pear, but there is little reason to expect differences. Fruit set, however, can be low in dense canopies with high light interception. Khemira et al. (1993) suggested >30% of full illumination was required for floral initiation of 'd'Anjou', as similarly reported for apple (Cain, 1971). Reducing incident radiation on 60-year-old 'd'Anjou' pear trees to 40%, near the purported period of floral initiation (i.e., 60 dafb, Westwood, 1993), decreased the number of fruit per tree the following year by 55% (Einhorn et al., 2012). Kappel and Neilsen (1994) showed significant linear correlations between the natural light environment of 'd'Anjou' and 'Bartlett' spurs and their return bloom over three seasons, but the relationship was only significant for 'Bartlett' in 1 year. The authors suggested that the biennial tendencies of 'Bartlett' exerted more influence over bloom than light.

Regulation of fruit abscission under low illumination is not well understood. Carbohydrate deficit presumably contributes to fruit abscission (Section 4.4). Apple fruit growth was significantly reduced by severe shade (≤50% full light), but shoot growth remained unaffected, indicating greater sink strength of shoots for carbohydrate under shade conditions (Bepete and Lakso, 1998). Reduced fruit growth rates from shading during cell division may precondition fruit for abscission. van Zyl and Strydom (1982) showed competition from vegetative sinks reduced pear fruit set. Fifteen days of 44 or 77% whole-canopy shading between petal fall and 22 mm fruit diameter reduced 'Bartlett' fruit set by 50 and 75%, respectively, of non-shaded canopies (Einhorn and Arrington, 2018). Systematic research is necessary to identify the magnitude and duration of shading necessary to elicit pear abscission. Byers et al. (1991) derived these thresholds for apple. Bangerth (2000) proposed that the shift in red:far red ratio in shaded canopies may alter IAA transport and promote abscission. Given the rapidly emerging canopy-imaging and light-detection technologies (i.e., lidar, structure from motion, hyperspectral sensing), generation of 3-D canopy models and maps to develop canopy and cropping indices is attractive. Precise estimation of the light environment of canopies is needed to determine the photosynthetic efficiency per unit leaf area and the leaf-to-fruit ratios required for fruit set, flower initiation, and fruit quality. Despite the potential for precision agriculture technologies to transform management of tree fruit systems, research has lagged. A survey of Washington, New York, and Michigan apple producers revealed that willingness to adopt precision agriculture technologies was greater when research was conducted by non-biased, applied research, and extension sources (Gallardo et al., 2019). Economies of size created an additional barrier to adoption. Structured, planar canopies composed of repeating fruiting units are necessary to facilitate precision agriculture technologies. In some parts of the world, pear systems clearly fit this model, but many regions still lack plant material (primarily rootstocks) to enable such a transition.

Low illumination after June drop, in combination with greater sink demand of fruit, reduced fruit growth of pear (Einhorn et al., 2012; Garriz et al., 1998; Zhang et al., 2016). In fact, shading mature 'd'Anjou' trees from 60 dafb to harvest reduced fruit set and yield the subsequent year by 46% of non-shaded canopies (Einhorn et al., 2012). Yield did recover to non-shaded levels, but required 1 year without artificial shade, indicating transitory rather than irreversible effects. Though not covered in this review, fruit from interior, shaded canopy positions are typically of poorer quality due to reduced soluble solids concentration compared to fruit in higher light environments (Einhorn et al., 2012; Kappel and Neilsen, 1994; Zhang et al., 2016). 'd'Anjou' pears from internal canopy positions were greener, had lower sorbitol and sucrose accumulation, and higher malic acid than fruit from exterior canopy positions (Rudell et al., 2017). The rapid starch accumulation in 'Abbe Fetel' pears between 50 and 100 dafb, and the simultaneous inverse pattern of decreasing leaf starch, underscores the sink strength of fruit during the linear growth phase (Mesa et al., 2016).

There are many ways to increase light interception and its distribution within canopies. New pear orchards can be established on dwarfing rootstocks and trained to modern, planar fruiting walls. Changing the light environment of mature canopies is more difficult, but achievable by most techniques that control vigor; however, several years may be required for trees to reach a new equilibrium, particularly when using root and canopy pruning. Reflective fabric caused the most rapid response; this author has observed this fact when applied to alleyways of 60-year-old, low-density (269 trees per hectare) 'd'Anjou' trees. Reflective fabric significantly increased PAR (17%), yield (27%), and fruit quality in the year of application (Einhorn et al., 2012). Over 4 years, higher fruit set resulted in a cumulative yield increase of 13% when fabric was maintained from bloom to harvest compared to 7% when fabric was removed 60 dafb. Carbohydrate reserves of fruiting units were not quantified, but the results suggest that whole-season fabric increased carbohydrate availability to fruiting spurs from the increased PAR and photosynthetic efficiency. Reflective fabric increased floral bud production by roughly two-fold for 12-year-old slender spindle 'Clara Frijs' pear trees by the third year (Bertelsen, 2005). Collectively, these observations illustrate the importance of light management for fruit set, productivity, and fruit quality and the relative plasticity of pear trees to respond to short-term, corrective horticultural procedures.

9.2 Temperature

Clearly, freeze events can limit flowering and fruit set. Pear flowers are relatively cold hardy with fully acclimated buds surviving minimum temperatures between −30°C and −40°C (Rajashekar et al., 1982). Buds progressively lose freeze resistance during deacclimation when development resumes in spring. Pear bud hardiness is achieved primarily by freeze avoidance mechanisms (e.g., supercooling), but water transport to extracellular ice occurs. Unlike stone fruit, whose dormant flowers are isolated from woody vascular tissues of spurs or stems, the vascular connections of pear flowers and subtending structures are intact. Thus, water migration from pear pistils through the pedicels of the raceme to extracellular space lowers the free water content and ice nucleation temperature of flowers (Fig. 9a–c). Water migration is facilitated by the higher vapor pressure of supercooled water compared to ice when both phases exist at the same temperature (Sakai and Larcher, 1987). Often, following severe winter freezes or episodic spring frosts, oxidative browning of the pith can be readily observed. This tissue is in stark contrast to the healthy phloem and xylem tissues of fruiting spurs and the floral meristems within the bud (Fig. 9d). Extensive exposure to subfreezing temperatures can cause dehydration stress, denaturation of proteins, and ultrastructural changes to membranes in extreme events. Differences in the critical freezing temperatures of flowers among European pear cultivars appear negligible (Ballard et al., 1993), when comparisons are made at equivalent phenology stages. The apparent frost sensitivity of some cultivars likely is due to advanced development and relatively earlier bloom timing. The use of PGRs to rescue fruit whose embryos were killed by frost was described earlier.

Woody tissues of pear are slightly less hardy than apple, but considerably hardier than most temperate-zone tree fruit species (Quamme, 1976). Characterization of tissue hardiness within dormant, pre-conditioned pear stems indicated greater freeze susceptibility of xylem (−30 to −35°C) than phloem (−35 to −50°C) (Quamme, 1976). Xylem freezing was associated with the tightly packed xylem ray parenchyma cells (Quamme et al., 1972). Moreover, a geographic distribution study of pear cultivation throughout North America related low-temperature exotherms of apple and pear xylem with the northern limits of production (Quamme, 1976). An examination of blackheart injury of apple trees and temperature records corroborate these correlations (Quamme and Hampson, 2004). Notwithstanding extreme midwinter anomalies, autumn freeze events tend to be responsible for freeze injury of pear stems, typically in years with limited low temperatures for acclimation. These conditions may increase susceptibility to infection by *Pseudomonas* spp., indirectly reducing flowering or fruit set by weakening trees. Exogenous ABA, which has been applied experimentally to initiate acclimation and dormancy of young apple trees and improve N withdrawal from leaves (Guak and Fuchigami, 2001, 2002), has not yet been studied for potential acclimation initiation in pear.

Far more genetic variation in freeze resistance occurs among rootstocks of *Pyrus* species and, certainly, quince. Lombard and Westwood (1987) summarized the relative hardiness of *Pyrus* rootstocks. Illustrative of the

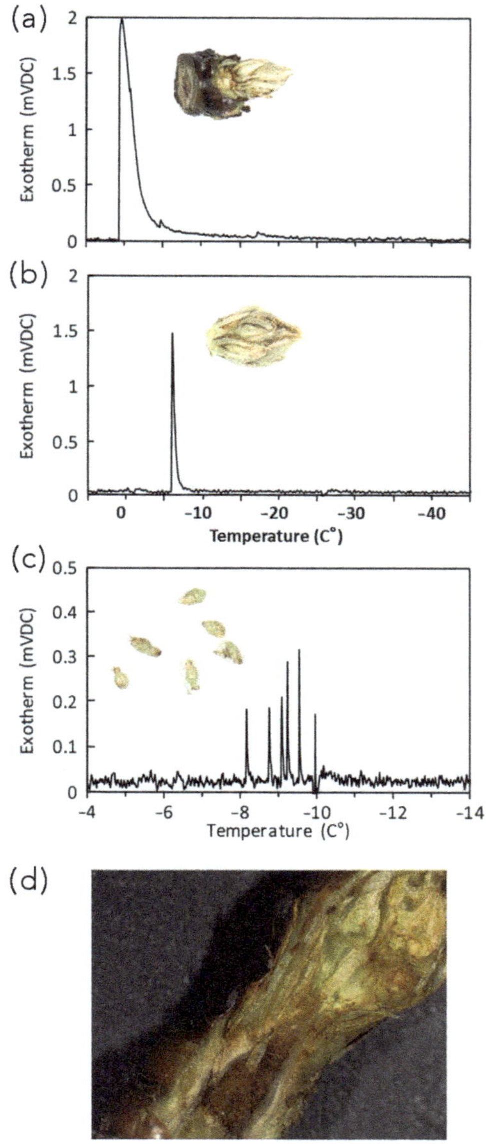

Figure 9 Microvolt signals produced by exothermic reaction from the freezing of extracellular water in tissues of a 'Bartlett' pear inflorescence: (a) with bud scales removed, but still attached to woody spur tissue; (b) intact inflorescence removed from subtending woody tissue; and (c) individual flowers isolated from the inflorescence. Exotherms were non-lethal in (a) and (b) and lethal in (c) given that water migration from the ovary through the vascular tissue of the inflorescence was eliminated when flowers were removed from the raceme; (d) oxidation of pith tissue in a 'Bartlett' fruiting spur. Extracellular freezing of the pith is an adaptive 'tolerance' strategy for ice nucleation. Photos and data: T. Einhorn and D. Gibeaut (unpublished).

complexity of hardiness, the native distribution *of Pyrus ussuriensis* is above the −40°C isotherm; hence, dormant *P. ussuriensis* can survive till −50°C. Since temperatures remain below 0°C for extended periods during winter, *P. ussuriensis* has a low chill requirement to break dormancy. Consequently, *P. ussuriensis* grown in mid-latitudes is highly susceptible to injury during late winter freeze–thaw cycles (Einhorn et al., 2011; Rajashekar et al., 1982). Roots, in general, are more susceptible to freeze damage than above-ground tissues, but benefit from the relatively greater heat sink of water and soil than aerial parts of the tree. Quince rootstocks generally are regarded to lack sufficient cold hardiness for use in northern regions (Wertheim, 1998); however, 'Bosc' on Quince A has shown no apparent cold injury in a 10-year evaluation in Geneva, New York (T. Robinson, personal communication). Cold hardiness of quince rootstocks was discussed previously (Section 7).

High temperature limits floral initiation, development, and fruit set in several ways. European pears require 800–1700 h of chilling at temperatures between 0 and 7°C to break endodormancy (Westwood, 1993). A lack of chill unit accumulation in lower latitude climates leads to developmental delay, often resulting in protracted bloom, bud abortion, and/or poor fruit set (Marodin et al., 2008). Higher-than-average temperatures during autumn (in a temperate environment) had no limiting effect on pear fruit set or yield the subsequent spring (Tromp and Borsboom, 1994). However, high temperatures during bloom reduced pear fruit set compared to cool. Despite a positive relationship between temperature and pollen tube growth, high temperatures shortened the EPP by accelerating maturation of the stigma and ovule (Sanzol and Herrero, 2001). The temperature responses of photosynthesis and respiration for European pear are not readily available. However, dark respiration in response to temperature was consistent for apple, peach, Asian pear, and grape (*Vitis vinifera*) (Higgins et al., 1992). European pear was not included in this work, but would likely have a similar response. The photosynthesis response to temperature, on the other hand, varied considerably among apple, peach, Asian pear, and grape, with Asian pear having higher temperature optima than apple (Higgins et al., 1992). It is notable, however, that maximum photosynthesis rates of 10 to 12 $\mu mol \cdot m^{-2} \cdot s^{-1}$ for Asian pear are substantially lower than for European pear (Einhorn et al., 2012; Martin et al., 1997). Regardless, higher carbohydrate loss to respiration with increasing temperature would be expected to increase fruit abscission because less carbohydrate would be available to meet demands of rapidly growing fruit. The temperature response of pear and apple fruit abscission was linear when evaluating the thinning efficacy of ethephon (Wertheim, 2000). Warm early season temperatures increased the cell division rate of 'Bartlett' pear (Rodríguez et al., 2011). Temperature explained 70–90% of the variability in fruit growth rate of 'Bartlett' pears between 0 and 60 dafb (Rodríguez et al., 2011). These authors showed a positive relationship between final fruit size and cell number, as demonstrated previously for apple (Goffinet et al., 1995). High temperature, in combination with high radiation, is related to the development of physiological disorders of fruit such as cork spot and sunburn, but will not be addressed in this review.

9.3 Water

Sustainable management of pear orchards depends on the conservation of finite resources such as water. Moreover, economic sustainability may be directly or indirectly enhanced by reduced irrigation withdrawals and improved management efficiency, respectively, the latter being due to reduced vigor of pear trees subjected to water deficit (Brun et al., 1985; Caspari et al., 1994; Chalmers, 1989; Chalmers et al., 1986; Mitchell et al., 1986, 1989). In the context of this review, however, rarely has water stress from deficit irrigation been shown to improve flowering of pear trees (Marsal et al., 2002; Mitchell et al., 1984). Replacement of 26% evaporation from a US Class A pan (Epan) over the first 60 dafb significantly increased floral initiation of 'Bartlett' pear trees (Mitchell et al., 1984). Measures of plant water stress and soil moisture were not provided, but presumably, the enhanced flowering was a result of improved canopy light distribution since pruning weights were one-third of fully irrigated trees and the deficit period would have overlapped with the timing of floral induction. Conversely, Caspari et al. (1994) showed lower return bloom of deficit-irrigated 'Hosui' Asian pear trees when regulated deficit irrigation was applied from 42 to 115 dafb. A late deficit irrigation treatment, applied 126–159 dafb, did not affect flowering. Fruit growth of several pear cultivars appears to be quite sensitive to water stress (Marsal et al., 2000, 2008; O'Connell and Goodwin, 2007; Pacheco et al., 2011; Shackel et al., 1997). Critical stem water potential for 'Conference' fruit growth was between −1 and −1.2 MPa, depending on cropload (Marsal et al., 2008). Daily fruit growth rate of the small-fruited 'Rocha' under non-irrigated conditions was ~50% of a maximum available water treatment or a mild, season-long deficit irrigation (~93% of maximum available water volume) when predawn leaf water potential reached values around −0.9MPa (Pacheco et al., 2011). Application of 15% of control irrigation (based on Penman ET_o) to containerized 'Bartlett' pear trees between 32 and 60 dafb lowered stem water potential to −1.4 MPa and reduced fruit growth rate (Marsal et al., 2000). Differences in mature fruit size among treatments was related to cell size and not a reduction in cell number, despite trees being returned to full irrigation at 60 dafb. O'Connell and Goodwin (2007) applied 50% of ET_c to 'Williams Bon Chretien' for the entire season using partial rootzone drying deficit irrigation in the semi-arid Goulburn Valley, Victoria, Australia. The treatment reduced stomatal conductance by ~50% and significantly lowered water potential, resulting in reduced yields due to increased fruit drop and reduced fruit size.

These results differ from Mitchell et al. (1984, 1986), which showed vigor-controlling benefits of early-season water deficits without deleterious effects on 'Bartlett' fruit growth or yield. In fact, Mitchell et al. (1986) reported significantly higher fruit growth rates when regulated deficit-irrigated 'Bartlett' trees were returned to control levels (~120% Epan). Jerie et al. (1989) showed a similar fruit growth response upon returning 'Williams Bon Chretien' trees irrigated at 25% E for ~8 weeks to 100% E. At the end of the deficit period, osmotic potential of RDI-treated fruit was −2.4 MPa compared to −1.7 MPa for control fruit, indicating favorable water transport and turgor maintenance from the energy gradient established by the lower osmotic potential of the deficit fruit.

Critical water savings are likely to be achievable in most regions simply by irrigating according to Kc values (ETc) suggested for pear (Allen et al., 1998; Marsal et al., 2002). Implementing deficit irrigation to gain greater savings appears difficult given the sensitivity of fruit growth to water potential. Water savings will depend on climatic factors, available soil moisture, rooting depth, percent ground cover, and required harvestable fruit size. Increasing bloom does not seem possible without sacrificing yield and fruit size.

10 Summary and future trends

The intention of this review was to address the critical factors limiting European pear productivity and propose techniques to improve sustainable production. Continued breeding and selection of dwarfing rootstocks are critical. In their absence, several viable, pragmatic solutions exist for managing vigor. Many of these, however, need to be recognized simply as tools, capable of repairing one deficiency, i.e., vigor, with the potential of creating others. Two examples are the application of P-Ca or root pruning to control vigor, resulting in floral inhibition or reduced fruit size, respectively. A systems approach to tree fruit management requires recognition of the interrelationship of separate components and their effect on the system as a whole. Development of a highly efficient tree is a dynamic process, requiring constant reassessment of inputs based on tree response and management objectives.

The focus on production efficiency limited attention to other, equally important topics - most notably fruit quality. Highly efficient, dwarf trees that produce abundant yields of poor-quality fruit are not profitable. Market demand for high-quality fruit will only increase. Achieving consistent fruit quality requires equal attention to extrinsic [blemish-free skin, freedom from russet or full russet (in some markets), and optimum fruit size] and intrinsic (sugars, acidity, flavor, texture, and freedom from internal disorders) attributes. While first-time purchases are dependent on external quality, repeat purchases are associated with intrinsic fruit quality. Pears, far more than alternative tree fruits, have unique market challenges due to their ripening requirements. For the consumer to have consistent eating experiences, pears must be of uniform quality and maturity, provided sufficient low temperature conditioning (i.e., chill), and be handled carefully throughout the entire postharvest supply chain to mitigate the appearance of surface maladies during ripening at retail markets or on home countertops. All of this must be integrated into a fast-paced world demanding greater supply of ready-to-eat products.

Homer described pear as one of the "gifts of the gods" in the ninth century BC literary work, The *Odyssey*. Surely, pear orchards have evolved greatly from the pear trees that graced the garden of Alcinöus. The future is bright for continued progress in pear systems biology and new, improved genetic material to ensure economic sustainability for producers of this wondrous temperate fruit. Breeding efforts toward development of novel germplasm, both scion and rootstocks, is imperative. Rootstocks that confer scion dwarfing and precocity need to replace genotypes that perform poorly in warm climates, calcareous soils, or have insufficient hardiness for northern production regions. New cultivar development is also essential to the economic sustainability of pear (see review by Brewer, 2011). New selections should have superior fruit quality. Several new and exciting cultivars have recently been released: three early maturing (pre-'Williams' timing) and one mid-season maturing variety with diverse quality attributes, PE1UNIBO 'Lucy Sweet'™, PE2UNIBO 'Early Giulia'™, PE3UNIBO 'Debby Green'™, and the red-skinned, PE4UNIBO 'Lucy Red', respectively, from the University of Bologna (Ancarani et al., 2018); the bi-color, crisp 'Gem' from the USA (Bell et al., 2014); the bi-color and red-skinned, crisp novel interspecific hybrids 'Piqareo'™ and PremP009 'Piqa Boo'™ from New Zealand (Brewer et al., 2011; Volz et al., 2008); and the bi-color 'Cheeky'™ from ARC Infruitec-Nietvoorbij, South Africa. The highly attractive, bi-color, and red-skinned pears have the potential to increase consumption by enticing that portion of consumers who remain disinterested in the standard cultivars that have occupied the same shelf space for decades.

Considerable progress has been made on the molecular mechanisms underlying flowering genes that promote (FLOWERING LOCUS T, FT) or repress (TERMINAL FLOWER 1, TFL1) reproductive developmental transitions and flowering (Wickland and Hanzawa, 2015). In combination with florigenic treatments, gene expression studies and transformation technology offer enormous potential to advance our understanding of these processes and inform breeding efforts. Genetic control of dwarfing is another vibrant research area. The recent identification of quantitative trait loci [*Dwarfing 1* (*Dw1*) and *Dwarfing 2* (*Dw2*)] associated with dwarfing in pear (Wang et al., 2014) and apple (Celton et al., 2009; Fazio et al., 2014; Foster et al., 2015) has the potential to markedly reduce the timeline for developing novel genetic material. Whether rootstock improvements are realized from traditional breeding efforts or biotechnological approaches, robust field performance trials are a necessary step to facilitate industry adoption and characterize genetic and environmental effects.

High-density pear plantings with structured canopies are a prerequisite for the integration of precision agricultural technologies. Use of unmanned aerial vehicles (UAVs or drones), cameras, and sensors to monitor plant stress, disease pressure, predict fruit size, yield estimation, etc. are all nascent technologies with potential for substantive gains in production efficiency. An understanding of spatial and temporal variations of flowering and fruiting within pear canopies, and imaging and mapping of fruit quality, is the next frontier for orchard management.

With new advances in dwarfing rootstocks and high-density plantings, far more emphasis will be placed on crop load management. Greater importance on thinning will require development of new thinning chemistries and predictive models for their success, much of which is rapidly developing in modern apple production systems. There will be a continual need for development of effective new PGR formulations to replace existing chemistries and provide alternatives in the event of future restrictions.

11 Where to look for further information

When appropriate, recommended review articles were provided within each section.

The International Society for Horticultural Science (ISHS) Division Physiology and Plant-Environment Interactions of Horticulture Crops in Field Systems (https://www.ishs.org/physiology-and-plant-environment-interactions-horticultural-crops-field-systems) and Division Temperate Tree Fruits (https://www.ishs.org/temperate-tree-fruits) list upcoming symposia and latest news. In addition, the peer-reviewed proceedings of these symposia, Acta Horticulturae, are valuable resources for additional information on all aspects of pear horticulture (https://www.ishs.org/acta-horticulturae).

Organizations that focus on orchard systems and management include the International Fruit Tree Association (IFTA) https://www.ifruittree.org/and the European Fruit Research Institutes Network (EUFRIN) working groups http://eufrin.eu/index.php?id6 The IFTA hosts summer study tours in addition to an annual meeting. The EUFRIN Thinning Working Group is of notable interest to cropload management of pear.

Organizations that focus on plant growth regulators include the International Plant Growth Substances Association (https://pages.wustl.edu/ipgsa) and the Plant Growth Regulation Society of America (http://www.pgrsa.org/).

12 References

Addicott, F. T. and Lyon, J. L. 1969. Physiology of abscisic acid and related substances. *Annu. Rev. Plant. Physiol.* 20(1), 139–64. doi:10.1146/annurev.pp.20.060169.001035.

Allen, R. G., Luis, S. P., Raes, D. and Smith, M. 1998. *Crop Evapotranspiration: Guidelines for Computing Crop Water Requirements*. FAO irrigation and drainage Paper No. 56. FAO, Rome.

Ancarani, V., Spinelli, F. and Musacchi, S. 2018. Produttività e sapore premium nelle pere dell'Università di Bologna. *Frutticoltura* 7, 38–47.

Arrington, M. 2015. Novel methods for crop load management of pear cultivars in the Pacific Northwest. MS thesis. Oregon State University, Corvallis.

Arrington, M., Pasa, M. S. and Einhorn, T. C. 2017. Postbloom thinning response of 'Bartlett' pears to abscisic acid. *HortScience* 52(12), 1765–71. doi:10.21273/HORTSCI12429-17.

Asín, L., Alegre, S. and Montserrat, R. 2007. Effect of paclobutrazol, prohexadione-Ca, deficit irrigation, summer pruning and root pruning on shoot growth, yield, and return bloom, in a 'Blanquilla' pear orchard. *Sci. Hort.* 113(2), 142–8. doi:10.1016/j.scienta.2007.02.008.

Atkinson, D. 1980. The distribution and effectiveness of the roots of tree crops. In: Janick, J. (Ed.), *Horticultural Reviews* (vol. 2). AVI Publishing Co. Inc., Westport, CT, pp. 424–90.

Ayala, M. and Lang, G. 2018. Current season photoassimilate distribution in sweet cherry. *J. Amer. Soc. Hort. Sci.* 143(2), 110–7. doi:10.21273/JASHS04200-17.

Baldini, E., Rossi, F., Baraldi, R. and Marangoni, B. 1984. The availability of radiant energy in mechanically pruned pear trees. *Acta Hort.* 161(161), 201–10. doi:10.17660/ActaHortic.1984.161.16.

Ballard, J. K., Proebsting, E. L. and Tukey, R. B. 1993. Critical temperatures for blossom buds, pears. Extension Bulletin 0978. Washington State University, Pullman, WA.

Bangerth, F. 2000. Abscission and thinning of young fruit and their regulation by plant hormones and bioregulators. *Plant Growth Regul.* 31(1–2), 43–59.

Bangerth, F. 2005. A more holistic view on the hormonal interactions among different organs in a fruit tree. *Acta Hort.* 671(671), 25–36. doi:10.17660/ActaHortic.2005.671.1.

Barritt, B. H. 1992. *Intensive Orchard Management: A Practical Guide to the Planting, Establishment and Management of High Density Apple Orchards*. Good Fruit Grower, Yakima, WA.

Batjer, L. P. and Thompson, A. H. 1949. Effects of boric acid sprays during bloom upon the set of pear fruits. *Proc. Amer. Soc. Hort. Sci.* 53, 141–2.

Bell, R., van der Zwet, T., Castagnoli, S., Einhorn, T., Turner, J., Spotts, R., Moulton, G. A., Reighard, G. L. and Shane, W. W. 2014. 'Gem' pear. *HortScience* 49(3), 361–3.

Bepete, M. and Lakso, A. N. 1998. Differential effects of shade on early season fruit and shoot growth rates in 'Empire' apple. *HortScience* 33(5), 823–5. doi:10.21273/HORTSCI.33.5.823.

Bertelsen, M. 2005. Reflective mulch improves fruit size and flower bud formation of pear cv 'Clara Frijs'. *Acta Hort.* 671, 87–95.

Brewer, L. R. and Palmer, J. W. 2011. Global pear breeding programs: goals, trends and progress for new cultivars and new rootstocks. *Acta Hort.* 909, 105–20.

Brown, P. H. 2001. Transient nutrient deficiencies and their impact on yield – a rationale for foliar fertilizers? *Acta Hort.* 564(564), 217–23. doi:10.17660/ActaHortic.2001.564.24.

Browning, G. and Watkins, R. 1991. Preliminary evaluation of new quince (*Cydonia oblonga* Miller) hybrid rootstocks for pear. *J. Hort. Sci.* 66(1), 35–42. doi:10.1080/00221589.1991.11516122.

Brun, C., Raese, J. T. and Stahly, E. 1985. Seasonal response of 'Anjou' pear trees to different irrigation regimes: I. Soil moisture, water relations, tree and fruit growth. *J. Amer. Soc. Hort. Sci.* 110, 830–4.

Bubán, T. 2000. The use of benzyladenine in orchard fruit growing a mini review. *Plant Growth Regul.* 32(2–3), 381–90.

Bubán, T. and Faust, M. 1982. Flower bud induction in apple trees: internal control and differentiation. *Hort. Rev.* 4, 174–203.

Bukovac, M. J., Sabbatini, P. and Schwallier, P. G. 2006. Modifying alternate bearing of spur-type 'Delicious' apple with ethephon. *HortScience* 41(7), 1606–11. doi:10.21273/HORTSCI.41.7.1606.

Byers, R. E., Carbaugh, D. H., Presley, C. N. and Wolf, T. K. 1991. The influence of low light on apple fruit abscission. *HortScience* 66(1), 7–17. doi:10.1080/00221589.1991.11516119.

Cain, J. C. 1971. Effects of mechanical pruning of apple hedgerows with a slotting saw on light penetration and fruiting. *J. Amer. Soc. Hort. Sci.* 5, 664–7.

Callan, N. W. and Lombard, P. B. 1978. Pollination effects on fruit and seed development in 'Comice' pear. *J. Amer. Soc. Hort. Sci.* 103, 496–500.

Carra, B., Pasa, M. S., da Silva, C. P., do Amaranted, C. V. T., Steffens, C. A., Barnicki, V. A., Ciotta, M. N., Mello-Farias, P. C. and Einhorn, T. 2018. Early spring inhibition of ethylene synthesis increases fruit set and yield of 'Rocha' pear trees in Southern Brazil. *Sci. Hortic.* 232, 92–6. doi:10.1016/j.scienta.2017.12.062.

Carrera, M. 1982. Floral initiation in pear varieties. *Acta Hort.* 124(124), 157–64. doi:10.17660/ActaHortic.1982.124.20.

Caspari, H. W., Hossein, M. H. and Chalmers, D. J. 1994. Water use, growth, and fruit yield of 'Hosui' Asian pears under deficit irrigation. *J. Amer. Soc. Hort. Sci.* 119(3), 383–8. doi:10.21273/JASHS.119.3.383.

Castagnoli, S. 2008. Evaluation of pea pear pollinizers for *Pyrus communis*. *Acta Hort.* 800(800), 179–82. doi:10.17660/ActaHortic.2008.800.18.

Celton, J. M., Tustin, D. S., Chagne, D. and Gardiner, S. E. 2009. Construction of a dense genetic linkage map for apple rootstocks using SSRs developed from *Malus* ESTs and *Pyrus* genomic sequences. *Tree Genet. Genomes* 5(1), 93–107. doi:10.1007/s11295-008-0171-z.

Chalmers, D. J. 1989. A physiological examination of regulated deficit irrigation. *N.Z. J. Agric. Sci.* 23, 44–8.

Chalmers, D. J., Burge, G., Jerie, P. H. and Mitchell, P. D. 1986. The mechanism of regulation of 'Bartlett' pear fruit and vegetative growth by irrigation withholding and regulated deficit irrigation. *J. Amer. Soc. Hort. Sci.* 111, 904–7.

Chan, B. G. and Cain, J. C. 1967. The effect of seed formation on subsequent flowering in apple. *Proc. Amer. Soc. Hort. Sci.* 91, 63–8.

Chitu, V., Chitu, E. and Braniste, N. 2008. Effects of GA3 and paclobutrazol treatment on fruit set and yield of 'Beurré Bosc' and 'Triumf' pear cultivars. *Acta Hort.* 800(800), 163–8. doi:10.17660/ActaHortic.2008.800.15.

Cline, J. A. and Bakker, C. J. 2017. Prohexadione-calcium, ethepon, trinexapac-ethyl, and maleic hydrazide reduce extension growth of apple. *Can. J. Plant Sci.* 97, 457–65.

Cline, J. A., Carter, K., Gunter, A., Bakker, C. and Green, A. C. 2018. Response of Bosc and Cold Snap pears to thinning with NAA, 6-BA, ACC, and s-ABA. *Can. J. Plant Sci.* 98(4), 830–43. doi:10.1139/cjps-2017-0258.

Cody, D. A., Larsen, F. E. and Fritts Jr., R. 1985. Stimulation of lateral branch development in tree fruit nursery stock with GA_{4+7}+BA. *HortScience* 20, 758–9.

Corelli-Grapadelli, L., Musacchi, S. and Magnanini, E. 2001. Single-leaf and whole canopy gas exchange of pear as affected by graft incompatibility. *Acta Hortic.* 557(557), 377–84. doi:10.17660/ActaHortic.2001.557.50.

Costa, G., Andreotti, C., Bucchi, F., Sabatini, E., Bazzi, C., Malaguti, S. and Rademacher, W. 2001. Prohexadione-ca (Apogee®): growth regulation and reduced fire blight incidence in pear. *HortScience* 36(5), 931–3. doi:10.21273/HORTSCI.36.5.931.

Costa, G., Sabatini, E., Spinelli, F., Andreotti, C., Spada, G. and Mazzini, F. 2004. Prohexadione-Ca controls vegetative growth and cropping performance in pear. *Acta Hortic.* 563(653), 127–32. doi:10.17660/ActaHortic.2004.653.16.

Crisosto, C. H., Vasilakakis, M. D., Lombard, P. B., Richardson, D. G. and Tetley, R. 1986. Effect of ethylene inhibitors on fruit set, ovule longevity and polyamine levels in 'Comice' pear. *Acta Hort.* 179(179), 229–36. doi:10.17660/ActaHortic.1986.179.25.

Crisosto, C. H., Lombard, P. B., Richardson, D. G. and Tetley, R. 1992. Putrescine extends effective pollination period in 'Comice' pear (*Pyrus communis* L.) irrespective of post-anthesis ethylene levels. *Sci. Hortic.* 49(3–4), 211–21. doi:10.1016/0304-4238(92)90158-9.

Cristoferi, G., Meconcelli, M. and Schena, B. 1981. Controllo chimico dell'allegagione e della fruttificazione nel pero 'Decana del Comizio'. *Riv. Ortoflorofruttic. Ital.* 65, 309–18.

Curetti, M., Rodríguez, R., Magdalena, C. and Rodríguez, A. 2010. Effect of concentration, application volume and addition of a surfactant on response to benzyladenine as thinning agent in 'Williams' pears. *Acta Hort.* 909, 395–401.

Curetti, M., Sanchez, E., Tagliavini, M. and Gioacchini, P. 2013. Foliar applied urea at bloom improves early fruit growth and nitrogen status of spur leaves in pear trees, cv. Williams Bon Chretien. *Sci. Hort.* 150, 16–21. doi:10.1016/j.scienta.2012.10.022.

Curry, E. A. 1991. NAA-induced ethylene and ACC in 'Delicious' spur tissues: changes with temperature and time. *J. Amer. Soc. Hort. Sci.* 116(5), 846–50. doi:10.21273/JASHS.116.5.846.

Dallabetta, N. 2014. Effect of training systems and pruning methods on fruit quality in apple. PhD dissertation thesis. Alma Mater Studiorum- University of Bologna, Italy.

Davies, W. J.and Zhang, J. 1989. Chemical regulation of growth and physiology in drying soil: the case for abscisic acid. *Curr. Top. Plant Biochem. Phys.* 8, 100–9.

Deckers, T. and Daemen, E. 1998. Pear growing in Belgium: production and problems. *Acta Hort.* 475, 49–58.

Deckers, T. and Schoofs, H. 2002. Improvement of fruit set on young pear trees cultivar conference with gibberellins. *Acta Hort.* 596(596), 735–43. doi:10.17660/ActaHortic.2002.596.126.

De Franceschi, P., Dondini, L. and Sanzol, J. 2012. Molecular bases and evolutionary dynamics of self-incompatibility in the Pyrinae (Rosaceae). *J. Exp. Bot.* 63(11), 4015–32. doi:10.1093/jxb/ers108.

Dejong, T. M. 2007. Canopy and light management. In: Mitcham, E. J. and Elkins, R. B. (Eds), *Pear Production and Handling Manual*. University of California, Agriculture and Natural Resources, Oakland.

De Maeyer, L. and Deckers, J. C. 1984. Flower bud quality in the pear cultivar Doyenné du Comice: 1982-1983 trials. *Acta Hort.* 149(149), 153-60. doi:10.17660/ActaHortic.1984.149.18.

Dennis, F. G. 2000. The history of fruit thinning. *Plant Growth Reg.* 31, 1-16.

Dussi, M. C. 2011. Sustainable use of plant bioregulators in pear production. *Acta Hortic.* 909(909), 353-67. doi:10.17660/ActaHortic.2011.909.40.

Dussi, M. C. and Huysamer, M. 1995. Severe postharvest summer pruning of mature 'Forelle' pear trees influences canopy light distribution and fruit and spur leaf characteristics in the following season. *J. S. Afr. Soc. Hort. Sci.* 5, 57-60.

Dussi, M. C. and Sugar, D. 2011. Fruit thinning and fruit size enhancement with 6-benzyladenine application to 'Williams' pear. *Acta Hortic.* 909(909), 403-7. doi:10.17660/ActaHortic.2011.909.45.

Duyvelshoff, C. and Cline, J. A. 2013. Ethephon and prohexadione-calcium influence the flowering, early yield, and vegetative growth of young 'Northern Spy' apple trees. *Sci. Hort.* 151, 128-34. doi:10.1016/j.scienta.2012.12.002.

Edgerton, L. J. 1971. Apple abscission. *HortScience* 6, 378-82.

Einhorn, T. C. and Arrington, M. 2018. ABA and shading induce 'Bartlett' pear abscission and inhibit photosynthesis but are not additive. *J. Plant Growth Regul.* 37(1), 300-8. doi:10.1007/s00344-017-9729-z.

Einhorn, T. C., Turner, J., Gibeaut, D. and Postman, J. D. 2011. Characterization of cold hardiness in quince: potential pear rootstock candidates for northern pear production regions. *Acta Hortic.* 909(909), 137-43. doi:10.17660/ActaHortic.2011.909.13.

Einhorn, T. C., Turner, J. and Laraway, D. 2012. Effect of reflective fabric on yield of mature 'd'Anjou' pear trees. *HortScience* 47(11), 1580-5. doi:10.21273/HORTSCI.50.11.1580.

Einhorn, T., Castagnoli, S., Smith, T., Turner, J. and Mielke, E. 2013. Summary of the 2002 Pacific Northwest Pear Rootstock Trials: performance of 'd'Anjou' and 'Golden Russet Bosc' pear on eight *Pyrus* rootstocks. *J. Amer. Pom. Soc.* 67(2), 80-8.

Einhorn, T. C., Pasa, M. S. and Turner, J. 2014. 'D'Anjou' pear shoot growth and return bloom, but not fruit size, are reduced by prohexadione-calcium. *HortScience* 49(2), 180-7. doi:10.21273/HORTSCI.49.2.180.

Elfving, D. C., Lombardini, L., McFerson, J. R., Drake, S. R., Faubion, D. F., Auvil, T. D., Van Ee, G. and Visser, D. B. 2003a. Effects of directed applications of prohexadione-calcium to tops of mature pear trees on shoot growth, light penetration, pruning and fruit quality. *J. Amer. Pom. Soc.* 57(2), 45-57.

Elfving, D. C., Lang, G. A. and Visser, D. B. 2003b. Prohexadione-ca and ethephon reduce shoot growth and increase flowering in young vigorous sweet cherry trees. *HortScience* 38(2), 293-8. doi:10.21273/HORTSCI.38.2.293.

Elkins, R., Bell, R. and Einhorn, T. 2012. Needs assessment for future US pear rootstock research directions based on the current state of pear production and rootstock research. *J. Amer. Pom. Soc.* 66(3), 153-63.

Ermel, F. F., Kervella, J., Catesson, A. M. and Poëssel, J. L. 1999. Localized graft incompatibility in pear/quince (*Pyrus communis/Cydonia oblonga*) combinations: multivariate analysis of histological data from 5-month-old grafts. *Tree Physiol.* 19(10), 645-54. doi:10.1093/treephys/19.10.645.

Espen, L., Cocucci, M. and Sacchi, G. A. 2005. Differentiation and functional connection of vascular elements in compatible and incompatible pear/quince internode micrografts. *Tree Physiol.* 25(11), 1419-25. doi:10.1093/treephys/25.11.1419.

Evans, J. R., Evans, R. R., Regusci, C. L. and Rademacher, W. 1999. Mode of action, metabolism, and uptake of BAS 125W, prohexadione-calcium. *HortScience* 34(7), 1200–1. doi:10.21273/HORTSCI.34.7.1200.

Fazio, G., Wan, Y., Kviklys, D., Romero, L., Adams, R., Strickland, D. and Robinson, T. 2014. *Dw2* a new dwarfing locus in apple rootstocks and its relationship to induction of early bearing in apple scions. *J. Amer. Soc. Hort. Sci.* 139(2), 87–98. doi:10.21273/JASHS.139.2.87.

Ferree, D. C. 1989. Growth and carbohydrate distribution of young apple trees in response to root pruning and tree density. *Hort.Sci.* 24(1), 62–5.

Ferree, D. C. 1990. Pros and cons of root pruning. *Compact Fruit Tree* 23, 100–1.

Ferree, D. C. 1992. Time of root pruning influences vegetative growth, fruit size, biennial bearing, and yield of 'Jonathan' apple. *J. Amer. Soc. Hort. Sci.* 117(2), 198–202. doi:10.21273/JASHS.117.2.198.

Ferree, D. C., Myers, S. C. and Schupp, J. R. 1992. Root pruning and root restriction of fruit trees-current review. *Acta Hort.* 322(322), 153–66. doi:10.17660/ActaHortic.1992.322.17.

Flaishman, M., Amihai Shargal, A. S. and Raphael, S. 2001. The synthetic cytokinin CPPU increases fruit size and yield of 'Spadona' and 'Costia' pear (*Pyrus communis* L.). *J. Hort. Sci. Biotech.* 76(2), 145–9. doi:10.1080/14620316.2001.11511341.

Flaishman, M. A., Shargal, A., Shlizerman, L. and Lev-Yadun, S. 2005. The synthetic cytokinins CPPU and TDZ prolong the phase of cell division in developing pear (*Pyrus communis* L.) fruit. *Acta Hort.* 671, 151–7.

Forshey, C. G., Elfving, D. C. and Stebbins, R. L. 1992. *Training and Pruning Apple and Pear Trees*. American Society for Horticultural Science, St. Joseph, MI.

Foster, T., Johnston, R. and Seleznyova, A. 2003. A morphological and quantitative characterization of early floral development in apple (*Malus* × *Domestica* Borkh.). *Ann. Bot.* 92(2), 199–206. doi:10.1093/aob/mcg120.

Foster, T. M., Celton, J. M., Chagne, D., Tustin, D. S. and Gardiner, S. E. 2015. Two quantitative trait loci, *Dw1* and *Dw2* are primarily responsible for rootstock-induced dwarfing in apple. *Hort. Res.* 2, 15001; doi:10.1038/hortres.2015.1.

Gallardo, R. K., Grant, K., Brown, D. J., McFerson, J. R., Lewis, K. M., Einhorn, T. and Sazo, M. M. 2019. U.S. fresh apple industry perceptions of precision agriculture technologies. *HortTechnology* 29(2), 151-62. doi:10.21273/HORTTECH04214-18.

Garriz, P. I., Alvarez, H. L. and Alvarez, A. J. 1997. Influence of altered irradiance on fruits and leaves of mature pear trees. *Biol. Plant.* 39(2), 229–34. doi:10.1023/A:1000344920479.

Garriz, P. I., Colavita, G. M. and Alvarez, H. L. 1998. Fruit and spur leaf growth and quality as influenced by low irradiance levels in pear. *Sci. Hort.* 77(3–4), 195–205. doi:10.1016/S0304-4238(98)00182-4.

Geisler, D. and Ferree, D. 1984. Response of plants to root pruning. *Hort. Rev.* 6, 155–88.

Gil, G. F., Griggs, W. H. and Martin, G. C. 1972. Gibberellin-induced parthenocarpy in 'Winter Nelis' pear. *HortScience* 7, 559-61.

Gimenez, G., Reeb, P., Dussi, M. C., Elosegui, F., Siviero, P., Fantaguzzi, S. and Sugar, D. 2009. Optimizing benzyladenine application timing in 'Williams' pear. *Acta Hort.* 884, 265–72.

Goffinet, M. C., Robinson, T. L. and Lakso, A. N. 1995. A comparison of 'Empire' apple fruit size and anatomy in unthinned and hand-thinned trees. *J. Hort. Sci.* 70(3), 375–87. doi:10.1080/14620316.1995.11515307.

Goldway, M., Takasaki, T., Sanzol, J., Mota, M., Zisovich, A. H., Stern, R. A. and Sansavini, S. 2009. Renumbering the S-RNase alleles of European pears (*Pyrus communis* L.) and cloning the S109 RNase allele. *Sci. Hort.* 119, 417-22.

Goren, R., Huberman, M. and Goldschmidt, E. E. 2004. Girdling: physiological and horticultural aspects. *Hort. Rev.* 30, 1-35.ü.

Greene, D. W. 2012. Influence of abscisic acid and bezyladenine on fruit set and fruit quality of 'Bartlett' pears. *HortScience* 47(11), 1607-11. doi:10.21273/HORTSCI.47.11.1607.

Griggs, W. H. 1953. Pollination requirements of fruit and nuts. Calif. Agr. Expt. Sta. Ext. Serv. Circ. 423.

Griggs, W. H. and Iwakiri, B. T. 1961. Effect of gibberellin and 2,4,5-trichlorophenoxypropionic acid sprays on 'Bartlett' pear trees. *Proc. Amer. Soc. Hort. Sci.* 77, 73-89.

Griggs, W. H., Martin, G. C. and Iwakiri, B. T. 1970. The effect of seedless versus seeded fruit development on flower bud formation in pear. *J. Amer. Soc. Hort. Sci.* 95, 243-24.

Guak, S. and Fuchigami, L. H. 2001. Effects of applied ABA on growth cessation, bud dormancy, cold acclimation, leaf senescence and N mobilization in apple nursery plants. *J. Hortic. Sci. Biotechnol.* 76(4), 459-64. doi:10.1080/14620316.2001.1151 1394.

Guak, S. and Fuchigami, L. H. 2002. Foliar applications of urea or ABA affect growth cessation, leaf senescence and abscission, cold acclimation and levels of reserve nitrogen and carbohydrates in nitrogen-treated apple nursery plants. *J. Hortic. Sci. Biotechnol.* 77(2), 137-42. doi:10.1080/14620316.2002.11511469.

Hansen, P. 1971. 14C-studies on apple trees. VII. The early seasonal growth in leaves, flowers and shoots as dependent upon current photosynthates and existing reserves. *Physiol. Plant.* 25(3), 469-73. doi:10.1111/j.1399-3054.1971.tb01475.x.

Hanson, E. J. 1991. Movement of boron out of tree fruit leaves. *HortScience* 26(3), 271-3. doi:10.21273/HORTSCI.26.3.271.

Hanson, E. J. and Breen, P. J. 1985. Xylem differentiation and boron accumulation in 'Italian' prune flower buds. *J. Amer. Soc. Hort. Sci.* 110, 566-70.

Harley, C. P., Moon, H. H. and Regeimbal, L. O. 1958. Evidence that postbloom apple thinning sprays of naphthaleneacetic acid increase blossom bud formation. *Proc. Amer. Soc. Hort. Sci.* 72, 52-6.

Higgins, S. S., Larsen, F. E., Bendel, R. B., Radamaker, G. K., Bassman, J. H., Bidlake, W. R. and Wir, A. A. 1992. Comparative gas exchange characteristics of potted, glasshouse-grown almond, apple, fig, grape, olive, peach and Asian pear. *Sci. Hort.* 52(4), 313-29. doi:10.1016/0304-4238(92)90032-8.

Hoad, G. V. 1978. The role of seed derived hormones in the control of flowering in apple. *Acta Hortic.* 80(80), 93-104. doi:10.17660/ActaHortic.1978.80.14.

Hoad, G. V. 1984. Hormonal regulation of fruit-bud formation in fruit trees. *Acta Hort.* 149, 13-24. doi:10.17660/ActaHortic.1984.149.1.

Iglesias, I. and Batlle, I. 2011. Agronomical performance and fruit quality of 'Conference' on some *Cydonia* and *Pyrus* rootstocks. *Acta Hort.* 909(909), 195-200. doi:10.17660/ActaHortic.2011.909.20.

Irisarri, P., Binczycki, P., Errea, P., Martens, H. J. and Pina, A. 2015. Oxidative stress associated with rootstock-scion interactions in Pear/Quince combinations during early stages of graft development. *J. Plant Physiol.* 176, 25-35. doi:10.1016/j.jplph.2014.10.015.

Ito, A., Yaegaki, H., Hayama, H., Kusaba, S., Yamaguchi, I. and Yoshioka, H. 1999. Bending shoots stimulates flowering and influences hormone levels in lateral buds of Japanese pear. *HortScience* 34(7), 1224–8. doi:10.21273/HORTSCI.34.7.1224.

Jacob, H. B. 2002. New pear rootstocks from Geisenheim, Germany. *Acta Hort.* 596(596), 337–44. doi:10.17660/ActaHortic.2002.596.52.

Jacyna, T. 1996. Induction of lateral branching in nursery pear and apple trees with plant growth regulators. *Fruit Var. J.* 50, 151–6.

Janssens, P., Deckers, T., Elsen, F., Elsen, A., Schoofs, H., Verjans, W. and Vandendriessche, H. 2011. Sensitivity of root pruned 'Conference' pear to water deficit in a temperate climate. *Agric. Water Manag.* 99(1), 58–66. doi:10.1016/j.agwat.2011.07.018.

Jerie, P. H., Mitchell, P. D. and Goodwin, I. 1989. Growth of Williams' Bon Chretien pear fruit under regulated deficit irrigation (RDI). *Acta Hort.* 240(240), 271–4. doi:10.17660/ActaHortic.1989.240.50.

Johnson, D., Evans, K., Spencer, J., Webster, T. and Adam, S. 2005. Orchard comparisons of new quince and *Pyrus* rootstock clones. *Acta Hort.* 671(671), 201–7. doi:10.17660/ActaHortic.2005.671.26.

Kappel, F. and Neilsen, G. H. 1994. Relationship between light microclimate, fruit growth, fruit quality, specific leaf weight and N and P content of spur leaves of 'Bartlett' and 'Anjou' pear. *Sci. Hort.* 59(3–4), 187–96. doi:10.1016/0304-4238(94)90012-4.

Kemp, H., Koskela, E., van Dieren, M. C. A. and Maas, F. 2008. Selected *Pyrus* genotypes as pollinizers for *Pyrus communis* cultivars. *Acta Hort.* 800, 189–97.

Ketchie, D. O., Fairchild, E. D. and Drake, F. R. 1996. Viability of different pear pollen and the effect on fruit set of 'Anjou' pear (*Pyrus communis* L.). *Fruit Var. J.* 50, 118–24.

Khan, Z. U., McNeil, D. L. and Samad, A. 1998. Root pruning of apple trees grown at ultra-high density affects carbohydrate reserves distribution in vegetative and reproductive growth. *N.Z. J. Crop and Hort. Sci.* 26, 291–7.

Khemira, H., Lombard, P. B., Sugar, D. and Azarenko, A. N. 1993. Hedgerow orientation affects canopy exposure, flowering, and fruiting of 'Anjou' pear trees. *HortScience* 28(10), 984–7. doi:10.21273/HORTSCI.28.10.984.

Knoche, M. and Petracek, P. D. 2014. Foliar uptake of PGRs: barriers, mechanisms, model systems, and factors. *Acta Hort.* 1042, 125–41. doi:10.17660/ActaHortic.2014.1042.16.

Kviklys, D. and Kvikliene, N. 2004. Pear rootstock effect on growth, productivity and fruit internal quality. *Acta Hortic.* 568(658), 359–63. doi:10.17660/ActaHortic.2004.658.52.

Lafer, G. 2008. Effects of different bioregulator applications on fruit set, yield and fruit quality of 'Williams' pears. *Acta Hortic.* 800(800), 183–8. doi:10.17660/ActaHortic.2008.800.19.

Lakso, A. N. 1994. Apple. In: Schaffer, B. and Anderson, P. C. (Eds), *Handbook of Environmental Physiology of Fruit Crops vol. 1: Temperate Crops*. CRC Press, Boca Raton.

Lakso, A. N. 2011. Early fruit growth and drop – the role of carbon balance in the apple tree. *Acta Hort.* 903(903), 733–42. doi:10.17660/ActaHortic.2011.903.102.

Lakso, A. N., Robinson, T. L. and Greene, D. W. 2006. Integration of environment, physiology and fruit abscission via carbon balance modeling: implications for understanding growth regulator responses. *Acta Hort.* 727(727), 321–6. doi:10.17660/ActaHortic.2006.727.38.

Lawes, G. S., Spence, C. B., Tustin, D. S. and Max, S. M. 1997. Tree quality and canopy management effects on the growth and floral precocity of young 'Doyenne Du

Comice' pear trees. *N. Z. J. Crop Hortic. Sci.* 25(2), 177-84. doi:10.1080/01140671.1997.9514004.

Lee, S. H., Kim, W. S. and Han, T. H. 2009. Effects of post-harvest foliar boron and calcium applications on subsequent season's pollen germination and pollen tube growth of pear (*Pyrus pyrifolia*). *Sci. Hort.* 122(1), 77-82. doi:10.1016/j.scienta.2009.04.006.

Le Lezec, M., Belouin, A. and Simard, M. H. 2005. A selection from *Pyrus betulaefolia* as a new pollinator for the main *Pyrus communis* cultivars. *Acta Hort.* 671(671), 253-5. doi:10.17660/ActaHortic.2005.671.36.

Lombard, P. B. and Richardson, D. G. 1982. Increase fruit set and cropping of `Comice' pear trees with an ethylene inhibitor, amino-ethoxyvinylglycine. *Acta Hort.* 124(124), 165-70. doi:10.17660/ActaHortic.1982.124.21.

Lombard, P. B. and Westwood, M. N. 1987. Pear rootstocks. In: Rom, R. and Carlson, R. (Eds), *Rootstocks for Fruit Crops*. John Wiley & Sons, New York.

Lombard, P. B., Westwood, M. N. and Thompson, M. M. 1971. Effective pollination-the facts of life behind uniform pear cropping. *Proc. Oregon Hort. Soc.* 62, 31-6.

Lombard, P. B., Williams, R. R., Stott, K. G. and Jeffries, C. J. 1972. Temperature effects on pollen tube growth in styles of Williams' pear with a note on pollination deficiencies of 'Comice' pear. In: *Compte rendue du symposium 'Culture du Poirier'*, pp. 265-79.

Lombard, P. B., Stebbins, R. L. and Westwood, M. N. 1984. Related genera and Pyrus species for pear rootstock to control size and yield. *Acta Hort.* 146, 197-202. doi:10.17660/ActaHortic.1984.146.21.

Lombard, P. B., Callan, N. W., Dennis, F. G., Jr., Looney, N. E., Martin, G. C., Renquist, A. R. and Mielke, E. A. 1988. Towards a standardized nomenclature, procedures, values, and units in determining fruit and nut tree yield performance. *HortScience* 23(5), 813-7.

Lordan, J., Robinson, T. L., Sazo, M. M., Cowgill, W., Black, B. L., Huffman, L., Grigg-McGuffin, K., Francescatto, P. and McArtney, S. 2017. Use of plant growth regulators for feathering and flower suppression of apple nursery trees. *HortScience* 52(8), 1080-91. doi:10.21273/HORTSCI11918-17.

Maas, F. 2005. Shoot growth, fruit production and return bloom in 'conference' and 'Doyenné Du Comice' treated with regalis (prohexadione-calcium). *Acta Hort.* 671, 517-24.

Maas, F. 2008. Evaluation of *Pyrus* and quince rootstocks for high-density pear orchards. *Acta Hort.* 800, 599-610.

Maas, F. 2015. Evaluation of yield efficiency and winter hardiness of quince rootstocks for 'conference' pear. *Acta Hort.* 1094, 93-102.

Maas, F. M., Kanne, H. J. and van der Steeg, P. A. H. 2010. Chemical thinning of 'Conference' pears. *Acta Hort.* 884, 293-304. doi:10.17660/ActaHortic.2010.884.34.

Malladi, A. and Johnson, L. K. 2011. Expression profiling of cell cycle genes reveals key facilitators of cell production during carpel development, fruit set, and fruit growth in apple (*Malus* × *domestica* Borkh.). *J. Exp. Bot.* 62(1), 205-19. doi:10.1093/jxb/erq258.

Marodin, G., Koller, O. C., Bender, R. J., Bergamaschi, H., Camelatto, D. and Kretzschmar, A. A. 2008. Index of bud abortion of 'Packham's Triumph' pear trees in the production areas of the Rio Grande do Sul state and Santa Catarina state. *Acta Hort.* 800(800), 215-20. doi:10.17660/ActaHortic.2008.800.24.

Marsal, J., Rapoport, H. F., Manrique, T. and Girona, J. 2000. Pear fruit growth under regulated deficit irrigation in container-grown trees. *Sci. Hort.* 85(4), 243-59. doi:10.1016/S0304-4238(99)00151-X.

Marsal, J., Mata, M., Arbonès, A., Rufat, J. and Girona, J. 2002. Water stress limits for vegetative and reproductive growth of 'Barlett' pears. *Acta Hort.* 596, 659–63.

Marsal, J., Mata, M., Arbones, A., Del Campo, J., Girona, J. and Lopez, G. 2008. Factors involved in alleviating water stress by partial crop removal in pear trees. *Tree Physiol.* 28(9), 1375–82. doi:10.1093/treephys/28.9.1375.

Martin, M. M., Larsen, F. E., Higgins, S. S., Ku, M. S. B. and Andrews, P. K. 1997. Comparative growth and physiology of selected one-year-old, red and green-fruited European pear cultivars. *Sci. Hort.* 71(3–4), 213–26. doi:10.1016/S0304-4238(97)00100-3.

McArtney, S. J. 2002. Ethylene evolution from detached apple spurs in response to chemical thinners. *HortScience* 37(4), 662–5. doi:10.21273/HORTSCI.37.4.662.

McArtney, S. J. and Belton, R. P. 1992. Apple shoot growth and cropping responses to root pruning. *N. Z. J. Crop Hort. Sci.* 20(4), 383–90.

McArtney, S., Unrath, D., Obermiller, J. D. and Green, A. 2007. Naphthaleneacetic acid, ethephon, and gibberellin A4 + A7 have variable effects on flesh firmness and return bloom of apple. *HortTechnology* 17, 32–8. doi:10.21273/HORTTECH.17.1.32.

McArtney, S. J., Obermiller, J. D. and Arellano, C. 2012. Comparison of the effects of metamitron on chlorophyll fluorescence and fruit set in apple and peach. *HortScience* 47(4), 509–14. doi:10.21273/HORTSCI.47.4.509.

McArtney, S., Greene, D., Schmidt, T. and Yuan, R. 2013. Naphthaleneacetic acid and ethephon are florigenic in the biennial apple cultivars Golden Delicious and York Imperial. *HortScience* 48, 742–6.

McArtney, S. J., Abrams, S. R., Woolard, D. D. and Petracek, P. D. 2014. Effects of S-abscisic acid and (+)-8′-acetylene abscisic acid on fruit set and stomatal conductance in apple. *HortScience* 49(6), 763–8.

McGrath, M. J., Koczan, J. M., Kennelly, M. M. and Sundin, G. W. 2009. Evidence that prohexadione-calcium induces structural resistance to fire blight infection. *Phytopathology* 99(5), 591–6. doi:10.1094/PHYTO-99-5-0591.

Mellenthin, W. M., Wang, C. Y. and Wang, S. Y. 1972. Influence of temperature on pollen tube growth and initial fruit development in 'd'Anjou' pear. *HortScience* 7, 557–9.

Mesa, K., Serra, S., Masia, A., Gagliardi, F. and Bucci, D. 2016. Seasonal trends of starch and soluble carbohydrates in fruits and leaves of 'Abbé Fétel' pear trees and their relationship to fruit quality parameters. *Sci. Hort.* 211, 60–9.

Miller, S. S. 1988. Plant bioregulators in apple and pear culture. *Hort. Rev.* 10, 309–401.

Mitchell, P. D., Jerie, P. H. and Chalmers, D. J. 1984. The effects of regulated water deficits on pear tree growth, flowering, fruit growth, and yield. *J. Amer. Soc. Hort. Sci.* 109, 544–7.

Mitchell, P. D., Chalmers, D. J., Jerie, P. H. and Burge, G. 1986. The use of initial withholding of irrigation and tree spacing to enhance the effect of regulated deficit irrigation on pear trees. *J. Am. Soc. Hort. Sci.* 111, 858–61.

Mitchell, P. D., van den Ende, B., Jerie, P. H. and Chalmers, D. J. 1989. Responses of 'Bartlett' pear to withholding irrigation, regulated deficit irrigation, and tree spacing. *J. Am. Soc. Hort. Sci.* 114, 5–19.

Moriya, Y., Yamamoto, K., Okada, K., Iwanami, H., Bessho, H., Nakanishi, T. and Takasaki, T. 2007. Development of a CAPS marker system for genotyping European pear cultivars harboring 17 S alleles. *Plant. Cell. Rep.* 26(3), 345–54. doi:10.1007/s00299-006-0254-y.

Morrison Baird, L. A. and Webster, B. D. 1979. The anatomy and histochemistry of fruit abscission. *Hort. Rev.* 1, 172–203.

Mota, M., Tavares, L. and Oliveira, C. M. 2007. Identification of S-alleles in pear (*Pyrus communis* L) cv. 'Rocha' and other European cultivars. *Sci. Hort.* 113, 13–9.

Müller, D. and Leyser, O. 2011. Auxin, cytokinin and the control of shoot branching. *Ann. Bot.* 107(7), 1203–12. doi:10.1093/aob/mcr069.

Musacchi, S. 2006. Recenti innovazioni dell'impiantistica e della tecnica colturale del pero. *Italus Hortus* 13(6), 24–31.

Musacchi, S. 2008. Bibaum: a new training system for pear orchards. *Acta Hort.* 800, 763–9.

Musacchi, S. 2011. Training system and management for a high density orchard of 'Abbé Fetel'. *Acta Hortic.* 909(909), 225–40. doi:10.17660/ActaHortic.2011.909.25.

Musacchi, S. 2015. https://www.youtube.com/watch?v=5h5aQ5DwYOo.

Musacchi, S. and Greene, D. 2017. Innovations in apple tree cultivation to manage crop load and ripening. In: Evans, K. (Ed.), *Achieving Sustainable Cultivation of Apples*. Burleigh Dodds Science Publishing, Cambridge, UK, pp. 195–237.

Musacchi, S., Serra, S. and Ancarani, V. 2011. Comparison among pear training systems and rootstocks for high density planting (HDP) of the cultivar 'Abbé Fetel'. *Acta Hort.* 909, 251–8.

Myers, S. C. and Ferree, D. C. 1983. Influence of time of summer pruning and limb orientation on growth and flowering of 'Delicious' apple trees. *J. Am. Soc. Hort. Sci.* 108, 634–8.

Neilsen, D., Millard, P., Neilsen, G. H. and Hogue, E. J. 1997. Sources of N for leaf growth in a high-density apple (*Malus domestica*) orchard irrigated with ammonium nitrate solution. *Tree Physiol.* 17(11), 733–9. doi:10.1093/treephys/17.11.733.

Nocito, F. F., Espen, L., Fedeli, C., Lancilli, C., Musacchi, S., Serra, S., Sansavini, S., Cocucci, M. and Sacchi, G. A. 2010. Oxidative stress and senescence-like status of pear calli co-cultured on suspensions of incompatible quince microcalli. *Tree Physiol.* 30(4), 450–8. doi:10.1093/treephys/tpq006.

O'Connell, M. and Goodwin, I. 2007. Water stress and reduced fruit size in micro-irrigated pear trees under deficit partial rootzone drying. *Aust. J. Agric. Res.* 85(7), 670–9.

Olenz, H. and Zielinski, Q. B. 1965. Pollen incompatibility between Bartlett and Seckel pears. *Proc. Amer. Soc. Hort. Sci.* 87, 117–22.

Pacheco, A., Azevedo, J. and Oliveira, C. M. 2011. Water potential and fruit growth as management indicators in trickle-irrigation in 'Rocha' pear. *Acta Hort.* 909, 331–7.

Palmer, J. W., Seymour, S. M. and Diack, R. 2011. Feathering of 'Doyenné Du Comice' pear in the nursery using repeat sprays of benzyladenine and gibberellins. *Sci. Hort.* 130(2), 393–7.

Pasa, M. and Einhorn, T. 2014. Heading cuts and prohexadione-calcium affect the growth and development of 'd'Anjou' pear shoots in a high-density orchard. *Sci. Hortic.* 168, 267–71.

Pasa, M. S. and Einhorn, T. 2017. Prohexadione calcium on shoot growth of 'Starkrimson' pear trees. *Pesq. Agropec. Bras.* 52(2), 75–83.

Picchioni, G. A. and Weinbaum, S. A. 1995. Retention and the kinetics of uptake and export of foliage-applied, labeled boron by apple, pear, prune, and sweet cherry leaves. *J. Amer. Soc. Hort. Sci.* 120(1), 28–35.

Plooy, P., Sadie, A., Jacobs, G. and Cook, N. 2002. Branching habit of 2-year-old pear branches classified on the basis of length and position of 1-year-old laterals. *Sci. Hort.* 95(3), 193–201.

Poni, S., Tagliavini, D., Neri, D., Scudellari, D. and Toselli, M. 1992. Influence of root pruning and water stress on growth and physiological factors of potted apple, grape, peach and pear trees. *Sci. Hort.* 52, 223–36.

Quamme, H. A. 1976. Relationship of the low temperature exotherm to apple and pear production in North America. *Can. J. Plant Sci.* 56(3), 493–500.

Quamme, H. A. and Hampson, C. L. 2004. Winter hardiness measurements on 15 new apple cultivars. *J. Amer. Pom. Soc.* 58, 98–107.

Quamme, H. A., Stushnoff, C. and Weiser, C. J. 1972. The relationship of exotherms to cold injury in apple stem tissues. *J. Amer. Soc. Hort. Sci.* 97, 608–13.

Quinet, M. and Jacquemart, A. L. 2015. Difference between pollination and parthenocarpy in the 'conference' pear production. *Acta Hortic.* 1094(1094), 359–66. doi:10.17660/ActaHortic.2015.1094.45.

Quinet, M., Kelecom, S., Raspé, O. and Jacquemart, A. 2014. S-genotype characterization of 13 Northwestern European pear (*Pyrus communis*) cultivars. *Sci. Hort.* 165, 1–4.

Rademacher, W. 1991. Biochemical effects of plant growth retardants. In: Gausman, H. W. (Ed.), *Plant Biochemical Regulators*. Marcel Dekker, New York.

Rademacher, W. 2000. Growth retardants: effects on gibberellin biosynthesis and other metabolic pathways. *Annu. Rev. Plant Physiol. Plant Mol. Biol.* 51, 501–31. doi:10.1146/annurev.arplant.51.1.501.

Rademacher, W., van Saarloos, K., Garuz Porte, J. A., Riera Forcades, F., Senechal, Y., Andreotti, C., Spinelli, F., Sabatini, E. and Costa, G. 2004. Impact of prohexadione-ca on the vegetative and reproductive performance of apple and pear trees. *Europ. J. Hort. Sci.* 69(6), 221–8.

Rajashekar, C., Westwood, M. N. and Burke, M. J. 1982. Deep supercooling and cold hardiness in genus *Pyrus*. *J. Amer. Soc. Hort. Sci.* 107, 968–72.

Reynolds, L. P., Jacobs, G. and Theron, K. I. 2005a. The effect of scoring during flower induction or the initiation phase on return bloom in *Pyrus communis* L. *Acta Hortic.* 671(671), 171–6. doi:10.17660/ActaHortic.2005.671.21.

Reynolds, L. P., Jacobs, G. and Theron, K. I. 2005b. Reproductive bud development in pears (*Pyrus communis* L.). with the emphasis on bourse shoot. *Acta Hort.* 671, 451–4.

Robbie, F. A., Atkinson, C. J., Knight, J. N. and Moore, K. G. 1993. Branch orientation as a factor determining fruit set in apple trees. *HortScience* 68(3), 317–25.

Robinson, T. L. 2011. High density pear production with *Pyrus communis* rootstocks. *Acta Hort.* 909, 259–69.

Robinson, T. L. and Dominguez, L. 2015. Yield and profitability of high-density pear production with *Pyrus* rootstocks. *Acta Hort.* 1094, 247–56.

Rodriguez, A., Sanchez, E. and de la Casa, A. 2011. Contributions of early season temperatures to *Pyrus communis* 'Bartlett' fruit growth. *Acta Hort.* 909, 657–64.

Rudell, D., Serra, S., Sullivan, N., Mattheis, J. and Musacchi, S. 2017. Survey of 'd'Anjou' pear metabolic profile following harvest from different canopy positions and fruit tissues. *HortScience* 52, 1501–10.

Sakai, A. and Larcher, W. 1987. *Frost Survival of Plants*. Springer-Verlag, Berlin.

Salisbury, F. B. and Ross, C. W. 1992. *Plant Physiology*. Wadsworth, Belmont, CA.

Sánchez, E. E. 2002. Nitrogen nutrition in pear orchard. *Acta Hort.* 596, 653–7.

Sánchez, E. E. 2015. Nutrition and water management in intensive pear growing. *Acta Hort.* 1094, 307–16.

Sánchez, E. E. and Righetti, T. L. 1990. Tree nitrogen status and leaf canopy position influence postharvest nitrogen accumulation and efflux from pear leaves. *J. Amer. Soc. Hort. Sci.* 115, 934–7.

Sánchez, E. E., Righetti, T. L., Sugar, D. and Lombard, P. 1990a. Seasonal differences and soil texture alter uptake of newly absorbed nitrogen in field-grown pear trees. *J. Hort. Sci.* 65, 395–400.

Sánchez, E. E., Righetti, T. L., Sugar, D. and Lombard, P. 1990b. Response of 'Comice' pear trees to a postharvest urea spray. *J. Hort. Sci.* 65, 541–6.

Sánchez, E. E., Righetti, T. L., Sugar, D. and Lombard, P. B. 1991. Recycling of nitrogen in field grown 'Comice' pears. *J. Hort. Sci.* 66(4), 479–86. doi:10.1080/00221589.1991.11516177.

Sánchez, E. E., Righetti, T. L., Sugar, D. and Lombard, P. 1992. Effects of timing of nitrogen application on nitrogen partitioning between vegetative, reproductive, and structural components of mature 'Comice' pears. *J. Hort. Sci.* 67, 51–8.

Sánchez, E. E., Khemira, H., Sugar, D. and Righetti, T. L. 1995. Nitrogen management in orchards. In: Bacon, P. (Ed.), *Nitrogen Fertilization in the Environment*. Marcel Dekker, Inc., New York, pp. 327–80.

Sanchez, E., Righetti, T. and Sugar, D. 1998. Partitioning and recycling of fall applied boron in comice pears. *Acta Hort.* 475(475), 347–54. doi:10.17660/ActaHortic.1998.475.41.

Sanchez, E., Curetti, M. and Retamales, J. 2011. Effect of AVG applications on fruit set, yield and fruit size in 'Abate Fetel' and 'Packham's Triumph' pears in a semi-commercial statistical trial. *Acta Hort.* 909, 135–440.

Sansavini, S. 1966. Caratteristiche produttive dei rami a frutto nelle diverse cultivar di pero. *Riv. Ortoflorofruttic. It.* 2, 153–71.

Sansavini, S. 1974. Mechanic pruning. The first findings of a three year research on pear, apple and peach trees. [Italian] Italia Agricola.

Sansavini, S. 1982. High density orchards of various crops. In: *Proceedings of the XXI International Hort. Congress*, pp. 182–97.

Sansavini, S. and Musacchi, S. 1994. Canopy architecture, training and pruning in the modern European pear orchards: an overview. *Acta Hort.* 367, 152–16.

Sansavini, S. and Zocca, A. 1968. Studies on fruit trees pruning systems, the rapidly trained palmette for pears. *Fruticoltura* 30, 917–38.

Sansavini, S., Cristoferi, G., Antonelli, M. and Montalti, P. 1981. Growth regulators in pear production. *Acta Hort.* 120, 143–8.

Sanyal, D. and Bangerth, F. 1998. Stress induced ethylene evolution and its possible relationship to auxin-transport, cytokinin levels, and flower bud induction in shoots of apple seedlings and bearing apple trees. *Plant Growth Reg.* 24, 127–34.

Sanzol, J. and Herrero, M. 2001. The 'Effective pollination period' in fruit trees. *Sci. Hort.* 90(1–2), 1–17.

Sanzol, J. and Robbins, T. 2008. Combined analysis of S-alleles in European pear by pollinations and PCR-based S-genotyping; correlation between S-phenotypes and S-RNase genotypes. *J. Amer. Soc. Hort. Sci.* 133(2), 213–24.

Sanzol, J., Rallo, P. and Herrero, M. 2003. Stigmatic receptivity limits the effective pollination period in 'Agua de Aranjuez' pear. *J. Amer. Soc. Hort. Sci.* 128, 458–62.

Sassa, H., Kakui, H., Miyamoto, M., Suzuki, Y., Hanada, T., Ushijima, K., Kusaba, M., Hirano, H. and Koba, T. 2007. S locus F-box brothers: multiple and pollen-specific F-box genes with S haplotype specific polymorphisms in apple and Japanese pear. *Genetics* 175(4), 1869–81. doi:10.1534/genetics.106.068858.

Sastry, K. K. S. and Muir, R. M. 1963. Gibberellin: effect on diffusible auxin in fruit development. *Science* 140(3566), 494–5. doi:10.1126/science.140.3566.494.

Saunders, R. C., Jacobs, G. and Strydom, D. K. 1991. Effect of pruning on fruit set and shoot growth of 'Packham's Triumph' pear trees. *Sci. Hort.* 47(3-4), 239–45.

Saure, M. C. 2007. Root pruning–a poorly understood management practice in fruit trees. *Int. J. Fruit Sci.* 7(2), 43-56.

Sazo, M. M., De Marree, A. and Robinson, T. 2010. The platform factor-Labor positioning machines producing good results for NY apple industry. *N. Y. Fruit Q.* 18, 5–9.

Schmidt, T., Elfving, D. C., McFerson, J. R. and Whiting, M. D. 2009. Crop load overwhelms effects of gibberellic acid and ethephon on floral initiation in apple. *HortScience* 44, 1900–6.

Schupp, J. R. and Ferree, D. C. 1987. Effect of root pruning at different growth stages on growth and fruiting of apple trees. *HortScience* 22, 387–90.

Schupp, J. R. and Ferree, D. C. 1988. Effects of root pruning at four levels of severity on growth and yield of 'Melrose'/M.26 apple trees. *J. Amer. Soc. Hort. Sci.* 113, 194–8.

Schupp, J. R. and Ferree, D. C. 1990. Influence of time of root pruning on growth, mineral nutrition, net photosynthesis and transpiration of young apple trees. *Sci. Hort.* 42, 299–306.

Seleznyova, A., Dayatilake, G., Watson, A. and Tustin, D. 2013. After initial invigoration by heading, young pear trees show reduction in axis vigour and increased propensity to flower. *Funct. Plant Biol.* 40(1), 34–43.

Shackel, K., Ahmadi, H., Biasi, W.,Buchner, R., Goldhamer, D., Gurusinghe, S., Hasey, J., Kester, D., Krueger, B., Lampinen, B., McGourty, G., Micke, W., Mitcham, E., Olson, B., Pelletrau, K., Philips, H., Ramos, D., Schwankl, L., Sibbett, S., Snyder, R., Southwick, S., Stevenson, M., Thorpe, M., Weinbaum, S. and Yeager, J. 1997. Plant water status as an index of irrigation need in deciduous fruit trees. *HortTechnology* 7, 23–9.

Silva, L., Herrero, M. and Oliveira, C. M. 2008. Effects of gibberellic acid and pollination on fruit set and fruit quality in 'Rocha' pear. *Acta Hort.* 800(800), 199–204. doi:10.17660/ActaHortic.2008.800.21.

Simard, M. H., Guisnel, R., Daguin, F., Demilly, D., Billy, B. and Honoré, D. 2011. Is dwarfing in pear rootstocks due to aneuploidy genetic structures. *Acta Hort.* 909, 59–66.

Smit, M., Meintjes, J. J., Jacobs, G., Stassen, P. J. C. and Theron, K. I. 2005. Shoot growth control of pear trees (*Pyrus communis* L.) with prohexadione-calcium. *Sci. Hort.* 106, 515–29.

Sosna, I. 2002. Reducing vegetative growth of 'Golden Delicious' apple trees by root pruning and ways of planting. *J. Fruit Ornament. Plant Res.* 10, 63–74.

Sousa, R. M., Calouro, F. and Oliveira, C. M. 2008. Influence of trunk girdling on growth and fruit production of 'Rocha'/BA29. *Acta Hortic.* 800(800), 319–24. doi:10.17660/ActaHortic.2008.800.39.

Southwick, S. M., Ingels, C., Hansen, R. and Glozer, K. 2004. The effects of Apogee® on shoot growth, secondary flowering, fire blight, fruit quality, yield and return bloom in 'Bartlett' pear growing in California. *J. Hort. Sci. Biotech.* 79(3), 380–9. doi:10.1080/14620316.2004.11511777.

Srivastava, L. M. 2002. *Plant Growth and Development: Hormones and Environment*. Academic Press, San Diego, CA.

Stanley, R. G. and Lichtenberg, E. A. 1963. The effect of various boron compounds *in vitro* on germination of pollen. *Physiol. Plant.* 16, 337–46.

Stebbins, R. L., Mellenthin, W. M. and Lombard, P. B. 1979. Pollination and fruit set of pear. Oregon State University Extension Circular 986.

Stephen, W. P. 1958. Pear pollination studies. Technical Bulletin No. 43. Oregon Agricultural Experimental Station, Corvallis, OR, 42p.

Sugar, D. and Basile, S. R. 2011. Performance of 'Comice' pear on quince rootstocks in Oregon, USA. *Acta Hort.* 909, 215–8.

Sugar, D., Elfving, D. C. and Mielke, E. A. 2004. Effects of prohexadione-calcium on fruit size and return bloom in pear. *HortScience* 39(6), 1305–8. doi:10.21273/HORTSCI.39.6.1305.

Tagliavini, M., Quartieri, M. and Millard, P. 1997. Remobilized nitrogen and root uptake of nitrate for spring leaf growth, flowers and developing fruits of pear (*Pyrus communis* L.) trees. *Plant Soil* 195(1), 137–42.

Theron, K. I. and Stein, W. J. 2008. Girdling: science behind the age-old technique. *Acta Hort.* 800(800), 51–60. doi:10.17660/ActaHortic.2008.800.2.

Theron, K. I., Chabikwa, T. G. and Lötze, G. F. A. 2011. Evaluation of 6-benzyladenine (BA) and naphthylacetamide (NAD) as post-bloom thinning compounds for 'Early Bon Chrétien' pear. *Acta Hort.* 909(909), 387–93. doi:10.17660/ActaHortic.2011.909.43.

Thimann, K. V. and Skoog, F. 1933. Studies on the growth hormone of plants III. The inhibitory action of the growth substance on bud development. *Proc. Nat. Acad. Sci. U.S.A.* 19, 714–6.

Tromp, J. 1982. Flower bud formation in apple as affected by various gibberellins. *J. Hort. Sci.* 57(3), 277–82. doi:10.1080/00221589.1982.11515053.

Tromp, J. and Borsboom, O. 1994. The effect of autumn and spring temperature on fruit set and on the effective pollination period in apple and pear. *Sci. Hort.* 60(1-2), 23–30.

Vansell, G. H. 1946. Bees and pear pollination. *Oregon State Hort. Soc. Proc.* 37, 51–3.

van Zyl, H. J. and Strydom, D. K. 1982. Some factors affecting fruit set of the 'Packham's Triumph' pear. *Acta Hort.* 124, 149–56.

Vasilakakis, M. and Porlingis, I. C. 1985. Effect of temperature on pollen germination, pollen tube growth, effective pollination period and fruit set of pear. *HortScience* 20, 733–5.

Vercammen, J. 2011. Comparison among different planting systems for 'Conference'. *Acta Hort.* 909, 271–6.

Vercammen, J. and Gomand, A. 2008. Fruit set of 'Conference': a small dose of gibberellins or Regalis. *Acta Hort.* 800(800), 131–8. doi:10.17660/ActaHortic.2008.800.11.

Vercammen, J., Van Daele, G. and Gomand, A. 2005. Root pruning: a valuable alternative to reduce the growth of 'Conference'. *Acta Hort.* 671, 533–7.

Vertucci, C. W. and Stushnoff, C. 1992. The state of water in acclimating vegetative buds from *Malus* and *Amelanchier* and its relationship to winter hardiness. *Phys. Plant.* 86(4), 503–11. doi:10.1111/j.1399-3054.1992.tb02162.x.

Vilardell, P., Pagés, M. J. and Asín, L. 2008. Effect of bioregulator applications on the fruit set in Abate Fetel pear trees. *Acta Hort.* 800, 169–74.

Volz, R. K., White, A. G. and Brewer, L. R. 2008. Breeding for red skin color in interspecific pears. *Acta Hort.* 800, 469–74.

Walsh, C. S., Swartz, H. J. and Edgerton, L. J. 1979. Ethylene evolution in apple following post-bloom thinning sprays. *HortScience* 14, 704–6.

Wang, Y., Bertelsen, M., Petersen, K., Andersen, M. and Liu, F. 2014. Effect of root pruning and irrigation regimes on leaf water relations and xylem ABA and ionic concentrations in pear trees. *Agric. Water Manag.* 135, 84–9.

Warmund, M. 2008. Leaf to fruit ratios: a balancing act for sustained apple production. University of Missouri, Integrated Pest Management. Available at: https://ipm.missouri.edu/meg/2008/3/Leaf-to-Fruit-Ratios-A-Balancing-Act-for-Sustained-Apple-Production/(accessed on 28 April 2019).

Webster, A. D. 1998. A brief review of pear rootstock development. *Acta Hort.* 475, 135–41.

Webster, A. D. 2002. Factors influencing the flowering, fruit set and fruit growth of European pears. *Acta Hort.* 596(596), 699–709. doi:10.17660/ActaHortic.2002.596.121.

Weinbaum, S. A., DeJong, T. M. and Maki, J. 2001. Reassessment of seed influence on return bloom and fruit growth in 'Bartlett' pear. *HortScience* 36, 295–7.

Wertheim, S. J. 1998. Rootstock guide; apple, pear, cherry, European plum. Publication no. 25. Fruit Research Station, Wilhelminadorp, The Netherlands.

Wertheim, S. J. 2000. Developments in the chemical thinning of apple and pear. *Plant Growth Regul.* 31(1–2), 85–100.

Wertheim, S. J. 2002. Rootstocks for European pear: a review. *Acta Hort.* 596, 299–309.

Westwood, M. N. 1993. *Temperate-Zone Pomology: Physiology and Culture*. Timber Press, Portland, OR.

Westwood, M. N. and Bjornstad, D. K. 1974. Fruit set as related to girdling, early cluster thinning and pruning of 'Anjou' and 'Comice' pear. *HortScience* 9, 342–4.

Westwood, M. N. and Grim, J. 1962. Effect of pollinizer placement on long term yield of Anjou, Bartlett and Bose pears. *Proc. Amer. Soc. Hort. Sci.* 81, 103–7.

Westwood, M. N., Lombard, P. W. and Brannock, L. D. 1968. Effect of seeded fruits and foliar-applied auxin on seedless fruit set of pear the following year. *HortScience* 3(3), 168–9.

Wickland, D. P. and Hanzawa, Y. 2015. The flowering locus t/terminal flower 1 gene family: functional evolution and molecular mechanisms. *Mol. Plant* 8(7), 983–97. doi:10.1016/j.molp.2015.01.007.

Williams, R. R. 1965. The effect of summer nitrogen applications on the quality of apple blossom. *J. Hort. Sci.* 40, 31–41.

Williams, M. W. 1972. Induction of spur and flower bud formation in young apple trees with chemical growth retardants. *J. Amer. Soc. Hort. Sci.* 97, 210–2.

Williams, M. W. 1981. Response of apple trees to aminoethoxyvinylglycine (AVG) with emphasis on apical dominance, fruit set, and mechanism of action of fruit thinning chemicals. *Acta Hort.* 120, 137–41.

Williams, M. W., Couey, H. M., Moffitt, H. and Coyier, D. L. 1978. Pear production. USDA Handbook No. 330, 53pp.

Wismer, P. T., Proctor, J. T. A. and Elfving, D. C. 1995. Benzyladenine affects cell division and cell size during apple fruit thinning. *J. Amer. Soc. Hort. Sci.* 120, 802–7.

Wojcik, P. and Wojcik, M. 2003. Effects of boron fertilization on 'Conference' pear tree vigor, nutrition, and fruit yield and storability. *Plant Soil* 256(2), 413–21. doi:10.1023/A:1026126724095.

Wünsche, J. N., Lakso, A. N., Robinson, T. L., Lenz, F. and Denning, S. S. 1996. The bases of productivity in apple production systems: the role of light interception by different shoot types. *J. Amer. Soc. Hort. Sci.* 121, 886–93.

Yehia, Taher A., Hegazia, A. A., Saleh, M. A. and Hassan, S. A. 2011. Effect of root pruning on flowering of 'Le Conte' pear trees. *J. Hort. Sci. Ornament. Plants* 3(3), 199–206.

Yoder, K. S., Miller, S. S. and Byers, R. E. 1999. Suppression of fire blight in apple shoots by prohexadione-calcium following experimental and natural inoculation. *HortScience* 34, 1202–4.

Yoder, K. S., Yuan, R., Combs, L. D., Byers, R. E., McFerson, J. and Schmidt, T. 2009. Effects of temperature and the combination of liquid lime sulfur and fish oil on pollen germination, pollen tube growth, and fruit set in apples. *HortScience* 44, 1277-83.

Yoder, K. S., Peck, G. M., Combs, L. D. and Byers, R. E. 2013. Using a pollen tube growth model to improve apple bloom thinning for organic production. *Acta Hort.* 1001(1001), 207–14. doi:10.17660/ActaHortic.2013.1001.23.

Zhang, C., Tanabe, K., Tamura, F., Itai, A. and Wang, S. 2005a. Spur characteristics, fruit growth, and carbon partitioning in two late-maturing Japanese pear (*Pyrus pyrifolia* Nakai) cultivars with contrasting fruit size. *J. Amer. Soc. Hort. Sci.* 130(2), 252–60.

Zhang, C., Tanabe, K., Tamura, F., Itai, A. and Wang, S. 2005b. Partitioning of 13C-photosynthate from spur leaves during fruit growth of three Japanese pear (*Pyrus pyrifolia*) cultivars differing in maturation date. *Ann. Bot.* 95(4), 685–93. doi:10.1093/aob/mci070.

Zhang, J., Serra, S., Leisso, R. and Musacchi, S. 2016. Effect of light microclimate on the quality of 'd'Anjou' pears in mature open-centre tree architecture. *Biosyst. Eng.* 141, 1–11.

Zielinski, Q. B. and Thompson, M. M. 1966. Pollen germination in *Pyrus* species and species hybrids. *Euphytica* 15, 195–8.

Zisovich, A. H., Stern, R. A., Shafir, S. and Goldway, M. 2009. Identification of seven haplotype-specific SFBs in European pear (*Pyrus communis*) and their use as molecular markers. *Sci. Hort.* 121, 49-53.

Zuccherelli, S., Tassinari, P., Broothaerts, W., Tartarini, S., Dondini, L. and Sansavini, S. 2002. S-allele characterization in self-incompatible pear (*Pyrus communis* L.). *Sex. Plant Reprod.* 15, 153–8.

Chapter 7

Advances and challenges in apple breeding

Amanda Karlström, NIAB EMR and University of Reading, UK; Magdalena Cobo Medina, NIAB EMR and University of Nottingham, UK; and Richard Harrison, NIAB EMR, UK

1 Introduction

1.1 Current production

Apple, *Malus × domestica* (Borkh), is one of the most important global fruit crops, both in terms of its cultural symbolism and with respect to the extent of production, which amounts to an average of 87 million tonnes of annual production globally for the years 2014–16 (FAOSTAT, 2018). Apple is primarily a temperate crop due to the chilling requirements for the initiation of blossoming (Heide and Prestrud, 2005), nevertheless there are varieties with lower requirements for chilling, which can be grown in subtropical climates (Labuschagné et al., 2002).

Global apple production has increased 117% between the mid-1980s and the mid-2010s (FAOSTAT, 2018). The growth in production is mainly attributed to a higher productivity per ha as the area utilised for apple production has increased by only 20% during the same period (FAOSTAT, 2018). As illustrated in Fig. 1, Asia surpassed Europe as the major apple-producing region of the world in the 1990s. Mainland China has been the major contributor to this large expansion of apple production in Asia and today represents 75% of the Asian production (Fig. 2). Any development towards a more sustainable cultivation

http://dx.doi.org/10.19103/AS.2018.0040.21
Published by Burleigh Dodds Science Publishing Limited, 2019.

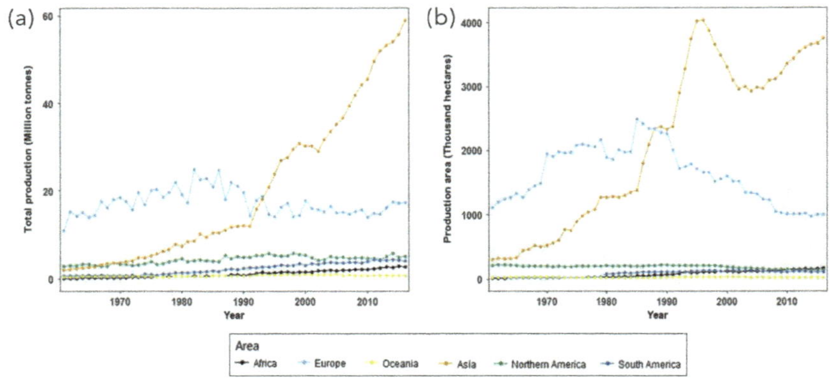

Figure 1 (a) The total annual production quantity of apple in million tonnes shown by continent. (b) Total area used for apple production in 1000 ha shown by continent. Source: data retrieved from the Food and Agriculture Organization of the United Nations (FAO) website, FAOSTAT (2018).

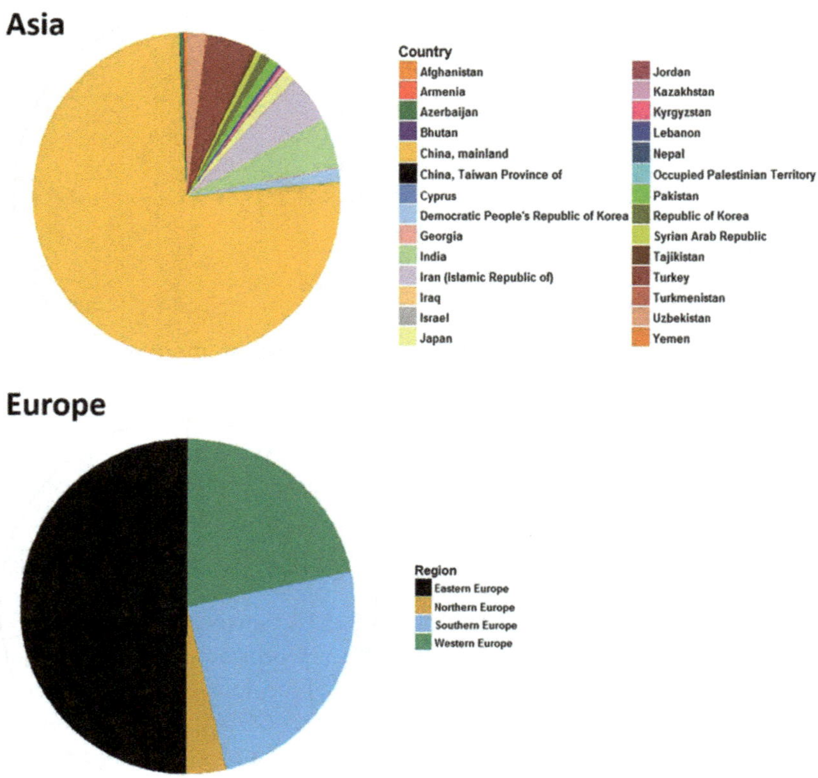

Figure 2 Breakdown of the total apple production quantity in 2016 by country and region in Asia and Europe, the two major apple producing regions of the world. Source: data retrieved from FAOSTAT (2018).

of apple, either genetic or agronomic, will therefore have the most significant impact if adopted by commercial production in Asia.

Modern apple production systems are characterised by trees grown on dwarfing or semi-dwarfing rootstocks which allows for high-density planting. Such production systems have the benefit of earlier cropping trees and a higher production efficiency per ha (Robinson, 2007). Two clonally propagated dwarfing rootstocks, the varieties M9 and M26 (the M denoting 'Malling'), dominate the European and North American apple industry, although both were commercially released before 1950 and susceptible to certain pests and diseases (Marini et al., 2009, 2014, 2017). In China, 80–90% of apple production systems use seedlings from wild species of *Malus* (*M. sieversii, M. baccata, M. prunifolia* and *M. hupehensis*) as rootstocks. Due to genetic variation among the seedling rootstocks these systems suffer from heterogeneity in vigour and a late onset of fruiting due to rootstock juvenility (Li et al., 2015).

While there is diversity in the apple scion varieties that are commercially grown, the genetic base from which they are derived is narrow (Bannier, 2011; Noiton and Alspach, 1996). A study which considered 500 apple varieties, predominantly with a Central European and/or US origin, found that 'Golden Delicious' occurred at least once in 51% of the pedigrees, whereas 'McIntosh' occurred in 35% of the varieties, 'Jonathan' in 31%, 'Cox's Orange Pippin' in 30%, 'Red Delicious' in 18% and 'James Grieve' in 15% (Bannier, 2011). The inclusion of these progenitor varieties and their derivatives in breeding programmes is principally attributed to their fruit quality, storability and cropping (Bannier, 2011). Because of the relatively narrow genetic pool of commercial apple production, negative traits originating from these founder varieties may be incorrectly attributed to be intrinsic to modern apple cultivation, rather than genetic weaknesses.

Novel apple varieties are released continuously, and there are currently 821 pending applications for *Malus domestica* variety registrations across 40 countries (CPVO, 2018). However, the adoption and spread of new varieties is limited due to a high degree of competition within the industry. Grower adoption of new varieties is also associated with a large financial risk as orchard establishment requires a high initial capital investment (Badiu et al., 2015) and there is therefore a need for growers to be reassured that there is a market demand for the variety once the trees begin to crop. For that reason the success of a novel variety tends to be reliant on the backing of grower associations and marketing organisations following extensive trialling. Many vigorously marketed apples are 'club' or 'managed' varieties for which exclusive rights to grow and sell the variety are licensed to specific grower organisations and nurseries. Such varieties are often granted patent and plant variety rights under a variety name or selection number and then trademarked and marketed under a separate name (Legun, 2015). Examples of club varieties are 'Cripps Pink' which is trademarked under the name Pink Lady® by Apple and Pear Australia Limited (APAL) and 'Scifresh', trademarked as Jazz® and owned and licensed by ENZAFRUIT International Ltd (Brown, 2009).

1.2 Current breeding of apple scions and rootstocks

As with other perennial tree crops, the breeding cycle of apple scions is long and the development of a new variety takes a minimum of 15–20 years (Fruitbreedomics, 2014a). This timeframe is further prolonged if germplasm with suboptimal fruit quality is used for the introduction of a new trait such as disease resistance (Flachowsky et al., 2011).

The most common breeding strategy in apple utilises controlled-pollinated crosses between elite selections or varieties to produce large full-sibling families, which are phenotypically evaluated in the field (Kumar, 2010). Genotypes which are considered to be outstanding are selected from the seedling trial and clonally propagated and planted in replicated advanced trials. The selected genotypes are thereafter assessed for fruit quality traits across multiple sites, years and harvests (Harshman et al., 2016). The breeding cycle is completed when elite clones, either from the advanced trials or after commercialisation, are used as breeding parents. The different stages of apple breeding are shown in Fig. 3. Due to the long juvenility phase (the period prior to reproductive maturity) of apple, a period of 4–7 years is required from when the initial cross is made until the progeny start to produce fruit and can be evaluated (Edge-Garza et al., 2015). The time and money invested in each

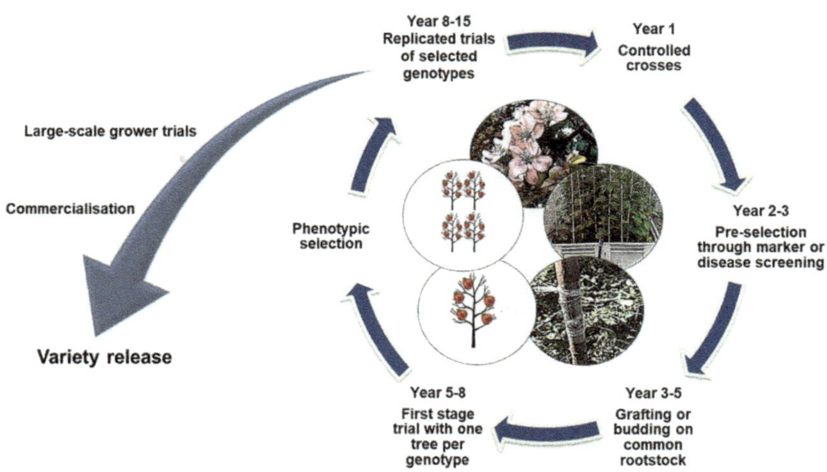

Figure 3 Illustration of a generalisation of the breeding cycle of apple, which can be seen as consisting of a population improvement cycle and a separate entity of variety development.

seedling that reaches cropping is therefore substantial (Edge-Garza et al., 2015; Ru et al., 2015). Hence, in breeding programmes with resistance to biotic stress as a major breeding objective, where possible, it is common to select for resistance at an early seedling stage (Kellerhals et al., 2017). A small percentage of individuals are moved forward to the advanced trialling stage, nevertheless the long period of trialling and the labour-intensive fruit phenotyping can contribute substantially to the costs of the breeding programme (Harshman et al., 2016).

Fruit quality traits, such as texture, attractiveness, aroma and storability, as well as tree productivity and consistent bearing are universally important breeding objectives of apple scion breeding programmes across the world (Centro de Pomaceas, 2019; Evans, 2011; Fruitbreedomics, 2012, 2013a,b,c; Johan Kriel, pers. comm.; Sansavini et al., 2004). In warmer regions with high exposure to sunlight, the development of varieties with low winter chilling requirement and fruits that are less prone to be damaged by sunburn are also priorities (Centro de Pomaceas, 2019; Johan Kriel, pers. comm.). Disease resistance objectives differ depending on region, with the major diseases being apple scab (*Venturia inaequalis*), powdery mildew, (*Podosphaera leucotricha*), fire blight (*Erwinia amylovora*) and European apple canker (*Neonectria ditissima*) (Fruitbreedomics, 2012, 2013a,b,c; Johan Kriel, pers. comm.; Sansavini et al., 2004). A major biotic constraint in apple growing regions in eastern Asia is Valsa canker, caused by the fungus *Valsa mali* (Abe et al., 2007). The genetic improvement of resistance levels to this pathogen is therefore a priority within this region (Abe et al., 2007; Tan et al., 2017a,b).

The breeding cycle for apple rootstocks can last even longer than for breeding for scion varieties due to the period of time required to replicate rooted material as well as extensive periods of trialling. The objectives of modern rootstock breeding programmes include vigour control of the scion (ranging from dwarfing to semi-invigorating depending on the target market), increased yield efficiency and precocity, good nursery performance (good production of liners in stoolbeds as well as adventitious rooting ability and grafting compatibility), resistance to woolly apple aphid (*Eriosoma lanigerum*), fire blight (*E. amylovora*), collar rot (*Phytophthora cactorum*), European apple canker (*N. ditissima*) and tolerance to the apple replant disease complex (Gregory et al., 2013).

1.3 Current genomic tools in apple

The first draft apple genome became available in 2010 (Velasco et al., 2010) and has since been a valuable resource of this economically important perennial crop and a crucial tool to advance the development of new and improved apple varieties. In 2017, a high-quality apple genome generated from a double haploid of 'Golden Delicious' was released, from a combination of long reads (PacBio) and short sequencing reads (Illumina). The new genome provides an excellent foundation for future genetic studies not only in apple but also for other species of Rosaceae (Daccord et al., 2017).

Simple sequence repeat (SSR) markers have been routinely utilised in apple research and breeding for decades. This is not surprising given their advantages: they are co-dominant, multi-allelic, abundant and, generally, are uniformly distributed (Gianfranceschi et al., 1998). A significant amount of progress in apple genetics has relied upon the utilisation of SSRs, for example, fine mapping of the rosy apple aphid resistance locus (Pagliarani et al., 2016), rootstock linkage map construction (Celton et al., 2009; Fernández-Fernández et al., 2012) as well as from the development of linkage maps of scion varieties (Liebhard et al., 2002, 2003; Kenis and Keulemans, 2005; N'Diaye et al., 2008; Ziya Motalebipour et al., 2015; Liu et al., 2016).

Several single nucleotide polymorphism (SNP) arrays have been developed specifically for apple. In 2012, the IRSC apple Infinium II 8K array became available. This was developed from a SNP-discovery panel consisting of 27 apple cultivars from around the world which were re-sequenced at low coverage (Chagné et al., 2012). Shortly after, a 20K Illumina apple array was developed using re-sequencing data from 13 apple cultivars and one crab apple species (Bianco et al., 2014). Both have been used to generate saturated linkage maps of an apple rootstock progeny (Antanaviciute et al., 2012), in genomic selection for fruit quality traits (Kumar et al., 2012) and in multiple studies to detect quantitative trait loci (QTL) for several traits including apple skin russeting (Falginella et al., 2015), sugar and soluble solid content (Guan et al., 2015) and budbreak and flowering time (Allard et al., 2016). These arrays have been useful for the generation of saturated genetic linkage maps and QTL detection. Nevertheless, these arrays also have certain limitations since the low-medium density of markers limits its applicability to genome-wide association studies (GWAS) (Bianco et al., 2016).

In 2016, the Axiom Apple 480K array was developed after high-depth re-sequencing of 63 different cultivars and two doubled haploids (Bianco et al., 2016). It became an important tool for GWAS, thanks to the high percentage of well-distributed and robust SNPs (Bianco et al., 2016). It was subsequently used for this purpose for the mapping of flowering and ripening periods in apples (Urrestarazu et al., 2017). There are also limitations to array-based genotyping, since the genotype calls often result in useless or unreliable data due to the high levels of off-target polymorphism in varieties which are absent from the initial discovery panel (Miller et al., 2013). Furthermore, as a closed genotyping system, limited by the discovery panel used, arrays often suffer from built-in ascertainment biases leading to limitations when applying some population genetic analyses (Albrechtsen et al., 2010).

There are several benefits of using genotyping-by-sequencing (GBS) technologies over microarray-based methods in apple, including a reduction of the ascertainment bias and as an opened genotyping system you can obtain genotypic data across the whole population (Elshire et al., 2011; Gardner et al., 2014). GBS methods provide a rapid tool to genotype breeding populations for several applications like GWAS and genomic selection (He et al., 2014). These sequencing technologies have been applied to map QTL for disease resistance to blue mould (Norelli et al., 2017), soft scald (McClure et al., 2016) and apple skin colour (Gardner et al., 2014), as well as for GWAS of scab and fruit quality traits (McClure et al., 2018). Some potential drawbacks of GBS are large proportion of missing data points due to low coverage of sequencing and management and analysis of a large amount of sequence data (Bhatia et al., 2013).

One of the latest technologies available to breeders is genome editing, a powerful method used for genome modification. The CRISPR/Cas9 system is one of the best-known techniques in this field since it allows for the targeted removal or addition of genes in a defined location. CRISPR/Cas9 has been applied in apple as a proof of concept to modify an apple phytoene desaturase (PDS) gene, essential for chlorophyll biosynthesis (Nishitani et al., 2016). PDS mutants showed the expected albino phenotype demonstrating the potential to modify the apple genome using CRISPR/Cas9 and how this could contribute to generate desired apple traits (Nishitani et al., 2016). CRISPR/Cas9 ribonucleoproteins have been used to silence DIPM-1, DIPM-2 and DIPM-4 in apple protoplasts to increase resistance to fire blight and proved that DNA-free genome editing can be performed on apple (Malnoy et al., 2016). Organisms obtained through novel mutagenesis techniques are not being regulated as genetically modified (GM) in the United States and Canada, while the Court of Justice of the European Union has ruled that they are to be classified as GMOs and therefore subject to the same regulation as transgenic organisms (Judgement of 25 July 2018, Case C-528/16, ECLI:EU:C:2018:583).

The numerous advances in genomic tools have contributed to improvements for apple breeding, but there are several challenges to face in order to take advantage of these tools. Future challenges are related to the improvement of bioinformatic tools to facilitate the extraction of information from raw data and the improvement of algorithms for genotype calling (Gardner et al., 2014). Another challenge is to reduce the cost of genotyping to make it more affordable. The effectiveness of genomic tools is proven and the choice of the method is dependent on many factors but one of the most important and decisive is the price.

For the remainder of this chapter we will discuss scientific achievements and challenges within three areas of apple breeding which we consider to be central in order to ensure sustainable production in the future: (1) shortening the breeding cycle and improving selection to deliver better varieties to growers, (2) increasing the orchard productivity and resource-use efficiency through genetic means and (3) mitigating pre- and post-harvest losses through improved genetic approaches.

2 Shortening the breeding cycle and improving selection

There is a great potential to increase genetic gains in apple through shortening the breeding cycle and making selection more effective. The current constraints are mainly due to the long period of seedling juvenility and extensive trialling of advanced selections in traditional apple breeding.

2.1 Accelerating breeding cycles

The need for a reduction in the time taken for novel trait introgression is more pressing than ever as environmental change and increased global movement increases the need to rapidly adapt germplasm to new abiotic and biotic stresses. The introgression of novel traits from other *Malus* species or non-elite material into commercial apple varieties is traditionally a slow process that can require five generations of backcrosses or more to recover enough of the domesticated apple genome in order to become a commercially viable variety (Joshi et al., 2009; Joshi, 2010; Flachowsky et al., 2011). An example of a gene introgression from a wild relative to *M.* × *domestica* is the *Rvi6* (previously denoted *Vf*) apple scab resistance gene from *Malus floribunda* 821. The initial crosses to introgress the *Rvi6* gene from *M. floribunda* 821 were made in 1914, but it was approximately 80 years before *Rvi6* varieties with a good fruit quality were released on the market (Joshi et al., 2009).

Several methods have been developed to decrease the generation time in apple, while trying to maintain an end product which does not have the regulatory status of a GMO (Flachowsky et al., 2009, 2011; Yamagishi et al., 2011). One approach is to introduce *trans*- and *cis*-genes that shorten the juvenile phase, and thereafter segregate out the *trans/cis* early flowering gene construct in the final population. Overexpression of the gene *BpMADS4* from silver birch (*Betula pendula*), a homologue of *Arabidopsis thaliana FRUITFULL,* in apple resulted in F1-seedlings that flowered within 3–10 months after seed germination, and produced viable offspring when used as a pollen parent (Flachowsky et al., 2009, 2011). Introgression of resistance from the variety 'Evereste' has reached the fifth generation of fire blight-resistant seedlings after only 7 years, using the *BpMADS4*-early flowering transgenic lines combined with marker-assisted selection (MAS; Schlathölter et al., 2018). The same method has been used to pyramid resistance genes to fire blight, apple powdery mildew and apple scab (Flachowsky et al., 2011).

Published by Burleigh Dodds Science Publishing Limited, 2019.

Cisgenesis refers to the genetic transformation of a plant with genetic material from a donor with which it is sexually compatible (Camacho et al., 2014). Some *cis*-genic approaches to shorten the juvenile phase include RNAi silencing of *MdTFL1 (TERMINAL FLOWER 1)* (Kotoda et al., 2005; Szankowski et al., 2009) and the constitutive expression of *MdFT (FLOWERING LOCUS T)* (Tränkner et al., 2010). However, these studies were conducted with promoters and selection markers that were non-native to apple, and can therefore not be considered completely *cis*-genic.

A plant virus vector-based method, known as virus-induced gene silencing, has also been utilised to transiently express *AtFT* from *A. thaliana* and silence *MdTFL1* by using the *Apple latent spherical virus* (ALSV) as a vector system. The use of an RNA virus vector to induce early flowering can be applied to a wide variety of genotypes and has the advantage that the viral genome seldom integrates into the host genome. However, plant viruses can be transmitted through seeds, and ALSV was detected through qRT-PCR in a small percentage of the progeny from an infected parent (Yamagishi et al., 2014). All progeny obtained from a virus-infiltrated parent would therefore have to be tested for presence of the ALSV-vector.

The null segregants of the above methods do not contain any trans- or cis-genes and could, depending on the method of transformation used, be exempt of GM regulation according to US legislation (Camacho et al., 2014). In Argentina, Brazil, Chile and Colombia the regulation of plant material derived from new breeding techniques is reviewed on a case-by-case basis, but if the final product is free from any transgene then the norm is that it is not regulated as a GMO (Eriksson et al., 2018). The legal status of null segregants within the European Union is not clearly defined (European Commission, 2001), but is likely to fall within the definition of GM as of the ruling of the Court of Justice of the European Union in 2018 (Court of Justice of the European Union, press release No.111/18).

Another approach to reduce the period of juvenility in apple seedlings is through the manipulation of cultural conditions (van Nocker and Gardiner, 2014). Promoting vigorous growth through optimal conditions in a glasshouse has been reported by several authors to lead to an earlier induction of flowering in apple seedlings, in some accounts as early as within 16 months after germination (Visser, 1964; Aldwinckle et al., 1976; Fischer, 1994). The transition to the adult phase has been postulated to occur when the seedlings reach a height of 75–80 nodes, whereafter wood from above this height can be grafted or budded on a dwarfing rootstock (Hanke et al., 2007). The use of intensive LED-light regimes in glasshouse conditions has successfully been used as a means to reduce generation time in several annual crops (Hickey et al., 2017; Watson et al., 2018). This method may also have the potential to improve early flowering in apple as there is evidence that apple seedlings grown under 16-h light conditions induced earlier flowering (Fischer, 1994).

2.2 Marker-assisted selection

The selection in apple breeding programmes has traditionally been based on phenotypic evaluations of performance (Ru et al., 2015; Ru, 2016). At the seedling trial stage, this requires the plant material to be evaluated to reach physical maturity before it can be assessed for fruit-related traits (Edge-Garza et al., 2015). The phenotypic selection of resistance to biotic stresses at this stage can also be highly laborious and costly depending on phenotyping protocol (Ru et al., 2015). Furthermore, the destructive nature of assessing tolerance to certain biotic stresses requires multiplying seedling plant material, which may not be economically feasible. Owing to these factors apple is a good candidate for the implementation of genotyping technologies as tools to reduce the cost of plant phenotyping (van Nocker and Gardiner, 2014).

MAS is to date the most widely utilised genetic tool applied for apple seedling selection. In 2011–12, 43% of the apple breeding programmes in the European Union stated that they deployed markers in some way in their breeding programme, however only six of these programmes used markers to do progeny selection (Fruitbreedomics, 2014b). The traits for which predictive marker tests are available include fruit storability/ethylene biosynthesis (Edge-Garza et al., 2010), fruit acidity (Bai et al., 2012; Verma, 2014), bitter pit susceptibility (Kumar et al., 2013; Buti et al., 2015), fruit sweetness/fructose content (Guan et al., 2015), fruit skin and flesh colour (Cheng et al., 1996), fruit crispness (Verma, 2014), fruit firmness (Zhu and Barritt, 2008; Longhi et al., 2013; Nybom et al., 2013; Verma, 2014), scab resistance (Cheng et al., 1998; Vinatzer et al., 2004), powdery mildew resistance (Markussen et al., 1995; Evans and James, 2003) and fire blight resistance (Gardiner et al., 2012).

In a comparison of selection strategies in simulated data, Ru (2016) showed that incorporating marker information in the selection decision tended to increase the genetic gain compared to phenotype-only selection in scenarios where the proportion of genetic variance explained by the marker was higher than the broad-sense heritability of the trait. Empirical comparisons of MAS to phenotypic seedling selection in apple have been conducted for scab resistance (Tartarini et al., 2000), fruit storability/ethylene biosynthesis (Edge-Garza et al., 2010) and fruit firmness, acidity and crispness (Ru, 2016). The breeding value, defined as the component of a trait phenotype constituted by additive genetic effects, is commonly used as a measure of the value of an individual that is transferable to its offspring. A comparison of average estimate breeding value between seedlings selected through MAS or phenotypic selection for the traits such as fruit firmness, acidity and crispness indicated that the genetic gain was similar between the two selection strategies, although the variance of breeding values for MAS seedlings for firmness and crispness was lower (Ru, 2016). As mentioned by Ru, the use of genetic gain, defined as the average increase in performance of all individuals per unit of time, might not be the best measure of response to selection in an apple breeding programme with the aim of selecting a few outstanding genotypes and not to improve the average population performance. In such a setting it might be preferred to look at the maximum breeding value achieved in a percentage of the highest performing progeny (Ru, 2016).

In calculations of the cost-efficiency of incorporating MAS in an apple breeding programme, Edge-Garza et al. (2015) concluded that maximum savings would be achieved if genotype-based culling of seedlings is performed at a stage where minimal routine seedling reductions are likely afterwards but before a large cost per seedling is incurred. The magnitude of savings in the cost scenarios of the same study were highly dependent on the proportion of seedlings that were culled after MAS. Traits that can be phenotyped at an early seedling stage, such as some disease screening, were not cost-effective to select for using markers (Edge-Garza et al., 2015).

2.3 Genomic selection

MAS is expected to be useful for traits where large-effect loci are present, however many agronomically important traits are highly polygenic and can therefore not be accurately predicted by markers at a few single loci (de Los Campos et al., 2013). Genomic selection (GS) overcomes this limitation by simultaneously estimating the effect of a large set of markers distributed across the genome, thereby enabling genetic predictions for highly quantitative traits controlled by multiple small effect loci (Meuwissen et al., 2001). GS utilises phenotypic and genotypic data from a training population to fit a prediction model which is then used to produce genomic estimated breeding values (GEBV) of non-phenotyped, test genotypes (Crossa et al., 2017). Because GS can be performed at an early seedling stage, the field phenotyping of the full population can be omitted and the selected progeny directly moved into replicated field trials (Crossa et al., 2017, see Fig. 4 for a depiction of a GS breeding cycle). GS therefore has the potential to accelerate the breeding cycle and lower the cost of selection (Kumar et al., 2013). As generations progress, the training population can be updated by adding data from a subset of the test population (e.g. the best performing genotypes) (Neyhart et al., 2017).

The predictiveness of GS in apple has been empirically trialled in four studies for a total of 23 fruit and one disease resistance trait (Kumar et al., 2012, 2013, 2015b; Muranty et al., 2015; McClure et al., 2018). Prediction accuracy, defined as the correlation between predicted GEBV and observed estimated breeding value for a test population, ranged between 0.15 (for juiciness) and 0.35 (for firmness) (Kumar, et al., 2015b), −0.14 (for type of colour) and 0.37 (per cent over-colour) (Muranty et al., 2015); 0.67 (for astringency) and 0.89

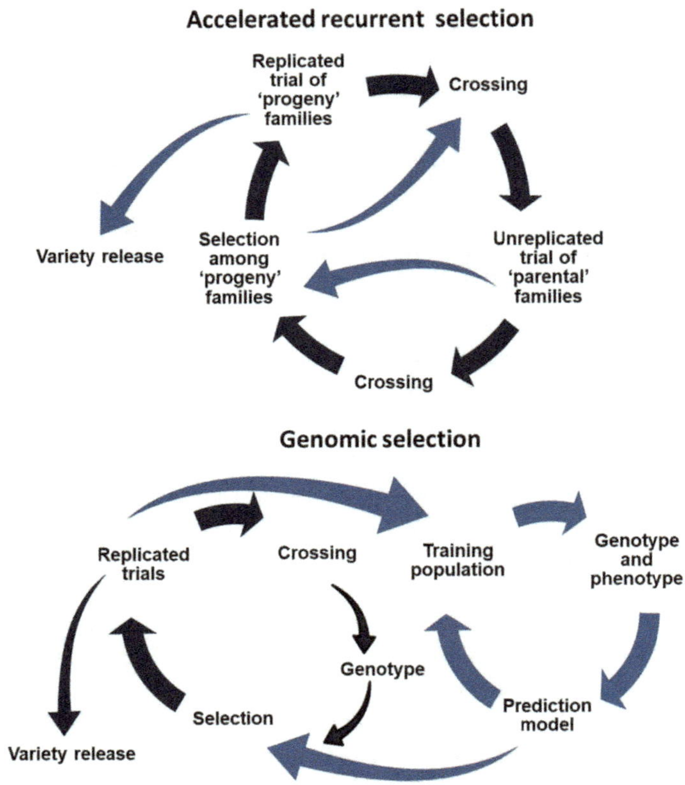

Figure 4 Depictions of two alternative breeding cycles for apple scion and rootstock which aim to cycle through generations more rapidly than the traditional breeding cycle shown in Fig. 3. Accelerated recurrent selection is a breeding scheme for rapid population improvement in which full-sibling families are trialled while simultaneously being crossed to each other in an incomplete or complete diallel design to produce progeny families. Selections among the progeny families are then made based on the performance of the trialled parental families. By this point the trees in the progeny generation may already have reached maturity and the selected families can therefore immediately be used for another round of crossing while outstanding genotypes within these families are moved on to replicated trials. In genomic selection a 'training population', comprising genetic material that is sufficiently related to all material of interest to the breeding programme, is genotyped and phenotyped for all traits relevant for selection. This data is then integrated to develop a prediction model which is used to make selections among cross progeny solely based on their genotype and without field-testing. The selected genotypes are thereafter fast-tracked directly to replicated trialling and further crossing.

(for soluble solid content) (Kumar et al., 2012) and 0.08 (for change in firmness between harvest and storage) and 0.72 (for scab resistance) (McClure et al., 2018). Both Kumar et al. (2015b) and Muranty et al. (2015) found that the accuracy of prediction was positively correlated to the narrow sense heritability of a trait.

The large discrepancies in prediction accuracy between studies, in some cases for the same traits, are likely due to differences in study design. Kumar et al. (2012) achieved high prediction accuracies, but used a random sample of the training population to make predictions, which meant that predictions were made within the same families as were present in the training population. The relatedness between a training population and test population affect the prediction ability, with higher predictions being achieved with closer relatedness (Crossa et al., 2017; Zhou et al., 2018). The prediction accuracies in Muranty et al. (2015) were low compared to the other studies, likely due to the use of single replicates of genotypes (possibly impacting on the phenotyping accuracy) and that predictions were made for subjectively scored traits across environments. In previous applications of genomic prediction, accuracies have been shown to be reduced when predictions are attempted in an environment other than in which the phenotypic data was collected (Beaulieu et al., 2014).

The genetic gain per year was higher for GS than conventional selection in apple when estimated by Kumar et al. (2012), with the assumption that GS would reduce the seedling selection cycle from 7 to 4 years. Due to the high prediction accuracies in this study, the relative efficiency of GS in apple might have been overestimated. Nevertheless, depending on selection cycle and strategy, GS could improve genetic gains per year even with low prediction ability (Longin et al., 2015). For practical application of GS in apple to be successful, the following considerations should be highlighted: (1) The optimum design of the training population; population size, genetic relationship between training and test individuals and the training population structure are important factors to consider to optimise prediction ability (further reading on population design can be found in Lorenz and Nice, 2017). (2) The heritability of traits under selection; genomic prediction accuracy increases with the heritability of a trait (Kumar et al., 2015b; Muranty et al., 2015). Measures to increase the trait heritability, such as increasing the number of replicates and using objective and accurate phenotyping, could therefore improve the predictions. (3) The choice of genomic prediction model may affect prediction ability, and measures to improve prediction models include incorporating genotype × environment interaction (Oakey et al., 2016; Cuevas et al., 2017) and integrating secondary traits which are correlated with the trait under selection into the model (Sun et al., 2017).

2.4 Allocation of resources to optimise genetic gain

A key feature of all breeding programmes is the allocation of limited resources to obtain a maximum response to selection. Design factors which affect the utilisation of resources include population size at different stages of trialling, number of replications and number of trial environments (Harshman et al., 2016; Lorenz and Nice, 2017).

The selection strategy at the seedling stage of many apple scion and rootstock breeding programmes relies upon mass selection based on phenotypic information from single individuals (Kumar et al., 2010; Ru, 2016). This strategy is usually adequate for traits with a medium to high heritability, such as many fruit quality characteristics (Oraguzie et al., 2001), but less effective for low heritability traits. Family-based selection, where either the family mean or the relative performance of an individual within its family informs the selection decision, offers the ability to make selections based on traits with a low heritability from the early stages of selection (Walsh and Lynch, 1999). Furthermore, if seedlings from within a family are split between multiple trial sites then genotype × environment interaction estimates can be obtained without the replication of individual genotypes (Kumar et al., 2010). In such schemes, best linear unbiased predictions of breeding values can be calculated based on models incorporating both individual and family performance as well as spatial data (Kumar et al., 2010; Walsh and Lynch, 1999). A recurrent selection scheme for apple based on a combination of within- and between-family selection was implemented in the Plant and Food Research apple breeding programme in New Zealand to broaden the genetic diversity in the programme and ensure long-term genetic gains (Kumar et al., 2010). Families within this programme were initially selected based on the general combining ability of the parents, followed by selecting the individual genotypes with the best performance within families.

An accelerated recurrent selection scheme, as proposed for sugarbeet by Mackay et al. (1999), has the potential to reduce the generation time in apple by half. In the accelerated recurrent selection, full-sibling families are evaluated in trials and while they are being trialled all individuals are also crossed in a complete or incomplete diallel mating design. Selections among the 'progeny' full-sibling families are thereafter made on the basis of the performance of parental families (Mackay et al., 1999). Since the crossing and trialling is conducted simultaneously the progeny families could potentially already be cropping when selections are made from the parental families, thereby halving the breeding cycle (see Fig. 4).

Harshman et al. (2016) evaluated different designs of the second phase trials in the Washington State University apple breeding programme. The evaluations were based on the current trialling system of the breeding programme, which assesses candidate selections in three locations for 3 years with three harvests per year and location. Fruit quality traits for these selections were assessed on fruit both at harvest and after 8 weeks of storage. The reduction of the number of replication units to two locations, 3 years and two harvests had only a small impact on the accuracy and response to selection, but the cost reduction incurred by the less-extensive trialling made it possible to increase the number of trialled selections from 10 to 16.

2.5 Improved phenotyping

In order to correctly make selections, plant breeders need to accurately and quickly measure large amounts of plant traits and this can be achieved through integration of non-destructive technologies into phenotyping protocols. The past few years have seen an increase in the development and application of imaging sensors and software for image analysis, nevertheless many breeding programmes are still heavily dependent on conventional phenotyping for making selections (Finkel, 2009; Fiorani et al., 2012; Fiorani and Schurr, 2013; Fahlgren et al., 2015).

Phenotyping in apple can be divided into two major groups according to the traits measured: tree phenotyping and fruit phenotyping. There are several technologies available for measuring plant traits and the most commonly used in apple are discussed below.

An imaging method used for fruit phenotyping of internal quality characteristics in apple is near-infrared reflectance (NIR) spectroscopy, which uses the near-infrared region of the electromagnetic spectrum (Lammertyn et al., 1998). This imaging technique has several advantages, including its non-destructive nature, high measurement speed and the ability to measure several traits at the same time (Lammertyn et al., 1998). NIR spectroscopy has been used to measure fruit firmness, soluble solid content (SSC) and dry matter content (DMC) in seedlings of apple (Kumar et al., 2015a). The prediction accuracy of NIR was high, with correlation coefficients between observed and predicted values being 0.91 and 0.95 for SSC and DMC, respectively (Kumar et al., 2015a). This technique was used to develop a calibration model for vitamin C, polyphenol and sugar content from a mix of apple varieties and breeding populations and the accuracy of the model was thereafter verified in reference samples with known concentrations, for which the accuracies were $R^2 = 0.8$, $R^2 = 0.94$ and $R^2 = 0.94$, respectively (Pissard et al., 2013). The results are promising, and NIR could be a suitable tool in fruit phenotyping of quality and nutritional traits. There are many more studies where NIR imaging has been used for non-destructive fruit phenotyping; however, these have not been carried out in genetically heterogeneous populations but on single varieties, for example, to measure phenolic compounds and DMC (Pissard et al., 2018), SSC and firmness (Xiaobo et al., 2007; Fan et al., 2009) and to determine apple storage time (Liu and Tang, 2015).

Thermal imaging cameras detect radiation in the long-infrared range, that is, they are able to distinguish the heat emitted by an object. Thermography has been used for both tree and fruit phenotyping to assess incidence of diseases, such as apple scab (Oerke et al., 2011; Belin et al., 2013) and water stress (Gómez-Candón et al., 2015, 2016), to estimate the number and diameter of fruits in field-grown trees (Stajnko et al., 2004) and to detect watercore and bruising in fruits (Varith et al., 2003; Baranowski et al., 2008, 2012).

Unmanned aerial vehicles (UAVs), also known as drones, are used with different imaging technologies attached to the vehicle, which allows for cheap and quick field phenotyping (Gómez-Candón et al., 2017). UAVs have been used in apple to detect water stress (Gómez-Candón et al., 2015, 2016) and to monitor water input and canopy cover (Bulanon et al., 2014). UAVs have also been used as a mean to estimate the number of apple flowers as a proxy for yield and precocity (Xiao et al., 2014) as well as estimating fruit number and yield in the field (Chen et al., 2017; Roy and Isler, 2017).

High-throughput phenotyping platforms (HTPPs) are automated phenotyping facilities commonly used to assess plants grown in highly controlled environments such as glasshouses or growth chambers. One such platform, PhenoArch, has been used with apple trees to phenotype water use efficiency, biomass and transpiration (Lopez et al., 2015, 2017a,b). The PhenoArch platform is available in a glasshouse at INRA in Montpellier (Bosquet et al., 2015). The main limitations of applying HTPPs to apple are: (1) there are only a few available for tree phenotyping since most are designed for broad acre crops, (2) the platforms are currently developed for phenotyping in glasshouses or cabinets which do not always correspond to phenotypes observed in the field, (3) they do not capture genotype × environment interaction (Fiorani and Schurr, 2013; Araus and Cairns, 2014).

Root phenotyping is essential to comprehend nutrient use efficiencies in apple trees, a target for rootstock breeding programmes, but still remains a challenge due to our limited ability to access growing roots through non-destructive methods (Fiorani and Schurr, 2013). Nowadays, several semi-automatic technologies are available for the analysis of root architecture like RootNav (Pound et al., 2013) or SmartRoot (Lobet et al., 2011). These tools often require pre-treatment, such as root tracing, which translates into additional manual work. Advances in root analysis software and in phenotyping systems are needed in order to reduce the analysis time and obtain better measurements of root architecture.

3 Increasing efficiency of production

There is a pressing need to establish integrated and sustainable approaches that will increase both the resource-use efficiency of crops and production per unit area to ensure global food security. To meet the projected demands, food production needs to increase 50% by 2030 and double by 2050 (Parry and Hawkesford, 2010).

3.1 Yield efficiency

One of the ways in which apple breeders try to improve efficiency of production is with the adoption of dwarfing and precocious rootstocks. In recent years, the industry has moved from traditional apple orchard systems (70–100 trees/ha) to planting high-density orchards with less vigorous, closely spaced trees (1000–6000 trees/ha) (Robinson, 2003). Dwarfing rootstocks reduce yield per tree but considerably enhance yield per unit area (Robinson, 2007; Robinson et al., 1991; Hampson et al., 2002). Rootstock-induced dwarfing has been associated in apple with two QTL, denoted *Dw1* and *Dw2*, on chromosomes 5 and 11, respectively (Pilcher et al., 2008; Fazio et al., 2014). Furthermore, a three-locus model for rootstock-induced dwarfing was developed by mapping the root bark ratio, a primary rootstock trait which measures the proportion of cortical cells in the roots. Two of the QTL identified colocalised with *DW1* and *DW2* regions and a third one was found on chromosome 13, a region not previously associated with dwarfing (Harrison et al., 2016).

A recent study carried out in 48 rootstock accessions grafted with the scion 'Fuji' has shown a high negative correlation between cumulative yield efficiency (CYE) and trunk cross-sectional area (TCSA), a commonly used proxy for tree size estimation (Reig et al., 2018). In addition, a moderate positive correlation ($r = 0.5$, $P \leq 0.01$) between CYE and leaf calcium concentration was found (Reig et al., 2018). Among the rootstocks with the highest yield efficiency it is worth mentioning the Geneva rootstocks CG.2034, CG4088, CG.2406 and the Malling rootstocks M9 and M26 (Reig et al., 2018). Data from NIAB EMR, UK, support this negative correlation between TCSA and CYE, confirming that dwarfing rootstocks have a better yield efficiency (see Fig. 5).

Precocious production is another desired trait for increasing the efficiency of apple orchards. The precocity of apple scions have been shown to be affected by which rootstock they are grown on, with dwarfing rootstocks leading to a more precocious production (Webster, 1995). Fazio et al. (2014) studied the effect of rootstocks on precocity and fruit bearing and found that the induction of early bearing was influenced by two loci (*Eb1* and *Eb2*) and that a major QTL for a number of fruits per tree colocalised with *Dw2*. Markers segregating with these loci could therefore be suitable for predicting the effect on precocity and fruiting of apple rootstock genotypes (Fazio et al., 2014). There is also genetic variation in the precocity of apple scion genotypes, as shown in an evaluation of the fruiting ability of a total of 756 seedlings from 23 apple scion progenies (Blažek et al., 2016). The crossing of genotype HL1737 × Pink Lady resulted in the most precocious offspring, with an average time till they start to bear fruit of 1.56 years, compared to the least precocious offspring which started bearing in the fifth year. There is some evidence that length of seedling juvenility is

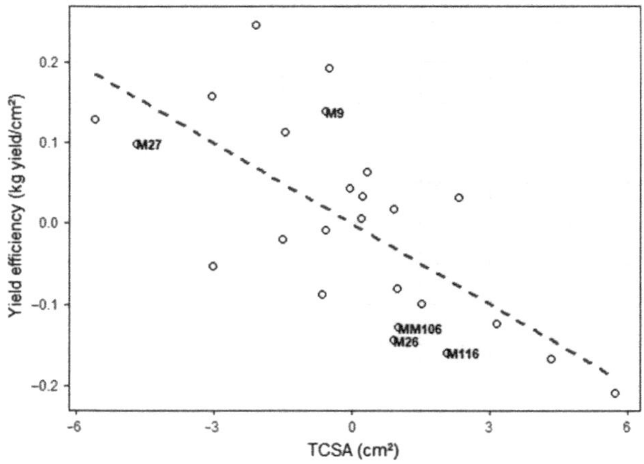

Figure 5 The correlation between best linear unbiased predictions of yield efficiency and trunk cross-sectional area (TCSA: a common proxy for tree volume) for rootstocks trialled in East Malling, UK. The graph showcases the negative relationship between tree size and yield efficiency in apple rootstocks. Source: adapted from Karlström, Lipska and Fernandez (unpublished results).

correlated with the precocity of mature plant material (Visser and de Vries, 1970), in which case it would be worthwhile to select for precocity in seedling material.

Many studies have been focused on the control of biennial bearing which is causing irregular yield at a mature stage Biennial bearing occurs when apple trees have a heavy fruit load one year which affects the floral formation for the next year (Jonkers, 1979; Monselise and Goldschmidt, 1982). A study carried out in flowering individuals from an F1 population, from a Starkrimson × Granny Smith cross, a strong biennial bearer and regular bearer respectively, showed that biennial bearing is heritable and segregates in an apple progeny (Guitton et al., 2012). Inflorescence traits exhibited the highest genotypic effect, and three QTLs explained 50% of the phenotypic variability for biennial bearing. The results suggest that biennial bearing is more closely linked to floral induction, rather than to pollination or fruit set, and this discovery was supported by the colocalisation of QTLs for annual yield with the position of genes related to metabolism, degradation and transport of gibberellins and auxins (Guitton et al., 2012). Subsequently, the same population was utilised for estimating the genetic value of segregating apple progenies for irregular bearing where relevant QTLs were detected (Durand et al., 2013). All these studies suggest that selecting new varieties with intrinsic regular bearing is a possible strategy through the development of specific molecular markers.

Although there is a focus on improving yields in breeding programmes of apple rootstocks, it has a lower priority in scion breeding, for example, in a survey by Sansavini et al. (2004) only 5 out of 25 European breeding programmes answered that productivity was a main breeding objective.

3.2 Resource-use efficiency

Breeding for water use efficiency (WUE) and drought resistance has also been of interest for breeders since feeding an increasing world population with reduced water resources has become a pressing issue (Hamdy et al., 2003; Tester and Langridge, 2010). WUE measures how much biomass is produced for a given amount of water transpired and is a good indicator for plant adaptation to drought conditions (Adiredjo et al., 2014; Lopez et al., 2015, 2017a,b). Selecting drought resistance genotypes could contribute to improving yield in adverse conditions as well as to growing apples in new climatic areas with low water availability.

To study the genetic variation and heritability of WUE, plant biomass and transpiration, 193 apple genotypes were evaluated under well-watered conditions at seedling stage (Lopez et al., 2017a). Plants were grown and phenotyped in a glasshouse with an HTPP. WUE showed a high heritability (H^2 = 0.73) but the majority of the variability in WUE (70%) was explained by plant biomass, indicating that indirect selection on biomass would be sufficient to select for improved water use under well-watered conditions. The evaluation of WUE in apple genotypes under drought conditions could further the understanding of the genetic control of WUE in apple and the potential of adapting germplasm to low water availability.

The effect of rootstock on the absorption and translocation of macro- and micronutrients to the scion has been studied but poorly translated into breeding and selection for increased nutrient availability in rootstock material (Fazio et al., 2013). A QTL mapping study to determine the genetic loci involved in the absorption and translocation of nutrients was conducted using progenies from an Ottawa 3 × Robusta 5 cross. The QTL for potassium concentration in leaves colocated with the dwarfing locus *DW1* (Fazio et al., 2013). Another study was carried out in an Ottawa 3 rootstock progeny which segregated for both dwarfing loci, *DW1* and *DW2* and showed that fruit nitrogen and sulphur concentration were higher in trees grafted on rootstocks that possessed either of the dwarfing alleles, revealing that segregation of these loci could have a significant effect on the nutrient levels of fruits and leaves (Fazio et al., 2017). In addition, tree size was positively correlated with leaf concentration of potassium, sulphur and calcium (Fazio et al., 2017). Improving the ability of apple rootstocks to increase nutrient availability to grafted scions could contribute positively to the general health and productivity of the tree as well as to increase the nutritional value, organoleptic quality and storage potential of apples (Autio, 1994; Fallahi, 2012).

4 Mitigating production losses and waste

Increases in apple yield can be greatly offset by the action of pests and diseases and through the development of physiological disorders. This is not only a concern in the pre-harvest stage of the production chain but also during post-harvest storage of the produce, as 8–10% of the total global apple production in 2011–13 was estimated to have been lost due to wastage between harvest and reaching the consumer (FAOSTAT, 2018). In order to reduce losses and waste, while at the same time lowering pesticide inputs, an emphasis will have to be placed on breeding varieties tolerant to biotic and abiotic stresses during all stages of production. Here we discuss biotic factors that are important in the commercial production of apple. The major focus of apple breeding and research has been on improving resistance to diseases occurring in the field, while relatively little attention has been given to tolerance to storage diseases (Sansavini et al., 2004).

4.1 Pre-harvest

Increasing the resilience to field diseases has been of large importance in many apple breeding programmes, both for scion and rootstock varieties. Breeding efforts have mainly been focused on improving resistance to the two fungal pathogens apple scab, *Venturia inaequalis,* and powdery mildew, *Podosphaera leucotricha,* as well as the fire blight causing bacterium *Erwinia amylovora*. Varietal tolerance to the two major wood pathogens, the fungus *Neonectria ditissima* and bacterium *Valsa mali*, is recognised as being of high importance, but has received relatively little attention up until recently (Abe et al., 2007, 2011; Ghasemkhani et al., 2015b; Garkava-Gustavsson et al., 2016; Gómez-Cortecero et al., 2016). Resistance to the woolly apple aphid, *Eriosoma lanigerum*, has been an important objective for rootstock breeding and varieties with resistance from *Malus robusta* 'Robusta 5' (used as a parent in the Malling Merton series) and the variety 'Northern Spy' (parent in the Geneva-series) have been released (Bus et al., 2008).

4.1.1 Apple scab

Apple scab, *Venturia inaequalis,* causes brown lesions on leaves and fruits, making them unmarketable, and is considered the most economically damaging disease in apple (Vaillancourt and Hartman, 2000). In the early twentieth century, genetic resistance was recognised as a good control measure of the fungus, which led breeders to initiate crosses with wild apple species exhibiting resistance to the disease (Joshi et al., 2009). The apple scab pathosystem was also one of the first host-pathogen interactions in which the gene-for-gene model was demonstrated (Flor, 1971). A total of 20 different major resistance loci (denoted *Rvi1-20*) have since been identified in *Malus* germplasm (Clark et al., 2014). Avirulence alleles corresponding to 5 out of the 20 *Rvi* loci have been described in isolates of *V. inaequalis* (see Clark et al., 2014) or sources of resistance and interactions. The genomic location of several of these resistance genes have been published and markers developed (Cheng et al., 1998; Tartarini et al., 2000; Gygax et al., 2004; Patocchi et al., 2009). The *Rvi6* (previously denoted *Vf*) resistance, originating from *Malus floribunda* 821, is by far the most utilised source of resistance in the development of scab-resistant varieties (Gessler and Pertot, 2012). Furthermore, the majority of the commercially released *Rvi6*-resistant varieties have ancestry in the same resistant selections produced by the PRI Cooperation Program (Purdue University, Rutgers University and the University of Illinois) (Gessler and Pertot, 2012). There is an understanding that the genetic base of resistance needs to be broadened to avoid the selection of *V. inaequalis* pathotypes able to overcome the *Rvi6* resistance (Parisi et al., 1993). There have therefore been attempts to use plant material with other, or polygenic resistance and to pyramid resistance genes in a single genotype (Kellerhals et al., 2009; Clark et al., 2014; Bastiaanse et al., 2016; Peil et al., 2007).

4.1.2 Apple powdery mildew

Apple powdery mildew, *Podosphaera leucotricha,* can be observed as a white layer of mycelia on leaves and shoots of apple. The pathogen can cause diminished shoot growth and flowering, fruit russeting and a general reduction in health due to reduced photosynthesis (Holb, 2017). Five sources of major resistance have been identified and markers segregating with the resistance developed: Pl_1 from *M. robusta* (Dunemann et al., 2007), Pl_2 from *Malus zumi* (Baumgartner et al., 2015), Pl_d from 'D12' (James et al., 2004), Pl_m from 'Mildew Immune Selection' (Bus et al., 2010) and Pl_w from the ornamental crab apple 'White Angel' (Evans and James, 2003). In addition, there is evidence of polygenic resistance to *P. leucotricha* (Calenge and Durel, 2006; Stankiewicz-Kosyl et al., 2005). Alternative approaches that can be used in breeding for reduced incidence of powdery mildew are selection of loss-of-function mutations in susceptibility genes. *Mildew resistance locus o* (*Mlo*) genes are known to confer susceptibility to powdery mildew in a multitude of crops (Pessina et al., 2016a,b; Bracuto et al., 2017; Kusch and Panstruga, 2017). More than 20 *Mlo* gene homologues have been identified in the 'Golden Delicious' genome (Pessina et al., 2014). A knock-down mutation of one of these *Mlo* genes, *MdMLO19,* was shown to lead to a significant reduction of powdery mildew in apple (Pessina et al., 2016b). This finding supports the hypothesis that targeted mutation of susceptibility genes could be a feasible method to increase resistance to apple powdery mildew, as long as there are no pleiotropic effects which affect normal plant development. Furthermore, natural variations in alleles of *Mlo* genes could be exploited for resistance breeding (Pessina et al., 2017).

4.1.3 Fire blight

Fire blight, *Erwinia amylovora,* is caused by a gram-negative bacterium which produces blossom, rootstock and shoot blight in host trees (Norelli et al., 2003). Resistance towards the pathogen has been identified in both cultivated apple varieties and in wild apple accessions (Calenge et al., 2005; Durel et al., 2009; Le Roux et al., 2010; Emeriewen et al., 2017; Harshman et al., 2017). One of the resistant accessions described is the crab apple *Malus robusta* 'Robusta 5', from which a major resistance locus to *E. amylovora* was discovered through linkage mapping (Peil et al., 2007). A putative resistance gene was subsequently identified within the locus, and designated *FB_MR5* (Fahrentrapp et al., 2013). The effect of *FB_MR5* on the incidence of *E. amylovora* was verified by cloning the resistance gene into the susceptible variety 'Gala Galaxy', which rendered transgenic lines with significantly lower symptom development after artificial inoculations compared to the untransformed equivalents (Broggini et al., 2014; Kost et al., 2015). Resistance loci from the species *Malus arnoldiana* (Emeriewen et al., 2017), *Malus fusca* (Emeriewen et al., 2014) and the variety 'Evereste' (Durel et al., 2009) have also been mapped. The application of these varieties in scion breeding has been limited due to their poor fruit quality, which means that several generations of pseudo-backcrossing would be necessary to obtain a commercial variety (Kellerhals et al., 2008). Nevertheless, QTL from several varieties of cultivated apple have also been mapped (Calenge et al., 2005; Durel et al., 2009; Le Roux et al., 2010; van de Weg et al., 2018), which has enabled the development of molecular markers linked to a major resistance QTL on linkage group 7 in the variety 'Fiesta' (Khan et al., 2007).

4.1.4 European apple canker

European apple canker, *Neonectria ditissima,* is mainly a wood pathogen which causes trunk cankers, stem and branch lesions and dieback (Weber, 2014). The disease has become increasingly significant in recent years, particularly in apple growing regions with maritime temperate climates, where the high susceptibility of many modern varieties and limited effective control measures add to its prevalence (Beresford and Kim, 2011; Weber, 2014; Gómez-Cortecero et al., 2016). To date, there has been no report of any apple variety or species of *Malus* that exhibits complete immunity to *N. ditissima.* There is however germplasm that has exhibited partial resistance in experiments with controlled inoculations (van de Weg, 1989; Gelvonauskienė et al., 2007; Ghasemkhani et al., 2015a; Gómez-Cortecero et al., 2016). Breeding efforts to improve resistance in the germplasm have been limited, probably due to the highly polygenic nature of identified resistance and the large influence of environment on disease expression (Gómez-Cortecero et al., 2016). Nonetheless, there is ongoing research to map resistance QTL from 'Golden Delicious', *Malus robusta* 'Robusta 5', as well as from a multiparental breeding population (Peter Braun, pers. comm.; Vincent Bus, pers. comm.; Amanda Karlström, unpublished results).

4.2 Post-harvest

Due to the year-round demand for apples, the storability of a variety is an important factor for its acceptability on the market. Consequently, it has also been a priority for apple breeders and researchers to understand genetic factors involved in physiological disorders and breakdown processes that occur during storage (Costa et al., 2008; Edge-Garza et al., 2010; Buti et al., 2015; McClure et al., 2016; Howard et al., 2018). Yet, genetically based tolerance to post-harvest disease is an underdeveloped research area. This could partly be due to the fact that susceptibility to several post-harvest diseases has been found to be correlated with harvest date and firmness which makes it difficult to separate other genetic components involved in resistance.

4.2.1 Blue mould

Blue mould, *Penicillium expansum,* is a fungal storage rot that can produce the mycotoxins patulin and citrinin (Tahir et al., 2015). In commercial apple cultivars, the variation in susceptibility to blue mould has been shown to be negatively correlated to harvest date and fruit firmness at harvest and positively correlated to loss of firmness during storage (Ahmadi-Afzadi et al., 2013). The current breeding practice of selecting for high fruit firmness at harvest which is maintained during storage will therefore indirectly select for an increased tolerance to *P. expansum*. The varieties surveyed in these two studies showed a variation in susceptibility to blue mould, but none of them exhibited complete immunity. Complete resistance to *P. expansum* has, however, been reported in a few accessions of *Malus sieversii* (Janisiewicz et al., 2008). A QTL, explaining 30% of the variation in resistance, was subsequently mapped in the progeny of a cross between the *M. sieversii* accession GMAL4593 × 'Royal Gala' (Norelli et al., 2014). GMAL4593 was however, not one of the accessions found to be completely immune in the study by Janisiewicz et al. (2008), but classified as moderately resistant.

4.2.2 Grey mould

Grey mould, *Botrytis cinerea,* and *Colletotrichum gloeosporioides* are the other major fungal causes of fruit decay in apple during storage worldwide (Davey et al., 2007; Xiao and Kim, 2008). A set of 30 varieties has been screened for resistance to these two pathogens by Auwerkerken et al. (2009). In the tested varieties, resistance to *B. cinerea* but not *C. gleosporioides* (formerly *Gloeosporium fructigenum*) was correlated with the harvest date of the cultivar. As in the case of *P. expansum,* none of the varieties were reported to be completely resistant to the two pathogens studied (Auwerkerken et al., 2009).

There are several other fungal species affecting apple fruit during post-harvest storage, such as *Neonectria ditissima, Mucor piriformis* and *Neofabrea* spp. (Spotts et al., 1999; Gariepy et al., 2005; Xu and Robinson, 2010). However, the knowledge on resistance in the apple germplasm to these post-harvest pathogens is limited (Noble and Drysdale, 1983; Spotts et al., 1999).

5 Summary

Research and development in perennial crops is crucial to ensure that the future of apple growing remains a profitable and sustainable method of delivering high-quality, nutritious food to consumers. This chapter has identified some traits which if introgressed into mainstream varieties will offer a significant improvement in apple production. This chapter also described a range of modern breeding and genomics approaches that can be used to more rapidly lead to genetic gain and the delivery of new cultivars to the market. However, for truly sustainable crop production and intensification, genetic improvements cannot be seen in isolation but will need to be integrated with agronomical developments. It has been proposed that nitrogen and water availability are larger limitations to yield increase than plant genetics (Sinclair and Rufty, 2012). Hence, to increase production without expanding agricultural land, there is a need to shift focus from genotype performance at optimal conditions to the interaction between genotype, management practices and environment. For apple breeding, which has traditionally focused on qualitatively controlled traits such as fruit quality and disease resistance, there is a need to increase efforts towards optimised tree architecture, vegetative growth and phenology. This could involve designing tree ideotypes for environments and management practices, to be used as a guidance for breeding targets. There is also a need to take into account future climates with potentially more extreme weather events, for which optimal phenotypes could be developed based on climate forecasting.

6 Future challenges

A future challenge for the long-term success of apple breeding is the accumulation of deleterious mutations in the genome which may hamper the progress of improving apple yields, quality, resilience or production efficiency. Genetic mutations can have either a positive, neutral or negative effect on evolutionary fitness, however there is evidence that the majority of mutations which lead to an amino acid substitution (i.e. non-synonymous) have weak detrimental effects (Eyre-Walker and Keightley, 2007). Deleterious mutations are considered to be a major factor in inbreeding depression (Charlesworth and Willis, 2009) and the complementation of recessive or partly recessive mutations an explanation of heterosis (Yang et al., 2017). Factors affecting the rate and degree of accumulation of deleterious mutations include (1) population bottlenecks, such as a domestication event, (2) purging of mutations through events of recombination and selection, (3) rate of mutation (Gaut et al., 2015). There is empirical evidence of a substantial genetic load in annual crop species (Mezmouk and Ross-Ibarra, 2013; Renaut and Rieseberg, 2015; Liu et al., 2017) as well as in cassava (Ramu et al., 2017), and data corroborating the effect of deleterious mutations on traits of agronomic importance (Yang et al., 2017). Previous studies have indicated that cultivated apple has not been subject to a domestication bottleneck (Gross et al., 2014; Duan et al., 2017), and the genetic load in the apple genome would thus be expected to be lower than for crop species with substantial bottlenecks (Lohmueller, 2014). However the current practice of using a genetically narrow gene pool for breeding for commercial production is likely to result in a bottleneck if material of a wider genetic diversity is not utilised in population improvement (Mezmouk and Ross-Ibarra, 2013). There is also a possibility that deleterious variants have accumulated in apple due to it being clonally propagated and therefore subject to fewer events of recombination, as well as recessive mutations being maintained in the population because of its self-incompatibility system and obligate outcrossing status (Gaut et al., 2015). In cassava (which similarly to apple is both clonally propagated and predominantly bred through outcrossing) a reduction in homozygous deleterious mutations was evident in domesticated genotypes compared to their progenitors, whereas the same material had a higher heterozygous genetic load (Ramu et al., 2017), thus indicating that recessive deleterious mutations have been kept masked through selection of genotypes heterozygous at these loci. There are several approaches which could be taken to purge perennial crop genomes from deleterious genetic burdens. One approach, which can be directly incorporated in breeding programmes, is to unmask partly recessive mutations through selfing followed by deselection of lower performing genotypes. The level of recombination per generation is limited due to constraints in the number of meiotic crossovers that can occur for each pair of chromosomes (Fernandes et al., 2017). Hence, methods to increase meiotic crossovers in low-recombination regions of the genome could be a means of improving the removal of detrimental genetic load. There is evidence that manipulating external factors such as temperature, nutrient availability, irradiation and pathogen infection can increase recombination (Lambing et al., 2017). Additionally, developments towards the understanding how meiotic crossover recombinations are genetically and epigenetically regulated are underway to deliver new techniques that could be used to alter recombination rates in breeding programmes. This has been demonstrated in *Arabidopsis* where transgenic approaches have been used to achieve up to eightfold increases in meiotic crossovers (Fernandes et al., 2017; Ziolkowski et al., 2017; Serra et al., 2018). There is potential to use such methods coupled with genomic prediction models that incorporate predictors of putatively deleterious SNPs to rapidly purge breeding material from genetic load (Yang et al., 2017). Genome editing of multiple deleterious alleles simultaneously has also been suggested as a feasible method to remove deleterious variants in breeding programmes (Johnsson et al., 2018).

7 Where to look for further information

For a more comprehensive description of apple physiology and cultivation, as well as breeding, see the publication *Achieving Sustainable Cultivation of Apples* edited by the apple breeder at Washington State University, Kate Evans (Evans, 2017).

Information and updates on available DNA marker tests and phenotyping protocols for apple are published on the RosBREED project website (https://www.rosbreed.org), where you also can find recent publications on trait-marker associations and advances in breeding methods. We would also like to recommend the Genome Database for Rosaceae (https://www.rosaceae.org) which provides data mining tools and publicly available genomics, genetics and breeding data for Rosaceous species, including apple.

The International Rosaceae Genomics Conference is a biennially held research conference with a focus on molecular genetics and genomics of Rosaceous species. In addition to apple this conference covers other horticulturally important fruit crops such as peach, cherry, pear, strawberry and raspberry.

8 References

Abe, K., Kotoda, N., Kato, H. and Soejima, J. 2007. Resistance sources to Valsa canker (*Valsa ceratosperma*) in a germplasm collection of diverse *Malus* species (*Valsa ceratosperma*). *Plant Breeding* 126(4), 449–53. doi:10.1111/j.1439-0523.2007.01379.x.

Abe, K., Kotoda, N., Kato, H. and Soejima, J. 2011. Genetic studies on resistance to Valsa canker in apple: Genetic variance and breeding values estimated from intra- and inter-specific hybrid progeny populations. *Tree Genetics and Genomes* 7(2), 363–72. doi:10.1007/s11295-010-0337-3.

Adiredjo, A. L., Navaud, O., Muños, S., Langlade, N. B., Lamaze, T. and Grieu, P. 2014. Genetic control of water use efficiency and leaf carbon isotope discrimination in sunflower (*Helianthus annuus* L.) subjected to two drought scenarios. *PLoS ONE* 9(7), e101218. doi:10.1371/journal.pone.0101218.

Ahmadi-Afzadi, M., Tahir, I. and Nybom, H. 2013. Impact of harvesting time and fruit firmness on the tolerance to fungal storage diseases in an apple germplasm collection. *Postharvest Biology and Technology* 82, 51–8. doi:10.1016/j.postharvbio.2013.03.001.

Albrechtsen, A., Nielsen, F. C. and Nielsen, R. 2010. Ascertainment biases in SNP chips affect measures of population divergence. *Molecular Biology and Evolution* 27(11), 2534–47. doi:10.1093/molbev/msq148.

Aldwinckle, H. S., Gustafson, H. L. and Lamb, R. C. 1976. Early determination of genotypes for apple scab resistance by forced flowering of test cross progenies. *Euphytica* 25(1), 185–91. doi:10.1007/BF00041544.

Allard, A., Bink, M. C. A. M., Martinez, S., Kelner, J. J., Legave, J. M., di Guardo, M., Di Pierro, E. A., Laurens, F., van de Weg, E. W. and Costes, E. 2016. Detecting QTLs and putative candidate genes involved in budbreak and flowering time in an apple multiparental population. *Journal of Experimental Botany* 67(9), 2875–88. doi:10.1093/jxb/erw130.

Antanaviciute, L., Fernández-Fernández, F., Jansen, J., Banchi, E., Evans, K. M., Viola, R., Velasco, R., Dunwell, J. M., Troggio, M. and Sargent, D. J. 2012. Development of a dense SNP-based linkage map of an apple rootstock progeny using the *Malus* Infinium whole genome genotyping array. *BMC Genomics* 13, 203. doi:10.1186/1471-2164-13-203.

Araus, J. L. and Cairns, J. E. 2014. Field high-throughput phenotyping: The new crop breeding frontier. *Trends in Plant Science* 19(1), 52–61. doi:10.1016/j.tplants.2013.09.008.

Autio, W. R. 1994. Rootstock affects apple ripening, quality and storability. *Compact Fruit Tree* 27, 41–7.

Auwerkerken, A., Davey, M. W., Van den Putte, A. and Keulemans, J. 2009. Screening for *Botrytis cinerea* resistant apple cultivars. *Acta Horticulturae* 814(814), 759–64. doi:10.17660/ActaHortic.2009.814.128.

Badiu, D., Arion, F., Muresan, I., Lile, R. and Mitre, V. 2015. Evaluation of economic efficiency of apple orchard investments. *Sustainability* 7(8), 10521-33. doi:10.3390/su70810521.

Bai, Y., Dougherty, L., Li, M., Fazio, G., Cheng, L. and Xu, K. 2012. A natural mutation-led truncation in one of the two aluminium-activated malate transporter-like genes at the Ma locus is associated with low fruit acidity in apple. *Molecular Genetics and Genomics: MGG* 287(8), 663–78. doi:10.1007/s00438-012-0707-7.

Bannier, H.-J. 2011. Moderne apfelzüchtung: Genetische verarmung und tendenzen zur inzucht. *Erwerbs-Obstbau* 52(3–4), 85–110. doi:10.1007/s10341-010-0113-4.

Baranowski, P., Lipecki, J., Mazurek, W. and Walczak, R. T. 2008. Detection of watercore in 'Gloster' apples using thermography. *Postharvest Biology and Technology* 47(3), 358–66. doi:10.1016/j.postharvbio.2007.07.014.

Baranowski, P., Mazurek, W., Wozniak, J. and Majewska, U. 2012. Detection of early bruises in apples using hyperspectral data and thermal imaging. *Journal of Food Engineering* 110(3), 345–55. doi:10.1016/j.jfoodeng.2011.12.038.

Bastiaanse, H., Bassett, H. C. M., Kirk, C., Gardiner, S. E., Deng, C., Groenworld, R., Chagné, D. and Bus, V. G. 2016. Scab resistance in 'Geneva' apple is conditioned by a resistance gene cluster with complex genetic control. *Molecular Plant Pathology* 17(2), 159–72. doi:10.1111/mpp.12269.

Baumgartner, I. O., Patocchi, A., Frey, J. E., Peil, A. and Kellerhals, M. 2015. Breeding elite lines of apple carrying pyramided homozygous resistance genes against apple scab and resistance against powdery mildew and fire blight. *Plant Molecular Biology Reporter* 33(5), 1573–83. doi:10.1007/s11105-015-0858-x.

Beaulieu, J., Doerksen, T. K., MacKay, J., Rainville, A. and Bousquet, J. 2014. Genomic selection accuracies within and between environments and small breeding groups in white spruce. *BMC Genomics* 15, 1048. doi:10.1186/1471-2164-15-1048.

Belin, É., Rousseau, D., Boureau, T. and Caffier, V. 2013. Thermography versus chlorophyll fluorescence imaging for detection and quantification of apple scab. *Computers and Electronics in Agriculture* 90, 159–63. doi:10.1016/j.compag.2012.09.014.

Beresford, R. M. and Kim, K. S. 2011. Identification of regional climatic conditions favorable for development of European canker of apple. *Phytopathology* 101(1), 135–46. doi:10.1094/PHYTO-05-10-0137.

Bhatia, D., Wing, R. A. and Singh, K. 2013. Genotyping by sequencing its implications and benefits. *Crop Improvement* 40(2), 101–11.

Bianco, L., Cestaro, A., Sargent, D. J., Banchi, E., Derdak, S., Di Guardo, M., Salvi, S., Jansen, J., Viola, R., Gut, I., et al. 2014. Development and validation of a 20K single nucleotide polymorphism (SNP) whole genome genotyping array for apple (*Malus × domestica* Borkh). *PLoS ONE* 9(10), e110377. doi:10.1371/journal.pone.0110377.

Bianco, L., Cestaro, A., Linsmith, G., Muranty, H., Denancé, C., Théron, A., Poncet, C., Micheletti, D., Kerschbamer, E., Di Pierro, E. A., et al. 2016. Development and validation of the Axiom(®) Apple480K SNP genotyping array. *The Plant Journal: for Cell and Molecular Biology* 86(1), 62–74. doi:10.1111/tpj.13145.

Blažek, J., Zelený, L. and Křelinová, J. 2016. Precocity and a long-term cropping in apple progenies grown on M 9 rootstock. *Horticultural Science* 43(4), 167–74. doi:10.17221/268/2015-HORTSCI.

Bosquet, L. C., Brichet, N., Fournier, C., Grau, A., Mineau, J., Negre, V., Neveu, P., Suard, B., Tireau, A., Welcker, C., et al. 2015. PHENOARCH, a multiscale phenotyping platform for plant architecture, growth rate, water use efficiency and radiation use efficiency. In: *Communication at EPPN Plant Phenotyping Symposium*, November 2015, Barcelona, Spain.

Bracuto, V., Appiano, M., Ricciardi, L., Göl, D., Visser, R. G. F., Bai, Y. and Pavan, S. 2017. Functional characterization of the powdery mildew susceptibility gene *SmMLO1* in eggplant (*Solanum melongena* L.). *Transgenic Research* 26(3), 323–30. doi:10.1007/s11248-016-0007-9.

Broggini, G. A. L., Wöhner, T., Fahrentrapp, J., Kost, T. D., Flachowsky, H., Peil, A., Hanke, M. V., Richter, K., Patocchi, A. and Gessler, C. 2014. Engineering fire blight resistance into the apple cultivar 'Gala' using the FB_MR5 CC-NBS-LRR resistance gene of *Malus × robusta* 5. *Plant Biotechnology Journal* 12(6), 728–33. doi:10.1111/pbi.12177.

Brown, S. K. 2009. Making sense of new apple varieties, trademarks and clubs: Current status. *New York Fruit Quarterly* 17(3): 9–12.

Bulanon, D., Horton, M., Salvadore, P. and Fallahi, E. 2014. Apple orchard monitoring using aerial multispectral imaging. ASABE Paper No. 1913165. American Society of Agricultural and Biological Engineers, pp. 1–7.

Bus, V. G. M., Chagné, D., Bassett, H. C. M., Bowatte, D., Calenge, F., Celton, J.-M., Durel, C.-E., Malone, M. T., Patocchi, A., Ranatunga, A. C., et al. 2008. Genome mapping of three major resistance genes to woolly apple aphid (*Eriosoma lanigerum* Hausm.). *Tree Genetics and Genomes* 4(2), 223–36. doi:10.1007/s11295-007-0103-3.

Bus, V. G. M., Bassett, H. C. M., Bowatte, D., Chagné, D., Ranatunga, C. A., Ulluwishewa, D., Wiedow, C. and Gardiner, S. E. 2010. Genome mapping of an apple scab, a powdery mildew and a woolly apple aphid resistance gene from open-pollinated mildew immune selection. *Tree Genetics and Genomes* 6(3), 477–87. doi:10.1007/s11295-009-0265-2.

Buti, M., Poles, L., Caset, D., Magnago, P., Fernández-Fernández, F., Colgan, R. J., Velasco, R. and Sargent, D. J. 2015. Identification and validation of a QTL influencing bitter pit symptoms in apple (*Malus × domestica*). *Molecular Breeding* 35(1). doi:10.1007/s11032-015-0258-9.

Calenge, F. and Durel, C.-E. 2006. Both stable and unstable QTLs for resistance to powdery mildew are detected in apple after four years of field assessments. *Molecular Breeding* 17(4), 329–39. doi:10.1007/s11032-006-9004-7.

Calenge, F., Drouet, D., Denancé, C., van de Weg, W. E., Brisset, M. N., Paulin, J. P. and Durel, C. E. 2005. Identification of a major QTL together with several minor additive or epistatic QTLs for resistance to fire blight in apple in two related progenies. *Theoretical and Applied Genetics* 111(1), 128–35. doi:10.1007/s00122-005-2002-z.

Camacho, A., Van Deynze, A., Chi-Ham, C. and Bennett, A. B. 2014. Genetically engineered crops that fly under the US regulatory radar. *Nature Biotechnology* 32(11), 1087–91. doi:10.1038/nbt.3057.

Celton, J.-M., Tustin, D. S., Chagné, D. and Gardiner, S. E. 2009. Construction of a dense genetic linkage map for apple rootstocks using SSRs developed from *Malus* ESTs and *Pyrus* genomic sequences. *Tree Genetics and Genomes* 5(1), 93–107. doi:10.1007/s11295-008-0171-z.

Centro de Pomaceas. 2019. Programa mejoramiento genético. Available at: http://pomaceas.utalca.cl/investigacion/programa-mejoramiento-genetico.

Chagné, D., Crowhurst, R. N., Troggio, M., Davey, M. W., Gilmore, B., Lawley, C., Vanderzande, S., Hellens, R. P., Kumar, S., Cestaro, A., et al. 2012. Genome-wide SNP detection, validation, and development of an 8K SNP array for apple. *PLoS ONE* 7(2), e31745. doi:10.1371/journal.pone.0031745.

Charlesworth, D. and Willis, J. H. 2009. The genetics of inbreeding depression. *Nature Reviews. Genetics* 10(11), 783–96. doi:10.1038/nrg2664.

Chen, S. W., Shivakumar, S. S., Dcunha, S., Das, J., Okon, E., Qu, C., Taylor, C. J. and Kumar, V. 2017. Counting apples and oranges with deep learning: A data-driven approach. *IEEE Robotics and Automation Letters* 2(2), 781–8. doi:10.1109/LRA.2017.2651944.

Cheng, F. S., Weeden, N. F. and Brown, S. K. 1996. Identification of co-dominant RAPD markers tightly linked to fruit skin color in apple. *Theoretical and Applied Genetics* 93(1–2), 222–7. doi:10.1007/BF00225749.

Cheng, F. S., Weeden, N. F., Brown, S. K., Aldwinckle, H. S., Gardiner, S. E. and Bus, V. G. 1998. Development of a DNA marker for *Vm*, a gene conferring resistance to apple scab. *Genome* 41(2), 208–14. doi:10.1139/g98-020.

Clark, M. D., Luby, J. J. and Bradeen, J. M. 2014. Characterizing the host response and genetic control of resistance in 'Honeycrisp' to apple scab (*Venturia inaequalis*). *European Journal of Plant Pathology* 140(1), 69–81.

Costa, F., van De Weg, W. E., Stella, S., Dondini, L., Pratesi, D., Musacchi, S. and Sansavini, S. 2008. Map position and functional allelic diversity of *Md-Exp7*, a new putative expansin gene associated with fruit softening in apple (*Malus* × *domestica* Borkh.) and pear (*Pyrus communis*). *Tree Genetics and Genomes* 4(3), 575–86. doi:10.1007/s11295-008-0133-5.

Court of Justice of the European Union. 2018. Press Release No 111/18, 25 July 2018. Judgment in case C-528/16. Available at: https://g8fip1kplyr33r3krz5b97d1-wpengine.netdna-ssl.com/wp-content/uploads/2018/07/CP180111EN.pdf.

CPVO. 2018. Community Plant Variety Office (CPVO) statistics database. Available at: http://cpvoextranet.cpvo.europa.eu/DENOMINATIONS_WEB/UK/PAGE_statistics.htm (Accessed on 14 June 2018).

Crossa, J., Pérez-Rodríguez, P., Cuevas, J., Montesinos-López, O., Jarquín, D., de Los Campos, G., Burgueño, J., González-Camacho, J. M., Pérez-Elizalde, S., Beyene, Y., et al. 2017. Genomic selection in plant breeding: Methods, models, and perspectives. *Trends in Plant Science* 22(11), 961–75. doi:10.1016/j.tplants.2017.08.011.

Cuevas, J., Crossa, J., Montesinos-López, O. A., Burgueño, J., Pérez-Rodríguez, P. and de Los Campos, G. 2017. Bayesian genomic prediction with genotype × environment interaction kernel models. *G3 (Bethesda, Md.)* 7(1), 41–53. doi:10.1534/g3.116.035584.

Daccord, N., Celton, J. M., Linsmith, G., Becker, C., Choisne, N., Schijlen, E., van de Geest, H., Bianco, L., Micheletti, D., Velasco, R., et al. 2017. High-quality de novo assembly of the apple genome and methylome dynamics of early fruit development. *Nature Genetics* 49(7), 1099–106. doi:10.1038/ng.3886.

Davey, M. W., Auwerkerken, A. and Keulemans, J. 2007. Relationship of apple vitamin C and antioxidant contents to harvest date and postharvest pathogen infection. *Journal of the Science of Food and Agriculture* 87(5), 802–13. doi:10.1002/jsfa.2777.

de Los Campos, G., Hickey, J. M., Pong-Wong, R., Daetwyler, H. D. and Calus, M. P. 2013. Whole-genome regression and prediction methods applied to plant and animal breeding. *Genetics* 193(2), 327–45. doi:10.1534/genetics.112.143313.

Duan, N., Bai, Y., Sun, H., Wang, N., Ma, Y., Li, M., Wang, X., Jiao, C., Legall, N., Mao, L., et al. 2017. Genome re-sequencing reveals the history of apple and supports a two-stage model for fruit enlargement. *Nature Communications* 8(1), 249. doi:10.1038/s41467-017-00336-7.

Dunemann, F., Peil, A., Urbanietz, A. and Garcia-Libreros, T. 2007. Mapping of the apple powdery mildew resistance gene *Pl1* and its genetic association with an NBS-LRR candidate resistance gene. *Plant Breeding* 126(5), 476–81. doi:10.1111/j.1439-0523.2007.01415.x.

Durand, J. B., Guitton, B., Peyhardi, J., Holtz, Y., Guédon, Y., Trottier, C. and Costes, E. 2013. New insights for estimating the genetic value of segregating apple progenies for irregular bearing during the first years of tree production. *Journal of Experimental Botany* 64(16), 5099–113. doi:10.1093/jxb/ert297.

Durel, C. E., Denancé, C. and Brisset, M. N. 2009. Two distinct major QTL for resistance to fire blight co-localize on linkage group 12 in apple genotypes 'Evereste' and *Malus floribunda* clone 821. *Genome* 52(2), 139–47. doi:10.1139/g08-111.

Edge-Garza, D. A., Peace, C. P. and Zhu, Y. 2010. Enabling marker-assisted seedling selection in the Washington apple breeding program. *Acta Horticulturae* 859(859), 369–73. doi:10.17660/ActaHortic.2010.859.44.

Edge-Garza, D. A., Luby, J. J. and Peace, C. 2015. Decision support for cost-efficient and logistically feasible marker-assisted seedling selection in fruit breeding. *Molecular Breeding* 35(12), 223. doi:10.1007/s11032-015-0409-z.

Elshire, R. J., Glaubitz, J. C., Sun, Q., Poland, J. A., Kawamoto, K., Buckler, E. S. and Mitchell, S. E. 2011. A robust, simple genotyping-by-sequencing (GBS) approach for high diversity species. *PLoS ONE* 6(5), e19379. doi:10.1371/journal.pone.0019379.

Emeriewen, O., Richter, K., Kilian, A., Zini, E., Hanke, M., Malnoy, M. and Peil, A. 2014. Identification of a major quantitative trait locus for resistance to fire blight in the wild apple species *Malus fusca*. *Molecular Breeding* 34(2), 407–19. doi:10.1007/s11032-014-0043-1.

Emeriewen, O. F., Peil, A., Richter, K., Zini, E., Hanke, M. and Malnoy, M. 2017. Fire blight resistance of *Malus* × *arnoldiana* is controlled by a quantitative trait locus located at the distal end of linkage group 12. *European Journal of Plant Pathology* 148(4), 1011–8. doi:10.1007/s10658-017-1152-6.

Evans, K. 2011. Apple breeding in the Pacific Northwest. In: *XIII Eucarpia Symposium on Fruit Breeding and Genetics*. Acta Horticulturae 976, 75–8. doi:10.17660/ActaHortic.2013.976.6.

Evans, K. 2017. *Achieving Sustainable Cultivation of Apples*. Cambridge, UK: Burleigh Dodds Science Publishing Limited.

Evans, K. M. and James, C. M. 2003. Identification of SCAR markers linked to *Pl-w* mildew resistance in apple. *Theoretical and Applied Genetics* 106(7), 1178–83. doi:10.1007/s00122-002-1147-2.

Eriksson, D., Kershen, D., Nepomuceno, A., Pogson, B. J., Prieto, H., Purnhagen, K., Smyth, S., Wesseler, J. and Whelan, A. 2018. A comparison of the EU regulatory approach to directed mutagenesis with that of other jurisdictions, consequences for international trade and potential steps forward. *New Phytologist*. doi:10.1111/nph.15627.

European Commission. 2001. Directive 2001/18/EC of the European Parliament and of the Council of 12 March 2001 on the deliberate release into the environment of genetically modified organisms and repealing Council Directive 90/220/EEC. *Official Journal of the European Communities*, L106, 1–39. Available at: europa.eu.int/eur-lex/pri/en/oj/dat/2001/l_106/l_10620010417en00010038.pdf.

Eyre-Walker, A. and Keightley, P. D. 2007. The distribution of fitness effects of new mutations. *Nature Reviews. Genetics* 8(8), 610–8. doi:10.1038/nrg2146.

Fahlgren, N., Gehan, M. A. and Baxter, I. 2015. Lights, camera, action: High-throughput plant phenotyping is ready for a close-up. *Current Opinion in Plant Biology* 24, 93–9. doi:10.1016/j.pbi.2015.02.006.

Fahrentrapp, J., Broggini, G. A. L., Kellerhals, M., Peil, A., Richter, K., Zini, E. and Gessler, C. 2013. A candidate gene for fire blight resistance in *Malus* × *robusta* 5 is coding for a CC-NBS-LRR. *Tree Genetics and Genomes* 9(1), 237–51. doi:10.1007/s11295-012-0550-3.

Falginella, L., Cipriani, G., Monte, C., Gregori, R., Testolin, R., Velasco, R., Troggio, M. and Tartarini, S. 2015. A major QTL controlling apple skin russeting maps on the linkage group 12 of 'Renetta Grigia di Torriana'. *BMC Plant Biology* 15, 150. doi:10.1186/s12870-015-0507-4.

Fallahi, E. 2012. Influence of rootstock and irrigation methods on water use, mineral nutrition, growth, fruit yield, and quality in 'Gala' apple. *HortTechnology* 22, 731–7. doi:10.21273/HORTTECH.22.6.731.

Fan, G., Zha, J., Du, R. and Gao, L. 2009. Determination of soluble solids and firmness of apples by Vis/NIR transmittance. *Journal of Food Engineering* 93(4), 416–20. doi:10.1016/j.jfoodeng.2009.02.006.

FAOSTAT. 2018. Food and Agriculture Organization of the United Nations. Available at: http://www.fao.org/faostat/en/(Accessed on 14 June 2018).

Fazio, G., Kviklys, D., Grusak, M. A. and Robinson, T. 2013. Phenotypic diversity and QTL mapping of absorption and translocation of nutrients by apple rootstocks. *Aspects of Applied Biology* 119, 37–50.

Fazio, G., Wan, Y., Kviklys, D., Romero, L., Adams, R., Strickland, D. and Robinson, T. 2014. *Dw2*, A new dwarfing locus in apple rootstocks and its relationship to induction of early bearing in apple scions. *Journal of the American Society for Horticultural Science* 139(2), 87–98. doi:10.21273/JASHS.139.2.87.

Fazio, G., Grusak, M. and Robinson, T. L. 2017. Apple rootstocks' dwarfing loci relationships with mineral nutrient concentration in scion leaves and fruit. *Acta Horticulturae* 1177(1177), 93–102. doi:10.17660/ActaHortic.2017.1177.10.

Fernandes, J. B., Seguéla-Arnaud, M., Larchevêque, C., Lloyd, A. H. and Mercier, R. 2017. Unleashing meiotic crossovers in hybrid plants. *Proceedings of the National Academy of Sciences* 115, 2431–6.

Fernández-Fernández, F., Antanaviciute, L., van Dyk, M. M., Tobutt, K. R., Evans, K. M., Rees, D. J. G., Dunwell, J. M. and Sargent, D. J. 2012. A genetic linkage map of an apple rootstock progeny anchored to the *Malus* genome sequence. *Tree Genetics and Genomes* 8(5), 991–1002. doi:10.1007/s11295-012-0478-7.

Finkel, E. 2009. Imaging. With 'phenomics', plant scientists hope to shift breeding into overdrive. *Science* 325(5939), 380–1. doi:10.1126/science.325_380.

Fiorani, F. and Schurr, U. 2013. Future scenarios for plant phenotyping. *Annual Review of Plant Biology* 64, 267–91. doi:10.1146/annurev-arplant-050312-120137.

Fiorani, F., Rascher, U., Jahnke, S. and Schurr, U. 2012. Imaging plants dynamics in heterogenic environments. *Current Opinion in Biotechnology* 23(2), 227–35. doi:10.1016/j.copbio.2011.12.010.

Fischer, C. 1994. Shortening of the juvenile period in apple breeding. In: Schmidt, H. and Kellerhals, M. (Eds), *Developments in Plant Breeding: Progress in Temperate Fruit Breeding*. London, UK: Kluwer Academic Publishers, pp. 161–4.

Flachowsky, H., Hanke, M.-V., Peil, A., Strauss, S. H. and Fladung, M. 2009. A review on transgenic approaches to accelerate breeding of woody plants. *Plant Breeding* 128(3), 217–26. doi:10.1111/j.1439-0523.2008.01591.x.

Flachowsky, H., Le Roux, P. M., Peil, A., Patocchi, A., Richter, K. and Hanke, M. V. 2011. Application of a high-speed breeding technology to apple (*Malus* × *domestica*) based on transgenic early flowering plants and marker-assisted selection. *New Phytologist* 192(2), 364–77. doi:10.1111/j.1469-8137.2011.03813.x.

Flor, H. H. 1971. Current status of the gene-for-gene concept. *Annual Review of Phytopathology* 9(1), 275–96. doi:10.1146/annurev.py.09.090171.001423.

Fruitbreedomics. 2012. DAY 1 - Stakeholder Day - Conferences 1, 2 - Welcome - Jaroslav Vacha and Jan Blazek (video file). 9 April 2012. Available at: https://vimeo.com/63414205.

Fruitbreedomics. 2013a. CONF. 15 D. 1 -New apple cultivars released at the research institute of horticulture Skierniewice. Mariusz Lewandowski (Poland) (video file). 5 April 2013. Retrieved from: https://vimeo.com/63414205.

Fruitbreedomics. 2013b. CONF. 10 D. 1 -Progress in Better3fruit's apple and pear breeding program- Inge de Wit – Better3fruit (Belgium). (video file). 5 April 2013. Available at: https://vimeo.com/63413153.

Fruitbreedomics. 2013c. CONF. 12 D. 1 -Advances in CRA-FRF pear and apple breeding- Gianluca Baruzzi – CRA-FRF (Italy). (video file). 5 April 2013. Available at: https://vimeo.com/63413152 12 D, 1.

Fruitbreedomics. 2014a. C5 D1 – Molecular markers on breeding: Apple – Dr. M. Kellerhals – (Agroscope, Switzerland) - FB 3rd A. Meeting - Zürich 2014. (video file). 10 April 2014. Available at: https://vimeo.com/91614678.

Fruitbreedomics. 2014b. C8 D1 – MAB Transfer – Dr. François Laurens (INRA, France) – FB 3rd A. Meeting – Zürich 2014. video file. Available at: https://vimeo.com/91614738.

Gardiner, S. E., Norelli, J. L., de Silva, N., Fazio, G., Peil, A., Malnoy, M., Horner, M., Bowatte, D., Carlisle, C., Wiedow, C., et al. 2012. Putative resistance gene markers associated with quantitative trait loci for fire blight resistance in *Malus* 'Robusta 5' accessions. *BMC Genetics* 13, 25. doi:10.1186/1471-2156-13-25.

Gardner, K. M., Brown, P., Cooke, T. F., Cann, S., Costa, F., Bustamante, C., Velasco, R., Troggio, M. and Myles, S. 2014. Fast and cost-effective genetic mapping in apple using next-generation sequencing. *G3 (Bethesda, Md.)* 4(9), 1681–7. doi:10.1534/g3.114.011023.

Gariepy, T. D., Rahe, J. E., Lévesque, C. A., Spotts, R. A., Sugar, D. L. and Henriquez, J. L. 2005. *Neofabraea* species associated with bull's-eye rot and cankers of apple and pear in the Pacific Northwest. *Canadian Journal of Plant Pathology* 27(1), 118–24. doi:10.1080/07060660509507202.

Garkava-Gustavsson, L., Ghasemkhani, M., Zborowska, A., Englund, J.-E., Lateur, M. and van de Weg, E. 2016. Approaches for evaluation of resistance to European canker (*Neonectria ditissima*) in apple. *Acta Horticulturae* 1127(1127), 75–82. doi:10.17660/ActaHortic.2016.1127.14.

Gaut, B. S., Díez, C. M. and Morrell, P. L. 2015. Genomics and the contrasting dynamics of annual and perennial domestication. *Trends in Genetics: TIG* 31(12), 709–19. doi:10.1016/j.tig.2015.10.002.

Gelvonauskienė, D., Sasnauskas, A. and Gelvonauskis, B. 2007. The breeding of apple tree resistant to European canker (*Nectria galligena* bres.) scientific works of the Lithuanian Institute of Horticulture and Lithuanian University of Agriculture. *Sodininkyste ir Darzininkyste* 26(3).

Gessler, C. and Pertot, I. 2012. *Vf* scab resistance of *Malus*. *Trees* 26(1), 95–108. doi:10.1007/s00468-011-0618-y.

Ghasemkhani, M., Liljeroth, E., Sehic, J., Zborowska, A. and Nybom, H. 2015a. Cut-off shoots method for estimation of partial resistance in apple cultivars to fruit tree canker caused by *Neonectria ditissima*. *Acta Agriculturae Scandinavica, Section B – Soil and Plant Science* 65(5), 412–21.

Ghasemkhani, M., Sehic, J., Ahmadi-Afzadi, M., Nybom, H. and Garkava-Gustavsson, L. 2015b. Screening for partial resistance to fruit tree canker in apple cultivars. *Acta Horticulturae* 1099(1099), 687–90. doi:10.17660/ActaHortic.2015.1099.84.

Gianfranceschi, L., Seglias, N., Tarchini, R., Komjanc, M. and Gessler, C. 1998. Simple sequence repeats for the genetic analysis of apple. *Theoretical and Applied Genetics* 96(8), 1069–76. doi:10.1007/s001220050841.

Gómez-Candón, D., Costes, E., Martinez, S., Labbé, S. and Regnard, J.-L. 2015. High-throughput field phenotyping of apple tree response to drought by UAV thermal imaging: Supervised image classification for unravelling intra-canopy temperature variation. In: *EPPN Plant Phenotyping Symposium*, November 2015, Barcelona, Spain.

Gómez-Candón, D., Virlet, N., Labbé, S., Jolivot, A. and Regnard, J. 2016. Field phenotyping of water stress at tree scale by UAV-sensed imagery: New insights for thermal acquisition and calibration. *Precision Agriculture* 17(6), 786–800. doi:10.1007/s11119-016-9449-6.

Gómez-Cortecero, A., Saville, R. J., Scheper, R. W. A., Bowen, J. K., Agripino De Medeiros, H., Kingsnorth, J., Xu, X. and Harrison, R. J. 2016. Variation in host and pathogen in the *Neonectria/Malus* interaction; toward an understanding of the genetic basis of resistance to European canker. *Frontiers in Plant Science* 7, 1365. doi:10.3389/fpls.2016.01365.

Gross, B. L., Henk, A. D., Richards, C. M., Fazio, G. and Volk, G. M. 2014. Genetic diversity in *Malus × domestica* (Rosaceae) through time in response to domestication. *American Journal of Botany* 101(10), 1770–9. doi:10.3732/ajb.1400297.

Gregory, P. J., Atkinson, C. J., Bengough, A. G., Else, M. A., Fernández-Fernández, F., Harrison, R. J. and Schmidt, S. 2013. Contributions of roots and rootstocks to sustainable, intensified crop production. *Journal of Experimental Botany* 64(5), 1209–22. doi:10.1093/jxb/ers385.

Guan, Y., Peace, C., Rudell, D., Verma, S. and Evans, K. 2015. QTLs detected for individual sugars and soluble solids content in apple. *Molecular Breeding* 35(6), 135. doi:10.1007/s11032-015-0334-1.

Guitton, B., Kelner, J. J., Velasco, R., Gardiner, S. E., Chagné, D. and Costes, E. 2012. Genetic control of biennial bearing in apple. *Journal of Experimental Botany* 63(1), 131–49. doi:10.1093/jxb/err261.

Gygax, M., Gianfranceschi, L., Liebhard, R., Kellerhals, M., Gessler, C. and Patocchi, A. 2004. Molecular markers linked to the apple scab resistance gene *Vbj* derived from *Malus baccata* jackii. *Theoretical and Applied Genetics* 109(8), 1702–9. doi:10.1007/s00122-004-1803-9.

Hamdy, A., Ragab, R. and Scarascia-Mugnozza, E. 2003. Coping with water scarcity: Water saving and increasing water productivity. *Irrigation and Drainage* 52(1), 3–20. doi:10.1002/ird.73.

Hampson, C. R., Quamm, H. A. and Brownlee, R. T. 2002. Canopy growth, yield, and fruit quality of Royal Gala apple trees grown for eight years in five tree training systems. *HortScience* 37, 627–31.

Hanke, M. V., Flachowsky, H., Peil, A. and Hättasch, C. 2007. No flower no fruit–Genetic potentials to trigger flowering in fruit trees. *Genes Genomes Genomics* 1(1), 1–20.

Harrison, N., Harrison, R. J., Barber-Perez, N., Cascant-Lopez, E., Cobo-Medina, M., Lipska, M., Conde-Ruíz, R., Brain, P., Gregory, P. J. and Fernández-Fernández, F. 2016. A new three-locus model for rootstock-induced dwarfing in apple revealed by genetic mapping of root bark percentage. *Journal of Experimental Botany* 67(6), 1871–81. doi:10.1093/jxb/erw001.

Harshman, J. M., Evans, K. M. and Hardner, C. M. 2016. Cost and accuracy of advanced breeding trial designs in apple. *Horticulture Research* 3, 16008. doi:10.1038/hortres.2016.8.

Harshman, J. M., Evans, K. M., Allen, H., Potts, R., Flamenco, J., Aldwinckle, H. S., Wisniewski, M. E. and Norelli, J. L. 2017. Fire blight resistance in wild accessions of *Malus sieversii*. *Plant Disease* 101, 1738–45.

He, J., Zhao, X., Laroche, A., Lu, Z. X., Liu, H. and Li, Z. 2014. Genotyping-by-sequencing (GBS), an ultimate marker-assisted selection (MAS) tool to accelerate plant breeding. *Frontiers in Plant Science* 5, 484. doi:10.3389/fpls.2014.00484.

Heide, O. M. and Prestrud, A. K. 2005. Low temperature, but not photoperiod, controls growth cessation and dormancy induction and release in apple and pear. *Tree Physiology* 25(1), 109–14. doi:10.1093/treephys/25.1.109.

Hickey, L. T., Germán, S. E., Pereyra, S. A., Diaz, J. E., Ziems, L. A., Fowler, R. A., Platz, G. J., Franckowiak, J. D. and Dieters, M. J. 2017. Speed breeding for multiple disease resistance in barley. *Euphytica* 213(3), 64. doi:10.1007/s10681-016-1803-2.

Holb, I. J. 2017. Categorization of apple cultivars based on seasonal powdery mildew disease progression in two disease management systems over 12 years. *Trees* 31(6), 1905–17. doi:10.1007/s00468-017-1595-6.

Howard, N. P., van de Weg, E., Tillman, J., Tong, C. B. S., Silverstein, K. A. T. and Luby, J. J. 2018. Two QTL characterized for soft scald and soggy breakdown in apple (*Malus* × *domestica*) through pedigree-based analysis of a large population of interconnected families. *Tree Genetics and Genomes* 14(1), 2. doi:10.1007/s11295-017-1216-y.

James, C. M., Clarke, J. B. and Evans, K. M. 2004. Identification of molecular markers linked to the mildew resistance gene *Pl-d* in apple. *Theoretical and Applied Genetics* 110(1), 175–81. doi:10.1007/s00122-004-1836-0.

Janisiewicz, W. J., Saftner, R. A., Conway, W. S. and Forsline, P. L. 2008. Preliminary evaluation of apple germplasm from Kazakhstan for resistance to postharvest blue mold in fruit caused by *Penicillium expansum*. *HortScience* 43, 420–6.

Johnsson, M., Gaynor, R. C., Jenko, J., Gorjanc, G., de Koning, D. J. and Hickey, J. M. 2018. Removal of alleles by genome editing – RAGE against the deleterious load. *bioRxiv*, 335497. doi:10.1101/335497.

Jonkers, H. 1979. Biennial bearing in apple and pear: A literature survey. *Scientia Horticulturae* 11(4), 303–17. doi:10.1016/0304-4238(79)90015-3.

Joshi, S. G. 2010. Towards durable resistance to apple scab using cisgenes. (Doctoral dissertation). Wageningen University.

Joshi, S. G., Soriano, J. M., Kortstee, A., Schaart, J. G., Krens, F. A., Jacobsen, E. and Schouten, H. J. 2009. Development of cisgenic apples with durable resistance to apple scab. *Acta Horticulturae* 839(839), 403–6. doi:10.17660/ActaHortic.2009.839.53.

Kellerhals, M., Patocchi, A., Duffy, B. and Frey, J. 2008. Modern approaches for breeding high quality apples with durable resistance to scab, powdery mildew and fire blight. In: Boos, M. (Ed.), *Ecofruit - 13th International Conference on Cultivation Technique and Phytopathological Problems in Organic Fruit-Growing: Proceedings to the Conference from 18–20 February 2008 at Weinsberg/Germany*, pp. 226–31.

Kellerhals, M., Spuhler, M., Duffy, B., Patocchi, A. and Frey, J. E. 2009. Selection efficiency in apple breeding. *Acta Horticulturae* 814(814), 177–84. doi:10.17660/ActaHortic.2009.814.22.

Kellerhals, M., Schütz, S. and Patocchi, A. 2017. Breeding for host resistance to fire blight. *Journal of Plant Pathology : An International Journal of the Italian Phytopathological Society* 99(Special Issue), 37–43.

Kenis, K. and Keulemans, J. 2005. Genetic linkage maps of two apple cultivars (*Malus domestica* Borkh.) based on AFLP and microsatellite markers. *Molecular Breeding* 15(2), 205–19. doi:10.1007/s11032-004-5592-2.

Khan, M. A., Durel, C. E., Duffy, B., Drouet, D., Kellerhals, M., Gessler, C. and Patocchi, A. 2007. Development of molecular markers linked to the 'Fiesta' linkage Group 7 major QTL for fire blight resistance and their application for marker-assisted selection. *Genome* 50(6), 568–77. doi:10.1139/g07-033.

Kost, T. D., Gessler, C., Jänsch, M., Flachowsky, H., Patocchi, A. and Broggini, G. A. 2015. Development of the first cisgenic apple with increased resistance to fire blight. *PLoS ONE* 10(12), e0143980. doi:10.1371/journal.pone.0143980.

Kotoda, N., Iwanami, H., Takahashi, S. and Abe, K. 2005. Antisense expression of *MdTFL1*, a *TFL1*-like gene, reduces the juvenile phase in apple. *Journal of the American Society for Horticultural Science* 131(1), 74–81. doi:10.21273/JASHS.131.1.74.

Kumar, S., Volz, R. K., Alspach, P. A. and Bus, V. G. M. 2010. Development of a recurrent apple breeding programme in New Zealand: A synthesis of results, and a proposed revised breeding strategy. *Euphytica* 173(2), 207–22. doi:10.1007/s10681-009-0090-6.

Kumar, S., Chagné, D., Bink, M. C. A. M., Volz, R. K., Whitworth, C. and Carlisle, C. 2012. Genomic selection for fruit quality traits in apple (*Malus × domestica* Borkh.). *PLoS ONE* 7(5), e36674. doi:10.1371/journal.pone.0036674.

Kumar, S., Garrick, D. J., Bink, M. C., Whitworth, C., Chagné, D. and Volz, R. K. 2013. Novel genomic approaches unravel genetic architecture of complex traits in apple. *BMC Genomics* 14, 393. doi:10.1186/1471-2164-14-393.

Kumar, S., McGlone, A., Whitworth, C. and Volz, R. 2015a. Postharvest performance of apple phenotypes predicted by near-infrared (NIR) spectral analysis. *Postharvest Biology and Technology* 100, 16–22. doi:10.1016/j.postharvbio.2014.09.021.

Kumar, S., Molloy, C., Muñoz, P., Daetwyler, H., Chagné, D. and Volz, R. 2015b. Genome-enabled estimates of additive and nonadditive genetic variances and prediction of apple phenotypes across environments. *G3 (Bethesda, Md.)* 5(12), 2711–8. doi:10.1534/g3.115.021105.

Kusch, S. and Panstruga, R. 2017. *mlo*-based resistance: An apparently universal 'weapon' to defeat powdery mildew disease. *Molecular Plant-Microbe Interactions: MPMI* 30(3), 179–89. doi:10.1094/MPMI-12-16-0255-CR.

Labuschagné, I. F., Louw, J. H., Schmidt, K. and Sadie, A. 2002. Genetic variation in chilling requirement in apple progeny. *Journal of the American Society for Horticultural Science* 127(4), 663–72. doi:10.21273/JASHS.127.4.663.

Lambing, C., Franklin, F. C. H. and Wang, C. R. 2017. Understanding and manipulating meiotic recombination in plants. *Plant Physiology* 173(3), 1530–42. doi:10.1104/pp.16.01530.

Lammertyn, J., Nicolaï, B., Ooms, K., De Smedt, V. and De Baerdemaeker, J. 1998. Non-destructive measurement of acidity, soluble solids and firmness of Jonagold apples using NIR-spectroscopy. *Transactions of the ASAE (USA)* 41(4), 1089–94.

Le Roux, P.-M. F., Khan, M. A., Broggini, G. A. L., Duffy, B., Gessler, C. and Patocchi, A. 2010. Mapping of quantitative trait loci for fire blight resistance in the apple cultivars 'Florina' and 'Nova Easygro'. *Genome* 53(9), 710–22. doi:10.1139/G10-047.

Legun, K. A. 2015. Club apples: A biology of markets built on the social life of variety. *Economy and Society* 44(2), 293–315. doi:10.1080/03085147.2015.1013743.

Li, H., Wan, Y., Wang, M., Han, M. and Huo, X. 2015. History, status and prospects of the apple industry in China. *Journal of the American Pomological Society* 69(4), 174–85.

Liebhard, R., Gianfranceschi, L., Koller, B., Ryder, C. D., Tarchini, R., van de Weg, E. and Gessler, C. 2002. Development and characterisation of 140 new microsatellites in apple (*Malus* × *domestica* Borkh.). *Molecular Breeding* 10(4), 217–41. doi:10.1023/A:1020525906332.

Liebhard, R., Koller, B., Patocchi, A., Kellerhals, M., Pfammatter, W., Jermini, M. and Gessler, C. 2003. Mapping quantitative field resistance against apple scab in a 'fiesta' × 'discovery' progeny. *Phytopathology* 93(4), 493–501. doi:10.1094/PHYTO.2003.93.4.493.

Liu, F. and Tang, X. 2015. Fuji apple storage time rapid determination method using Vis/NIR spectroscopy. *Bioengineered* 6(3), 166–9. doi:10.1080/21655979.2015.1038001.

Liu, Z., Bao, D., Liu, D., Zhang, Y., Ashraf, M. A. and Chen, X. 2016. Construction of a genetic linkage map and QTL analysis of fruit-related traits in an F1 Red Fuji × Hongrou apple hybrid. *Open Life Sciences* 11(1). doi:10.1515/biol-2016-0063.

Liu, Q., Zhou, Y., Morrell, P. L. and Gaut, B. S. 2017. Deleterious variants in Asian rice and the potential cost of domestication. *Molecular Biology and Evolution* 34(4), 908–24. doi:10.1093/molbev/msw296.

Lobet, G., Pages, L. and Draye, X. 2011. A novel image analysis toolbox enabling quantitative analysis of root system architecture. *Plant Physiology* 157(1), 29–39. doi:10.1104/pp.111.179895.

Lohmueller, K. E. 2014. The distribution of deleterious genetic variation in human populations. *Current Opinion in Genetics and Development* 29, 139–46. doi:10.1016/j.gde.2014.09.005.

Longhi, S., Hamblin, M. T., Trainotti, L., Peace, C. P., Velasco, R. and Costa, F. 2013. A candidate gene based approach validates Md-PG1 as the main responsible for a QTL impacting fruit texture in apple (*Malus* × *domestica* Borkh). *BMC Plant Biology* 13, 37. doi:10.1186/1471-2229-13-37.

Longin, C. F. H., Mi, X. and Würschum, T. 2015. Genomic selection in wheat: Optimum allocation of test resources and comparison of breeding strategies for line and hybrid breeding. *Theoretical and Applied Genetics* 128(7), 1297–306. doi:10.1007/s00122-015-2505-1.

Lopez, G., Pallas, B., Martinez, S., Lauri, P. É, Regnard, J. L., Durel, C. É and Costes, E. 2015. Genetic variation of morphological traits and transpiration in an apple core collection under well-watered conditions: Towards the identification of morphotypes with high water use efficiency. *PLoS ONE* 10(12), e0145540. doi:10.1371/journal.pone.0145540.

Lopez, G., Pallas, B., Martinez, S., Lauri, P. E., Regnard, J. L., Durel, C. E. and Costes, E. 2017a. Heritability and genetic variation of plant biomass, transpiration, and water use efficiency for an apple core-collection. *Acta Horticulturae* 1172(1172), 317–22. doi:10.17660/ActaHortic.2017.1172.59.

Lopez, G., Pallas, B., Martinez, S., Lauri, P. E., Regnard, J. L., Durel, C. E. and Costes, E. 2017b. High-throughput phenotyping of an apple core collection: Identification of genotypes with high water use efficiency. *Acta Horticulturae* 1150(1150), 335–40. doi:10.17660/ActaHortic.2017.1150.48.

Lorenz, A. and Nice, L. 2017. Training population design and resource allocation for genomic selection in plant breeding. In: Varshney, R. K., Roorkiwal, M. and Sorrells, M. E. (Eds), *Genomic Selection for Crop Improvement*. Cham: Springer International Publishing, pp. 7–22.

Mackay, I. J., Caligari, P. D. S. and Gibson, J. P. 1999. Accelerated recurrent selection. *Euphytica* 105(1), 43–51. doi:10.1023/A:1003428430664.

Malnoy, M., Viola, R., Jung, M. H., Koo, O. J., Kim, S., Kim, J. S., Velasco, R. and Nagamangala Kanchiswamy, C. 2016. DNA-free genetically edited grapevine and apple protoplast using CRISPR/Cas9 ribonucleoproteins. *Frontiers in Plant Science* 7, 1904. doi:10.3389/fpls.2016.01904.

Marini, R. P., Black, B. and Crassweller, R. M. 2009. Performance of 'golden delicious' apple on 23 rootstocks at 12 locations: A five-year summary of the 2003 NC-140 dwarf rootstock trial. *Journal of the American Pomological Society* 63(3), 115–27.

Marini, R. P., Black, B., Crassweller, R. M., Domoto, P. A., Hampson, C., Moran, R., Robinson, T., Stasiak, M. and Wolfe, D. 2014. Performance of 'Golden Delicious' apple on 23 rootstocks at eight locations: A ten year summary of the 2003 NC-140 dwarf apple rootstock trial. *Journal of the American Pomological Society* 68(2), 54–68.

Marini, R. P., Crassweller, R. M., Autio, W., Moran, R., Robinson, T. L., Cline, J., Wolfe, D. and Parra-Quezada, R. 2017. The interactive effects of early season temperatures, crop density and rootstock on average fruit weight of 'Golden Delicious' apple. *Acta Horticulturae* 1177(1177), 189–94. doi:10.17660/ActaHortic.2017.1177.26.

Markussen, T., Kruger, J., Schmidt, H. and Dunemann, F. 1995. Identification of PCR-based markers linked to the powdery-mildew-resistance gene *Pl1* from *Malus robusta* in cultivated apple. *Plant Breeding* 114(6), 530–4. doi:10.1111/j.1439-0523.1995.tb00850.x.

McClure, K. A., Gardner, K. M., Toivonen, P. M., Hampson, C. R., Song, J., Forney, C. F., DeLong, J., Rajcan, I. and Myles, S. 2016. QTL analysis of soft scald in two apple populations. *Horticulture Research* 3, 16043. doi:10.1038/hortres.2016.43.

McClure, K. A., Gardner, K. M., Douglas, G. M., Song, J., Forney, C. F., DeLong, J., Fan, L., Du, L., Toivonen, P. M. A., Somers, D. J., et al. 2018. A genome-wide association study of apple quality and scab resistance. *The Plant Genome* 11(1). doi:10.3835/plantgenome2017.08.0075.

Meuwissen, T. H., Hayes, B. J. and Goddard, M. E. 2001. Prediction of total genetic value using genome-wide dense marker maps. *Genetics* 157(4), 1819–29.

Mezmouk, S. and Ross-Ibarra, J. 2013. The pattern and distribution of deleterious mutations in maize. *G3: Genes, Genomes, Genetics* 4, 163–71.

Miller, A. J., Matasci, N., Schwaninger, H., Aradhya, M. K., Prins, B., Zhong, G. Y., Simon, C., Buckler, E. S. and Myles, S. 2013. *Vitis* phylogenomics: Hybridization intensities from a SNP array outperform genotype calls. *PLoS ONE* 8(11), e78680. doi:10.1371/journal.pone.0078680.

Monselise, S. P. and Goldschmidt, E. E. 1982. Alternate bearing in fruit trees. In: Janick, J. (Ed.), *Horticultural Reviews*. Hoboken, NJ: John Wiley & Sons, Inc., pp. 128–73.

Muranty, H., Troggio, M., Sadok, I. B., Rifaï, M. A., Auwerkerken, A., Banchi, E., Velasco, R., Stevanato, P., van de Weg, W. E., Di Guardo, M., et al. 2015. Accuracy and responses of genomic selection on key traits in apple breeding. *Horticulture Research* 2, 15060. doi:10.1038/hortres.2015.60.

N'Diaye, A., van de Weg, W. E., Kodde, L. P., Koller, B., Dunemann, F., Thiermann, M., Tartarini, S., Gennari, F. and Durel, C. E. 2008. Construction of an integrated consensus map of the apple genome based on four mapping populations. *Tree Genetics and Genomes* 4(4), 727–43. doi:10.1007/s11295-008-0146-0.

Neyhart, J. L., Tiede, T., Lorenz, A. J. and Smith, K. P. 2017. Evaluating methods of updating training data in long-term genomewide selection. *G3 (Bethesda, Md.)* 7(5), 1499-510. doi:10.1534/g3.117.040550.

Nishitani, C., Hirai, N., Komori, S., Wada, M., Okada, K., Osakabe, K., Yamamoto, T. and Osakabe, Y. 2016. Efficient genome editing in apple Using a CRISPR/Cas9 system. *Scientific Reports* 6, 31481. doi:10.1038/srep31481.

Noble, J. P. and Drysdale, R. B. 1983. The role of benzoic acid and phenolic compounds in latency in fruits of two apple cultivars infected with *Pezicula malicorticis* or *Nectria galligena*. *Physiological Plant Pathology* 23(2), 207–16. doi:10.1016/0048-4059(83)90003-6.

Noiton, D. A. M. and Alspach, P. A. 1996. Founding clones, inbreeding, coancestry, and status number of modern apple cultivars. *Journal of the American Society for Horticultural Science* 121(5), 773–82. doi:10.21273/JASHS.121.5.773.

Norelli, J. L., Jones, A. L. and Aldwinckle, H. S. 2003. Fire blight management in the twenty-first century: using new technologies that enhance host resistance in apple. *Plant Disease* 87(7), 756–65.

Norelli, J. L., Wisniewski, M. and Droby, S. 2014. Identification of a QTL for postharvest disease resistance to *Penicillium expansum* in *Malus sieversii*. *Acta Horticulturae* 1053(1053), 199–203. doi:10.17660/ActaHortic.2014.1053.21.

Norelli, J. L., Wisniewski, M., Fazio, G., Burchard, E., Gutierrez, B., Levin, E. and Droby, S. 2017. Genotyping-by-sequencing markers facilitate the identification of quantitative trait loci controlling resistance to *Penicillium expansum* in *Malus sieversii*. *PLoS ONE* 12(3), e0172949. doi:10.1371/journal.pone.0172949.

Nybom, H., Ahmadi-Afzadi, M., Sehic, J. and Hertog, M. 2013. DNA marker-assisted evaluation of fruit firmness at harvest and post-harvest fruit softening in a diverse apple germplasm. *Tree Genetics and Genomes* 9(1), 279–90. doi:10.1007/s11295-012-0554-z.

Oakey, H., Cullis, B., Thompson, R., Comadran, J., Halpin, C. and Waugh, R. 2016. Genomic selection in multi-environment crop trials. *G3 (Bethesda, Md.)* 6(5), 1313-26. doi:10.1534/g3.116.027524.

Oerke, E.-C., Fröhling, P. and Steiner, U. 2011. Thermographic assessment of scab disease on apple leaves. *Precision Agriculture* 12(5), 699–715. doi:10.1007/s11119-010-9212-3.

Oraguzie, N. C., Hofstee, M. E., Brewer, L. R. and Howard, C. 2001. Estimation of genetic parameters in a recurrent selection program in apple. *Euphytica* 118(1), 29–37. doi:10.1023/A:1004056423344.

Pagliarani, G., Dapena, E., Miñarro, M., Denancé, C., Lespinasse, Y., Rat-Morris, E., Troggio, M., Durel, C. E. and Tartarini, S. 2016. Fine mapping of the rosy apple aphid resistance locus *Dp-fl* on linkage group 8 of the apple cultivar 'Florina'. *Tree Genetics and Genomes* 12(3), 56. doi:10.1007/s11295-016-1015-x.

Parisi, L., Lespinasse, Y., Guillaumes, J. 1993. A new race of *Venturia inaequalis* virulent to apples with resistance due to the *Vf* gene. *Phytopathology* 83(5), 533–7. doi:10.1094/Phyto-83-533.

Parry, M. A. J. and Hawkesford, M. J. 2010. Food security: Increasing yield and improving resource use efficiency. *Proceedings of the Nutrition Society* 69(4), 592–600. doi:10.1017/S0029665110003836.

Patocchi, A., Frei, A., Frey, J. E. and Kellerhals, M. 2009. Towards improvement of marker assisted selection of apple scab resistant cultivars: *Venturia inaequalis* virulence surveys and standardization of molecular marker alleles associated with resistance genes. *Molecular Breeding* 24(4), 337–47. doi:10.1007/s11032-009-9295-6.

Peil, A., Garcia-Libreros, T., Richter, K., Trognitz, F. C., Trognitz, B., Hanke, M.-V. and Flachowsky, H. 2007. Strong evidence for a fire blight resistance gene of *Malus robusta* located on linkage group 3. *Plant Breeding* 126(5), 470–5. doi:10.1111/j.1439-0523.2007.01408.x.

Pessina, S., Pavan, S., Catalano, D., Gallotta, A., Visser, R. G., Bai, Y., Malnoy, M. and Schouten, H. J. 2014. Characterization of the *MLO* gene family in Rosaceae and gene expression analysis in *Malus domestica*. *BMC Genomics* 15, 618. doi:10.1186/1471-2164-15-618.

Pessina, S., Lenzi, L., Perazzolli, M., Campa, M., Dalla Costa, L., Urso, S., Valè, G., Salamini, F., Velasco, R. and Malnoy, M. 2016a. Knockdown of *MLO* genes reduces susceptibility to powdery mildew in grapevine. *Horticulture Research* 3, 16016. doi:10.1038/hortres.2016.16.

Pessina, S., Angeli, D., Martens, S., Visser, R. G., Bai, Y., Salamini, F., Velasco, R., Schouten, H. J. and Malnoy, M. 2016b. The knock-down of the expression of MdMLO19 reduces susceptibility to powdery mildew (*Podosphaera leucotricha*) in apple (*Malus domestica*). *Plant Biotechnology Journal* 14(10), 2033–44. doi:10.1111/pbi.12562.

Pessina, S., Palmieri, L., Bianco, L., Gassmann, J., van de Weg, E., Visser, R. G. F., Magnago, P., Schouten, H. J., Bai, Y., Riccardo Velasco, R., et al. 2017. Frequency of a natural truncated allele of *MdMLO19* in the germplasm of *Malus domestica*. *Molecular Breeding: New Strategies in Plant Improvement* 37(1), 7. doi:10.1007/s11032-016-0610-8.

Pilcher, R. L. R., Celton, J. M., Gardiner, S. E. and Tustin, D. S. 2008. Genetic markers linked to the dwarfing trait of apple rootstock 'Malling 9'. *Journal of the American Society for Horticultural Science* 133(1), 100–6. doi:10.21273/JASHS.133.1.100.

Pissard, A., Fernández Pierna, J. A., Baeten, V., Sinnaeve, G., Lognay, G., Mouteau, A., Dupont, P., Rondia, A. and Lateur, M. 2013. Non-destructive measurement of vitamin C, total polyphenol and sugar content in apples using near-infrared spectroscopy. *Journal of the Science of Food and Agriculture* 93(2), 238–44. doi:10.1002/jsfa.5779.

Pissard, A., Baeten, V., Dardenne, P., Dupont, P. and Lateur, M. 2018. Use of NIR spectroscopy on fresh apples to determine the phenolic compounds and dry matter content in peel and flesh. *Biotechnology, Agronomy and Society and Environment* 22(1), 3–12.

Pound, M. P., French, A. P., Atkinson, J. A., Wells, D. M., Bennett, M. J. and Pridmore, T. 2013. RootNav: Navigating images of complex root architectures. *Plant Physiology* 162(4), 1802–14. doi:10.1104/pp.113.221531.

Ramu, P., Esuma, W., Kawuki, R., Rabbi, I. Y., Egesi, C., Bredeson, J. V., Bart, R. S., Verma, J., Buckler, E. S. and Lu, F. 2017. Cassava haplotype map highlights fixation of deleterious mutations during clonal propagation. *Nature Genetics* 49(6), 959-63. doi:10.1038/ng.3845.

Reig, G., Lordan, J., Fazio, G., Grusak, M. A., Hoying, S., Cheng, L., Francescatto, P. and Robinson, T. 2018. Horticultural performance and elemental nutrient concentrations on 'Fuji' grafted on apple rootstocks under New York State climatic conditions. *Scientia Horticulturae* 227, 22–37. doi:10.1016/j.scienta.2017.07.002.

Renaut, S. and Rieseberg, L. H. 2015. The accumulation of deleterious mutations as a consequence of domestication and improvement in sunflowers and other compositae crops. *Molecular Biology and Evolution* 32(9), 2273–83. doi:10.1093/molbev/msv106.

Robinson, T. L. 2003. Apple-orchard planting system. In: Ferree, D. C. and Warrington, I. J. (Eds), *Apples*. CABI Publishing, Wallingford, UK, pp. 345–407.

Robinson, T. L. 2007. Effects of tree density and tree shape on apple orchard performance. *Acta Horticulturae* 732(732), 405–14. doi:10.17660/ActaHortic.2007.732.61.

Robinson, T. L., Lakso, A. N. and Carpenter, S. G. 1991. Canopy development, yield, and fruit quality of 'Empire' and 'Delicious' apple trees grown in four orchard production systems for ten years. *Journal of the American Society for Horticultural Science Jashs* 116(2), 179–87.

Roy, P. and Isler, V. 2017. Vision-based apple counting and yield estimation. In: Kulić, D., Nakamura, Y., Khatib, O. and Venture, G. (Eds), *2016 International Symposium on Experimental Robotics*. Cham: Springer International Publishing, pp. 478–87.

Ru, S. 2016. Theoretical and empirical evaluation of efficiency of marker-assisted seedling selection in Rosaceae tree fruit breeding. (Doctoral dissertation). Washington State University.

Ru, S., Main, D., Evans, K. and Peace, C. 2015. Current applications, challenges, and perspectives of marker-assisted seedling selection in Rosaceae tree fruit breeding. *Tree Genetics and Genomes* 11(1), 8. doi:10.1007/s11295-015-0834-5.

Sansavini, S., Donati, F., Costa, F. and Tartarini, S. 2004. Advances in apple breeding for enhanced fruit quality and resistance to biotic stresses: New varieties for the European market. *Journal of Fruit and Ornamental Plant Research* 12.

Schlathölter, I., Jänsch, M., Flachowsky, H., Broggini, G. A. L., Hanke, M. V. and Patocchi, A. 2018. Generation of advanced fire blight-resistant apple (*Malus × domestica*) selections of the fifth generation within 7 years of applying the early flowering approach. *Planta* 247(6), 1475–88. doi:10.1007/s00425-018-2876-z.

Serra, H., Lambing, C., Griffin, C. H., Topp, S. D., Nageswaran, D. C., Underwood, C. J., Ziolkowski, P. A., Séguéla-Arnaud, M., Fernandes, J. B., Mercier, R., et al. 2018. Massive crossover elevation via combination of *HEI10* and *recq4a recq4b* during *Arabidopsis meiosis*. *Proceedings of the National Academy of Sciences* 115, 2437–42.

Sinclair, T. R. and Rufty, T. W. 2012. Nitrogen and water resources commonly limit crop yield increases, not necessarily plant genetics. *Global Food Security* 1(2), 94–8. doi:10.1016/j.gfs.2012.07.001.

Spotts, R. A., Cervantes, L. A. and Mielke, E. A. 1999. Variability in postharvest decay among apple cultivars. *Plant Disease* 83(11), 1051–4. doi:10.1094/PDIS.1999.83.11.1051.

Stajnko, D., Lakota, M. and Hočevar, M. 2004. Estimation of number and diameter of apple fruits in an orchard during the growing season by thermal imaging. *Computers and Electronics in Agriculture* 42(1), 31–42. doi:10.1016/S0168-1699(03)00086-3.

Stankiewicz-Kosyl, M., Pitera, E. and Gawronski, S. W. 2005. Mapping QTL involved in powdery mildew resistance of the apple clone U 211. *Plant Breeding* 124(1), 63–6.

Sun, J., Rutkoski, J. E., Poland, J. A., Crossa, J., Jannink, J. L. and Sorrells, M. E. 2017. Multitrait, random regression, or simple repeatability model in High-throughput phenotyping data improve genomic prediction for wheat grain yield. *The Plant Genome* 10(2). doi:10.3835/plantgenome2016.11.0111.

Szankowski, I., Waidmann, S., El-Din Saad Omar, A., Flachowsky, H., Hättasch, C. and Hanke, M.-V. 2009. RNAi-silencing of *MdTFL1* induces early flowering in apple. *Acta Horticulturae* 839(839), 633–6. doi:10.17660/ActaHortic.2009.839.86.

Tahir, I. I., Nybom, H., Ahmadi-Afzadi, M., Røen, K., Sehic, J. and Røen, D. 2015. Susceptibility to blue mold caused by *Penicillium expansum* in apple cultivars adapted to a cool climate. *European Journal of Horticultural Science* 80(3), 117–27. doi:10.17660/eJHS.2015/80.3.4.

Tan, Y., Shen, F., Chen, R., Wang, Y., Wu, T., Li, T., Xu, X., Han, Z. and Zhang, X. 2017a. Candidate genes associated with resistance to Valsa canker identified via quantitative trait loci in apple. *Journal of Phytopathology* 165(11–12), 848–57.

Tan, Y., Lv, S., Liu, X., Gao, T., Li, T., Wang, Y., Wu, T., Zhang, X., Han, Y., Korban, S. S., et al. 2017b. Development of high-density interspecific genetic maps for the identification of QTLs conferring resistance to *Valsa ceratosperma* in apple. *Euphytica* 213(1), 10. doi:10.1007/s10681-016-1790-3.

Tartarini, S., Sansavini, S., Vinatzer, B., Gennari, F. and Domizi, C. 2000. Efficiency of marker assisted selection (MAS) for the *vf* scan resistance gene. *Acta Horticulturae* 538, 549–52.

Tester, M. and Langridge, P. 2010. Breeding technologies to increase crop production in a changing world. *Science* 327(5967), 818–22. doi:10.1126/science.1183700.

Tränkner, C., Lehmann, S., Hoenicka, H., Hanke, M. V., Fladung, M., Lenhardt, D., Dunemann, F., Gau, A., Schlangen, K., Malnoy, M., et al. 2010. Over-expression of an *FT-homologous* gene of apple induces early flowering in annual and perennial plants. *Planta* 232(6), 1309–24. doi:10.1007/s00425-010-1254-2.

Urrestarazu, J., Muranty, H., Denancé, C., Leforestier, D., Ravon, E., Guyader, A., Guisnel, R., Feugey, L., Aubourg, S., Celton, J. M., et al. 2017. Genome-wide association mapping of flowering and ripening periods in apple. *Frontiers in Plant Science* 8, 1923. doi:10.3389/fpls.2017.01923.

Vaillancourt, L. J. and Hartman, J. R. 2000. Apple scab. *Plant Health Instructor*.

van de Weg, W. E. 1989. Screening for resistance to *Nectria galligena* Bres. in cut shoots of apple. *Euphytica* 42(3), 233–40. doi:10.1007/BF00034459.

van de Weg, E., Di Guardo, M., Jänsch, M., Socquet-Juglard, D., Costa, F., Baumgartner, I., Broggini, G. A. L., Kellerhals, M., Troggio, M., Laurens, F., et al. 2018. Epistatic fire blight resistance QTL alleles in the apple cultivar 'Enterprise' and selection X-6398 discovered and characterized through pedigree-informed analysis. *Molecular Breeding* 38(1), 5. doi:10.1007/s11032-017-0755-0.

van Nocker, S. and Gardiner, S. E. 2014. Breeding better cultivars, faster: Applications of new technologies for the rapid deployment of superior horticultural tree crops. *Horticulture Research* 1, 14022. doi:10.1038/hortres.2014.22.

Varith, J., Hyde, G. M., Baritelle, A. L., Fellman, J. K. and Sattabongkot, T. 2003. Non-contact bruise detection in apples by thermal imaging. *Innovative Food Science and Emerging Technologies* 4(2), 211–8. doi:10.1016/S1466-8564(03)00021-3.

Velasco, R., Zharkikh, A., Affourtit, J., Dhingra, A., Cestaro, A., Kalyanaraman, A., Fontana, P., Bhatnagar, S. K., Troggio, M., Pruss, D., et al. 2010. The genome of the domesticated apple (*Malus* × *domestica* Borkh.). *Nature Genetics* 42(10), 833–9. doi:10.1038/ng.654.

Verma, S. 2014. A predictive genetic knowledge for apple 'fresh sensation' provides information to increase breeding efficiency in Washington. (Doctoral dissertation). Available from ProQuest Dissertations and Theses database. (UMI No. 3628891).

Vinatzer, B. A., Patocchi, A., Tartarini, S., Gianfranceschi, L., Sansavini, S. and Gessler, C. 2004. Isolation of two microsatellite markers from BAC clones of the *Vf* scab resistance region and molecular characterization of scab-resistant accessions in *Malus* germplasm. *Plant Breeding* 123(4), 321–6. doi:10.1111/j.1439-0523.2004.00973.x.

Visser, T. 1964. Juvenile phase and growth of apple and pear seedlings. *Euphytica* 13(2), 119–29.

Visser, T. and De Vries, D. P. 1970. Precocity and productivity of propagated apple and pear seedlings as dependent on the juvenile period. *Euphytica* 19(2), 141–4. doi:10.1007/BF01902937.

Walsh, B. B. and Lynch, B. 1999. Evolution and selection of quantitative traits. Available at: http://nitro.biosci.arizona.edu/zbook/volume_2/vol2.html.

Watson, A., Ghosh, S., Williams, M. J., Cuddy, W. S., Simmonds, J., Rey, M. D., Asyraf Md Hatta, M., Hinchliffe, A., Steed, A., Reynolds, D., et al. 2018. Speed breeding is a powerful tool to accelerate crop research and breeding. *Nature Plants* 4(1), 23–9. doi:10.1038/s41477-017-0083-8.

Weber, R. W. S. 2014. Biology and control of the apple canker fungus *Neonectria ditissima* (syn. *N. galligena*) from a Northwestern European perspective. *Erwerbs-Obstbau* 56(3), 95–107. doi:10.1007/s10341-014-0210-x.

Webster, A. D. 1995. Rootstock and interstock effects on deciduous fruit tree vigour, precocity, and yield productivity. *New Zealand Journal of Crop and Horticultural Science* 23(4), 373–82. doi:10.1080/01140671.1995.9513913.

Xiao, C. L. and Kim, Y. K. 2008. Postharvest fruit rots in apples caused by *Botrytis cinerea*, *Phacidiopycnis washingtonensis*, and *Sphaeropsis pyriputrescens*. *Plant Health Progress* 9(1). doi:10.1094/PHP-2008-0919-01-DG.

Xiao, C., Zheng, L., Sun, H., Zhang, Y. and Li, M. 2014. *Estimation of the Apple Flowers Based on Aerial Multispectral Image*. Montreal, Quebec Canada: American Society of Agricultural and Biological Engineers, pp. 1–8.

Xiaobo, Z., Jiewen, Z., Xingyi, H. and Yanxiao, L. 2007. Use of FT-NIR spectrometry in non-invasive measurements of soluble solid contents (SSC) of 'Fuji' apple based on different PLS models. *Chemometrics and Intelligent Laboratory Systems* 87(1), 43–51. doi:10.1016/j.chemolab.2006.09.003.

Xu, X.-M. and Robinson, J. D. 2010. Effects of fruit maturity and wetness on the infection of apple fruit by *Neonectria galligena*. *Plant Pathology* 59(3), 542–7. doi:10.1111/j.1365-3059.2009.02232.x.

Yamagishi, N., Sasaki, S., Yamagata, K., Komori, S., Nagase, M., Wada, M., Yamamoto, T. and Yoshikawa, N. 2011. Promotion of flowering and reduction of a generation time in apple seedlings by ectopical expression of the *Arabidopsis thaliana FT* gene using the Apple latent spherical virus vector. *Plant Molecular Biology* 75(1–2), 193–204. doi:10.1007/s11103-010-9718-0.

Yamagishi, N., Kishigami, R. and Yoshikawa, N. 2014. Reduced generation time of apple seedlings to within a year by means of a plant virus vector: A new plant-breeding technique with no transmission of genetic modification to the next generation. *Plant Biotechnology Journal* 12(1), 60–8. doi:10.1111/pbi.12116.

Yang, J., Mezmouk, S., Baumgarten, A., Buckler, E. S., Guill, K. E., McMullen, M. D., Mumm, R. H. and Ross-Ibarra, J. 2017. Incomplete dominance of deleterious alleles contributes substantially to trait variation and heterosis in maize. *PLoS Genetics* 13(9), e1007019. doi:10.1371/journal.pgen.1007019.

Zhou, L., Mrode, R., Zhang, S., Zhang, Q., Li, B. and Liu, J. F. 2018. Factors affecting GEBV accuracy with single-step Bayesian models. *Heredity* 120(2), 100–9. doi:10.1038/s41437-017-0010-9.

Zhu, Y. and Barritt, B. H. 2008. Md-ACS1 and Md-ACO1 genotyping of apple (*Malus* × *domestica* Borkh.) breeding parents and suitability for marker-assisted selection. *Tree Genetics and Genomes* 4(3), 555–62. doi:10.1007/s11295-007-0131-z.

Ziolkowski, P. A., Underwood, C. J., Lambing, C., Martinez-Garcia, M., Lawrence, E. J., Ziolkowska, L., Griffin, C., Choi, K., Franklin, F. C., Martienssen, R. A., et al. 2017. Natural variation and dosage of the HEI10 meiotic E3 ligase control *Arabidopsis* crossover recombination. *Genes and Development* 31(3), 306–17. doi:10.1101/gad.295501.116.

Ziya Motalebipour, E., Kafkas, S., Özongun, Ş. and Atay, A. N. 2015. Construction of dense genetic linkage maps of apple cultivars Kaşel-41 and Williams' Pride by simple sequence repeat markers. *Turkish Journal of Agriculture and Forestry = Turk tarim ve ormancilik dergisi* 39, 967–75.

Chapter 8

Advances and challenges in sustainable apple cultivation

Pierre-Éric Lauri and Sylvaine Simon, INRA, France

1 Introduction

Apple (*Malus domestica*) is one of the most produced temperate fruits worldwide, reaching 84.6 MT in 2014, among which 41 MT are produced by China (FAO, 2017). Originating from forests in central Asia, apple progressively spread towards various areas of Asia, the Mediterranean region and all of Europe thousands of years ago, with remains of apple identified in historic sites dating to 6500 BC (Luby, 2003). Apple was brought to the Americas by the first European colonists during the sixteenth and seventeenth centuries. During this spread, intense crop-to-wild and wild-to-crop gene flows among the various *Malus* species occurred, though it is usually considered that the wild Central Asian species *Malus sieversii* is a main contributor to the genome of the cultivated apple, with the wild European crab apple *Malus sylvestris* being a major secondary contributor (Cornille et al., 2014, 2015).

Apple is cultivated commercially in both the northern and the southern hemispheres in most temperate climatic zones, roughly within a latitude range of 25° to 50°. However, provided cultural techniques are adapted appropriately, apple is also grown at higher latitudes in warmer microclimatic zones, close to significant bodies of water, for example Norway at ca. 60°N (Palmer et al., 2003), and even within the intertropical zone at high elevation, for example Brazil at ca. 27°S above 1000 m (Petri and Leite, 2004) and the island of Java at ca. 8°S above 1300 m (Suryanata, 1994).

http://dx.doi.org/10.19103/AS.2018.0040.22

In past decades, and especially following the end of the Second World War, apple productivity (i.e. yield per unit of land area) has increased considerably due to optimization of the production system that combines various elements. A first element encompasses the strongly related 'orchard system components': plant material (rootstock and cultivar), tree arrangement, planting density, support system, tree training (defined as the global directions imposed to the branches and to the tree) and pruning (defined as cutting away branch parts) (Barritt, 1992; Robinson, 2003). The other elements are water and nutrient supply, fossil-fuel-based mechanization and control of weeds, pests and diseases. Intensive apple production, similar to agriculture in general, is now questioned because of its high dependence on non-renewable energy resources together with the loss of biodiversity and soil erosion, and above all, the high reliance on chemical fertilizers and pesticides that can increase environmental pollution and negatively impact human and animal health (Altieri, 1999). This disconnection between the food industry, especially agriculture, and the health industry raises the question of agricultural sustainability not only in terms of resource management, but also in terms of the environment and health (IPES-Food, 2016).

This chapter addresses how the sustainability issue may be handled for apple cultivation, focusing on plant biology and plant and system design and management. Genetic and breeding aspects will not be developed, although they present new challenging objectives for geneticists and breeders with regard to productivity and reducing susceptibility to multiple pests and diseases (Litrico and Violle, 2015). A growing body of evidence from current research in apple, as well as in other agricultural fields (e.g. organic agriculture) and ecology (e.g. agroecology, agroforestry), indicates that building more sustainable apple production systems may require complementary aspects to be tightly adjusted. The first is the apple tree itself, with the idea that better knowledge of the genetic variability of tree architecture and physiology would optimize fruit-oriented tree manipulation. The second is the design of the agrosystem in which apples are grown with the idea that efficient management of biodiversity, in terms of the choice of species and their spatial and temporal arrangement, would optimize desirable ecosystem services (provisioning, regulating, supporting and cultural; see Section 4.1).

2 Apple tree growth and fruiting

2.1 Tree architecture and shape typology

Tree architecture is a discipline of botany that was initially developed during the 1960s and 1970s on existing trees by Hallé and co-workers (Hallé and Oldeman, 1970; Hallé et al., 1978). Tree architecture analysis considers that the tree first establishes its 'architectural model'. Model determination is based

on morphological observations and considers four main qualitative criteria: growth process (e.g. rhythmic vs. continuous), branching process (e.g. immediate or sylleptic vs. delayed or proleptic), morphological differentiation of axes (e.g. orthotropy or erect vs. plagiotropy or horizontal) and position of reproductive organs (lateral or axillary vs. terminal). The combination of these criteria led to the determination of 23 architectural models (Barthélémy and Caraglio, 2007; but see the recent description of 12 new architectural models in paleozoic plants, Chomicki et al., 2017), each being characterized by a unique combination of specific modalities of each of the qualitative criteria, and named after a botanist who has worked on a species considered as a type for this model. Research has shown that if some species can be rather strictly defined by a single model, others combine characteristics of two or more models, often in relation to changes of branching development during tree lifespan (Bell, 1991). For example, the apple tree combines features of two architectural models, Rauh and Scarrone. These two models are characterized by orthotropic trunk and branches, and rhythmic growth and branching. However, although branches are monopodial with lateral flowering in Rauh, they are sympodial with terminal flowering in Scarrone. Indeed, unlike peach (*Prunus persica*), apricot (*Prunus armeniaca*) or sweet and sour cherry (*Prunus avium*, *Prunus cerasus*) that have lateral flowering – therefore belonging to the Rauh architectural model – apple has both lateral and terminal flowering (Lauri and Térouanne, 1995; Lauri and Laurens, 2005; Costes et al., 2006).

Once the tree has established its architectural model, it may reiterate the whole model or a part of it, referred to as total or partial reiteration, respectively. Here, the term 'reiteration' designates both the process and its end product, the shoot, possibly with branching. The reiteration phenomenon was initially described by Oldeman (1974) on tropical trees, and it was soon established that the ability to reiterate is highly related to the species. Some species, such as the coconut palm (*Cocos nucifera*), only seldom reiterate, whereas the apple tree may reiterate as soon as during the first years after germination. Reiteration occurs naturally during tree growth, often related to changes of light distribution within the canopy, or in reaction to branch bending. In this latter case, a 'water sprout' or an 'epicormic shoot' (Meier et al., 2012) developed on the upper side of a bent branch typically corresponds to a partial or total reiteration. Reiteration is well known by fruit tree growers because it is usually provoked by pruning, referred to as 'traumatic reiteration'. Therefore, from an architectural point of view, the management of the pruned tree, such as a V- or Y-trellis system (Robinson, 2003), is fundamentally the management of a population of reiterations. It also has to be mentioned that although strong reiterations often indicate a lack of balance within the tree canopy, that is usually indicating vigorous vegetative growth to the detriment of flowering and fruiting, some cultivars may maintain strong flowering in the apical position on vigorous reiterations, suggesting that this trait is a relevant criterion to consider for both genetic selection and management of tree training and pruning (Lauri, 2002; Lauri et al., 2011).

It has long been known that beyond the architectural model that is qualitative by definition, apple cultivars differ widely in tree shape and fruiting behaviour. Cultivars can be categorized in two main 'growth habits': (1) the 'upright and spur-type' canopy with fruiting in terminal position on spurs (short shoots) or (2) the 'umbrella-type' canopy with fruiting in terminal position on 'downswept branches' (Forshey et al., 1992). A finer typology of apple tree growth and fruiting was established by Bernhard (1961) combining both tree growth orientation (upright to drooping branches) and position of fruiting on the structural framework (high frequency in apical position on spurs vs. high frequency in apical position on long shoots). This typology was extended by Lespinasse (1992) considering four 'ideotypes': 'columnar' as type I (e.g. cv. MacWijcik, a mutant of the McIntosh variety which was not described by Bernhard, 1961), and types II ('spur'; e.g. cv. Oregon Spur Delicious), III ('standard'; e.g. cv. Golden Delicious) and IV ('tip-bearing'; e.g. cv. Pink Lady). Lespinasse (1977, 1980) and later Lespinasse and Delort (1986) were likely the first to establish a relationship between these 'tree ideotypes' and the bearing pattern. Indeed, type I and type IV cultivars are generally characterized by an irregular and a regular bearing pattern, respectively.

As with other angiosperm trees, the molecular bases that predetermine apple tree architecture are progressively being deciphered (Hollender and Dardick, 2015). However, the determination of apple tree architecture based on the qualitative criteria established by Hallé et al. (1978) and the analyses of the quantitative variations of architectural traits such as the genetic variability in the topological distribution of branching along the parent branch (Lauri and Térouanne, 1998; Costes et al., 2003) and in the regularity of bearing (Lauri et al., 1995, 1997a, 2014; Lauri and Laurens, 2005) have essentially been developed in the temperate climate zone. Under such a climate, branching and phenology are strongly affected by winter cold temperatures (Cook and Jacobs, 1999; Legave et al., 2008). These architectural traits vary quantitatively when apple trees are grown in climates characterized by a lack of chilling. This is typically the case for budburst frequency in spring that involves differences in both the topological distribution along the parent shoot and the date of budburst (Dutra Schmitz et al., 2014, 2015; Legave et al., 2015; Malagi et al., 2015).

2.2 From descriptive to functional architecture

2.2.1 Tree architecture, living meristems and the extinction phenomenon

The whole-tree approach, developed in the conceptual framework of tree architecture analysis, has a clear descriptive interest, especially for breeders.

For example, the ideotype definition developed by Lespinasse (1992) is promoted by the International Board for Plant Genetic Resources (IBPGR; Watkins and Smith, 1997) to describe the 'bearing habit' of the apple tree by combining the branching pattern and the distribution of fruiting within the tree canopy. However, the variability of apple tree growth and fruiting patterns within each apple ideotype indicates that such a global and qualitative approach to tree architecture can hardly shed light on the physiological mechanisms underlying tree function. This is typically the case for some 'Fuji' clones that belong to type IV, according to Lespinasse typology, with terminal fruiting on long shoots, but may be characterized by a strong irregular bearing pattern (Lauri et al., 1997a). On the other hand, as shown by Barritt et al. (1997), the bearing pattern of a given cultivar may vary depending on the rootstock, with vigorous rootstocks decreasing the incidence of irregular bearing of spur-type cultivars but increasing irregular bearing of the standard and tip-bearing cultivars. More generally, the competitive relationship between crop and shoot growth, which is a classical principle in fruit tree biology and management textbooks (Forshey et al., 1992), may be true for spur-type cultivars, but is questioned for standard and tip-bearing cultivars (Lauri, 2002; Dennis, 2011; Lauri et al., 2011).

In the early 1990s, more detailed studies were developed on the behaviour of individual fruit-bearing shoots of various lengths, including spurs, with the objective to decipher the relative contributions of the functioning of the fruit-bearing shoot itself and of the branch, viewed as a population of shoots, in the regularity of bearing (Lauri et al., 1995, 1997a,b). These studies led to three main results. First, cultivars are differentiated by the frequency of axillary flower clusters whose bourse-shoots die, entailing the disappearance of shoots that would otherwise have developed from these flower clusters. This phenomenon, called 'extinction', gives rise to portions of branches without branching ('bare wood') in the following years (Fig. 1a). It is a specific case of cladoptosis, that is the shedding of a single shoot or of a whole-branch complex (Bell, 1991), related here to purely internal physiological mechanisms (Lauri, 2009a). More fundamentally, this finding showed that focusing only on growing organs, as done in classic architectural analyses, hides the fact that tree architecture results also from interactions between organs that are growing and organs that are not growing, being either in a latent stage (e.g. latent buds) or that have grown but later died (extinction) (Lauri, 2009a).

Second, cultivars can also be characterized by the frequency of bourse-over-bourse formation, that is the ability to string flower clusters over consecutive years on the same shoots, with values ranging from 20% to 90% depending on the cultivar (Fig. 1b) (Lauri et al., 1995). Third, there is a positive relationship between extinction and bourse-over-bourse across cultivars, supporting the idea that cultivars with a low frequency of branching (resulting from high extinction) also have an endogenous ability for regular bearing (Lauri et al., 1995).

However, the actual bearing pattern of a tree not only results from the behaviour of individual shoots as described earlier, but also involves interactions between the shoots at branch and tree scales. Three main strategies have been identified: (1) a high bourse-over-bourse frequency clearly indicates the ability of the cultivar to have a regular-bearing pattern; (2) a low bourse-over-bourse may be related to high irregular-bearing if the shoots are synchronized in their bearing and non-bearing years (Fig. 1c); or (3) a low bourse-over-bourse may be related to more regular-bearing if shoots are desynchronized (Fig. 1d) (Lauri et al., 1995, 1997a,b, 2014; Lauri and Laurens, 2005). These basic concepts have been integrated recently in the quantitative analysis of the genetic determination of the flowering behaviour of apple (Durand et al., 2013, 2017). All of the above-mentioned studies were generally carried out on trees or branches with little training and pruning operations. Although the concepts of bourse-over-bourse, and synchronization vs. desynchronization, are still valid from a fundamental point of view, it is clear that the actual bearing pattern of a tree results from both its endogenous potential and the way it reacts to the training and pruning procedures (e.g. Breen et al., 2015).

Figure 1 Architectural traits of apple. Extinction (a, yellow circles, cv. Granny Smith), bourse-over-bourse (b, cv. Granny Smith), and branches with synchronization (c, cv. Braeburn) and desynchronization (d, cv. Belrène) of flowering. Credit: PE Lauri, INRA.

2.2.2 The relationships between growth and flowering and fruiting

There is a complex network of interactions between vegetative growth and flowering (Looney and Lane, 1984; Forshey et al., 1992) that also affects fruit-set and eventually fruiting (Dennis, 2011). Indeed, 'spur-type' cultivars with a strong hierarchy between the trunk or main scaffolds and the spurs usually have an irregular-bearing pattern, whereas 'terminal-bearing' cultivars with a more polyarchic architecture, that is with no such differences between vegetative and fruiting shoots, are more regular in fruit production (Lauri et al., 1995) (Fig. 2).

The relationships between vegetative growth and flowering and fruiting have been explored in two ways. The first is the frequency of lateral flowering, that is on one-year-wood (1YW). This varies widely across cultivars, with consequences for tree management: a high frequency of flowering on 1YW may, in the case of high fruit-set, lead to a heavy crop load that year. This increases the risk of irregular-bearing as early as in the first years in the orchard (Lauri, 2002; Lauri et al., 2011). The frequency of flowering on 1YW also affects how the branch responds to bending, with generally an increase of the number of flower buds on bent branches of cultivars with high 1YW flowering frequency, and an increase in the length of vegetative laterals on cultivars with high 1YW vegetative bud frequency (Lauri and Lespinasse, 2001).

The second aspect is the relationship between the length of a shoot and its contribution to yield, with the literature usually separating the behaviours of spurs and long shoots. Although there is a positive relationship between the whole-canopy leaf area index (LAI) and yield, this relationship is in fact mostly driven by the LAI and the light interception of the spur canopy, and not of the long shoot canopy (Wünsche and Lakso, 2000). This prominent role

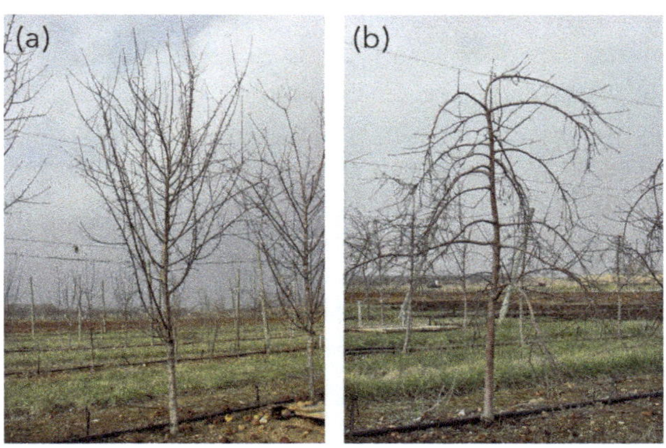

Figure 2 Whole-tree architecture of apple – 3-year-old trellised own-rooted apple hybrids belonging to the hierarchic or 'spur-type' (a) and to the polyarchic or 'terminal bearing type' (b) architectures. Credit: P. E. Lauri, INRA.

of the spurs to fruiting is usually related to the fact that the spur, compared to the long shoot, has a faster carbon autotrophy and therefore has a faster carbon export to the flowering and growing sites at the beginning of the growing season (Johnson and Lakso, 1986), therefore playing a positive role in fruit set and early growth before long shoots could export photosynthates to the fruit (Quinlan and Preston, 1971; Zhang and Tanabe, 2008). From a functional morphology point of view, it has been hypothesized that the higher frequency of flowering on spurs is in part due to the higher ratio of leaf to stem biomass, that is a lower 'axialisation' of the spur compared to the long shoot (Lauri and Kelner, 2001; Lauri and Normand, 2017). Other studies carried out at the scale of the individual shoot have shown that the relationship between the length of a shoot established in a given year and the frequency of apical position flowering in the following spring strongly differs depending on the cultivar (Lauri and Trottier, 2004). Interestingly, the relationship between shoot length and its fruiting frequency was parabolic for the two cultivars in the experiments of Lauri and Trottier (2004), illustrating the strong dependence of fruit-set on the supporting shoot growth during the previous year (Lauri and Correlli-Grappadelli, 2014).

3 Bases for sustainable apple training and pruning management

The agronomic efficiency of an apple orchard depends on several factors that need to be tightly adjusted. Apart from the soil, nutrient and water resource management that have to be well adapted to the site-specific characteristics (Neilsen et al., 2017), we develop here the specific aspect of tree training and pruning. Other aspects related to the plant material (cultivar and rootstock) and the design of the system (tree arrangement and density, support system, cultural practices) are detailed in Section 4.

More than 30 canopy training systems for apple have been described so far around the world, and these can be grouped in four basic canopy shapes: spheric, conic, flat-fan (vertical plane) and Y or V (angled dual plane) (Robinson, 2003). However, many more systems coexist worldwide, reflecting the ability of each individual grower to adapt the canopy system to his/her specific environmental and socio-economical context. From a biological point of view, this also means that the whole-tree shape is of minor importance when training and pruning procedures are managed in a way that promotes the growth of fruiting shoots and reduces undesirable vegetative growth reactions (Lauri, 2018). Past experiences that aimed to standardize tree shape, for example indiscriminate heading-back using mechanical hedging and topping (Forshey et al., 1992) or the management of old central leader trees (Warrington et al., 1995), often led to severe problems such as dense and vigorous vegetative reiterations with low flowering frequency and heavy shade within the tree canopy, resulting in poor fruiting. Therefore, tree canopy management may be optimized when manipulation is based on known endogenous architectural and physiological mechanisms to develop sustainable apple systems (Lespinasse, 1977, 1980; Lauri et al., 2009b). More specifically, regarding the classical vision of tree training and pruning, efficient manipulations of physiological sources (water, nutrients, carbohydrates) and sinks (shoot vs. fruit growth) in the tree are more important than maintaining a standardized shape (Lauri and Corelli-Grappadelli, 2014).

Based on apple architectural studies in past decades, we have proposed that sustainable apple tree management is that which minimizes undesirable vegetative reactions and promotes flower induction and good fruiting over consecutive years, through appropriate training and pruning procedures (Lauri and Lespinasse, 2000). To reach this objective, there is a crucial need to develop more knowledge of the endogenous potential of each cultivar to grow, flower and fruit, for example relationships between shoot length and fruiting (Lauri and Trottier, 2004; Lauri et al., 2011) and also on how the tree reacts to the cultural practices, for example reaction to bending (Lauri and Lespinasse, 2001) or to artificial spur extinction (Breen et al., 2016a). This can be illustrated by the training and pruning concepts that have been progressively developed based on architectural analyses (Lauri, 2016). The finding of a positive relationship between extinction and bourse-over-bourse formation across apple cultivars and genotypes suggests that the thinning out of the young fruiting spurs on spur-bound cultivars can be a means to increase canopy porosity to light and air, which is a main factor to optimize light interception within the whole canopy (Corelli-Grappadelli and Lakso, 2007). This procedure is called, in various ways, 'artificial extinction', 'extinction pruning', 'spur extinction' (Lauri et al., 1997b, 2004; Lauri, 2009b) or 'artificial spur extinction' (ASE; Tustin et al., 2012). It is carried out in all parts of the canopy where there is a lack of light and consequently a weak development of fruiting spurs, and is efficient at two levels. First, it reduces the number of inflorescences on each branch to reach as closely as possible the target commercial fruit number for that tree (Tustin et al., 2009, 2011, 2012, 2015; Breen et al., 2014, 2015, 2016a). Second, it improves light penetration within the tree canopy (Willaume et al., 2004; Breen et al., 2016b), optimizing fruit-set (Breen et al., 2014, 2015, 2016a), return-bloom (Lauri et al., 2004) and fruit quality (Lauri et al., 2007). Although artificial extinction has to be applied on well-balanced trees, that is without strong heading-back on vigorous shoots, it is successfully applied on various tree shapes, whether with a single trunk such as the tall spindle (Breen et al., 2015) or centrifugal training (Tustin et al., 2009; Lauri et al., 2004, 2007) systems, or with several reiterative trunks such as the Salsa system (Lauri et al., 2016a).

4 Challenging the apple production agroecosystem

Current apple orchards are intensively managed. They are comprised of high-density plantings of single cultivars and require a lot of external inputs, especially pesticides, to produce high fruit yields. Rethinking apple cultivation systems to improve the long-term sustainability of orchards should then reconsider plant material, temporal orchard design and management practices.

4.1 From simplified to complex productive and sustainable agroecosystems

The Millennium Ecosystem Assessment initiative (Millennium Ecosystem Assessment, 2005) established four 'ecosystem services' defined as the 'benefits people obtain from ecosystems': (1) 'provisioning' that provides goods such as food or timber to the people; (2) 'regulating' that affects climate, floods, disease and so on; (3) 'supporting' such as photosynthesis or nutrient cycling; and (4) 'cultural' that provides recreational, aesthetic and spiritual benefits. Each service can be considered as the output of one to several functions. For example, fruit yield depends not only on carbon allocation and partitioning within the plant, but also on canopy light interception and root uptake of water and nutrients from the soil. The concept of ecosystem service and the subtending physiological processes undoubtedly foster fundamental and applied research not only on the functioning of natural ecosystems, but also on anthropized ecosystems (e.g. agroecosystems). In the latter case, the idea is promoted that such systems can be engineered, meaning that they can be sustainably designed and managed using ecological principles that integrate human societies with their natural environment for the benefit of both (Lescourret et al., 2015; Barot et al., 2017a,b).

A great amount of research in recent years has shown that increasing plant diversity in agricultural systems improves the overall production of crops, forage and wood, and also improves several regulating and supporting ecosystem services (Storkey et al., 2015; Isbell et al., 2017). The concept of ecological intensification has progressively emerged based on the 'use of biological regulation to manage agroecosystems, at field, farm and landscape scales' (Doré et al., 2011) with the main objective to achieve food security (Bommarco et al., 2013). As perennial crops and complex habitats, orchards are an opportunity to develop such approaches. However, research on the subject is scarce, with most studies on the functioning of agroecosystems and ecological intensification developed on annual crops (Barot et al., 2017a). In a review on ecosystem services in orchards, Demestihas et al. (2017) showed that orchard agroecosystems (including plant diversity in groundcover and hedgerows) also have a high potential to provide ecosystem services other than provisioning, provided adequate agricultural practices are adopted by the grower. Regarding the ecosystem service of pest suppression, although there is no strict relationship between plant diversity and reduction of the incidence of pests and diseases, various mechanisms can explain the effects of plant diversity on pests and diseases, including the conservation of natural enemies and facilitation of their action against aerial pests, as well as direct and indirect architectural and/or physical effects (Ratnadass et al., 2012). Soil management procedures also may be a way to improve ecosystem support services. The utilization of groundcovers and composts in organic apple orchards to stimulate biological activity in the soil and enhance mineralization is generally advocated, as well as the use of legumes such as white clover (*Trifolium repens*) or alfalfa (*Medicago sativa*) as a source of nitrogen (Granatstein et al., 2016; Demestihas et al., 2017).

Standard apple orchards typically include a single cultivar in a simplified, input-based agroecosystem designed to maximize fruit production. The challenge is to redesign the whole system to produce fruit by mobilizing ecosystem services rather than external inputs. With that perspective, interest is increasing in the concept of 'multifunctional woody polycultures'. In examples in the United States and France detailed by Lovell et al. (2017), these systems combine apple, nut species, timber trees and other species of trees (e.g. chestnut, *Castanea* spp.) or shrubs (e.g. black currant, *Ribes nigrum*), possibly with annuals (e.g. corn–soybean rotation). These systems are 'agroforestry systems' combining agriculture and forestry, with various perennial and annual crops that provide different products: timber, pip- or stone-fruit, berries, vegetables, grains or tubers (Van Noordwijk et al., 2016). Agroforestry is considered as an agroecological practice encompassing both the mixture of species and the spatial and temporal organization of the system (Wezel et al., 2014).

4.2 Towards apple agroecosystems

4.2.1 Agroecological practices to foster ecosystem services for apple production

Apple-Tree-based AgroForestry Systems (AT-AFS) are comprised of mixtures of woody plants, namely apple and potentially other trees, and herbaceous crops. In AT-AFS, as for all cropping and farming systems, the first service to address is the provision of products. However, if the priority product is apple fruit, it is worth considering that combining various crop species in a same land area is also a way to increase the global productivity per land unit. A common metric to quantify the productivity of a multispecies system is the land equivalent ratio (LER), that is, the area needed to produce the same outputs as one unit of land with a pattern of sole cropping. For example, a two-species system LER is computed as follows: LER = (mixed yield A/pure yield A) + (mixed yield B/pure yield B) (Malézieux et al., 2009). Silvo-arable systems that combine timber trees and annual crops (typically cereals in rotation) have LER estimates of 1.3–1.6

(Lovell et al., 2017). LER values greater than 1 indicate that mixed systems may be advantageous compared to the sum of individual pure cultures on separate land areas.

To our knowledge, there is no published LER for AT-AFS to date. Considering other fruit trees, an LER of ~1.3 has been estimated for a cherry-walnut agroforestry system (Dupraz et al., 2005). This result suggests that growing apple with other herbaceous or perennial crops with short (e.g. vegetables, cereals) or longer (e.g. timber) terms could result in a favourable AT-AFS global provisioning service. Gao et al. (2013) studied two intercropping systems combining apple with either soybean (*Glycine max*) or peanut (*Arachis hypogaea*) on the Loess plateau in China. Although competition for water between apple and groundcover was evident in this experiment, intercropping was considered to be an effective method to increase land use efficiency and economic return across the whole plot.

Beside production, other ecosystem services can be fostered by practices and plant diversification in AT-AFS. At the tree scale, apple canopy management, through training and pruning, can have a significant (although partial) effect on reducing pest infestation and disease infection. Simon et al. (2006) compared the effect of apple canopy management systems on the most serious foliar pests (aphids, Aphididae; mites, Acari, mainly Tetranychidae) and diseases (apple scab due to *Venturia inaequalis*) in an organic orchard over 3 years. The interactions between canopy factors and infestations or infections were related to four main features: the direct effects of pruning method on (1) the removal of pests or inoculum, and (2) shoot density and distribution in the canopy, and the indirect effects of the severity of pruning on (3) the within-tree microclimate and (4) the shoot growth dynamics, possibly promoting the presence of attractive or susceptible leaves later in the season. For example, the degree of branching was identified as a main factor driving the within-branch development of the rosy apple aphid (*Dysaphis plantaginea*) during its wingless phase in spring. This finding could explain why Centrifugal trained trees, with a higher degree of branching in response to artificial spur extinction (i.e. the removal of branches stimulates branching on remaining ones), tended to be less infested than the Solaxe trained trees that does not integrate this type of pruning (Simon et al., 2012).

At the orchard scale, the increase in biodiversity, either cultivated or not, is a way to promote the number and diversity of natural enemies of pests. Stamps and Linit (1998) noticed that increasing plant diversity in temperate agroforestry, and especially the combination of trees and crops, provided greater niche partitioning that favoured arthropod diversity in both time and space than did monoculture of annual crops. Simon et al. (2010) reviewed the literature on the effects of manipulating plant diversity and habitats on the control of pests by arthropod and bird communities in apple, pear (*Pyrus communis*) and peach orchards. They found that the effect was mostly positive (50% of the studies), but could also be negative (15% of the studies). For example, with fresh (Wyss, 1995) and cider (Albert et al., 2017) apples, flower strips increased natural enemy abundance in the vicinity of the orchards, improving biological control of rosy apple aphid and/or green aphid (*Aphis* spp.). In organic fresh apple orchards, Miñarro and Prida (2013) identified local floral resources that could provide benefits for farmers by improving both pollination and the biological control of pests. As a whole, according to Smith et al. (2016), agroforestry with careful selection and, if possible, mixed planting of low-susceptibility apple cultivars seemed to reduce disease levels compared to pure cultivar orchards.

Intercropping apple trees with non-nitrogen-fixing annual grass usually leads to a decline in apple yield due to competition for resources (Robinson and O'Kennedy, 1978). In a 2-year experiment in an organic apple orchard, Goh et al. (1995) showed that biological nitrogen fixation was significantly correlated with clover dry matter production and maintenance of soil fertility. Further, in an experiment combining 17-year-old hybrid walnut trees (*Juglans nigra* × *Juglans regia* L.) and alfalfa in the inter-row, Querné et al. (2017) showed that shading improved the apparent light use efficiency (above-ground biomass produced per unit of global radiation) of alfalfa by 35%. Also, biological nitrogen fixation was stimulated close to the trees. Both results indicated that compensative and facilitative mechanisms developed when trees and alfalfa plants were intercropped and this was not observed when these species were grown in pure stands. Interestingly, the pest regulation and the nutrients availability services can also be enhanced together. Brown and Tworkoski (2004) showed in apple orchards that compost mulch application provided the tree with nutrients, increased soil fertility and promoted increased populations of detritivore species such as those of Collembola order that subsequently serve as prey of other organisms, thus maintaining predatory species in the food chain.

4.3 Designing an apple-tree-based agroforestry system

An AT-AFS designed to maximize a range of ecosystem services (e.g. pest suppression, soil fertility, pollination) must tightly integrate several components that go beyond the standard apple orchard system decisions described by Barritt (1992) or Hoying and Robinson (2000). Choices must also include the diversified plant species in the system and how to assemble them in space and time. Tree-based multiple cropping systems usually belong to 'intercropping' where two or more crops are grown at the same time, and to 'sequential' or 'relay cropping' where the crops are grown on the same land but one following the other (Gliessman, 1985; Gallaher, 2009; Parotta et al., 2015). The empirical design of fruit-tree-based agroforestry systems in temperate Europe likely was traditionally practised until the beginning of the twentieth century (Herzog, 1998). These systems were known as 'Joualle', 'Verger haute tige' or 'Pré-verger' in France (Dupraz and Liagre, 2011), 'Streuobst' in Germany (Herzog, 1998) or 'Dehesa' in Spain (Campos et al., 2007). These were multilayered (herbaceous, fruit bushes, trees) and multifunctional with plants (cereals, vegetables, fruits, timber, forage) and generally animals (grazing livestock).

Designing AT-AFS involves major changes in the spatial arrangement of the system compared to a standard orchard system. Changes are related to the plants that are assembled and the way they are managed. In past decades, most AT-AFS were developed as 'two-layered', including grass or flower strips between the apple rows at the lower stratum, and apple trees at the upper stratum, with various trees and shrubs in lines of hedgerows (Demestihas et al., 2017; Simon et al., 2017). Current thought takes inspiration from cocoa- and coffee (*Coffea sp.*)-based agroforestry systems in the tropics (Andres et al., 2016; Tscharntke et al., 2011), resulting in a multilayered AT-AFS (Lauri et al., 2016b; Lauri, 2018). Designing such systems is complex and needs to take into consideration the plant materials (apple, companion plants) and the design of the whole system, and seasonal management (planting choices and general design, cultural practices, and processes or approaches) (see Fig. 3).

This can be summarized considering three complementary aspects.

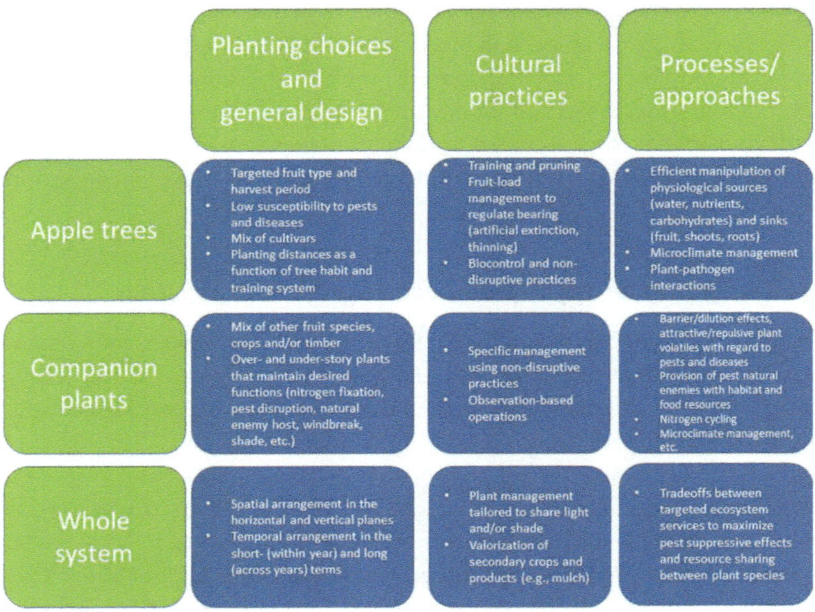

Figure 3 Elements for the design framework of an Apple-Tree-based Agroforestry System (AT-AFS).

4.3.1 Plant assemblage

The apple plant material needs to fit the objectives not only in terms of fruit production as the main objective, but also in terms of the other targeted services, for example pest control. If the choice of apple cultivar is determined primarily by the market, it should also be considered that some main pests and diseases can be controlled better on susceptible cultivars provided they are grown with other resistant or low-susceptibility cultivars. For example, in a study on scab (*Venturia inaequalis*) control in apple orchards, Didelot et al. (2007) showed that the within-row mixture of resistant and susceptible cultivars significantly reduced disease incidence on the susceptible cultivar over the 2-year experiment, compared to the monoculture of the susceptible cultivar alone.

Rootstock choice is also of primary importance to optimize vegetative growth and fruiting in agroforestry conditions where *a priori* strong below- and above-ground competitions exist among species. Semi-vigorous rootstocks (e.g. MM106 or M7; Ferree, 1988) or even growing own-rooted cultivars (Maguylo and Lauri, 2007) often are good candidates. Some rootstocks, such as those released by the USDA-ARS/Cornell University or Budagovski breeding programmes, provide a good balance between vegetative vigour and productivity, and resistance or tolerance to fire blight and/or crown rot (Lordan et al., 2017). The choice of companion species for the flower strips, grass alleys and/or hedgerows surrounding the apple trees must first consider expected ecosystem functions (e.g. windbreak, predation, nitrogen fixation). To focus on pest suppression, various processes, including push-pull and barrier and dilution effects, and the conservation of natural enemies of pests (Ratnadass et al., 2012), are important to design a 'pest suppressive' AT-AFS. The plant traits involved in desired functions such as emission of attractive or repulsive plant volatiles, or provision of pollen and nectar as alternative food resources to natural enemies, are a good basis to select these companion plants.

4.3.2 Plant arrangement

The spatial arrangement of the various species needs to be tightly adapted to the endogenous characteristics of each species and their expected functions. A first aspect of the spatial design is the distance between plants that should be large enough to decrease competition for light, water and nutrients, and also to permit easy access to the trees for manual tasks (fruit thinning, harvest). Differences in phenology among the agroecosystem species are also important to allow temporal complementarity and low competition. This is typically the case when trees and annual crops grow at different times of the year, avoiding competition for light as well as water and nutrients during their respective growing seasons, such as winter wheat combined with late-blooming walnut (Lovell et al., 2017). However, the situation is different in the case of an AT-AFS where the apple bloom-to-harvest season lasts from early spring to fall, that is overlapping most of the growing season for the companion species. The arrangement must also consider possible effects on pest infestation dynamics and infection epidemics. For example, Albert et al. (2017) showed that the presence of various arthropods as phytophagous and natural enemies around hedgerows and flower strips where they primarily develop varies consistently depending on the arthropod species, with measurable positive or negative effects on apple.

Similarly, distances between apple trees may have a significant effect on within-tree and within-row scab infections, resulting in different infection patterns depending on the year (Didelot et al., 2007). Another aspect is the relative position (height) of the species of the system. The success of growing a species in the understory is highly dependent on its intrinsic ability to grow in shade. Unlike some shade-tolerant tropical species, such as coffee and cocoa (*Theobroma cacao*) (Tscharntke et al., 2011), apple trees perform better in full sun due to significant negative effects of shade on photosynthates, fruit growth and colouring (Correlli-Grappadelli, 2003). However, there are no simple relations between light interception and leaf function, and shade leaves are in general more efficient than sun leaves at utilizing within-canopy low light levels or sun flecks (Correlli-Grappadelli, 2003). Moreover, if light interception by the apple canopy is considered to be critical, it should also be noticed that the relationships between light interception and productivity are not straightforward; these two features may be completely independent, for example in the case of an alternate-bearing apple tree in an 'off year'. In regions with hot and dry summers, apple trees may be subjected to excess solar radiation during the summer, with known deleterious effects on fruit quality (sunburn) and leaf function (photo-oxidative damage) (Correlli-Grappadelli and Lakso, 2007; Racsko and Schrader, 2012). In such climatic conditions, the design of multilayer agroecosystems in which the apple trees are overhung by other tree species may provide convenient shade for the apples, provided the overstory trees are carefully pruned (see below). This could be a good alternative to shade nets that are commonly used in monocropping apple orchards. Additionally, the system is optimized if these overstory trees provide other services such as provisioning (nuts, timber) or supporting (legume trees improving soil fertility through nitrogen fixation, greater organic matter production and recycling of nutrients; Ong et al., 1991).

4.3.3 Spatial and temporal agroecological engineering

Once the initial components of the AT-AFS are decided, the management of both overstory and understory trees, and of herbaceous plants, during consecutive years is primordial. This would include actions such as pollarding, repeated every 2–5 years (Le Bec et al., 2016), conducted on the overstory trees to improve light transmission to the apples and other plants underneath. In this low-light context, training and pruning of the apple tree is still a main means to increase light penetration within the apple tree canopy. Concepts such as those developed in Section 3 – 'Bases for sustainable apple training and pruning management' – may serve as guides to develop methods suitable for the management not only of apple trees but also of the whole system. Growing apples in an AT-AFS requires managing an agroecosystem, that is not only apple trees but a variety of functional plants. All plants are grown within and/or around the apple trees for a specific role in supporting apple production, and this spatio-temporal design and management is at the core of AT-AFS.

As concerns the temporal layout, all the plants to be assembled may not be planted at the outset of the system, but must be carefully planned and managed depending on the grower's objectives for production and the other expected services. For example, in a two-woody species system that combines apple trees and fruit bushes, all plants can be planted in the establishment year, with the goal to achieve bush fruit production before the apple trees reach full production, with the possible removal of fruit bushes thereafter. Similarly, in an apple tree–overstory tree system, all trees can be planted in the system establishment year, or alternatively, the overstory trees can be planted a few years later when apple trees are near production maturity. Various AT-AFS designs at the spatial and temporal levels address different ecosystem services.

5 Conclusion and future trends

Current strategies for sustainable apple production often focus on maintaining, but reducing, an array of traditional inputs, such as reduced nitrogen fertilization and the use of narrower spectrum pesticides to reduce unintentional impacts on the environment. However, as evidenced from the most recent literature on organic fruit tree culture, agroecology and more largely in the field of ecology and ecosystem services, it is logical for alternative designs for sustainable apple cultivation systems to consider not only the composite (rootstock and cultivar) apple tree itself, but also its biotic and abiotic environments. Designing such agroecosystems production strategies is complex and must draw upon lessons from the many concepts and experiments developed around the world on temperate and tropical-tree-based production systems, many of which include fruit trees (Parotta et al., 2015). Three main aspects are outlined in relation to work implications, the system approach and the moving dimension of AT-AFS.

First, as observed for agroforestry systems in general (Rapidel et al., 2015), and for organic apple orchards in particular, it is likely that implementing AT-AFS would increase the workload and complexity of the work needed for managing the agroecosystem, compared to standard apple orchards. Indeed, the spatial design of the whole system can increase the time needed to move from one part of the system to another and/or can require the combination of several tasks into a single operation. Therefore, consideration of ergonomics has to be included in the *a priori* evaluation step of the design process as well as the diversity of tasks, for which no simple and standardized rule or directive can be established. Considering the socioeconomic context of apple-tree-based agroforestry, Smith et al. (2016) concluded such systems 'could work well in a diverse, potentially small-scale system such as a market garden, where apples could contribute to direct marketing channels such as vegetable box schemes or farm shops'. However, since these systems are still experimental, it would be premature to confine them to a unique production and commercialization system. In rotational crops, precision agriculture has proved that both small-scale tailored practices and productivity could meet.

Second, increasing plant diversity to improve natural regulation of pests and diseases, as well as nitrogen delivery to the apple tree, also needs a system approach considering that each component of the spatial and temporal design, and its management, has to be consistent with the whole system (Debaeke et al., 2009). For example, if optimizing the spatial and temporal mixing of species is the basis of future AT-AFS, the control of pests and diseases, especially in the context of climate change, will likely need to be combined with innovative methods of biocontrol (Granatstein et al., 2016) to avoid the use of synthetic pesticides that could otherwise disrupt natural enemies of pests (Dib et al., 2016).

Finally, AT-AFS are perennial and the design of such systems relies heavily on both the ontogenetic changes the apple tree goes through during its lifetime (e.g. training and pruning rules vary depending on tree age) and the carry-over effects of cultural practices across consecutive years (e.g. crop load management one year has effects in the following years). These changes also affect the other species in the system composition, entailing a gradual change in the internal environment of the system (Simon et al., 2017). This is well illustrated by the concept of dynamic cocoa agroforestry systems that may include year-to-year changes in both the planting density and the management practices for the various species (Andres et al., 2016).

Tackling the design of innovative AT-AFS is a unique 'learn-by-doing' opportunity founded in interdisciplinarity and diversification of knowledge sources, including both advances in plant sciences, lessons from the functioning of natural ecosystems and local farmers' knowledge (Doré et al 2011; Lauri et al., 2016c). Therefore, the agroecological engineering that is embedded in the concept of AT-AFS is an open avenue for more fundamental and applied research aimed at increasing the sustainability of apple production.

6 Where to look for further information

There are only scarce websites entirely dedicated to sustainable apple production. Much of the information in English on sustainable fruit tree cultivation is available on numerous websites of land-grant universities in the United States (e.g. IPM at Washington State University: https://ipm.wsu.edu/). Main aspects of integrated fruit production and organic farming with examples in apple can be found on websites of international organizations (The International Organisation for Biological and Integrated Control, IOBC, https://www.iobc-wprs.org/; Organic Farming Research Foundation, http://ofrf.org/home), or of countries (Canada: Ministry of Agriculture, food and rural affairs, http://www.omafra.gov.on.ca/english/crops/facts/09-077.htm; European Commission: https://ec.europa.eu/agriculture/organic/organic-farming/what-is-organic-farming_en; China: https://www.rare.org/china).

Organic Eprints (http://orgprints.org/) is an international open access archive for papers and projects; some of them are on apple, related to research in organic food and farming.

Updated information on agroecology is available on the FAO website (http://www.fao.org/agroecology/en/). In the context of the European Environment Agency (EEA), the Common International Classification of Ecosystem Services (CICES; https://cices.eu/) gives an update of ecosystem services based on a review of the relevant scientific literature.

Several international programmes have been developed on agroforestry. A recent one is AGFORWARD (AGroFORestry that Will Advance Rural Development; https://www.agforward.eu) with several leaflets on innovation and best practices, some of them on apple (e.g. https://www.agforward.eu/index.php/en/grazed-orchards-in-france.html).

The journal *Agroforestry Systems* (https://link.springer.com/journal/10457) contains original and review articles on agroforestry, with some of them being on apple.

7 References

Albert, L., Franck, P., Gilles, Y. and Plantegenest, M. (2017). Impact of agroecological infrastructures on the dynamics of *Dysaphis plantaginea* (Hemiptera: Aphididae) and its natural enemies in apple orchards in Northwestern France. *Environmental Entomology* 46, 528–37.

Altieri, M. A. (1999). The ecological role of biodiversity in agroecosystems. *Agriculture, Ecosystems and Environment* 74, 19–31.

Andres, C., Comoé, H., Beerli, A., Schneider, M., Rist, S. and Jacobi, J. (2016). Chapter 3: Cocoa in monoculture and dynamic agroforestry. In: *Sustainable Agriculture Reviews* (E. Lichtfouse, ed.), volume 19. Springer International Publishing, Switzerland, pp. 121–53. DOI:10.1007/978-3-319-26777-7_3.

Barot, S., Allard, V., Cantarel, A., Enjalbert, J., Gauffreteau, A., Goldringer, I., Lata, J. C., Le Roux, X., Niboyet, A. and Porcher, E. (2017a). Designing mixtures of varieties for multifunctional agriculture with the help of ecology. A review. *Agronomy for Sustainable Development* 37, 13. DOI 10.1007/s13593-017-0418-x
Barot, S., Yéb, L., Abbadie, L., Blouin, M. and Frascaria-Lacoste, N. (2017b). Ecosystem services must tackle anthropized ecosystems and ecological engineering. *Ecological Engineering* 99, 486–95.
Barritt, B. H. (1992). *Intensive Orchard Management*. Good Fruit Grower and Washington State Fruit Commission, Yakima and Washington DC.
Barritt, B. H., Konishi, B. S. and Dilley, M. A. (1997). Tree size, yield and biennial bearing relationships with 40 apple rootstocks and three scion cultivars. *Acta Horticulturae* 451, 105–12.
Barthélémy, D. and Caraglio, Y. (2007). Plant architecture: A dynamic, multilevel and comprehensive approach to plant form, structure and ontogeny. *Annals of Botany* 99, 375–407.
Bell, A. (1991). *Plant Form an Illustrated Guide to Flowering Plant Morphology*. Oxford University Press, Oxford, UK.
Bernhard, R. (1961). Mise à fleur et alternance chez les arbres fruitiers. *Congrès pomologique, INRA, Paris, octobre* 1960, pp. 91–116.
Bommarco, R., Kleijn, D. and Potts, S. G. (2013). Ecological intensification: harnessing ecosystem services for food security. *Trends in Ecology and Evolution* 28(4), 230–8.
Breen, K. C., van Hooijdonk, B. M., Tustin, D. S., Wilkie, J. D., Bound, S. A., Middleton, S. G. and Close, D. C. (2014). Changes in fruit set of 'Gala' apple in response to environment and artificial spur extinction. *Acta Horticulturae* 1058, 77–83.
Breen, K. C., Tustin, D. S., Palmer, J. W. and Close, D. C. (2015). Method of manipulating floral bud density affects fruit set responses and productivity in apple. *Scientia Horticulturae* 197, 244–53.
Breen, K. C., Tustin, D. S., Palmer, J. W., Hedderley, D. I. and Close, D. C. (2016a). Effects of environment and floral intensity on fruit set behaviour and annual flowering in apple. *Scientia Horticulturae* 210, 258–67.
Breen, K. C., Palmer, J., Tustin, D. S. and Close, D. C. (2016b). Artificial spur extinction alters light interception by 'Royal Gala' apple trees. *Acta Horticulturae* 1130, 265–72.
Brown, M. W. and Tworkoski, T. (2004). Pest management benefits of compost mulch in apple orchards. *Agriculture Ecosystems and Environment* 103, 465–72.
Campos, P., Caparrós, A., Cerdá, E., Huntsinger, L. and Standiford, R. B. (2007). Modeling multifunctional agroforestry systems with environmental values: dehesa in Spain and woodland ranches in California. In: *Handbook of Operations Research in Natural Resources* (A. Weintraub, T. Bjorndal, R. Epstein and C. Romero, eds). Springer Publishers, Dordrecht, pp. 1–16.
Chomicki, G., Coiro, M. and Renner S. S. (2017). Evolution and ecology of plant architecture: Integrating insights from the fossil record, extant morphology, developmental genetics and phylogenies. *Annals of Botany* 120, 855–91.
Cook, N. C. and Jacobs, G. (1999). Suboptimal winter chilling impedes development of acrotony in apple shoots. *HortScience* 34, 1213–16.
Cornille, A., Giraud T., Smulders, M. J. M., Roldàn-Ruiz, I. and Gladieux, P. (2014). The domestication and evolutionary ecology of apples. *Trends in Genetics* 30(2), 57–65.
Cornille, A., Feurtey, A., Gélin, U., Ropars, J., Misvanderbrugge, K., Gladieux, P. and Giraud, T. (2015). Anthropogenic and natural drivers of gene flow in a temperate wild fruit

tree: A basis for conservation and breeding programs in apples. *Evolutionary Applications* 8, 373–84.

Correlli-Grappadelli, L. (2003). Light relations. In: *In Apples - Botany, Production and Uses* (D. C. Ferree and I. J. Warrington, eds). CABI Publishing, Wallingford, UK, pp. 195–216.

Correlli-Grappadelli, L. and Lakso, A. N. (2007). Is maximizing orchard light interception always the best choice? *Acta Horticulturae* 732, 507–18.

Costes, E., Sinoquet, H., Kelner, J. J. and Godin, C. (2003). Exploring within-tree architectural development of two apple tree cultivars over 6 years. *Annals of Botany* 91, 91–104.

Costes, E., Lauri, P.É. and Regnard, J. L. (2006). Analysing fruit tree architecture. Implication for tree management and fruit production. *Horticultural Reviews* 32, 1–61.

Debaeke, P., Munier-Jolain, N., Bertrand, M., Guichard, L., Nolot, J. M., Faloya, V. and Saulas, P. (2009). Iterative design and evaluation of rule-based cropping systems: Methodology and case studies. A review. *Agronomy for Sustainable Development* 29, 73–86.

Demestihas, C., Plénet, D., Génard, M., Raynal, C. and Lescourret, F. (2017). Ecosystem services in orchards. A review. *Agronomy for Sustainable Development* 37, 12. DOI 10.1007/s13593-017-0422-1.

Dennis, F. G. (2011). Factors affecting yield in apple, with emphasis on 'Delicious'. *Horticultural Reviews* 1, 396–422.

Dib, H., Sauphanor, B. and Capowiez, Y. (2016). Effect of management strategies on arthropod communities in the colonies of rosy apple aphid, *Dysaphis plantaginea* Passerini (Hemiptera: Aphididae) in south-eastern France. *Agriculture, Ecosystems and Environment* 216, 203–6.

Didelot, F., Brun, L. and Parisi, L. (2007). Effects of cultivar mixtures on scab control in apple orchards. *Plant Pathology* 56, 1014–22.

Doré, T., Makowski, D., Malézieux, É., Munier-Jolain, N., Tchamitchian, M. and Tittonell, P. (2011). Facing up to the paradigm of ecological intensification in agronomy: Revisiting methods, concepts and knowledge. *European Journal of Agronomy* 34, 197–210.

Dupraz, C. and Liagre, F. (2011). *Agroforesterie, des arbres et des cultures*. Éditions France-Agricole, Paris, France.

Dupraz, C., Burgess, P., Gavaland, A., Graves, A., Herzog, F., Incoll, L. D., Jackson, N., Keesman, K., Lawson, G., Lecomte, I., Liagre, F., Mantzanas, K., Mayus, M., Moreno, G., Palma, J., Papanastasis, V., Paris, P., Pilbeam, D. J., Reisner, Y., Van Noordwijk, M., Vincent, G. and Werf Van der, W. (2005). Synthesis of the Sylvoarable Agroforestry for Europe (SAFE) project. INRA UMR System, Montpellier, France. http://www1.montpellier.inra.fr/safe/english/results/final-report/SAFEFinalSynthesisReport.pdf (accessed 15 August 2017).

Durand, J. B., Guitton, B., Peyhardi, J., Holtz, Y., Guédon, Y., Trottier, C. and Costes, E. (2013). New insights for estimating the genetic value of segregating apple progeniesfor irregular bearing during the first years of tree production. *Journal of Experimental Botany* 64, 5099–113.

Durand, J. B., Allard, A., Guitton, B., van de Weg, E., Bink, M. C. A. M. and Costes, E. (2017). Predicting flowering behavior and exploring its genetic determinism in an apple multi-family population based on statistical indices and simplified phenotyping. *Frontiers in Plant Science* 8, 858. DOI:10.3389/fpls.2017.00858.

Dutra Schmitz, J., Guédon, Y., Gilberto Herter, F., Leite B. G. and Lauri, P. É. (2014). Exploring bud dormancy completion with a combined architectural and phonological analysis: The case of apple trees in contrasted winter temperature conditions. *American Journal of Botany* 101(3), 398–407.

Dutra Schmitz, J., Bonhomme, M., Cochard, H., Herter, F. G., Leite, B. G., Regnard, J. L. and Lauri, P. É. (2015). Are the effects of winter temperatures on spring budburst mediated by the bud water status or related to a whole-shoot effect? Insights in the apple tree. *Trees - Structure and Function* 29, 675–82.

FAO (2017). http://www.fao.org/faostat/en/#data (accessed 12 August 2017).

Ferree, D. C. (1988). Role of rootstocks and spur-type scions for controlling vegetative growth of apple and peach trees. *HortScience* 23, 464-7.

Forshey, C. G., Elfving, D. C. and Stebbins, R. L. (1992). *Training and Pruning Apple and Pear Trees*. American Society for Horticultural Science, Alexandria, VA.

Gallaher, R. N. (2009). Multiple cropping systems. In: *Management of Agricultural, Forestry and Fisheries Enterprises - Vol. I, Multiple Cropping Systems. Encyclopedia of Life Support Systems*. http://eolss.net/eolss-sampleallchapter.aspx.

Gao, L., Xu, H., Bi, H., XI, W., Bao, B., Wang, X., Bi, C. and Chang, Y. (2013). Intercropping competition between apple trees and crops in agroforestry systems on the loess plateau of China. *PLoS ONE* 8(7), e70739. doi:10.1371/journal.pone.0070739.

Gliessman, S. R. (1985). Chapter 5 - Multiple cropping systems: A basis for developing an alternative agriculture. In: *Innovative Biological Technologies for Lesser Developed Countries - Workshop proceedings*. OTA, Washington DC, pp. 69–83.

Goh, K., Ridgen, G. and Daly, M. (1995). Understorey biomass production and biological nitrogen fixation in an organic apple orchard in Canterbury, New Zealand. *Communications in Soil Science and Plant Analysis* 26, 3261–73.

Granatstein, D., Kirby, E., Ostenson, H. and Willer, H. (2016). Global situation for organic tree fruits. *Scientia Horticulturae* 208, 3–12.

Hallé, F. and Oldeman, R. A. A. (1970). *Essai sur l'architecture et la dynamique de croissance des arbres tropicaux*. Masson, Paris, France.

Hallé, F., Oldeman, R. A. A. and Tomlinson, P. B. (1978). *Tropical Trees and Forests*. Springer-Verlag, Berlin.

Herzog, F. (1998). Streuobst: A traditional agroforestry system as a model for agroforestry development in temperate Europe. *Agroforestry Systems* 42, 61–80.

Hollender, C. A. and Dardick, C. (2015). Molecular basis of angiosperm tree architecture. *New Phytologist* 206, 541–56.

Hoying, S. A. and Robinson, T. L. (2000). The apple orchard planting system puzzle. *Acta Horticulturae* 513, 257–60.

IPES-Food. (2016). From uniformity to diversity: a paradigm shift from industrial agriculture to diversified agroecological systems. *International Panel of Experts on Sustainable Food Systems*. www.ipes-food.org (accessed 15 August 2017).

Isbell, F., Adler, P. R., Eisenhauer, N., Fornara, D., Kimmel, K., Kremen, C., Letourneau, D. K., Liebman, M., Polley, H. W., Quijas, S. and Scherer-Lorenzen, M. (2017). Benefits of increasing plant diversity in sustainable agroecosystems. *Journal of Ecology* 105, 871–9.

Johnson, R. S. and Lakso, A. N. (1986). Carbon balance model of a growing apple shoot: II. Simulated effects of light and temperature on long and short shoots. *Journal of the American Society for Horticultural Science* 111, 164–9.

Lauri, P. É. (2002). From tree architecture to tree training - An overview of recent concepts developed in apple in France. *Journal of the Korean Society for Horticultural Science* 43(6), 782-8.

Lauri, P. É. (2009a). Does plant architecture only result from growing meristems? Atlan's principle of life and death as regulated morphogenetic processes. In: *Tree Growth: Influences, Layers and Types* (W. P. Karam, Ed.). Nova Science Publishers, Inc., New York, NY, pp. 45-55.

Lauri, P. É. (2009b). Developing a new paradigm for apple training. *Compact Fruit Tree* 42, 17-19.

Lauri, P. É. (2016). Concepts in fruit tree training and pruning, is there old versus new paradigms? Viewpoint of a biologist. In: *Des fruits d'ici et d'ailleurs- Regards sur l'histoire de quelques fruits consommés en Europe* (M. P. Ruas, Coord.). Omniscience, collection Histoire des savoirs, Paris, France, pp. 361-70. ISBN 13: 9782916097473.

Lauri, P. É. (2018). Apple tree architecture and cultivation - A tree in a system. *First International Apple Symposium*. Yangling, Shaanxi, China, 10-16 October 2016. *Acta Horticulturae* (in press).

Lauri, P. É. and Corelli-Grappadelli, L. (2014). Tree architecture, flowering and fruiting. Thoughts on training, pruning and ecophysiology. *Acta Horticulturae* 1058, 291-8.

Lauri, P. É. and Laurens, F. (2005). Architectural types in apple (*Malus x domestica* Borkh.). In: *Crops: Growth, Quality and Biotechnology* (R. Dris, Ed.). World Food Limited, Helsinki, Finland, pp. 1300-14.

Lauri, P. É. and Lespinasse, J. M. (2000). The vertical axis and solaxe systems in France. *Acta Horticulturae* 513, 287-96.

Lauri, P. É. and Lespinasse, J. M. (2001). Genotype of apple trees affects growth and fruiting responses to shoot bending at various times of year. *Journal of the American Society for Horticultural Science* 126(2), 169-74.

Lauri, P. É. and Kelner, J. J. (2001). Shoot type demography and dry matter partitioning. A morphometric approach in apple. *Canadian Journal of Botany* 79, 1270-3.

Lauri, P. É. and Normand, F. (2017). Are leaves only involved in flowering? Bridging the gap between structural botany and functional morphology. *Tree Physiology*. https://doi.org/10.1093/treephys/tpx068

Lauri, P. É. and Térouanne, É. (1995). Analyse de la croissance primaire de rameaux de pommier (*Malus × domestica* Borkh.) au cours d'une saison de végétation. *Canadian Journal of Botany* 73, 1471-89.

Lauri, P. É. and Térouanne, É. (1998). The influence of shoot growth on the pattern of axillary development on the long shoots of young apple trees (*Malus domestica* Borkh.). *International Journal of Plant Sciences* 159(2), 283-96.

Lauri, P. É. and Trottier, C. (2004). Patterns of size and fate relationships of contiguous organs in the apple (*Malus domestica* Borkh.) crown. *New Phytologist* 163, 533-46.

Lauri, P. É., Térouanne, É., Lespinasse, J. M., Regnard, J. L. and Kelner, J. J. (1995). Genotypic differences in the axillary bud growth and fruiting pattern of apple fruiting branches over several years - an approach to regulation of fruit bearing. *Scientia Horticulturae* 64, 264-81.

Lauri, P. É., Lespinasse, J. M. and Térouanne, É. (1997a). Relationship between the early development of apple fruiting branches and the regularity of bearing - An approach to the strategies of various cultivars. *Journal of Horticultural Science* 72, 519-30.

Lauri, P. É., Lespinasse, J. M. and Térouanne, É. (1997b). Vegetative growth and reproductive strategies in apple fruiting branches - An investigation into various cultivars. *Acta Horticulturae* 451, 717–24.

Lauri, P. É., Willaume, M., Larrive, G., and Lespinasse, J. M. (2004). The concept of centrifugal training in apple aimed at optimizing the relationship between growth and fruiting. *Acta Horticulturae* 636, 35–42.

Lauri, P. É., Crété, X. and Ferré, G. (2007). Centrifugal training in apple - Appraisal of a two-year experiment on cv. 'Galaxy' in southeast France. *Acta Horticulturae* 732, 391–6.

Lauri, P. É., Costes, E., Regnard, J. L., Brun, L., Simon, S., Monney, P. and Sinoquet H. (2009). Does knowledge on fruit tree architecture and its implications for orchard management improve horticultural sustainability? An overview of recent advances in the apple. *Acta Horticulturae* 817, 243–9.

Lauri, P. É., Hucbourg, B., Ramonguilhem, M. and Méry, D. (2011). An architectural-based tree training and pruning – identification of key features in the apple. *Acta Horticulturae* 903, 589–96.

Lauri, P. É., Combe, F. and Brun, L. (2014). Regular bearing in the apple – Architectural basis for an early diagnosis on the young tree. *Scientia Horticulturae* 174, 10–16.

Lauri, P. É., Hucbourg, B., Ramonguilhem, M. and Joannin, R. (2016a). A step further – Optimizing the natural apple tree habit with the Salsa concept. *Acta Horticulturae* 1130, 151–6.

Lauri, P. É., Simon, S., Navarrete, M., Parrot, L., Normand, F., Lesueur-Jannoyer, M. and Deguine, J. P. (2016b). Interdisciplinary and multiagent interactions for innovations in horticulture - paradigms beyond the words. *Acta Horticulturae* 1137, 255–65.

Lauri, P. É., Mézière, D., Dufour, L., Gosme, M., Simon, S., Gary, C., Jagoret, P., Wery, J. and Dupraz, C. (2016c). Fruit-trees in agroforestry systems - Review and prospects for the temperate and mediterranean zones. In: *Proceedings of the 3rd European Agroforestry Conference* (M. Gosme, et al., Eds). Montpellier, France, May 2016. ISBN: 978-2-87614-717-1. EAN: 9782876147171, pp. 106–9. https://www.repository.utl.pt/bitstream/10400.5/11690/1/EURAF2016%20-%20Book%20of%20Abstracts%20final.pdf

Le Bec, J., Bailly, A., Brossier, B. and Dupraz, C. (2016). Trunk-crown growth trade-off in pollarded trees: influence on wood production. In: *Proceedings of the 3rd European Agroforestry Conference* (M. Gosme, et al., Eds). Montpellier, France, May 2016. ISBN: 978-2-87614-717-1. EAN: 9782876147171, pp. 111–13. https://www.repository.utl.pt/bitstream/10400.5/11690/1/EURAF2016%20-%20Book%20of%20Abstracts%20final.pdf

Legave, J. M., Farrera, I., Almeras, T. and Calleja, M. (2008) Selecting models of apple flowering time and understanding how global warming has had an impact on this trait. *Journal of Horticultural Science & Biotechnology* 83(1), 76–84.

Legave, J. M., Guédon, Y., Malagi, G., El Yaacoubi, A. and Bonhomme, M. (2015). Differentiated responses of apple tree floral phenology to global warming in contrasting climatic regions. *Frontiers in Plant Science* 6, 1054. doi:10.3389/fpls.2015.01054.

Lescourret, F., Dutoit, T., Rey, F., Côte, F., Hamelin, M. and Lichtfouse, É. (2015). Agroecological engineering. *Agronomy for Sustainable Development* 35(4), 1191–8.

Lespinasse, J. M. (1977). *La conduite du Pommier. I - Types de fructification. Incidence sur la conduite de l'arbre*. INVUFLEC, Paris, France.

Lespinasse, J. M. (1980). *La conduite du pommier. II - L'axe vertical, la rénovation des vergers (2ème partie)*. Ctifl, Paris, France.

Lespinasse, Y. (1992). Breeding apple tree: aims and methods. In: *Proceedings of the Joint Conference of the EAPR Breeding and Varietal Assessment Section and the EUCARPIA Potato Section* (F. Rousselle-Bourgeois and P. Rousselle, Eds), Janvier 1992, Landerneau, France. INRA, Ploudaniel, France, pp. 103–10.

Lespinasse, J. M. and Delort, J. F. (1986). Apple tree management in VERTICAL AXIS: Appraisal after ten years of experiments. *Acta Horticulturae* 160, 120–55.

Litrico, I. and Violle, C. (2015). Diversity in plant breeding: A new conceptual framework. *Trends in Plant Science* 20(10), 604–13.

Looney, N. E. and Lane, W. D. (1984). Spur-type growth mutants of McIntosh apple: A review of their genetics, physiology and field performance. *Acta Horticulturae* 146, 31–46.

Lordan, J., Fazio, G., Francescatto, P. and Robinson, T. (2017). Effects of apple (*Malus* × *domestica*) rootstocks on scion performance and hormone concentration. *Scientia Horticulturae* 225, 96–105.

Lovell, S. T., Dupraz, C., Gold, M., Jose, S., Revord, R., Stanek, E. and Wolz, K. (2017). Temperate agroforestry research: Considering multifunctional woody polycultures and the design of long-term field trials. *Agroforestry Systems*. DOI: 10.1007/s10457-017-0087-4.

Luby, J. J. (2003). Taxonomic classification and brief history. In: *Apples - Botany, Production and Uses* (D. C. Ferree and I. J. Warrington, Eds). CABI Publishing, Wallingford, UK, pp.1–14.

Maguylo, K. and Lauri, P. É. (2007). Growth and fruiting characteristics of eight untrained genotypes assessed on both M.9 and own-rooted trees in Southern France. *Acta Horticulturae* 732, 93–9.

Malagi, G., Sachet, M. R., Citadin, I., Herter, F. G., Bonhomme, M., Regnard, J. L. and Legave, J. M. (2015). The comparison of dormancy dynamics in apple trees grown under temperate and mild winter climates imposes a renewal of classical approaches. *Trees* 29, 1365–80.

Malézieux, É., Crozat, Y., Dupraz, C., Laurans, M., Makowski, D., Ozier-Lafontaine, H., Rapidel, B., de Tourdonnet, S. and Valantin-Morison, M. (2009). Mixing plant species in cropping systems: Concepts, tools and models. A review. *Agronomy for Sustainable Development* 29, 43–62.

Meier, A. R., Saunders, M. R. and Michler, C. H. (2012). Epicormic buds in trees: A review of bud establishment, development and dormancy release. *Tree Physiology* 32, 565–84.

Millennium Ecosystem Assessment (2005). *Ecosystems and Human Well-Being: Synthesis*. Island Press. https://www.millenniumassessment.org/documents/document.356.aspx.pdf

Miñarro, M. and Prida, E. (2013). Hedgerows surrounding organic apple orchards in north-west Spain: Potential to conserve beneficial insects. *Agricultural and Forest Entomology* 15, 382–90.

Neilsen, G. H., Neilsen, D and Forge, T. (2017). Advances in soil and nutrient management in apple cultivation. In: *Achieving Sustainable Cultivation of Apples* (K. Evans, Ed.). Number 018, Burleigh Dodds Series in Agricultural Science, Cambridge, UK, pp. 239–78.

Oldeman, R. O. O. (1974). *L'architecture de la forêt guyanaise*. Mémoires 73, ORSTOM, Paris, France.

Ong, C. K., Corlett, J. E., Singh, R. P. and Black, C. R. (1991). Above and below ground interactions in agroforestry systems. *Forest Ecology and Management* 45, 45-57.

Palmer, J., Privé, J. P. and Tustin D. S. (2003). Temperature. In: *Apples - Botany, Production and Uses* (D. C. Ferree and I. J. Warrington, eds). CABI Publishing, Wallingford, UK, pp. 217-36.

Parrotta, J. A., Dey de Pryck, J., Darko, O. B., Padoch, C., Powell, B. and Sandbrook, C. (2015). The historical, environmental and socio-economic context of forests and tree-based systems for food security and nutrition. In: *Forests, Trees and Landscapes for Food Security and Nutrition. A Global Assessment Report*, pp. 51-85. https://www.researchgate.net/publication/276950253.

Petri, J. L. and Leite, G. B. (2004). Consequences of insufficient winter chilling on apple tree bud-break. *Acta Horticulturae* 662, 53-60.

Querné, A., Battie-laclau, P., Dufour, L., Wery, J. and Dupraz, C. (2017). Effects of walnut trees on biological nitrogen fixation and yield of intercropped alfalfa in a Mediterranean agroforestry system. *European Journal of Agronomy* 84, 35-46.

Quinlan, J. D. and Preston, A. P. (1971). The influence of shoot competition on fruit retention and cropping of apple trees. *Journal of Horticultural Science* 46, 525-34.

Racsko, J. and Schrader, L. E. (2012). Sunburn of apple fruit: Historical background, recent advances and future perspectives. *Critical Reviews in Plant Sciences* 31, 455-504.

Rapidel, B., Ripoche, A., Allinne, C., Metay, A., Deheuvels, O., Lamanda, N., Blazy, J. M., Valdés-Gómez, H. and Gary, C. (2015). Analysis of ecosystem services trade-offs to design agroecosystems with perennial crops. *Agronomy for Sustainable Development* 35, 1373-90.

Ratnadass, A., Fernandes, P., Avelino, J. and Habib, R. (2012). Plant species diversity for sustainable management of crop pests and diseases in agroecosystems: A review. *Agronomy for Sustainable Development* 32, 273-303.

Robinson, T. L. (2003). Apple-orchard planting systems. In: *Apples - Botany, Production and Uses* (D. C. Ferree and I. J. Warrington, Eds). CABI Publishing, Wallingford, UK, pp. 345-408.

Robinson, D. W. and O'Kennedy, N. D. (1978). The effect of the overall herbicide systems of soil management on the growth and yield of apple trees 'Golden Delicious'. *Scientia Horticulturae* 9, 127-36.

Simon, S., Lauri, P. É., Brun, L., Defrance, H. and Sauphanor, B. (2006). Does fruit-tree architecture manipulation affect the development of pests and pathogens? - A case study in apple orchard. *Journal of Horticultural Science & Biotechnology* 81(4), 765-73.

Simon, S., Bouvier, J. C., Debras, J. F. and Sauphanor, B. (2010). Biodiversity and pest management in orchard systems. A review. *Agronomy for Sustainable Development* 30, 139-52.

Simon, S., Morel, K., Durand, E., Brévalle, G., Girard, T. and Lauri, P. É. (2012). Aphids at crossroads: When branch architecture alters aphid infestation patterns in the apple tree. *Trees - Structure and Function* 26, 273-82.

Simon, S., Lesueur-Jannoyer, M., Plénet, D., Lauri, P. É. and Le Bellec, F. (2017). Methodology to design agroecological orchards: Learnings from on-station and on-farm experiences. *European Journal of Agronomy* 82, 320-30.

Smith, J., Crossland, E. M. and Wolfe, M. S. (2016). Silvoarable agroforestry: An alternative approach to apple production? In: *Proceedings of the International Farming Systems*

Association (IFSA), Europe Group, for 2016, 12–15 July 2016, Newport, UK. http://archive.harper-adams.ac.uk/events/ifsa-conference/papers/3/3.3%20Smith.pdf

Stamps, W. T. and Linit, M. J. (1998). Plant diversity and arthropod communities: Implications for temperate agroforestry. *Agroforestry Systems* 39, 73–89.

Storkey, J., Döing, T., Baddeley, J., Collins, R., Roderick, S., Jones, H. and Watson, C. (2015). Engineering a plant community to deliver multiple ecosystem services. *Ecological Applications*, 25(4), 1034–43.

Suryanata, K. (1994). Fruit trees under contract: Tenure and land use change in Upland Java, Indonesia. *World Development* 22(10), 1567–78.

Tscharntke, T., Clough, Y., Bhagwat, S. A., Buchori, D., Faust, H., Hertel, D., Hölscher, D., Juhrbandt, J., Kessler, M., Perfect, I., Scherber, C., Schroth, G., Veldkamp, E. and Wanger, T. C. (2011). Multifunctional shade-tree management in tropical agroforestry landscapes - a review. *Journal of Applied Ecology* 48, 619–29.

Tustin, D. S., Dayatilake, G. A., Breen, K. C., and Oliver, M. J. (2009). Improving fruit size and productivity of 'Scifresh' apple: Effects of regulating whole-tree floral bud density using Centrifugal Training. A report prepared for ENZA Ltd, The New Zealand Institute for Plant & Food Research Ltd, Private Bag 92 169, Auckland 1142, New Zealand.

Tustin, D. S., Dayatilake, G. A., Henriod, R. E., Breen, K. C. and Oliver, M. (2011). Changes in fruiting behaviour and vegetative development of 'Scifresh' apple in response to artificial spur extinction using centrifugal training. *Acta Horticulturae* 903, 603–10.

Tustin, D. S., Dayatilake, G. A., Breen, K. C. and Oliver, M. J. (2012). Fruit set responses to changes in floral bud load - a new concept for crop load regulation. *Acta Horticulturae* 932, 195–202.

Tustin, D. S., Van Hooijdonk, B. and Breen, K. (2015). Precision management of crop load in apples. Simple metrics to optimize orchard productivity, fruit quality, crop value. *The New Zealand Institute for Plant & Food Research Ltd (PFR) and Hawke's Bay Research Center, Havelock North, The Orchadist*, April 2015, pp. 34–40.

van Noordwijk, M., Coe, R. and Sinclair, F. (2016). Central hypotheses for the third agroforestry paradigm within a common definition. Working paper 233, World Agroforestry Centre (ICRAF) Southeast Asia Regional Program, Bogor, Indonesia. DOI:http://dx.doi.org/10.5716/WP16079.PDF

Warrington, I. J., Stanley, C. J., Julian, J. F., Tustin, D. S., Hirst, P. M. and Cashmore, W. (1995). Pruning strategies for restructuring top-dominant central leader 'Granny Smith' apple trees. *New Zealand Journal of Crop and Horticultural Science* 23, 315–22.

Watkins, R. and Smith, R. A. (1997). Descriptor list for apple. International Board for Plant Genetic Resources (IBPGR). https://cgspace.cgiar.org/bitstream/handle/10568/72794/Apple_descriptors_320.pdf?sequence=1 (accessed 16 august 2017).

Wezel, A., Casagrande, M., Celette, F., Vian, J. F., Ferrer, A. and Peigné, J. (2014). Agroecological practices for sustainable agriculture. A review. *Agronomy for Sustainable Development* 34, 1–20.

Willaume, M., Lauri, P.É. and Sinoquet, H. (2004). Light interception in apple trees influenced by canopy architecture manipulation. *Trees - Structure and Function* 18, 705–13.

Wünsche, J. N. and Lakso, A. N. (2000). The relationship between leaf area and light interception by spur and extension shoot leaves and apple orchard productivity. *HortScience* 37, 1202–6.

Wyss, É. (1995). The effects of weed strips on aphids and aphidophagous predators in an apple orchard. *Entomologia Experimentalis et Applicata* 75, 43–9.

Zhang, C. and Tanabe, K. (2008). Partitioning of 13C-photosynthates from different current shoots neighboring with fruiting spur in later-maturing Japanese pear during the period of rapid fruit growth. *Scientia Horticulturae* 117, 142–50.

Chapter 9

Sustainable plum and apricot cultivation

Mihai Botu, University of Craiova, Romania

1 Introduction

The explosive growth of the Earth's population and its implications for food security raise many challenges for science (Badgley et al., 2007; Sumner, 2008). Sustainability has been defined as "meeting the needs of today without compromising the ability of future generations to meet their needs" (United Nations World Commission on Environment and Development, 1987). Achieving sustainability in food production is recognized as a major challenge facing agriculture (Granatstein and Kupferman, 2008). Sustainable agriculture needs to produce sufficient food without depleting the Earth's resources (e.g. through soil erosion) or polluting the environment (Tan and Koksal, 2004; Lal, 2008; Bertschinger et al., 2009). To achieve this goal, sustainable agriculture integrates three main factors: environmental health, economic profitability, and social and economic equity (Malkina-Pykh and Pykh, 2003; cited by Gunduz et al., 2011; Xiloyannis et al., 2010).

Sustainable fruit cultivation requires the right growing conditions, cultivars and rootstocks suited to these conditions, and modern crop-cultivation techniques that minimize the use of inputs such as fertilizers or pesticides (Botu and Botu, 2003). Sustainability also implies promotion of biodiversity, improvement of microbiological processes in the soil, and protection of the environment (e.g. by avoiding pollution from fertilizer and pesticide runoff). It has been estimated that fruit production may need to increase two- to three-fold

http://dx.doi.org/10.19103/AS.2018.0040.24

to meet future demand, but this needs to be achieved in a sustainable way. This chapter explores ways this might be achieved for plum and apricot production.

2 Plum and apricot production

Of the more than 230 species of *Prunus* (Jukovski, 1947; cited by Cociu et al., 1997), a few dozen are known as 'plums' and 'apricots'. Not all of these species are used as fruit crops. Of the so-called 'plum' species, the most important are *Prunus domestica*, *P. insititia* (which is considered a subspecies of *P. domestica*), *P. salicina* (including some interspecific hybrids), and *P. cerasifera*. Other 'plum' species of minor importance are given in Table 1. The most important species of apricot are *P. armeniaca* and *P. mume*.

Prunus domestica (European plums) cultivars are generally classified into two main types: (1) 'plums' (mainly used for fresh consumption) and (2) 'prunes' (mainly used for drying and processing). *Prunus salicina* (Japanese, Chinese, Chino-Japanese, or Oriental plums) and interspecific hybrid cultivars are used mainly for fresh consumption and represent more than 40% of the total world plum production. Fruits of *P. americana*, *P. hortulana*, *P. maritima*, and *P. cerasifera* are used primarily for processing, with some for fresh consumption in local markets.

Plums and apricots are grown for their high nutritional value (e.g. as an important source of carbohydrates, proteins, lipids, minerals, and vitamins). The energy value of fresh plums is 46 kcal/100 g fruit and 240 kcal/100 g dried fruits. For apricots, the energy value is 48 kcal/100 g fruit and 241 kcal/100 g dried fruits. Particular focus is also given now to the potential nutraceutical benefits of these fruits (Igwe and Charlton, 2016; Fatima et al., 2018).

World plum and apricot production in 2016 was 12 050 800 t and 3 881 204 t, respectively (Tables 2 and 3). The world supply of prunes is about 300 000 t, including production and existing stocks (Amblard, 2015). These figures are relatively small compared to the production of other fruit tree crops such as apples and peaches. Plum production has increased by almost 11% since 2012, while apricot production has remained stable. If demand increases with a growing global population, there will likely be a need for significant and sustainable improvements in production (Liu, 2007).

The major production areas are shown in Tables 4 and 5. The first nine plum-producing countries account for ~78% of the total production and ~85% of the area harvested. China is the leading plum-producing country (55% of the world production and 72% of the total area harvested) (Table 2). China grows mostly *P. salicina*. The leading European plum producer is Romania with a modest 4% share of the world production. The first nine apricot-producing countries account for ~66% of the total production and ~62% of the area harvested.

Table 1 Taxonomy of the main *Prunus* species

Genus	Subgenus	Section	Species	Number of chromosomes (2n)	Common name
Prunus	*Prunophora*	*Euprunus*	*P. domestica* L.	48	European plum
			P. cerasifera Ehrh.	16, 24, 32	Myrobalan or cherry plum
			P. insititia L.	48	Damson plum or bullace
			P. spinosa L.	32	Blackthorn
			P. cocomilla Ten	16	Italian plum
			P. salicina Lindl	16	Japanese plum or Chinese plum
			P. simonii Carr	16	Apricot plum or Simon plum
		Prunocerasus	*P. americana* Marsh	16	American plum
			P. nigra Ait	16	Canada plum, Canadian plum or Black plum
			P. hortulana Bailey	16	Wild goose plum or Hortulan plum
			P. maritima Marsh	16	Beach plum
			P. angustifolia Marsh	16	Sand plum, Chickasaw plum or Cherokee plum
		Armeniaca	*P. armeniaca* L.	16	Apricot
			P. sibirica L.	16	Siberian apricot
			P. mume Sieb et Zuce	16	Japanese apricot

Source: adapted from Rehder (1940), Ingram (1948), and Maraux (1995).

Table 2 World production of plums and the main producing countries

	Plum production (t)						
	Year						World share
Country	2012	2013	2014	2015	2016	Mean	(%)
Chile	313400	312379	294416	294605	294873	301935	2.6
China	6014485	6176655	6338825	6500995	6663165	6338825	55.3
France	188969	167431	199594	165123	211269	186477	1.6
India	215000	220000	225000	227556	261903	229892	2.0
Iran	174408	192816	264921	285000	269113	237252	2.1
Romania	424068	512459	495287	496468	512975	488251	4.3
Serbia	297446	568840	401452	344612	463115	415093	3.6
Turkey	300046	305393	265490	279761	297589	289656	2.5
USA	511681	330068	409864	387747	392537	406379	3.6
World	*10873542*	*11325235*	*11441155*	*11630094*	*12050800*	*11464165*	*100.00*

Source: adapted from FAO Stat Database (2018).

Plum and apricot yields vary widely among countries and are significantly lower than other tree fruit crops such as apple and pear, where orchard densities are higher and cultivation techniques are more intensive. The average world plum yield reached 4.6 t/ha in 2016 (FAO Stat Database, 2018). Some countries have significantly higher plum yields such as Chile (17.1 t/ha), France (14.3 t/ha), Turkey (14.3 t/ha), and USA (13.6 t/ha). The average world apricot yield was 6.8 t/ha in 2016 (FAO Stat Database, 2018). Countries with significantly higher yields include Italy (12.5 t/ha), Greece (11.8 t/ha), and Uzbekistan (11.8 t/ha).

Table 3 World production of apricots and the main producing countries

	Apricot production (t)						
	Year						
Country	2012	2013	2014	2015	2016	Mean	World share (%)
Algeria	269308	319784	216941	293486	256771	271258	7.1
France	175228	127158	175760	159375	110850	149674	3.9
Greece	79457	74718	90038	93510	79070	83359	2.2
Iran	309908	380032	241569	252000	306115	297925	7.8
Italy	247146	198290	222690	217569	237021	224543	5.9
Pakistan	178489	177630	170504	172933	177658	175443	4.6
Spain	118114	131800	136446	153667	125335	133072	3.5
Turkey	760000	780000	278210	696100	730000	648862	17.0
Uzbekistan	426000	480000	547000	606000	662123	544225	14.2
World	*3865025*	*4093243*	*3338757*	*3965406*	*3881204*	*3828727*	*100.00*

Source: adapted from FAO Stat Database (2018).

Table 4 World plum area harvested and the main producing countries

	Plum area harvested (ha)						
	Year						
Country	2012	2013	2014	2015	2016	Mean	World share (%)
Chile	18701	18554	17408	17340	17278	17856	0.7
China	1816620	1853749	1889350	1924956	1959869	1888909	72.5
France	17454	16946	16045	15213	14741	16080	0.6
India	27000	27500	27600	27817	31410	28265	1.1
Iran	25342	19464	40685	41000	33881	32074	1.2
Romania	68478	68008	66552	65671	65114	66765	2.6
Serbia	150000	180000	77949	77949	77949	112769	4.3
Turkey	19330	19726	20027	20452	20811	20069	0.8
USA	31404	28571	27721	27203	28875	28755	1.1
World	*2565813*	*2663137*	*2550515*	*2602782*	*2651122*	*2606674*	*100.00*

Source: adapted from FAO Stat Database (2018).

Table 5 World apricot area harvested and the main producing countries

	Apricot area harvested (ha)						
	Year						
Country	2012	2013	2014	2015	2016	Mean	World share (%)
Algeria	47376	46893	38590	38857	38239	41991	7.6
France	12743	12176	12207	12014	11872	12202	2.2
Greece	6259	6518	7022	7530	6685	6803	1.2
Iran	35461	53205	53624	54500	54392	50236	9.0
Italy	19186	18999	19093	18718	18917	18983	3.4
Pakistan	27536	28578	26950	25746	26461	27054	4.9
Spain	18542	20300	18451	18822	18353	18894	3.4
Turkey	114052	115613	117907	123176	123805	118911	21.4
Uzbekistan	42300	46511	50250	53473	56206	49748	9.0
World	*532322*	*558924*	*554657*	*563742*	*568014*	*555532*	*100.00*

Source: adapted from FAO Stat Database (2018).

3 Genetic resources

Most plum and apricot species are diploid (2n = 2x = 16 chromosomes). Exceptions include *P. domestica* and *P. insititia*, which are hexaploid (2n = 6x = 48 chromosomes), *P. spinosa*, which is tetraploid (2n = 4x = 32 chromosomes), and *P. cerasifera*, which can be found in diploid (2n = 2x = 16 chromosomes), triploid (2n = 3x = 24 chromosomes), and tetraploid forms (2n = 4x = 32 chromosomes).

The genetic resources available for plum and apricot are key to improving future production. Mehlenbacher et al. (1990), Ramming and Cociu (1992), and Botu et al. (2006, 2012) have highlighted the value of plum and apricot biodiversity, the importance of conservation, and the use of genetic resources in future breeding programs. Creating new cultivars of plum and apricot is key to ensuring sustainable production able to meet increasing demand. Modern breeding techniques can increasingly combine molecular markers and conventional breeding techniques (Lecouls et al., 2004).

The range of plum and apricot species possesses valuable genes that can be useful for breeding programs to develop new cultivars. As an example, collections of plum genetic resources contain potential genes with resistance to various diseases such as brown rot, caused by *Monilinia* sp.; red leaf blotch, caused by *Polystigma rubrum* (Pers.) DC.; bacterial leaf spot, caused by *Xanthomonas arboricola* pv. *pruni* (Smith); and even plum pox virus (PPV or Sharka). Very good field resistance/tolerance to PPV has been observed in the case of 'Andreea', a plum cultivar obtained by crossing (Botu, 2000). Kegler et al. (2001) and Hartmann (2002, cited by Neumüller, 2011) mention another type of resistance to PPV, hypersensitivity resistance, which has its origin in 'Jojo', a cultivar obtained by Prof. Walter Hartmann at the University of Hohenheim (Germany) from a cross between 'Ortenauer' and Stanley'. The first genetically engineered (GMO) plum cultivar resistant to PPV was obtained in 1991 in the United States, initially called 'C5' and patented as 'HoneySweet' (Scorza et al., 2016). There are over 91 collections in the world with 42 514 *Prunus* accessions (Botu, 2000), including 5547 accessions of *P. domestica*, 802 accessions of *P. salicina*, and 6947 accessions of *P. armeniaca*. A 2010 study of 20 European countries found 22 779 *Prunus* accessions stored in national germplasm collections. Out of those, 4632 were European plums, 491 were Japanese plums, and 3510 were apricots (Benediková and Giovannini, 2013).

3.1 Plum cultivars and breeding

There are more than 6000 commercial plum cultivars from Europe, Asia, and America, derived from 19 to 40 species or hybrids. These are used mainly for the fresh market, but also for processing, canning, drying, and freezing. Despite the number of cultivars existing worldwide, there is a continuing need to develop new cultivars to meet the changing requirements of the market and the industry (Milošević and Milošević, 2018). Worldwide, there are/have been European plum genetic improvement programs in countries as diverse as Canada, Belarus, Bulgaria, Czech Republic, France, Germany, Italy, Latvia, Norway, Republic of Moldova, Poland, Romania, Russia, Serbia, Sweden, Switzerland, and the United States. More than 170 new plum cultivars have been named in Europe in the last two decades: 38 in Romania, 28 in Germany, 15 in

Belarus, 14 in Serbia, 13 in Italy, 12 in Sweden, 11 in Bulgaria, 10 in Republic of Moldova, 9 in Latvia, 6 in France, 3 in United Kingdom, 3 in Czech Republic, and 1 in Norway (Butac et al., 2013).

Plum breeding programs are focused on obtaining European and Japanese cultivars and various interspecific hybrids between diploid species. In the case of European plums, breeding objectives are focused mainly on characteristics such as self-fertility and improved ripening times, yield, adaptation to environmental stresses, resistance to major diseases (PPV, red leaf blotch, *Monilinia* rot, rust – *Tranzschelia pruni-spinosae*, bacterial canker – *Pseudomonas syringae* pv. *morsprunorum*, etc.) as well as fruit quality for fresh consumption, drying, and for processing (Okie and Weinberger, 1996). The main breeding objectives for Japanese plums are similar: self-fertility, ripening, yield, environmental adaptation, better fruit quality, better storage ability, and resistance to major diseases (bacterial leaf and fruit spot, bacterial canker, plum leaf scorch, PPV, etc.) (Okie and Weinberger, 1996).

Most breeding programs are focused on fresh or dessert use of fruits and less on drying and processing. Some new cultivars bred for fresh consumption now have a fruit mass of over 50 g (with some over 80 g, like the Romanian cultivars 'Record', 'Vâlcean', and 'Alutus', and the UK ones 'Avalon' and 'Excalibur'). These are also noted for their taste and appearance, from dark blue to blue-reddish as well as red and yellow. Popular European plum cultivars include 'Agen' (different clones), 'Tardicotes', 'Quetche', 'Stanley', 'Bluefre', 'Anna Späth', 'Sugar', 'Avalon', 'Victoria', 'Jubileum', 'Reeves', 'Opal', 'Reine Claude d'Althan', 'President', 'Damson', 'California Blue', 'Valor', 'Vision', 'Nancy Mirabelle', 'Green Gage', 'Čačanska lepotica', 'Čačanska najbolja', 'Čačanska rana', 'Čačanska rodna', 'Valjevka', 'Jojo', 'Haganta', 'Hanita', 'Haroma', 'Elena', 'Katinka', 'Presenta', 'Tophit', 'Topend Plus', 'Topfive', 'Empress', 'Centenar', 'Tuleu gras', and 'Record'. Some European plum cultivars also are recommended for integrated fruit production: 'Hanita', 'Katinka', 'Elena', 'Jojo', 'Top 2000', 'Topfirst', 'Toptaste', 'Tophit', and 'Čačanska najbolja' (Mezey, 2009; cited by Milošević and Milošević, 2018).

Dried plum or 'prune' breeding programs seek to improve on current standard cultivars. These include the 'Common Prune', known also as 'Požegača' in Serbia; 'Vinete', 'Vânăt românesc', or 'Bistriţene' in Romania; 'Besztercei' in Hungary; 'Kiustendilska' in Bulgaria; 'Vengerka' in former USSR countries; 'Hauszwetsche' or 'German Plum' in Germany; 'Casalinga' in Italy; and 'Quetsche' in France. The most used cultivar still used for drying is 'Agen', known also as 'Prune d'Ente' or 'French Prune'. The last few decades have seen several new cultivars for drying. Examples include 'Primacotes', 'Lorida', 'Tardicotes', and 'Spurdente' from France (Renaud, 1994); 'Sutter' and 'Muir Beauty' which compete with the dominant cultivar 'French Improved' in California (USA); and 'Čačanska Rodna', 'Čačanski Šećer', 'Valjevka', 'Krina',

'Pozna Plava', 'Nada', and 'Mildora' in Serbia (to replace the main cultivars used for drying: 'Požegača' and 'Stanley') (Milošević et al., 2016); and 'Andreea' and 'Agent' in Romania.

In recent decades, breeding of Japanese plums (*P. salicina* and diploid interspecific hybrids) has intensified in countries such as the United States, Brazil, South Africa, Australia, New Zealand, Japan, China, South Korea, Thailand, Taiwan, Russia, Italy, and France (Okie and Ramming, 1999). The most widely used cultivars are 'Friar', 'Angeleno', 'Black Amber', 'Santa Rosa', 'Frontier', 'Calita', 'Laroda', 'Black Beaut', 'Obilnaya', 'Ozark Premier', 'Shiro', 'Methley', 'Golden Japan', 'Red Beauty', 'Formosa', 'Casselman', 'Satsuma', 'Sungold', 'Sorriso di Primavera', 'T.C. Sun', and 'Burbank'. Cultivars of interspecific origin have been grouped as Plumcot, Pluot®, Aprium®, Cherry plums, and Color-cot® (Table 6).

3.2 Apricot cultivars and breeding

Key objectives for apricot breeding include tree size and shape, speed of ripening, fruit quality for fresh consumption and processing, productivity, resistance to abiotic stresses, and resistance to major diseases (PPV, brown rot and fruit rot, *Cytospora*, bacterial canker and leaf spot, and European stone fruit yellow phytoplasma (formerly known as Chlorotic Leaf Roll Phytoplasma and

Table 6 The main plum and apricot cultivars of interspecific origin

Group and Genetic Origin	Cultivars
Plumcot *P. salicina* × *P. armeniaca*	'Flavorella', 'Flavor Queen', 'Plum Parfait', 'Red Velvet', 'Royal Velvet', 'Spring Satin', 'Scarlet Sunrise'
Pluot® (*P. salicina* × *P. armeniaca*) x *P. salicina*	'Amigo 1', 'Amigo 2', 'Black Kat', 'Candy Stripe', 'Coparose', 'Crimson Heart', 'Crimson Rose', 'Crimson Royale', 'Dapple Dandy', 'Dapple Fire', 'Dapple Jack', 'Dapple Supreme', 'Early Dapple', 'Ebony Rose', 'Emerald Gem', 'Fall Fiesta', 'Festival Red', 'Flavor Blast', 'Flavor Delight', 'Flavor Grenade', 'Flavor Jewel', 'Flavor Heart', 'Flavor King', 'Flavor Rouge', 'Flavor Royale', 'Flavor Supreme', 'Flavorfall', 'Flavorite', 'Flavorosa', 'Gypzee', 'Honey Punch', 'Ruby Kat', 'Spring Flavor', 'Sunset Delight'
Aprium® (*P. salicina* × *P. armeniaca*) x *P. armeniaca*	'Autumn Sprite', 'Betty Cot', 'Cot-N-Candy White', 'Coral Cot', 'Country Cot', 'Escort', 'Flavor Delight', 'Hermosa', 'Honey Rich', 'Huntington', 'Kylese Cot', 'Late Brittney', 'Leah Cot', 'Macy Cot', 'Tasty Rich', 'Trinidad', 'Summer Delight', 'Wescot', 'Weston'
Cherry plum (*P. avium* × *P. salicina*) × *P. salicina*	'Delight', 'Sprite'
Color-cot® *P. armeniaca* × *P. salicina*	'Bella Cerise', 'Bella Gold', 'Bella Jean', 'Bella Rose', 'Bella Royale', 'Bella Sun'

Source: adapted from Dave Wilson Nursery Web Page (2018).

Apricot Decline Syndrome) (Bassi and Audergon, 2006, cited by Zhebentyayeva et al., 2012; Ledbetter, 2008; Krška, 2018).

There are 3000 apricot cultivars recorded worldwide (Zhebentyayeva et al., 2012), many of which are specific to particular regions and climatic conditions. Around 80% of global production is based on fewer than 30 cultivars (Bailey and Hough, 1975; Mehlenbacher et al., 1991; Lichou and Audubert, 1992; cited by Ruiz, 2011). These comprise a narrow genetic base, which makes apricot production vulnerable to threats such as disease. To address this problem, 563 new apricot cultivars have been introduced between 1980 and 2007, as well as releases of 61 hybrids (apricot × plum and plum × apricot) (Fideghelli and Della Strada, 2010; cited by Benediková and Giovannini, 2013). According to a survey carried out by Fideghelli and Della Strada (2010), almost all of these cultivars belong to *P. armeniaca*, 17 to *P. mume*, and two to *P. sibirica*. The cultivars were released in 32 countries; 74 in the United States, 70 in France, 62 in Italy, 47 in China, 29 in Czech Republic, and 26 in Spain.

Apricot cultivars are generally divided into six ecogeographical groups based on their geographical origin: Dzhungar-Zailij, East Chinese, European Iranian-Caucasian, Middle-Asian, and North Chinese (Layne et al., 1996). This classification is becoming more and more complex, given that new cultivars are from various crosses between genotypes in different groups (Faust et al., 1998; Halász et al., 2007; cited by Herrera et al., 2018).

Common and new cultivars in apricot-growing countries include (Mehlenbacher et al., 1990; Ogawa and Southwick, 1995; Bassi, 1999; cited by Hormaza et al., 2007; Indreiaş, 2011):

- Turkey: 'Aprikoz', 'Cataloglu', 'Cologlu', 'Darende', 'Hacihaliloglu', 'Hasanbey', 'Kabaasi', 'Sekerpare', 'Soganci', 'Tokaloglu', and 'Yegen'
- Spain: 'Boryana', 'Roxana', 'Albena' in Bulgaria; 'Bulida', 'Canino', 'Galta Rocha', 'Mauricio', 'Moniqui', 'Palabras', 'Paviot', 'Pepitos', and 'Real Fino' (new cultivars include: 'Rojo Pasión', 'Selene', 'Murciana', 'Dorada', 'Estrella', and 'Sublime')
- France: 'Bergeron', 'Canino', 'Earlyblush', 'Fantasme', 'Goldrich', 'Helena du Roussillon', 'Ivresse', 'Luizet', 'Malice', 'Orange Red', 'Rouge de Roussillon', and 'Tomcot' (new cultivars include 'Solédane', 'Florilège', and 'Bergarouge')
- Italy: 'Baracca', 'Bella di Imola', 'Cafona', 'Canino', 'Ceccona', 'Fracasso', 'Goldrich', 'Pallumella', 'Portici', and 'Vitillo' (new cultivars include 'Boreale', 'Ardore', 'Pieve', 'Angela', 'Gheriana', and 'Silvana')
- USA: 'Royal' (syn. 'Bleinheim'), 'Castlebrite', 'Goldrich', 'Katy', 'Modesto', 'Patterson', 'Tilton', and 'Tomcot'
- Greece: 'Bebeco', 'Tirynthos', and 'Luizet'

- Romania: 'Callatis', 'Comandor', 'Dacia', 'Excelsior', 'Favorit', 'Litoral', 'Mamaia', 'Olimp', 'Neptun', 'Saturn', 'Selena', 'Sulmona', 'Sulina', 'Tudor', and 'Venus' (new cultivars include 'Mamaia', 'Traian', 'Tudor', 'Cristal', 'Danubiu', 'Auraş', 'Amiral', 'Fortuna', 'Orizont', 'Augustin', 'Histria', 'Euxin', 'Ceres', 'Ovidius', and 'Elmar')
- Canada: 'Goldcot', 'Goldrich', 'Harcot', 'Harglow', 'Hargrand', 'Harlayne', 'Harogem', 'Veecot', 'Velvaglo', and 'Vivagold'
- Hungary: 'Ceglédi bíborkajszi', 'Ceglédi óriás', 'Gönci magyar kajszi', 'Magyar kajszi ('Hungarian Best'), and 'Mandula kajszi'
- Pakistan: 'Damavand', 'Lasgherdi', 'Malayer', 'Tabarza', 'Tokbam', and 'Shakarpara'
- Algeria and Morocco: 'Canino' and 'Amor Leuch'

Other new cultivars include (Zhebentyayeva et al., 2012):

- Australia: 'River Ruby', 'Riverbrite', 'Rivergold', and 'Rivergem'
- Tunisia: 'Asli', 'Atef', 'Fakher', 'Meziane', 'Ouafer', and 'Raki'
- China: 'Luotuohuang', 'Chuanzhihong', 'Fenren', 'Guoren', and 'Shajinhong'
- Japan: 'Hachirou' and 'Kagajizou'
- USA: 'OrangeRed', 'Robada', 'Helena', and 'Apache'
- Czech Republic: 'Lejuna', 'Leskora', 'Lebela', 'Ledana', 'Leala', and 'Lerosa', as well more recent cultivars with larger fruits and resistance to PPV such as 'Lemeda', 'Marlen', 'Minaret', 'Palava', 'Svatava', and 'Betinka' (Krška et al., 2006)
- Slovakia: 'Vesna', 'Vegama', 'Veharda', 'Velbora', 'Vemina', 'Velita', 'Vesprima', 'Barbora', 'Vestar', and 'Veselka' (these cultivars are characterized by late blossoming, better fruit quality, prolonged fruit maturation, and resistance to PPV (Benediková, 2010).

3.3 Rootstocks

All plum and apricot cultivars are currently budded/grafted onto rootstocks, a key element in sustainable orchard production. Plum and apricot rootstocks provide physical support for the tree and take up water and nutrients from the soil. Therefore, rootstocks need to be adaptable to different soil types. They play a key role in determining tree growth and canopy density, depending on the vigor of the scion cultivar. Many plum cultivars have been budded/grafted on Myrobalan (*P. cerasifera*) seedling rootstocks, which are relatively easy to obtain and induce strong and rapid growth of trees, but which are relatively late in coming into bearing fruit (Cociu et al., 1997). In England, Hatton (1914) selected A, B, C, and D types of *P. cerasifera*. Of these, 'Myrobalan B' has been used extensively in many countries. Bauer (1924) and Berndt and Dorner (1941) obtained rootstocks from *P. cerasifera* brought from Turkey (Adana). Interesting results also have been obtained in Germany ('Phätz' clones), France ('GF 1380'), California – the United States ('Myrobalan 29C'), and Romania ('Oteşani 8'). Trees grafted on Myrobalan seedlings have reduced vigor and productivity in heavy, clay, and wet soils. *P. cerasifera* seedling rootstocks also present problems of graft incompatibility with some valuable plum cultivars like 'Renclod D'Althan', 'Tuleu gras', and 'Vânăt românesc').

The rootstocks for European plums currently include:

- *P. domestica* (e.g., 'Oteşani 8', 'Oteşani 11', 'Pixy', 'Cirule 43', 'Saint Julien A', 'Ackerman', 'Roşior văratic', 'Voineşti B', 'Brompton', 'Damas 1869', 'Scolduş', 'Wagenheim', 'Reine Claude 1380')
- *P. cerasifera* (e.g., 'Myrobalan B', 'Miroval', 'C.163', 'Mirobolan 29C', 'Hamyra', 'Myruni', 'Myrabi', 'Myrocal')
- *P. cerasifera* × *P. munsoniana* hybrids (e.g., 'Marianna GF 8-1', 'Marianna 2624')
- *P. besseyi* × *P. cerasifera* hybrids (e.g., 'Ferlenain'); *P. tomentosa* × *P. cerasifera* hybrids ('VVA-1' named now 'Krymsk 1')

The rootstocks for Japanese plums currently include 'Myrobalan 29C', 'Citation', 'Nemaguard', and 'Marianna 2624' (Norton and Coates, 2012).

A study on the performance of 15 European plum cultivars grafted on 20 rootstocks in Romania assessed the vigor of trees 18 to 20 years after planting (Botu et al., 2002). The main plum cultivars used in the trial, which measured trunk cross-sectional area (TCSA), canopy area (CA), and crown volume (CV), were 'Stanley', 'Anna Späth', 'Agen', 'Centenar', 'Carpatin', 'Andreea', 'Minerva', 'Tuleu gras', 'Vâlcean', 'Renclod D'Althan', and 'Vânăt românesc'. The significant variations in growth were classified and rootstocks were assigned accordingly:

1. *dwarf vigor* (<20%) rootstocks: *P. besseyi*, *P. besseyi* × P. americana (CV<30 m^3; CA<10 m^2; TCSA<100 cm^2);
2. *semidwarf vigor* (20–40%) rootstocks: 'Pixy', 'Oteşani 8', 'St. Julien A', 'Voineşti B' (CV = 30–80 m^3; CA = 10–20 m^2; TCSA = 100–300 cm^2);
3. *medium vigor* (40–60%) rootstocks: 'Roşior văratic', 'Molive', 'Oteşani 11', 'Troianu 9', 'Galbene', 'Grase' (CV = 80–120 m^3; CA = 20–30 m^2; TCSA = 300–400 cm^2);
4. *high vigor* (60–80%) rootstocks: 'Mirobolan galben', 'Mirobolan GT_1', 'Mirobolan T_{7C}', 'Pepenii', 'Gogoşele negre' (CV = 120–200 m^3; CA = 30–40 m^2; TCSA = 400–500 cm^2);
5. *very high vigor* (80–100%) rootstocks: 'Marianna GF 8-1', 'Mirobolan de Horezu', 'Mirobolan de Sâmbureşti' (CV = 200 m^3; CA = 40 m^2; TCSA = 500 cm^2).

As with plum, rootstock selection for apricot depends on cultivar vigor (Botu, 1978). Apricot tree vigor is medium to high and planting distances generally range from 6 m × 6 m to 4 m × 2 m, resulting in tree densities of 277 to 1,250 tree/ha (Cociu, 1993; Norton and Coates, 2012). Regardless of the cultivar and rootstock, the larger the distances between trees, the higher is the vigor and the later the fruit bearing. Root competition can reduce tree vigor by 20–30% at higher planting densities, but yields decrease with aging trees (after 6–7 years). The most common apricot rootstocks are seedlings that tend to be vigorous: 'Blenheim' in USA; 'Zarzăr' and 'Constanta 14' in Romania; 'Mech-Mech' in Morocco; 'Manicot' in France; and 'Haggith' in Canada. Rootstocks originating from *P. cerasifera* also are used, such as 'Myrobalan B', 'Myrobalan GF-3.1', 'Dzhanska 4', and 'Aydemiroka' (Botu, 1978; Achim et al., 2006; Southwick et al., 2011; Marinov and Dimitrova, 2006). Other rootstocks include *P. insititia*, such as 'Pollito' in Spain and 'Buburuz' and 'Brompton' in Romania, as well as *P. persica* such as 'Montelar', 'Hygama', 'Rubira', and 'Nemared'. In the United States, the most used rootstocks are 'Myrobalan 29C', 'Citation', 'Marianna 2624', and 'Nemaguard' (Norton and Coates, 2012).

4 Environmental factors affecting plum and apricot cultivation

Maintaining and improving plum and apricot yields depend both on cultivar and the edaphic and climatological conditions in which they are grown. The main regions for European plum cultivation are between 30°–50° N and 30°–45° S and those for Japanese plum are between 25°–40° N and 20°–30° S (Cociu et al., 1997; Botu, 2000). Plum orchards have expanded particularly in Asia (especially with European plum) as well as in Europe, but also in Argentina, Chile, South Africa, Australia, and New Zealand. The areas where plums are grown are fragmented into individual microzones, each with specific ecological conditions. The main regions for apricot cultivation are located between 20°–50° N and 20°–35° S (Gautier, 1988; Cichi, 2005).

Botu (2000) evaluated the potential of an area for growing plums based on a method used by De Groot (1999), taking into account the following factors: geography; climate; cost of land, plant materials, labor and capital; infrastructure; knowledge and tradition of orchard cultivation; and potential markets and distribution network. He identified the following areas as having particular potential: Bosnia, France (Lot et Garonne), Serbia (Čačak – Valjeva), Romania (Oltenia), and the United States (California). Areas for apricot cultivation included Greece (Thessaloniki), the United States (California), Romania (Dobrogea), and Spain (Murcia). Within these overall zones, it is important to identify microzones with the most favorable growing conditions (based on location, soil, relief, etc.).

Plum and apricot orchards can be characterized as anthropogenic fruit ecosystems. Maintaining sustainable production depends on understanding these ecosystems and optimizing the way they function. Failure to do so has led to over-reliance on inputs such as fertilizers and pesticides, high input and energy costs, declining soil heath, environmental pollution, and loss of biodiversity (e.g., in numbers of pollinators and predatory species, which help to control pests). Over time, this compromises yields and profitability. The implementation of a sustainable system for these fruit crops greatly increases the productive orchard lifetime and can optimize yields in the long term.

4.1 Temperature

Resistance to low temperatures during winter and frost hardiness is very important for plum and apricot cultivars in order to produce good fruit yields. The limit of low temperature resistance during winter for European plums is −30°C to −32°C, while for Japanese plums the limit is −22°C to −24°C. The most low temperature-resistant species are from *P. nigra*, *P. ussuriensis,* and *P. americana*, which can resist damage at temperatures as low as −50°C (Cociu et al., 1997). Apricot cultivars are cold-hardy to −25°C to −27°C during winter, the most sensitive organs in this stage being the floral buds (Cociu, 1993). It is important to note that in the first 3–4 years after planting in the orchard when the trees are young, plum and apricot resistance to low winter temperatures is less than when mature. Flowers and fruit of young plum and apricot trees can be damaged at −2.2°C to −4.0°C.

In temperate zones, where plums and apricots are mainly grown, summer temperatures can reach 32°C to 40°C. These temperatures influence plant physiological processes, especially the intensity and rate of processes such as photosynthesis and transpiration and can negatively affect tree and fruit growth (Vişanu, 2011). Research work by Topor and Zamfir (2006) and Petre et al. (2006) on differing apricot cultivars found that climate changes, particularly the low temperatures in winter and during flowering, are limiting factors that significantly influence fruit yields. There were significant differences in the way cultivars in the study reacted to lower temperatures, either through higher or lower fruit yields or by biennial or alternate bearing. The most resilient cultivars were 'Goldrich', 'Manitoba', 'Vivagold', 'Sulmona', 'Comandor', and NJA 42, which maintained yields between 12.9 and 17.6 t/ha.

4.2 Rainfall

Insufficient or excessive moisture levels can significantly affect yields. Excess water is a particular risk in heavy, clay, undisturbed soils. Excessive moisture levels cause anaerobiosis followed by production of cyanogenic glycoside (prunasin), subsequently hydrolyzed into cyanhydric acid. At this point, growth ceases, epinastic bending of leaves occurs, and wilting, yellowing, and premature defoliation follow, all with negative effects on production. The phenomenon has been observed on plum cultivars grafted on 'Mirobolan galben' rootstock and grown on clay soils (Botu and Botu, 2003). Between 1960 and 1980, attempts at plum cultivation on clay soils in countries such as Romania, Republic of Moldova, Serbia, and Bulgaria failed. Subsequently, plum orchards were located on lighter soils, with access to irrigation.

Apricot is considered to be a drought-resistant fruit tree crop, while plum has medium resistance. The main measure of the water requirement of a crop is the transpiration coefficient. The value of this coefficient varies depending on factors such as cultivar and rootstock, age, productive capacity, and water content of the soil. In the case of stone fruit crops, the coefficient of transpiration is typically 250–500 L/kg of dry matter. This measure helps to predict potential water stress and irrigation requirements in the warmer, more arid areas where apricots are grown (e.g., the Mediterranean, California, and Central Asia).

4.3 Soil

Soil quality has a major impact on plum and apricot cultivation. Voiculescu (1999) and Păltineanu et al. (2015) list the main elements that influence the soil–tree relationship in the case of apricot and plum. These include physical properties such as texture, amount of humus, water-holding capacity and drainage as well as chemical properties such as pH and salinity. Nandwani and Nwosisi (2016) concluded that a soil with a high content of organic matter increases water-retention capacity and reduces the risk of soil erosion. As a result, use of organic fertilizers (manure) in orchards had a positive effect on fruit production. Voiculescu (1999) found that soils with 5–12% active calcium carbonate ($CaCO_3$) at a 60–80 cm depth inhibit orchard establishment, while soils with more than 12% active $CaCO_3$ at a 60–80 cm depth should not be used. Studies suggest that the optimal pH value for plum is 6.2–8.2 (minimum 5.5 and maximum 8.5) while, for apricot, the optimal pH value is 6.2–8.3 (minimum 6.0 and maximum 8.7) (Gyuro, 1974; cited by Botu and Botu, 2003).

5 Plum cultivation

Research suggests that under optimal conditions, plum yields can be as high as 70 t/ha, e.g., as in 'Centenar' grown in the southern Oltenia region of Romania (Cociu et al., 1997). However, average yields have been estimated at 25–40 t/ha for European plum cultivars and 30–50 t/ha for apricot cultivars. Improved yields have largely been a function of improved cultivars and higher density orchard systems. Recent research has documented (Table 7) the yield

Table 7 Yields of several European plum cultivars in different orchard systems in the tenth year after planting in the Oltenia region of Romania

		Orchard systems					
Cultivars	Rootstock	Classic (6 × 5m) with 333 trees/ha	Intensive (5 × 4m) with 500 trees/ha	Intensive (4 × 2m) with 1250 trees/ha	Intensive in buried containers with 1665 trees/ha	Intensive in unburied containers with 1665 trees/ha	Average yield per cultivar (t/ha)[a]
'Andreea'	'Mirobolan galben'	14.0	19.8	28.5	12.6	4.2	15.8
'Minerva'		12.4	18.1	24.2	11.8	3.3	14.0
'Tuleu gras'		12.7	17.8	20.5	12.1	3.8	13.4
'Stanley'		15.6	21.2	30.3	13.7	4.6	17.1
Mean (t/ha)[b]		13.7 100 (%)	19.2 (140%)	25.9 (189%)	12.6 (92%)	4.0 (29%)	
Mean (kg/tree)		41.1	38.4	20.7	7.6	2.4	

[a] $P < 0.05 = 2.27$.
[b] $P < 0.05 = 2.5$.
Source: adapted from Botu et al. (2013).

potential of some plums grown in the Oltenia region (Botu et al., 2013). The most productive orchard system was the intensive system with a density of 1,250 trees/ha, averaging 25.9 t/ha, 89% higher than the classical system (333 trees/ha), which produced 13.7 t/ha. It is important to note that this yield was achieved in the first 10 years after planting. After this point, the strong vigor of trees grafted on 'Mirobolan galben' began to shade each other, reducing subsequent yields. The 500 trees/ha-intensive orchard yielded 19.2 t/ha, 40% higher than in the classic system. Yields of very high density (1665 trees/ha) containerized plum trees (used for germplasm collections) varied from 12.6 t/ha (buried containers) to only 4.0 t/ha (unburied containers, above soil level). Under the study conditions, containerized plum cultivation yields were comparable to that obtained in the classical system. Individual plum cultivars achieved average fruit yields from 13.4 t/ha ('Tuleu gras') to 17.1 t/ha ('Stanley'). The most productive plum cultivars in the study were 'Stanley' and 'Andreea' (Botu et al., 2013).

Research also focused on two rootstocks, high vigor 'Mirobolan galben' and semidwarfing 'Oteşani 8'. Plums grafted on 'Oteşani 8' had 23% lower vigor compared to 'Mirobolan galben' (Botu and Botu, 2017). Yields varied by rootstock and cultivar, with mean yields beginning in the 3rd year after planting at 1.4 t/ha on 'Oteşani 8', but only at 0.3 t/ha on 'Mirobolan galben' (Table 8). Efficient yields occurred in the 4th year after orchard establishment. The mean yield oscillated from 10.0 t/ha (in the 4th year) to 39.7 t/ha (11th year) on 'Oteşani 8' and from 5.4 t/ha to 27.0 t/ha on 'Mirobolan galben'. The cumulative fruit yields of different cultivars, from the 3rd to 11th year after planting, varied from 211.5 t/ha on 'Oteşani 8' to 161.1 t/ha on 'Mirobolan galben'.

Across cultivars, the highest yields were obtained with 'Andreea' (26.2 t/ha), 'Stanley' (25.5 t/ha), and 'Centenar' (24.4 t/ha), all on 'Oteşani 8'. Yields were lower for 'Centenar' (19.2 t/ha), 'Andreea' (19.1 t/ha), and 'Stanley' (18.7 t/ha) on 'Mirobolan galben'. 'Andreea', 'Centenar', and 'Stanley' were the most productive cultivars (Botu and Botu, 2017). For all cultivars grafted on 'Oteşani 8', the average yields were 22.8 t/ha (at 500 trees/ha) and 26.3 t/ha (at 1250 trees/ha), and on 'Mirobolan galben', the average yields were 18.0 t/ha (at 500 trees/ha) and 20.1 t/ha (at 1250 trees/ha) (Table 7) (Botu and Botu, 2017).

These results can be compared with the study by Turcu et al. (1998) of 10 plum cultivars grown in the Vâlcea region of Romania over a period of 14 years. The mean plum yields varied as 26.1 t/ha ('Centenar'), 22.3 t/ha ('Stanley'), 19.9 t/ha ('Pescăruş'), 19.5 t/ha ('Minerva'), 10.3 t/ha ('Early Rivers' and 'Gras ameliorat'), and 7.9 t/ha ('Vânăt românesc cl.4'). In terms of dry matter yield (t/ha/year), the highest values were obtained for 'Centenar' (3.7 t/ha), 'Stanley' (3.5 t/ha), 'Agen 707' and 'Minerva' (3.0 t/ha), and 'Tuleu gras' (2.7 t/ha). There were no significant differences between the mean fruit sizes and weights related to the growing system and the rootstocks used. In both studies, the

Table 8 Yield of five plum cultivars grafted on two rootstocks planted in different orchard systems

		Yield (t/ha)										
		Year after planting										
Cultivar	Rootstock	3rd	4th	5th	6th	7th	8th	9th	10th	11th	Mean (3rd to 11th year)	Mean (4th to 11th year)
'Stanley'	'Oteşani 8'	1.5	11.0	30.3	32.2	15.4	30.3	28.7	39.6	41.5	25.6	28.6
'Andreea'	'Oteşani 8'	1.6	10.2	31.5	33.7	16.2	28.2	30.1	39.5	44.7	26.2	29.3
'Tuleu gras'	'Oteşani 8'	0.8	7.3	20.5	22.6	14.2	26.4	25.1	32.2	37.1	20.7	23.2
'Centenar'	'Oteşani 8'	1.9	12.7	29.9	31.8	18.5	23.5	27.2	34.8	39.6	24.4	27.2
'Minerva'	'Oteşani 8'	1.3	8.7	24.2	26.1	13.4	20.4	25.6	30.7	35.6	20.7	23.1
Mean ('Oteşani 8')		1.4	10.0	27.3	29.3	15.5	25.8	27.3	35.4	39.7	23.5	26.3
'Stanley'	'Mirobolan galben'	0.5	6.2	16.7	21.4	15.4	27.5	24.6	29.6	26.5	18.7	21.0
'Andreea'	'Mirobolan galben'	0.4	4.8	14.5	20.9	15.6	28.1	30.2	28.6	28.7	19.1	21.4
'Tuleu gras'	'Mirobolan galben'	0.2	3.6	11.2	18.7	14.3	21.0	23.4	22.5	24.3	15.5	17.4
'Centenar'	'Mirobolan galben'	0.2	7.7	17.2	21.7	18.2	25.7	29.3	25.6	26.8	19.2	21.6
'Minerva'	'Mirobolan galben'	0.2	4.5	14.1	19.6	15.6	21.7	27.6	24.4	25.1	17.0	19.1
Mean ('Mirobolan galben')		0.3	5.4	14.7	20.5	15.8	24.8	27.0	26.1	26.3	17.9	20.1
Difference between rootstocks		1.1	4.6	12.5	8.8	−0.3	1.0	0.3	9.2	13.4	5.6	6.2

Source: adapted from Botu and Botu (2017).

most important variable was the cultivar (Botu and Botu, 2017). Other research studies also confirm that production capacity depends on plum species and cultivar (Vişanu et al., 2012).

Research carried out in the northern area of Oltenia in Romania on 8 cultivars of European plum showed an average yield of 16.0 t/ha (in a non-irrigated orchard 15–18 years old, with trees planted at 5 × 4 m). The mean yield differences between cultivars fluctuated between 10 t/ha ('Vânăt românesc cl. 4') and 23.3 t/ha ('Centenar'). For five cultivars of *P. insititia*, the average yield reached 17.3 t/ha (varying between 13.0 t/ha and 23.0 t/ha). Four cultivars belonging to *P. cerasifera* produced an average yield of 17.7 t/ha (varying from 14.8 t/ha to 20.4 t/ha). In the case of *P. spinosa*, the average yield recorded was 5.0 t/ha. In the same study, low yields were achieved for four cultivars of Japanese plum (8.8 t/ha). One of the cultivars, 'Black Amber', yielded 10.5 t/ha. The low yield in this group correlated with the ecological conditions specific to the area (relatively low temperatures during winter and spring frosts), which were less suitable for these cultivars.

Milošević et al. (2012) studied several plum cultivars grown in Čačak, Serbia, including 'Opal', 'Valor', 'Jubileum', 'Hanita', and 'Victoria', grafted on 'St. Julien A' rootstock and planted at a distance of 5.0 × 1.5 m (1,334 trees/ha). Yields ranged between 21.9 t/ha ('Opal') and 15.2 t/ha ('Jubileum'). Average weights per fruit varied from 61.5 g ('Jubileum') to 31.8 g ('Opal'). Another study in Serbia by Lukic et al. (2012) analyzed the performance of 'Čačanska lepotica', Čačanska rodna', and 'Stanley' grafted on Myrobalan seedlings and planted at 5.0 × 4.0 m. This study found that 11 years after planting, the yield reached 45.8 t/ha ('Stanley') and 39.3 t/ha ('Čačanska lepotica'). The open vase tree training system proved to be higher yielding than the pyramid system (44.4 t/ha vs. 40.6 t/ha).

Blažek and Pištěková (2012) studied the productivity of 25 plum cultivars in a high-density orchard, grafted on 'St. Julien A' rootstock and planted at 4.0 × 1.0 m in Holovousy, Czech Republic. Very high yields were recorded for 'Topend Plus' (48.2 t/ha), 'Topstar' (47.5 t/ha), 'Topstar Plus' (42.5 t/ha), and 'Topfirst' (40.3 t/ha). The lowest yields were for 'Katinka' (13.1 t/ha), 'Ruth Gerstetter' (10.3 t/ha), and 'Top 2000' (10.0 t/ha).

6 Prune and apricot cultivation

The International Prune Association (IPA) was established in Agen (France) in January 1990 with the participation of France, Chile, Italy, Australia, Argentina, South Africa, and the United States. Other countries such as Serbia and Azerbaijan joined later. In 2015, the IPA estimated the total production of prunes across the globe to be at about 300 000 t, including both the harvested crop and existing stocks. The main prune-producing countries were the United

States (California), Chile, France, Argentina, Serbia, Australia, Italy, and South Africa (IPA Web Page, 2018).

About 99% of US prune production and 40% of the world's supply come from California (California Dried Plums Board Web Page, 2018). The main production areas are located in the Sacramento and San Joaquin Valleys. California exports prunes to more than 50 countries. The dried plums are sold pitted, whole, and as juice. However, in a total cultivated area of 35 000 ha, the average yield of dehydrated plums is only 4.0 t/ha, equivalent to 16 000-16 500 kg/ha fresh weight. About 96.5% of California's prune production comes from 'Agen' type plums, originally introduced in 1850 (Connell, 2012). Other cultivars like 'Imperial', 'Robe de Sergeant', and 'Sugar' are grown, but cultivars related to the original 'Petit d'Agen' are still dominant (California Dried Plums Board Web Page, 2018). Plum yields vary between 4–5 t/ha and 45 t/ha, the fresh consumption cultivars reaching yields of 18–45 t/ha and those for industry between 8 and 16 t/ha. The cultivars for processing have higher soluble solids (16–24%). Little innovation has occurred in prune-production systems over the past few decades, as higher density orchards are limited by the spacing required for access to trunk-shaker machine harvest (Tables 9 and 10).

Apricot yields are dependent on a multitude of factors including cultivar, environment, and method of production. Norton and Coates (2012) calculated average yields in California (which accounts for 95% of the US apricot production) to be at 24 t/ha, with 30 t/ha sometimes achieved, depending on the cultivar – 'Blenheim' and 'Patterson' are used for dehydrated fruits and 'Early Golden', 'Castlebrite', 'Flavor Giant', 'Flora Gold', and 'Nugget' are used for fresh fruit.

Topor and Zamfir (2006) analyzed production of 31 apricot cultivars grafted on 'Zarzăr' rootstock in the Dobrogea region of Romania, planted at 4.0 × 3.5 m (714 trees/ha). The early ripening cultivars produced 15.7–16.4 t/ha, midseason ripening cultivars yielded 12.8–17.3 t/ha, and the late ripening cultivars yielded 13.1–15.0 t/ha. The highest annual average yields (during 9 years) were obtained with 'Dacia' (17.3 t/ha), VT 10/79 (16.4 t/ha), and 'Selena' (14.7 t/ha). The authors also found that changing patterns of rainfall had a significant impact on production. Petre et al. (2006) studied 15 apricot cultivars planted at 4.0 × 3.5 m in the Moldova region of Romania. Average yield during 9 years varied from 12.9 to 17.6 t/ha (Table 11). The most productive cultivars were 'Precoce de Italia' (17.6 t/ha), 'Goldrich' (14.6 t/ha), and 'Manitoba' (13.6 t/ha). Frosts had a negative influence on yields.

Norton and Coates (2012), Xiloyannis et al. (2010), and Monastra and De Salvador (1995) have found that the most suitable densities for apricot cultivation range from 600 to 1300 trees/ha, depending on cultivar, rootstock, and orchard site. At these densities, yields of 18–30 t/ha can be achieved by the 5th year from planting. Irrigation is required since annual rainfall in most

Table 9 Mean 11-year production differences for five plum cultivars grafted on two rootstocks in orchard systems planted at two tree densities

		Average production at 500 trees/ha		Average production at 1250 trees/ha			
Cultivar	Rootstock	kg/tree	t/ha	kg/tree	t/ha	Difference between systems in t/ha	Significance
'Stanley'	'Oteşani 8'	19.60	24.5	22.88	28.6	4.1	
'Andreea'	'Oteşani 8'	19.84	24.8	23.44	29.3	4.5	
'Tuleu gras'	'Oteşani 8'	15.76	19.7	18.56	23.2	3.5	
'Centenar'	'Oteşani 8'	20.00	25.0	21.76	27.2	2.2	
'Minerva'	'Oteşani 8'	15.92	19.9	18.48	23.1	3.2	
Mean ('Oteşani 8')		18.22	22.8 (126.7%)	21.04	26.3	3.5 *	LSD5% = 2.78
'Stanley'	'Mirobolan galben'	38.4	19.2	42.0	21.0	1.8	
'Andreea'	'Mirobolan galben'	38.0	19.0	42.8	21.4	2.4	
'Tuleu gras'	'Mirobolan galben'	31.2	15.6	34.8	17.4	1.8	
'Centenar'	'Mirobolan galben'	37.4	18.7	43.2	21.6	2.9	
'Minerva'	'Mirobolan galben'	35.2	17.6	38.2	19.1	1.5	
Mean ('Mirobolan galben')		36.0	18.0 (100%)	40.2	20.1 (100%)	2.1* (111.7%)	LSD5% = 2.05
Difference between rootstocks		-	4.8	-	6.2	1.4	LSD5% = 2.82
Difference between growing systems		-	20.4	-	23.2	2.8 *	LSD5% = 2.66

Source: adapted from Botu and Botu (2017).

Table 10 Fruit characteristics of five plum cultivars grafted on two rootstocks in orchard systems planted at two tree densities

	Rootstocks							
	'Oteşani 8'				'Mirobolan galben'			
	500 tree/ha (5 × 4 m)		1250 tree/ha (4 × 2 m)		500 tree/ha (5 × 4 m)		1250 tree/ha (4 × 2 m)	
Cultivars	Fruit size index (mm)	Average fruit weight (g)	Fruit size index (mm)	Average fruit weight (g)	Fruit size index (mm)	Average fruit weight (g)	Fruit size index (mm)	Average fruit weight (g)
'Stanley'	37.6	34.5	37.9	35.6	37.9	37.4	38.2	37.1
'Andreea'	38.4	39.5	39.2	40.2	42.3	42.3	42.3	43.5
'Tuleu gras'	37.9	34.8	37.8	35.0	34.0	34.0	36.9	34.9
'Centenar'	37.2	40.5	37.4	38.4	38.6	38.6	38.8	37.7
'Minerva'	36.2	34.7	36.4	34.9	36.3	36.3	36.7	35.5
Mean	37.5	36.8	37.7	36.8	37.7	37.7	38.6	37.7
Difference (%)	100	100	100.7	100.05	100	100	100.05	100.5

Source: adapted from Botu and Botu (2017).

Table 11 Yields of some apricot cultivars in the Iaşi area of Romania

	Fruit yields (kg/tree)					9 year mean		Difference to	
Cultivars	1991	1992	1993	1995	2000	(kg/tree)	(t/ha)	mean (kg/tree)	Significance
'Precoce de Italia'	17.7	33.5	25.2	23.1	44.1	24.6	17.6	10.2	*
'Goldrich'	6.4	25.5	23.9	9.7	42.1	20.5	14.6	6.1	
'Manitoba'	9.2	5.8	21.5	19.0	37.5	19.0	13.6	4.6	
'Vivagold'	7.2	20.9	8.7	18.7	50.8	18.1	12.9	3.7	
'Sirena'	12.0	22.0	32.5	19.7	20.5	17.2	12.3	2.8	
'Umberto'	10.6	23.5	31.5	22.6	15.6	15.0	10.7	0.6	
Mean						14.4	10.3	-	
'Comandor'	8.0	13.0	20.0	18.6	13.8	13.3	9.5	−0.4	
'Băneasa 23/4'	3.5	17.9	11.1	25.6	28.6	12.3	8.9	−2.1	
'Sulmona'	7.3	12.9	14.2	23.9	15.5	11.1	7.9	−3.3	
'Favorit'	4.5	4.1	16.8	25.9	15.0	9.7	6.9	−4.7	
'Venus'	2.9	8.2	6.8	15.3	25.0	9.6	6.6	−4.8	
'Selena'	3.0	12.0	27.2	19.1	12.7	8.8	6.3	−5.6	
'Litoral'	4.0	5.0	11.3	10.7	18.0	8.4	6.0	−6.0	
'Sulina'	5.3	9.1	6.3	17.6	17.9	8.0	5.7	−6.4	
'Excelsior'	-	3.2	12.3	11.7	21.5	7.1	5.1	−7.3	
						LSD 5% = 9.4	LSD 1% = 12.4	LSD 0.1% = 16.0	

Source: adapted from Petre et al. (2006).

apricot-cultivation regions ranges between 500 and 600 mm. Other factors affecting apricot yields are diseases such as PPV, apricot apoplexy (Egea et al., 2012; Milatovich, 2013), and bacterial spot. Gunduz et al. (2011) analyzed the province of Malatya in Turkey, which produces 55% of the country's apricots and 20% of the world production, with over 50 000 apricot orchards. He identified the need for better support from the government to help growers select better cultivars and rootstocks as well as optimize production practices, in order for yields to increase. Unfortunately, vigor-controlling rootstocks and high-density intensive orchard systems are in only the earliest stages of development for apricot production.

7 Conclusion and future trends

Science has a central role to play in achieving more sustainable plum and apricot cultivation. A key goal of research is genetic improvement to develop new cultivars that are more productive, better adapted to the diverse climates of cultivation zones, and resistant or tolerant to the main diseases (PPV, bacterial spot, bacterial canker, etc.) while maintaining fruit quality. Another fundamental element is the choice of suitable regions and microzones for cultivation and the selection of cultivars, rootstocks, and cultivation techniques to match these zones. Extension specialists also have to be more active to train growers on the correct elements for sustainable plum and apricot cultivation systems.

One solution for sustainable cultivation is a production system that integrates conventional and organic techniques. This does not exclude the use of synthetic fertilizers and pesticides, but minimizes their use within a more integrated production strategy that better optimizes the agro-ecosystem within which cultivation takes place. A more integrated approach matches cultivar and rootstock selection to ecosystem as well as optimizes cultivation practices (e.g. planting densities). This maximizes fruit quality and sustainable yields by minimizing fruit contamination and wider pollution risks from excessive chemical use, promoting soil health, and encouraging biodiversity to support beneficial insects to help in pollination and pest control.

Future research needs include preserving and using genetic resources from wild relatives, including reversing trends in the narrow genetic base associated with much modern breeding, especially in the last 60–80 years. Genetic diversity and modern molecular techniques make it easier to identify genotypes with improved resistance to the most damaging diseases of plum and apricot. At the same time, new cultivars need to be characterized by other traits necessary for sustainable cropping systems (e.g. tolerance of abiotic stresses). In recent decades, more interspecific cultivars have been released. The combination of genetic material from different species has been shown to improve fruit taste, appearance, and shelf-life. Other research needs to focus on better understanding and improving soil-plant relations, in particular, the range of physical, chemical, and biological properties that best promote soil health and plant growth. These areas of research will become even more important in ensuring more resilient cultivation that is able to adapt to climate change without contributing to environmental degradation or global warming.

8 Where to look for further information

More information regarding current approach and future trends in sustainable plum and apricot cultivation can also be found in actual and future *Acta Horticulturae* publications of the International Society for Horticultural Sciences (ISHS), which contain proceedings of ISHS Symposia and the International Horticultural Congress:

- XVI International Symposium on Apricot Breeding and Culture (https://www.actahort.org/books/1214/index.htm)
- XVII International Symposium on Apricot Breeding and Culture - to be organized in 2019 in Malatya, Turkey (https://www.ishs.org/symposium/619)
- XII International Symposium on Plum and Prune Genetics, Breeding and Pomology - to be organized in 2020 in Zlatibor, Serbia (https://www.ishs.org/symposium/686)

9 References

Achim, Gh., Botu, I., Botu, M., Preda, S. and Giorgota, A. 2006. Comportarea în livadă a unor portaltoi noi în relaţie directă cu soiurile de prun în zona Vâlcea. *Lucr. Ştiinţifice "Prezent şi perspectivă în cercetarea pomicolă"* (vol. II). SCDP Constanţa, Ed. Cartea Universitară, Bucureşti, pp. 221-30.

Amblard, C. 2015. World prune production: statistical update. In: *15th World Prune Conference Sirmione*, Italy. International Prune Association. Available at: http://www.ipaprunes.org/congress/2015/2601%20Amblard%20World%20Stats.pdf.

Badgley, C., Moghtader, J., Quintero, E., Zakem, E., Chappell, M. J., Avilés-Vázquez, K., Samulon, A. and Perfecto, I. 2007. Organic agriculture and the global food supply. *Renewable Agriculture and Food Systems* 22(2), 86-108. doi:10.1017/S1742170507001640.

Benediková, D. 2010. Retrospective analysis of apricot breeding in Slovak Republic. *Acta Horticulturae* 862(862), 33-8. doi:10.17660/ActaHortic.2010.862.2.

Benediková, D. and Giovannini, D. 2013. Review on genetic resources in the ECPGR Prunus Working Group. *Acta Horticulturae* 981(981), 43-51. doi:10.17660/ActaHortic.2013.981.2.

Bertschinger, L., Baur, R., Carlen, C. and Doruchowski, G. 2009. Sustainability of horticulture in Europe (environmental, social, economic): examples from the pre- and the post-harvest food chain. *Acta Horticulturae* 817(817), 187-94. doi:10.17660/ActaHortic.2009.817.18.

Blažek, J. and Pištěková, I. 2012. Initial results from the evaluation of plum cultivars grown in a very dense planting. *Acta Horticulturae* 968(968), 99–108. doi:10.17660/ActaHortic.2012.968.13.

Botu, I. 1978. Studiul biotipurilor autohtone de prun şi comportarea lor ca portaltoi pentru prun şi cais. Teza doctorat, Universitatea din Craiova.

Botu, I. and Botu, M. 2003. *Pomicultura modernă şi durabilă*. Ed. Conphys, Rm.Vâlcea, România.

Botu, I., Achim, G., Botu, M., Godeanu, I. and Baciu, A. 2002. The evaluation and classification of growth vigor of the plum cultivars grafted on various rootstocks. *Acta Hortic*. 577, 299–306. doi: 10.17660/ActaHortic.2002.577.51.

Botu, M. 2000. Cercetări pentru stabilirea unor genotipuri şi soiuri valoroase de prun pentru zona subcarpatică a Olteniei. Teză de doctorat, Universitatea din Craiova.

Botu, M. and Botu, I. 2017. Evaluation of intensive plum culture system in the north of Oltenia – Romania. *Acta Horticulturae* 1175(1175), 55–60. doi:10.17660/ActaHortic.2017.1175.11.

Botu, M., Botu, I. and Preda, S. 2006. Metode de conservare a resursele genetice la speciile pomicole din genurile *Prunus*, *Juglans* şi *Corylus*. *Lucr. Ştiinţifice "Prezent şi perspectivă în cercetarea pomicolă"* (vol. II). SCDP Constanţa, Ed. Cartea Universitară, Bucureşti, pp. 41–52.

Botu, M., Tomic, L., Cvetković, M., Gjamovski, V., Jemric, T., Lazović, B., Ognjanov, V., Pintea, M., Sevo, R., Achim, G., Bozovic, D., Carka, F., Čiček, D., Fruk, G., Jaćimović, V., Kiprijanovski, M., Juraveli, A. and Hjalmarsson, I. 2012. *Balkan Pomology – Plums*. SEEDNet's WG for Fruit and Vitis. Hjalmarsson, I. and Tomić, L. (Eds). Exaktaprinting AB, Alnarp, 170pp.

Botu, M., Mitră Amza, M., Botu, I. and Papachatzis, A. 2013. European plum (*Prunus domestica* L.) cultural systems trial. *Acta Horticulturae* 985(985), 169–73. doi:10.17660/ActaHortic.2013.985.20.

Butac, M., Bozhkova, V., Zhivondov, A., Milosevic, N., Bellini, E., Nencetti, V., Blazek, J., Balsemin, E., Lafarque, B., Kaufmane, E., Gravite, I., Vasiljeva, M., Pintea, M., Juraveli, A., Webster, T., Hjalmarsson, I., Trajkovski, V. and Hjeltnes, S. H. 2013. Overview of plum breeding in Europe. *Acta Horticulturae* 981(981), 91–8. doi:10.17660/ActaHortic.2013.981.9.

California Dried Plums Board Web Page. 2018. Available at: https://www.californiadriedplums.org.

Cichi, M. 2005. Contributions regarding improvement of apricot assortment for South Zone of Oltenia. *Lucrări Ştiinţifice USAMV Iaşi*, Seria Horticultură, Anul XLIX vol. 1 (48), Ed. Ion Ionescu de la Brad, Iaşi, pp. 93–6.

Cociu, V. 1993. *Caisul*. Editura Ceres, Bucureşti, pp. 1–438.

Cociu, V., Botu, I., Minoiu, N., Pascu, I. and Modoran, I. 1997. *Prunul*. Ed. Conphys, Rm. Vâlcea.

Connell, J. 2012. The prune industry in California. In: Buchner, R. P. (Ed.), *Prune Production Manual*. Publication 3507, University of California, Division of Agriculture and Natural Resources, pp. 3–8.

Dave Wilson Nursery Web Page. 2018. Available at: http://www.davewilson.com/product-information-general/zaiger-interspecifics.

De Groot, N. S. P. 1999. Floriculture worldwide trade and consumption patterns. *Acta Horticulturae* 495(495), 101–22. doi:10.17660/ActaHortic.1999.495.4.

Egea, J., Dicenta, F., Burgos, L., Martínez-Gómez, P., Rubio, M., Campoy, J. A., Ortega, E., Patiño, J. L., Nortes, L., Molina, A., Molina Jr., A. and Ruiz, D. 2012. New promising high quality, Sharka resistant apricot cultivars from CEBAS-CSIC (Murcia, Spain) breeding program. *Acta Horticulturae* 940(940), 53–6. doi:10.17660/ActaHortic.2012.940.5.

FAO Stat Database Web Page. 2018. Available at: www.fao.org.

Fatima, T., Bashir, O., Gani, G., Bhat, T. A. and Jan, N. 2018. Nutritional and health benefits of apricots. *International Journal of Unani and Integrative Medicine* 2(2), 5–9.

Fideghelli, C. and Della Strada, G. 2010. The breeding activity on apricot in the world from 1980 through today. *Acta Horticulturae* 862(862), 93–8. doi:10.17660/ActaHortic.2010.862.13.

Gautier, M. 1988. *La culture fruitiere* (vol. 2). Edit. Lavoisier, Paris.

Granatstein, D. and Kupferman, E. 2008. Sustainable horticulture in fruit production. *Acta Horticulturae* 767(767), 295–308. doi:10.17660/ActaHortic.2008.767.31.

Gunduz, O., Ceyhan, V., Erol, E. and Ozkaraman, F. 2011. An evaluation of farm level sustainability of apricot farms in Malatya Province of Turkey. *Journal of Food, Agriculture and Environment* 9, 700–5.

Herrera, S., Lora, J., Hormaza, J. I., Herrero, M. and Rodrigo, J. 2018. Optimizing production in the new generation of apricot cultivars: self-incompatibility, *S-RNase* allele identification, and incompatibility group assignment. *Frontiers in Plant Science* 9, 527. doi:10.3389/fpls.2018.00527.

Hormaza, J. I., Yamane, H. and Rodrigo, J. 2007. Apricot. In: Kole, C. (Ed.), *Genome Mapping and Molecular Breeding in Plants, Volume 4: Fruits and Nuts*. Springer-Verlag Berlin Heidelberg, pp. 171–87. Chapter 7.

Igwe, E. and Charlton, K. 2016. A systematic review of the health effects of plums (*Prunus domestica* and *Prunus salicina*). *Phytotherapy Research* 30(5), 2032. doi:10.1002/ptr.5881.

Indreiaş, A. 2011. Breeding work on apricot and peach at Research Station for Fruit Growing Constanţa (personal communication).

Ingram, C. 1948. *Ornamental Cherries*. County Life, Limited, London.

International Prune Association Web Page. 2018. Available at: http://www.ipaprunes.org.

Krška, B. 2018. Genetic apricot resources and their utilisation in breeding. In: Soneji, J. R. and Nageswara-Rao, M. (Eds), *Breeding and Health Benefits of Fruit and Nut Crops*. IntechOpen, pp. 63–82. Chapter 4. doi:10.5772/intechopen.77125. Available at: https://www.intechopen.com/books/breeding-and-health-benefits-of-fruit-and-nut-crops/genetic-apricot-resources-and-their-utilisation-in-breeding.

Krška, B., Vachůn, Z. and Nečas, T. 2006. The apricot breeding programme at the horticulture faculty in Lednice. *Acta Horticulturae* 717(717), 145–8. doi:10.17660/ActaHortic.2006.717.29.

Lal, R. 2008. Sustainable horticulture and resource management. *Acta Horticulturae* 767(767), 19–44. doi:10.17660/ActaHortic.2008.767.1.

Layne, R. E. C., Bailey, C. and Hough, L. F. 1996. Apricots. In: Janick, J. and Moore, J. N. (Eds), *Fruit Breeding, Volume I: Tree and Tropical Fruits*. John Wiley & Sons, Inc., New York, NY, pp. 79–111.

Lecouls, A. C., Bergougnoux, V., Rubio-Cabetas, M. J., Bosselut, N., Voisin, R. and Sallesses, G. 2004. Marker assisted selection for the wide spectrum resistance to root-knot nematodes conferred by *Ma* gene from Myrobalan plum (*P. cerasifera*) in interspecific *Prunus* material. *Molecular Breeding* 13, 113–24. doi:10.1023/B:MOLB.0000018758.56413.cf.

Ledbetter, C. A. 2008. Apricots. In: Hancock, J. F. (Ed.), *Temperate Fruit Crop Breeding: Germplasm to Genomics*. Springer Science + Business Media B.V., Dordrecht, pp. 39–82. doi:10.1007/978-1-4020-6907-9_2.

Liu, W. 2007. Plum production in China. *Acta Horticulturae* 734(734), 89–92. doi:10.17660/ActaHortic.2007.734.9.

Lukic, M., Mitrovic, M., Milosevic, N. and Glissic, I. 2012. Biological proprieties of some plum cultivars grown under different training systems. *Acta Horticulturae* 968, 227–32.

Malkina-Pykh, I. G. and Pykh, Y. A. 2003. *Sustainable Food and Agriculture. The Sustainable World* (vol. 4). WIT Press, Southampton, Boston.

Maraux, E. 1995. *Contribution a la mise en point de techniques d' electrophorese en vue de la distinction varietate des Prunus*. Univ. Paul Sabatier, Toulouse and INRA, Bordeaux.

Marinov, P. and Dimitrova, M. 2006. Mineral balance of the nutrition of young apricot trees. *Lucr. Ştiinţifice "Prezent şi perspectivă în cercetarea pomicolă"* (vol. II). SCDP Constanţa, Ed. Cartea Universitară, Bucureşti, pp. 97–101.

Mehlenbacher, S. A., Cociu, V. and Hough, L. F. 1990. Apricots. In: Moore, J. N. and Ballington, R. J. (Eds), *Genetic Resources of Temperate Fruit and Nut Crops*. ISHS, Wagenigen, pp. 65–107.

Milatovic, D. 2013. *Kajsija*. Naucno vocarsko drustva Serbija, Čačak.

Milošević, T. and Milošević, N. 2018. Plum (*Prunus* spp.) breeding. In: Al-Khayri, J., Jain, S. and Johnson, D. (Eds), *Advances in Plant Breeding Strategies: Fruits*. Springer, Cham, pp. 165–215.

Milošević, N., Milincovic, V. and Mitrovic, M. 2012. Productive traits of some newly introduced plum cultivars grown under environment condition of Čačak (Western Serbia). *Acta Horticulturae* 968, 87–92.

Milošević, N., Glišić, I. and Mitrović, O. 2016. Plum cultivars from Fruit Research Institute Čačak intended for drying. Traditional Food Network. Available at: https://www.trafoon.org/sites/trafoon.org/files/info/news/1244/olsztyn_nebojsa_milosevic_20160913.pdf.

Monastra, F. and De Salvador, F. R. 1995. Apricot: present and future. *Acta Horticulturae* 384(384), 401–14. doi:10.17660/ActaHortic.1995.384.64.

Nandwani, D. and Nwosisi, S. 2016. Global trends in organic agriculture. In: Nandwani, D. (Ed.), *Organic Farming for Sustainable Agriculture*. Sustainable Development and Biodiversity, vol 9. Springer International Publishing, Cham, Switzerland.

Neumüller, M. 2011. Fundamental and applied aspects of plum (*Prunus domestica*) breeding. *Fruit, Vegetable and Cereal Science and Biotechnology* 5(1), 139–56.

Norton, M. and Coates, W. 2012. Growing apricots in California: an overview. University of California, USA. Available at: https://ucanr.edu/datastoreFiles/391-603.pdf.

Okie, W. R. and Ramming, D. W. 1999. Plum breeding worldwide. *HortTechnology* 9(2), 162–76. doi:10.21273/HORTTECH.9.2.162.

Okie, W. R. and Weinberger, J. H. 1996. Plums. In: Janick, J. and Moore, J. N. (Eds), *Fruit Breeding, volume 1: Tree and Tropical Fruits*. Wiley, New York, pp. 559–607.

Păltineanu, C., Chiţu, E., Septar, L., Gavăţ, C. and Opriţă, A. Vl. 2015. *Piersicul şi caisul în sistemul sol-plantă - atmosferă, în Dobrogea*. Editura Estfalia, Bucureşti, pp. 1-221.

Petre, L., Corneanu, G. and Cârdei, E. 2006. Comportarea unor soiuri de cais în condiţiile zonei Iaşi - România. *Lucr. Ştiinţifice "Prezent şi perspectivă în cercetarea pomicolă"* (vol. II). SCDP Constanţa Ed. Cartea Universitară, Bucureşti, pp. 53–60.

Ramming, D. W. and Cociu, V. 1992. Plums (*Prunus*). In: *Genetic Resources of Temperate Fruit and Nut Crops*. ISHS, Wagenigen, pp. 233–88.

Rehder, A. 1940. *Prunus*. In: *Manual of Cultivated Trees and Shrubs Hardy in North America* (2nd edn.). MacMillan Company, New York, NY, pp. 452-82.

Renaud, R. 1994. Fermareine - a new European plum variety. *Acta Horticulturae* 359(359), 93–6. doi:10.17660/ActaHortic.1994.359.12.

Ruiz, D., Martínez-Gómez, P., Rubio, M., Petri, C., Larios, A., Campoy, J. A. and Egea, J. 2011. Application of biotechnology tools to apricot breeding. *Fruit, Vegetable and Cereal Science and Biotechnology* 5(1), 101–17.

Scorza, R., Ravelonandro, M., Callahan, A., Zagrai, I., Polak, J., Malinowski, T., Cambra, M., Levy, L., Damsteegt, V., Krška, B., Cordts, J., Gonsalves, D. and Dardick, C. 2016. 'HoneySweet' (C5), the first genetically engineered plum pox virus-resistant plum (*Prunus domestica* L.) cultivar. *HortScience* 51(5), 601–3.

Southwick, S. M., Doyle, J. F. and Buchner, R. P. 2011. Prune rootstocks. In: *Prune Production Manual*. Publication 3507, University of California, Division of Agriculture and Natural Resources, pp. 37–40.

Sumner, J. 2008. Sustainable horticulture and rural development: more than just organic production. *Acta Horticulturae* 767(767), 111–22: doi:10.17660/ActaHortic.2008.767.11.

Tan, S. and Köksal, H. 2004. *Sustainable Agriculture*. Agricultural Economy and Research Institute. TEAE-BAKIŞ, Number: 5, Issue: 2, 4p (in Turkish).

Topor, E. and Zamfir, Gh. 2006. Cercetări privind dinamica producţiei de caise şi calitatea acesteia sub influenţa factorilor biotici şi abiotici. *Lucr. Ştiinţifice "Prezent şi perspectivă în cercetarea pomicolă"* (vol. II). SCDP Constanţa Ed. Cartea Universitară, Bucureşti, pp. 61–8.

Turcu, E., Botu, I. and Botu, M. 1998. Evaluation of the production capacity of some plum cultivars grown in Romania. *Acta Horticulturae* 478(478), 179–86: doi:10.17660/ActaHortic.1998.478.28.

United Nations World Commission on Environment and Development. 1987. Our Common Future: Report of the World Commission on Environment and Development (WCED). *The Brundtland Report*.

Vişanu (Ştefan), F. M. 2011. Cercetări privind principalele caracteristici ale creşterii şi fructificării la unele specii ale genului Prunus. Teză doctorat, Universitatea din Craiova.

Vişanu, F., Botu, I. and Baciu, A. 2012. Assessment of production capacity for some plum, myrobalan and sloe cultivars and selections grown in Northern Oltenia. *Acta Horticulturae* 968(968), 75–9. doi:10.17660/ActaHortic.2012.968.9.

Voiculescu, N. 1999. *Enciclopedia pomilor fructiferi*. Edit. Academiei Române, Bucureşti.

Xiloyannis, C., Dichio, B. and Montanaro, G. 2010. Sustainable apricot orchard management to improve soil fertility and water use efficiency. *Acta Horticulturae* 862(862), 419–24. doi:10.17660/ActaHortic.2010.862.64.

Zhebentyayeva, T., Ledbetter, C., Burgos, L. and Llacer, G. 2012. Apricot. In: Badenes, M. L. and Byrne, D. H. (Eds), *Fruit Breeding: Handbook of Plant Breeding* (vol. 8). Springer, Boston, MA, pp. 415–58.

Part 2

Berry fruits

Chapter 10

Advances and challenges in strawberry genetic improvement

Chris Barbey and Kevin Folta, University of Florida, USA

1 Introduction

The commercial strawberry (*Fragaria* × *ananassa*) is among a relatively small set of foods that are both nutritious and widely appreciated for flavour. Strawberry production in the United States has an annual value of US$2.4 billion (USDA, 2018), and trends of increased consumption are expected to continue (Rabobank, 2015). Despite its popularity, consumers feel that there is opportunity for improvement, primarily in terms of improved flavours and shelf life. Strawberry production is the foundation of regional economies, and a central crop of many farms. However, while consumption of fruits and vegetables is encouraged and consumers crave strawberry in the diet, the challenges of sustainable production are tremendous and only increasing.

Strawberries are produced from plants that yield fruits with specific seasonal timing. The June-bearing plants begin yielding fruit in spring, and continue through the longest days of summer. Day-neutral strawberries flower regardless of seasonal cues, and short-day varieties have been adopted for winter production in Florida, Mexico, Spain, Turkey, Morocco and other locales of similar climate. The thousands of acres of each modern variety originate from a single maternal plant that is then clonally propagated by runners bearing daughter plants. Plants for transplant are propagated in off-season nurseries in California or Canada and then transported to production areas.

http://dx.doi.org/10.19103/AS.2018.0040.25

The strawberry industry runs on two major production systems. Plasticulture is the dominant production method in commercial strawberry production (Fernandez et al., 2001; Poling, 2005). Soil is mounded into long rows under a meter wide, a drip tape is installed to provide fertilizer and water, and then plastic film is stretched over the top. Fumigants are injected under the plastic or via the drip line to suppress nematodes, fungi and weeds. After the fumigants dissipate the soil is nearly sterile, and the new plants are inserted through slits in the plastic. These production plants are established from vegetative, typically cold-treated daughter plants that grow quickly and begin flowering. The annual replacement of plants mitigates the carryover of pests and diseases (Poling, 2005). In some cases commercial strawberries are planted in matted rows, where their runners provide new plants for future seasons. This approach is more appropriate for colder regions with less pest and pathogen pressure (Tabatabaie and Murthy, 2016).

On the production side, regulations around water and fertilizer use, pesticide application, water availability, phase out of soil fumigants and changing labour practices constrain previously mainstream cultural practices. In strawberry, fertilizer, plastic and fuel are the resource inputs most impacting environmental sustainability (Tabatabaie and Murthy, 2016). A number of 'life cycle assessments' have been performed to analyse the major impacts on sustainable production in a variety of production locations, such as Iran (Khoshnevisan et al., 2013), Australia (Gunady et al., 2012), Italy (Girgenti et al., 2014), Spain and the United Kingdom (Williams et al., 2008). The consensus of these reports identified fuel and plastic as primary sources of greenhouse gas emissions, and a central environmental impact in field production. In 2012 it is estimated that 43 000 tons of polyethylene mulch were used in strawberry production in Italy alone (ISPRA, 2013), most of which ends up in landfills (Razza et al., 2012). New practices using biodegradable plastics have been proposed (Girgenti et al., 2014). Strawberry production in greenhouses requires extensive electrical energy input for growth lighting and climate control, which in some cases can be more environmentally adverse than field-based impacts (Khoshnevisan et al., 2013).

The issues of sustainable strawberry production encompass all three pillars of sustainability – economic, social and environmental. U.S. strawberry growers anticipate increasing competitive and regulatory challenges to profitable domestic production. On the other hand, strawberry production is highly taxing on the environment, and the myriad social aspects of labour, fumigant use, water consumption, equity and distribution remain long-standing issues. The way to satisfy all three pillars of sustainability going forward will lie in agricultural innovations that leverage technological advantages in automated production in synergy with new genetic resources.

Strawberry genetic improvement is achieved primarily through traditional breeding. These hybridization methods have provided outstanding varieties for sustainable production. New varieties satisfy basic production demands. Most modern strawberry production varieties produce many large, red, uniformly shaped fruits, and are more resilient to disease. Modern production varieties are, for the most part, genetically optimized to fruit during ideal regional production windows. Fruits from contemporary strawberry cultivars also must survive long-distance shipping and arrive at retail stores without physical damage or substantial natural decay. Agile strawberry genetic improvement is hindered by the complexity of its octoploid genome. Not only is this genome formidable, scientists are still unravelling the exact constituents of the resident subgenomes, a task that began with cytological studies almost a century ago (reviewed in Folta and Davis, 2006). Breeding of new varieties carries the burden of subgenome dominance and linkage drag, two processes that hinder the ability to consolidate genetic gain into advanced breeding selections.

The purpose of this chapter is to highlight some of the newest innovations in production, while explaining the current methods of genetic improvement. Because breeding has been a central barrier, the opportunities in streamlined variety improvement are the main focus of the chapter. Production has already been optimized within current regulations and limits, so discussions in production are more forward thinking and fantastical. Genetic solutions are available now, are being implemented and stand to place a better quality plant in any production system.

2 Threats and solutions to sustainable production

2.1 Competition

The United States is regarded as the second largest strawberry producer in the world; however imports have significantly decreased market prices and depressed domestic industry expansion over the last decade (FAOSTAT, 2016; Zhang et al., 2016). Florida strawberry growers perceive competition from Mexico as the most serious threat to the U.S. domestic strawberry industry, ahead of government regulation, labour shortages and far ahead of pest and disease control (Guan et al., 2015). Fresh imported strawberries to the United States come almost exclusively from Mexico, as imports have increased fourfold since 1997 with the passing of the North American Free Trade Agreement (NAFTA). Mexico has dramatically increased production in recent years, placing significant pressure on domestic growers particularly in the Florida region which relies on high wintertime market prices (Wu and Guan, 2015).

Other complications, including domestic labour shortages, make it increasingly difficult to compete with Mexican producers. The compounding challenges of depressed prices, and high labour cost or low labour availability have led to harvest becoming economically infeasible in certain situations. It is expected that the strawberry industry will become increasingly stressed by

Mexican competition, and further antagonized by various production challenges. It is possible that domestic strawberry market share may continue to erode, with some growers shifting to other crops seeking a greater competitive advantage against imports. However, California's strawberry industry maintained its 2016 acreage into 2017, breaking several consecutive years of decline. Per capita consumption of strawberries in the United States has continued to increase over two decades, indicating a growing domestic market coincident with increasing competition, perhaps buffering the perceived impact on American growers.

The impact of potential changes in free trade policies on the domestic strawberry industry are difficult to predict. Driscoll's, the largest American strawberry producer with large operations in both California and Mexico, has signalled that tariffs on Mexican imports would not fundamentally alter the trajectory of the strawberry industry (Charles, 2017). The trends towards Mexican production began in the early 1990s, prior to NAFTA. Overall, favourable labour costs and growing climates in Mexican strawberry agriculture present advantages compared to some aspects of U.S. strawberry production, allowing competition for the fresh domestic market.

Increasing global demand for strawberries, particularly in China, represents a potential avenue for growth for U.S. producers via exportation. Access to Chinese markets has recently been granted after over a decade of negotiations with the USDA. The California Strawberry Commission has indicated confidence in a growing export market to China, where strawberry production halts during the summer months. New domestic varieties that are adapted for Chinese export will be required to maximize U.S. competitiveness in strawberry production. This is similar to the shift towards higher yielding and summer-ripening U.S. varieties, in response to early Mexican competition. Additionally, the domestic U.S. strawberry industry may benefit from establishing cultivars that are distinctive to retail consumers, which may allow competition via differentiation in a market where strawberry cultivars are treated as generic by the consumer (Suh et al., 2017). Sustainable growing practices compatible with the standards of the import country, and rigorous certification processes ensuring pest and pathogen-free imports and exports, are critical to sustainable trade. Overall, the domestic U.S. industry faces significant challenges which will require considerable investment and innovation to overcome.

2.2 Labour, and the advent of mechanical harvesting

Strawberry is almost always harvested by hand. The work is performed by low-wage workers, typically immigrants on temporary agricultural work visas. The process typically involves ungloved workers bending to remove partially-ripened fruits from strawberry plants, placing them directly into plastic 'clamshell' cases, which are then stacked in flats for cooling and shipping. Hand harvested strawberry in the United States is unlikely to be sustained in the future, as improved standards for labourers encourage existing trends towards automation and production outsourcing to countries with a competitive advantage for inexpensive labour.

The loss of human labour will give rise to innovations in mechanical harvesting, which has been projected for over fifty years (Booster and Kirk, 1968; Denisen and Buchele, 1967). Methods of robotic harvesting have been explored for decades, with machines designed to identify the peduncle or fruit with machine vision (Tiezhong and Tianjuan, 2004) and then remove appropriately ripe berries (Hayashi et al., 2010). Examination of several cultivars and mechanical methods suggests that any solution will have to take the variety into consideration (Aliasgarian et al., 2015), which opens the door for breeding specifically for mechanical harvesting. Such varieties may present longer peduncles or thinner canopies, allowing machine vision detection of appropriately ripe fruits.

Reducing labour costs is critical for the competitiveness of the domestic U.S. industry, especially for Florida which is in direct competition with Mexican growers. While it may take considerable time and investment to establish a robust mechanized harvesting solution for strawberry, such a system would neutralize Mexico's labour advantage and allow the domestic U.S. industry to compete effectively in the future (Suh et al., 2017).

2.3 Improved chemical controls

Strawberries are prone to fungal and bacterial disease, as well as assault from insect pests and nematodes. Crop protection strategies are required to limit damage to plants before and during production, and additional strategies are needed at the onset of fruiting. Most of the critical diseases in strawberry are fungal, resulting in lower yields due to infections of the crown or petioles. Other pathogens affect fruit appearance or accelerate post-harvest decay. Strawberry remains as one of the most input-dependent crops, and improved management strategies have been devised. New approaches are pest and pathogen specific, are applied in small amounts and are applied less often. Still, there would be benefit in decreasing inputs in the interest of economic and environmental sustainability.

The emergence of pathogen populations that are resistant to important fungicides in strawberry is becoming commonplace. Resistant grey mould (*Botrytis cinerea*) populations in the United States (Fernández-Ortuño et al., 2012), Germany (Leroch et al., 2013), Spain (Dianez et al., 2002) and elsewhere have recently been reported. Quinone-outside Inhibitor (QoI; strobilurin) fungicides have begun to fail to control *Colletotrichum acutatum* populations in the United States (Forcelini et al., 2016), just as succinate dehydrogenase inhibitor fungicides have similarly failed to control *Botrytis cinerea* (Leroux et al., 2010). The emergence of adapted pathogens have provoked changes in strawberry cultural practices, including more judicious usage of fungicides, simultaneous application of alternative controls and shifts towards alternative but less efficacious fungicides (Leroch et al., 2013; Fernández-Ortuño et al., 2015). These strategies and others will help increase the sustainability of the current suite of chemical fungicides going into the future.

However, it is critical that new fungicides, and new fungicide classes, be developed and deployed. It is important that new fungicides be ecologically sustainable, have adequate target specificity and effectively control damaging pathogens. New and efficacious derivatives of existing fungicides for *Botrytis cinerea* have been created that show greater levels of control on resistant populations (Takagaki et al., 2014). Research into the molecular mechanisms of fungicidal compounds, and identifying differences in target protein structure between pathogen populations, is critical to the development of effective derivatives (Xiong et al., 2015). The integration of -omics tools applied to pathogens is expected to accelerate the rational design of new fungicides with novel targets (Cools and Hammond-Kosack, 2013). It is critical that responsible chemical usage and integrated pest management (IPM) be used in tandem to preserve these important compounds and improve environmental sustainability.

Grey mould and *Rhizopus* rot are the principal pathogens responsible for post-harvest decay. Post-harvest treatment with chitosan, benzothiadiazole and other organic compounds is often performed in strawberry agriculture to elicit strawberry defence genes during storage and shipping (Romanazzi et al., 2013). The elicitors acibenzolar-*S*-methyl and harpin protein, when applied prior to harvest, can also induce resistance in strawberry and suppress spider mites (Tomazeli et al., 2016). Further research in defence elicitors, and their induced resistance genetics, will improve strawberry sustainability and reduce reliance on broad, non-selective chemical controls.

New molecular technologies for selective pest and pathogen control are under development, and hold the promise to substantially affect sustainable pest and pathogen management. Spray-induced gene silencing (SIGS) presents a number of exciting advantages over traditional chemical controls, if it can be developed into an effective commercial solution (Majumdar et al., 2017). SIGs involves the exogenous application of double-stranded DNA (dsDNA) to silence genes in pests and pathogens. These dsDNAs are processed into small RNAs, which interfere with the expression of critical pathogen-derived genes and potentially lead to a loss of virulence or pathogen death (Wang and Jin, 2017). Effective pathogen resistance via transgenic RNAi has been demonstrated for a broad number of agricultural pests and pathogens. These include *Botrytis cinerea* (Wang et al., 2016), *Fusarium graminearum* (Machado et al., 2017), and various arthropods (Bolognesi et al., 2012) and nematodes (Huang et al., 2006).

In SIGS, dsDNAs designed against pathogen transcripts are applied to the crop exogenously, rather than transgenically, potentially balancing the need for biotechnological solutions with a consumer preference against genetic engineering, in some markets. Also, SIGS is likely to be durable if deployed correctly. It is hypothesized to be very unlikely for pathogens to evolve effective novel resistance to SIGS, as a variety of sequence targets per pathogen may be simultaneously exploited to maximize the effects (Majumdar et al., 2017). If resistance to a specific small inhibitory RNA were to evolve, it would be simple to design and deploy new dsRNAs targeting against the new mutation. In other words, SIGS represents a relatively durable and updatable strategy for pathogen control once a suitable RNA target for SIGS has been identified. While off-target silencing is measurable in some RNAi feeding studies, RNA-RNA interactions are generally much more predictable and testable than enzyme-chemical interactions (Tan et al., 2016). Thus, RNAi-based strategies for pest and pathogen control is more straightforward relative to pesticidal chemicals, as the potential impact on non-target organisms and the environment can be easily monitored (Bachman et al., 2016). Improved technologies for efficient uptake of exogenous dsDNA, as well as the identification of ideal target transcripts, remains an area of active research both in academia and industry (Chu et al., 2014).

SIGS represents a potential form of pesticidal control that is more specific, sustainable and amenable to rational design (Wang and Jin, 2017). The transgenic event MON87411, approved recently by Canada and the United States represents the first commercial application of RNAi against an insect pathogen, and works via silencing of the vacuolar sorting transcript *DvSnf7* in western corn rootworm (Head et al., 2017). Research into effective application methods and efficacious target transcripts is critical for the implementation of SIGS strategies widely in commercial agriculture.

2.4 Disease-resistant strawberry cultivars

Genetic resistance in cultivated strawberry is highly advantageous, as it would provide passive protection against commercially destructive diseases without additional management requirements or chemical inputs. Specific genetic resistance to most strawberry pathogens has been observed in various *Fragaria* × *ananassa* genotypes (Antanaviciute et al., 2015; Vining et al., 2015), as well as in various wild accessions of diploid *Fragaria* (Calis et al., 2015; Paynter, 2015). However, this resistance has not been extensively consolidated or well-optimized into elite production lines. Much work remains for elucidating the genetic basis for resistance at the molecular level. Marker-assisted selection for disease resistance, involving the selection of genetic elements closely associated with resistance, would provide strawberry breeders the ability to rapidly create disease-resistant cultivars for sustainable production. The development of strongly associated disease resistance loci has been reported for a variety of diseases, such as *Phytophthora fragariae* (Haymes et al., 1997) and *Phytophthora cactorum* (Mangandi et al., 2017). The elimination of methyl bromide use in strawberry production has been cited as a strong motivator for the identification of novel genetic sources of disease resistance (Whitaker et al., 2016). However, significant challenges exist for identifying and selecting disease-resistance genetics in strawberry.

Cultivated strawberry possesses a complex auto-allooctoploid genome, which substantially impedes efforts to identify and reliably consolidate desirable genetics via traditional selection (Folta and Davis, 2006). Additionally, disease resistance is often observed to be closely linked to alleles that are unfavourable for commercial production. Linkage drag is a substantial problem in strawberry breeding, as recombinants that have broken linkage are often still of low utility for breeding as favourable combinations of alleles are seldom to arise in an octoploid context (Stegmeir et al., 2010). Therefore, the ability to screen a very large number of progeny based on genetics, at an early developmental stage, would be highly advantageous for the strawberry breeder. If the causal genetics discriminating disease resistance can be resolved to the allelic level, horizontal transfer tools such as transgenics or gene editing may be used to quickly create production-ready cultivars without the penalties of linkage drag.

Individual genes added to strawberry via transgenesis have been shown to confer resistance disease, including chitinase genes (Asao et al., 1997; Vellicce et al., 2006; Chalavi et al., 2003), thaumatin genes (Schestibratov and Dolgov, 2005), *Npr1* (Silva et al., 2015), β-glucanase genes (Mercado et al., 2015) and the *ELONGATOR* 3 and 3 genes (Silva et al., 2017). Unfortunately, many resistance-associated traits in strawberry are themselves unsuitable for commercialization. For example, increased cell wall thickness is associated with disease resistance due to decreased permeability of pathogenic effectors, but also an undesirable fruit texture (Cantu et al., 2007). Selection for disease resistance in strawberry must be a holistic choice with consideration given to many areas of priority. A refocus on classes of disease-resistance genes which do not typically incur penalties to other traits may assist in this effort.

Several specific genes in strawberry have been associated with disease resistance, such as the Mannose-binding lectin protein associated with *C. acutatum* resistance (Guidarelli et al., 2014). Using new tools in molecular genetics, including targeted sequence capture and QTL analysis in pedigreed populations, several new large-effect QTL resolved to the subgenomic level. These include resistance to angular leaf spot caused by *Xanthomonas fragariae* (*RXf1*), crown and root rot caused by *Phytophthora cactorum* (*Pc1*) and crown rot caused by *Colletotrichum gloeosporioides* (*Cg1*) (Whitaker et al., 2016). These loci provide a high level of resistance, and represent an opportunity to substantially improve genetic resilience to biotic stress. These new methodologies for identifying the genetics of strawberry disease resistance, and other traits, are discussed in-depth in the 'Next steps in genetics' section.

2.5 Emerging pests

A wide variety of animal pests routinely affect strawberry agriculture, including the two-spotted spider mite, spotted wing *Drosophila*, flower thrips, fall armyworm and various secondary pests such as aphids, seed bugs, whiteflies and root weevils. Soil fumigants and foliar insecticides are used extensively in strawberry agriculture to combat these pests. Many pesticidal products have been approved for controlling these various strawberry agricultural pests, and are commonly used proactively and in combination to achieve a broad degree of control. Considerable research effort has been invested in developing sustainable and efficacious pesticides to enhance or replace the existing suite of chemical controls. For example, the elimination of the environmentally toxic methyl bromide has led to the adoption of less damaging alternatives, including metam, dimethyl disulphide, dichloropropene and particularly chloropicrin (Ziedan and Farrag, 2016). These compounds are frequently alternated in use to help prevent the evolution of biological resistance to these important chemical pest controls.

Alternative non-fumigant-based methods have been devised. Steam can be used to sterilize soil, however at high expense (Xu et al., 2017). Biofumigants have been proposed. Amending soil with mustards or arugula or mustard seed meal (Mazzola et al., 2017) also has been shown to have the capacity to suppress disease pressure.

A key area of growth is the development of new and highly selective pesticidal compounds to complement non-chemical strategies for broadly sustainable strawberry production. As described in the previous section, pest control via sprayable small RNAs has been demonstrated to control a number of damaging insect and nematode species. This represents a potentially selective and durable form of agricultural pest control, and is an area of highly active research.

IPM is an increasingly critical component of sustainable pest management in strawberry, and decline in the use of chemical pesticides (Strand, 2008). The use of predatory mites to control spider mites, for example, prevents about four applications of miticide each season for the 15% of Florida strawberry farms using this IPM strategy (Mossler, 2015). Common non-chemical practices include planting strawberries that are certified pest-free, however this does not guarantee that significant pest problems will not develop in the field. The intentional destruction of strawberry fields post-harvest is also commonplace, to remove potential food sources for nematode pests that also could harbour insects. Crop rotations and early field fallowing can reduce pest load, and have been shown specifically to reduce sting and root-knot nematode species (Noling, 1999). Sustainable pest management generally must be continual, prophylactic and adaptive. New strategies for IPM will increase actual yields by mitigating loss to disease, and sustainably decrease the ecological and chemical footprint of strawberry agriculture. New pesticides with increased specificity will help mitigate unintended damage to beneficial insects and the environment (Mossler and Nesheim, 2007). Decreasing yield loss to pests and other biotic stresses indirectly translates into more efficient and sustainable use of agricultural inputs (Pradhan et al., 2015).

Enhancing and optimizing strawberry genetics will promote sustainable agricultural practices through passive resistances to plant pests. The insecticidal Bt crystallized protein is applied in organic pest management of strawberry, and is also used notably as a sustainable transgenic strategy for diverse insect control in various other crops over the past twenty years (Bolda et al., 2003; Carrière et al., 2015). Research into transgenic Bt protein expression in strawberry may lead to cultivars with enhanced passive resistance to pests and potentially a concomitant decrease in pesticide application, as has been noted for other crops according to the U.S. National Academy of Sciences report on genetic engineering in agriculture (National Academies of Sciences, Engineering, and Medicine, 2016). Other transgenic strategies, such as the heterologous expression of trypsin inhibitors, may be used to protect strawberries against a host of pests without the application of exogenous pesticide and significantly advantaging growers and the environment. Native strawberry genes associated with pest resistance have also been identified, and are likely to be gradually introgressed into production cultivars through traditional and marker-assisted breeding selection (Roach et al., 2016). Genetic tolerance to mites has been noted in certain strawberry accessions, as have a host of natural inhibitory and repellent compounds produced under biotic stress conditions (Giménez-Ferrer et al., 1994). Many notable volatile organic compounds produced in the ripe strawberry fruit have antimicrobial and animal repellent properties, including the grape-like methyl anthranilate whose biochemical origins have recently been identified (Pillet et al., 2017).

Sustainable IPM must include all material and strategic sources of pest resistance, including optimized cultural practices, sustainable and selective pesticide application, transgenics, and the introgression and consolidation of favourable native genetics into high-yielding production cultivars.

2.6 Emerging pathogens

Diseases in cultivated strawberry are a major limiting factor for production, and are associated with various challenges in strawberry including high production costs, intensive management requirements, adverse ecological impact, sustainability and impact on human health (Maas, 2004). Various microbial and viral pathogens affect nearly all organs of the strawberry plant, including the fruit, foliar tissue and meristems. Common diseases negatively impacting strawberry production include grey mould caused by *Botrytis cinerea*, various *Colletotrichum* diseases, powdery mildew, angular leaf spot, leaf spot, leaf blight, leaf scorch, charcoal rot and *Phytophthora* crown rot (Peres, 2015). Highly intensive cultural practices are used routinely to mitigate pathogen load, including exclusive planting of certified stock, use of plasticulture, high-dose application of natural and synthetic microbicides, sanitation, replanting and crop rotation (Maas, 2004). Different pathogens may predominate depending on cultivar genetics, local climate and production practices.

Despite rigorous and diverse cultural strategies for pathogen control in strawberry, chemical microbicides remain essential for commercial production (Cordova et al., 2017). In particular, fungicides are key for effective control of strawberry diseases on a commercial scale (Will and Krüger, 1999). The use of synthetic chemical controls, even within the safe label usage, is controversial among broad sections of the potential consumer base (Guthman, 2016). Strawberry is regularly cited by non-governmental environmental organizations as being among the crops 'most contaminated' by trace residue of pesticides (Gatewood, 2017). In another aspect, the severity of negative health consequences caused by chronic and acute fungal mycotoxin contamination in fresh fruit produce, even at low doses, is becoming increasingly recognized (Fernández-Cruz et al., 2010; Smith et al., 2017). Fungal-infected strawberry fruits and jams demonstrate accumulation of toxic fungal metabolites, including Patulin and Ochratoxin A, and aflatoxins (Pensala et al., 1978). Considering the high production challenges posed by strawberry's broad pathogen susceptibility, and the environmental and human health considerations posed by chemical control of pathogens and potentially by the pathogens themselves, more effective control strategies are highly desired. Improved control of pathogens for sustainable strawberry production will arise in three principal areas: improved management strategies, improved chemical controls and improved genetic resistance in production cultivars.

2.7 Improved management strategies

2.7.1 Weather-based crop protection

One of the strongest initiatives towards sustainable strawberry production is the Agroclimate Strawberry Advisory System, based in Florida (Pavan et al., 2011). This web-based portal (www.agroclimate.org) analyses data from weather stations all around Florida, and provides guidance to farmers about the likelihood of disease pressure. The agroclimate algorithms provide a disease model output that informs growers of when conditions are most conducive for disease development. This guidance allows farmers to apply the appropriate fungicide with the right timing, resulting in a 50% decrease in the number of applications (Cordova et al., 2017). This resource has been helpful curbing the use of fungicides typically used to combat *Colletotrichum acutatum* and *Botrytis cinera*, two strawberry pathogens which frequently lead to disease in permissive conditions.

2.7.2 Controlled environments

Strawberry production is especially amenable to production in protected environments, such as greenhouses and artificially lit and heated environments. Such environments may rely on hydroponic production. These production methods are known to decrease the cost of transportation and support local economies with high-value fresh fruit (Mok et al., 2014). Dryland areas are particularly projected to benefit from hydroponic indoor fruit production, especially in terms of water usage and greenhouse gas emissions (de Anda and Shear, 2017). Such shifts to greenhouse production mirror the advances that have already taken place in The Netherlands and Belgium, where restrictions on soil fumigants and herbicides have motivated a shift towards protected cultivation. As recently as 2012, there were over 1600 ha under glass (Lieten, 2013), and that acreage likely stands to increase.

The advantages of protected cultivation are consistent with the ideals of sustainable production. Within a protected environment, strawberries may be grown on benches or in gutters, providing ease of harvest and year-round production that matches grower attractive market windows. Artificial lighting can alter flowering time to match market windows (Yoshida et al., 2012). At the same time these systems present new challenges, as the pests and pathogens irrelevant in the field may become an issue when plants are grown in enclosed space (Elad et al., 1996). Specific cultivars have been shown to perform better in these conditions, especially with respect to disease presentation (Fang et al., 2012). Various strategies have been implemented to combat angular leaf spot in controlled environments (Braun and Hildebrand, 2013). Light also has been used to alter disease pressure. UV-B radiation has been shown to suppress the growth of pathogens, including powdery mildew in strawberry. UV-B radiation in strawberry induced the transcription of a suite of known strawberry defence genes, including strawberry phenylalanine ammonia lyase and chalcone synthase, and improving fruit quality via anthocyanin induction (Kanto et al., 2014). Green LED light has been shown to induce disease resistance generally in a variety of crops including strawberry, suppressing spider mites, improving plant vigour and increasing the accumulation of antimicrobial compounds (Kudo and Yamamoto, 2016). Further research into the interactions of controlled light and defence physiology in strawberry is warranted, and may lead to economically viable solutions for pathogen control for indoor production systems and post-harvest storage. While the industry supporting indoor strawberry cultivation is growing in Europe and the United States, new crop protection strategies, production methods and genetically adapted cultivars will be needed to fit the environmental challenges and opportunities.

3 Post-harvest quality

Sustainability must take into consideration the issues of carbon footprint for transport, as well as post-harvest treatment, storage and waste. Post-harvest quality in strawberry is highly dependent upon fruit firmness – the ability to resist compression and mechanical damage during shipping and retail. One critical post-harvest method involves decreasing the ambient storage temperature to near freezing, quickly after harvest. This is done by a process called forced air cooling or hydrocooling, where berries are placed into a chilled water circulator that may contain sanitizing compounds (Tokarskyy et al., 2015). Many other approaches have been used to sustain fruit quality throughout the supply chain, including treatment with ozone (Pérez et al., 1999), essential oils (Sivakumar and Bautista-Baños, 2014), 1-methylcyclopropene (Ku et al., 1999), chitosan (Hernández-Muñoz et al., 2008, Reddy et al., 2000) and aloe vera (Sogvar et al., 2016). These treatments all have proven effects on slowing pathogen growth and maintaining fruit quality.

Improved metabolic content and flavour may provide a source of natural pest resistance that could also benefit healthful consumption. Volatile organic compounds (VOCs) are a critical component of strawberry flavour, along with sugars and acids (Bood and Zabetakis, 2002). These VOC flavour compounds are frequently shown to contribute to pest and pathogen defence as well as human health (Chambers et al., 2013; Vaughn et al., 1993; Goff and Klee, 2006). While appearance in strawberry has improved notably in larger fruit with more attractive colour, a perceived decline in flavour and aroma has been noted by consumers, breeders and growers for decades (Fletcher, 1917; Folta and Klee, 2016). Generally, the aroma intensity of modern cultivars is perceived as inferior to that of wild strawberries (Ulrich and Olbricht, 2014). Flavour has historically been considered a secondary goal in strawberry breeding, however consumer demands for improved flavour and increasing need for non-fungicidal controls in pathogen management have elevated flavour as a priority in many breeding programmes (Faedi et al., 2000; Whitaker, 2011). As discussed previously, product differentiation through the development of cultivars with distinctive and favourable flavours could help increase the competitiveness of the domestic U.S. industry (Suh et al., 2017).

For these reasons, the introgression of genes encoding enzymes required for the production of desirable VOCs has become a major point of focus for modern cultivar improvement (Vandendriessche et al., 2013). However, the specific introgression of desired flavour compounds in strawberry has been difficult for strawberry breeders. For example, the rare strawberry VOC methyl anthranilate was identified as a desired trait in the 1990s due to a broad range of positive qualities including flavour and antifungal activity (Olbricht et al., 2008). The instability of this trait in breeding populations proved to be a major issue, as unpredictable segregation ratios often delayed breeding efforts using traditional methods (Pillet et al., 2017). Despite considerable effort, progress towards introgressing this VOC into commercial varieties has remained slow, and the genetic mechanisms controlling its biosynthesis have not been elucidated.

Recently, a combination of modern assays in genetics, genomics and biochemistry were applied to octoploid strawberry to discover a necessary gene required for this polygenic trait. Using molecular approaches such as these, genetic markers are being designed to facilitate the introgression of post-harvest traits into advanced selections and commercial cultivars. Success in this area will benefit both consumers and strawberry producers.

4 Next steps in genetics

Associating traits with their causative genetics is crucial for developing more robust and sustainable strawberry cultivated varieties. The deductive process of gene-trait association is challenging in cultivated strawberry due to its peculiarly complex auto-allooctoploid genome. Strawberry's genome has been a substantial challenge for selecting and maintaining desired traits simultaneously in breeding programmes. Unexpected segregation behaviours may emerge due to a confluence of autopolyploid/diploid-like segregation of alleles, while other traits (or elements of a polygenic trait) may segregate with allopolyploid behaviour. Second-generation sequencing technologies resulted in rapid advances for simple crop genomes and model organisms. However, these genomics approaches typically provided insufficient resolution to discriminate octoploid strawberry's numerous allelic variants, and as a result, only situationally allowed the discrimination of genetics related to important traits. In the case of strawberry, it is difficult to achieve the necessary resolution for discovery, as numerous paralogous sequence variants occur both within and between homoeologous chromosomes. As such, successful association studies have often been limited to discrete traits underpinned by simple genetics, such as the deciphering of strawberry γ-decalactone biosynthesis via differential and categorical expression of single transcript. While RNAseq, AFLP and other reductive approaches have demonstrated success in delineating functional genetics in strawberry, the majority of agronomically important traits do not follow simple genetic behaviour and therefore must be studied in a more complete genomic context. Recently developed tools for genetic analysis in strawberry are now allowing these challenges to be addressed efficiently. The technologies discussed below will enable a marked increase in the delineation of gene function relating to marketable traits.

4.1 High-throughput genotyping and QTL analysis

High-density and inexpensive genotyping in octoploid strawberry is highly desired to facilitate association genetics approaches for gene discovery. The Affymetrix IStraw35 Axiom® SNP array, first released in 2015, was designed in collaboration with RosBREED and Affymetrix. The array interrogates over 90 000 subgenomic sequence variations identified within the germplasm of several major international breeding institutions. Subsequently, the IStraw35 array was recently released at lower cost (~$50/sample), and consists of the most useful subset of markers found in the IStraw90 array. These SNP/INDEL arrays can rapidly genotype octoploid germplasm, with informative marker densities typically in the tens of thousands. These genotyping platforms have been successfully used to identify many subgenomic QTL in octoploid strawberry. The demonstrated ability to rapidly and cheaply genotype large populations with high-density subgenomic markers promises to revolutionize gene/trait discovery in strawberry. Importantly, the entire process is relatively accessible to inexperienced users and does not require extremely technical or specific knowledge, nor does it rely on proprietary tools or techniques. A basic association genetics pipeline in strawberry can be described simply: isolated genomic DNA is submitted for IStraw90 or IStraw35 genotyping, and data is returned as a single human-readable text file. This file contains one of four possible calls for each marker (homozygous AA, heterozygous AB, homozygous BB or no-call NC). Basic quality control steps may be performed to remove uninformative markers (e.g. homozygous in all genotypes) and calls with impossible results (e.g. AA progeny from AA/BB parents). Basic genome-wide association studies (GWAS) may be performed using freely available tools to correlate genotypes with user-supplied phenotype scores. Positions of IStraw90 and IStraw35 markers in the diploid physical map are publically available, as are several high-quality octoploid genetic maps. The GWAS software can associate hundreds of phenotypes in hundreds of individuals with tens of thousands of markers in minutes, using only a typical consumer laptop. Pedigree-based QTL association offers a superior degree of association strength than GWAS, and may be executed using software such as FlexQTL, albeit with more demanding computational and technical requirements. Major loci controlling important traits for disease resistance and flavour have been described recently in strawberry using these technologies, and the number of discoveries should increase enormously in coming years. Additionally, identified

QTL regions may be directly sequenced via capture-based sequencing and/or third-generation sequencing. This additional resolution may define causal mutations, which would be important for developing perfect genetic markers, determining sequences for genetic engineering or genome editing, and for furthering basic biological understanding.

4.2 Targeted capture sequencing

Targeted capture sequencing represents an important and complementary technique for high-resolution genomic analysis of strawberry. This technique involves the bench-top enrichment of a desired genomic or transcriptomic subset, prior to sequencing. This enrichment is accomplished by the design and application of a panel of single-stranded and tagged probes, complementary to a desired nucleic acid sequence. The synthetic probes hybridize to their complementary sequences, which can then be co-purified with the tagged probe to create an enriched library for sequencing. Desired targets may be sequenced with high depth and coverage using this method, for example wide genomic QTL regions or specific gene and transcript families. For example, enriched R-gene sequencing (Renseq) has demonstrated success in expanding the repertoire of R-genes in wild potato, and has assisted in the isolation and cloning of novel disease-resistant genes. In the era of third-generation sequencing, capture techniques such as these are ready to be applied usefully to complex polyploids such as strawberry. In conjunction with QTL analysis, modern sequencing strategies will further enable rapid and high-resolution gene-trait associations. In some contexts, well-designed sequence capture experiments may allow for novel discovery via bulk segregants analysis alone.

4.3 Third-generation sequencing

Third-generation sequencing, typically involving long reads (>1 kb) of medium or high quality, represents a significant advancement for overcoming the sequence assembly challenges posed by auto-allooctoploid strawberry. The ability to sequence long contiguous sequences allows for relatively facile assembly in comparison to short reads from second-generation sequencing platforms. In the case of auto-allooctoploid strawberry, homeologous alleles may be more easily resolved from highly similar paralogs via comparisons of intergenic sequence contained in the same read. High-quality polyploid reference genomes derived from third-generation sequencing technologies, including that of auto-allooctoploid strawberry, will soon be widely and publicly available. Assembly of other octoploid strawberry genomes and transcriptomes will be made much easier by this octoploid reference. Additionally, in some contexts, less expensive second-generation reads may enjoy emergent feasibility for octoploid resequencing, via read-mapping to the octoploid reference. The application of third-generation sequencing on transcriptomes, or Iso-Seq, can provide direct and conclusive evidence for alternative splice variants, as well as the proportion of these splice variants. High-quality fruit transcriptomes, in combination with dense genetic maps, will also enable Expression QTL (eQTL) studies to determine the genetic basis for differential gene expression, between individuals.

4.4 Expression QTL

eQTL studies present the ability to determine the genetic basis of transcriptomic differences. A successful eQTL thus presents the strawberry breeder with the ability to control the transcription of specific genes via marker-assisted selection. Discoveries made with this approach may be very powerful in strawberry, when the genetic associations are derived from fruit transcriptomes of cultivars of dissimilar quality. Additionally, differentially expressed transcripts *not* associated to any genomic region may represent transcripts that are environmentally and/or developmentally regulated. This is a useful determination both for basic biological understanding and for filtering candidate genes for intentional crop improvement. Transcripts that demonstrate eQTL have an improved likelihood of representing an impactful genetic difference between cultivars, as their observed differential expression is not the result of a stochastic biochemical effect or common environmental response, but instead an intentional result of evolved and potentially selected genetic reregulation. Transcripts which demonstrate eQTL may be regarded as superior targets for study and potential genetic improvement.

5 High-throughput phenotyping

With the advent of rapid genomic and transcriptomic data generation on a grand scale, the bottleneck for gene-trait association increasingly falls to limitations in quantifying plant phenotypes (Araus and Cairns, 2014). High-throughput phenotyping solutions are becoming increasingly necessary, both to improve breeding through traditional selection and for integration with genomics data for discovery. Typical concepts for high-throughput and automated phenotyping involve the use of high-quality cameras and trained AI for image analysis. A proof-of-concept phenotyping device in strawberry was recently developed, involving a 3D camera system coupled with built-for-purpose image analysis software (He et al., 2017). This system reliably determined strawberry achene counts, calyx size, colour, height, length, width and volume. Continued improvement in 3D imaging, AI and foreseeable advances in robotics may soon automate large portions of the gene discovery and crop breeding pipeline. Additionally, high-throughput metabolomics phenotyping has advanced with the advent of sophisticated algorithms for accurately comparing raw GC-MS and LC-MS data (Tikunov et al., 2012; Lommen, 2009). High-quality comparisons of full-scan mass spectrometry data will assist breeders in identifying desirable biochemical phenotypes, and facilitate the association of strawberry metabolomics with genetics.

6 Future trends in research

Sustainable strawberry production must satisfy economic needs while meeting or exceeding social and environmental demands. Recent innovations in production and genetics are poised to meet this challenge, and provide even higher-quality fresh fruit with lower costs to producers and less environmental impact.

Decreasing production costs will come from implementation of precision agricultural innovations, such as automated drones or rovers to scout for pests and pathogens. Upon detection, highly localized treatments of specific crop protection strategies will be deployed, confining pests or diseases to a small part of the plot. Pests and pathogens will be detected and neutralized before symptoms are present, based on powerful sensors that detect threats before they become a problem. These approaches limit the spread of disease and pests, and preclude the need for prophylactic treatments with insecticides or antimicrobials, decreasing production costs for farmers and adverse impacts on the environment.

Automated harvesting is an area of very high potential economic value and is currently the subject of considerable research effort. Labour costs at harvest are substantial and only increasing, threatening economic sustainability. At the same time, the work is repetitive, physical and can negatively affect the health of farm labourers, especially over many seasons. The use of automation can speed harvests, eliminates the threat of labour shortages and high labour costs. Automation also meets the social sustainability need of not requiring people to perform hard physical labour in often inhospitable environments. Automation requires the development of specialized mechanical harvesters, but also varieties amenable to mechanical harvest. The co-evolution of these technologies will eliminate the need for manual labour in strawberry production.

The other major production trends will change the way strawberries are grown in and around urban centres. In The Netherlands, parts of Scandinavia and China, strawberries are grown in extensive greenhouse complexes that permit production throughout the year. These scenarios allow growers to produce fruit during high-value windows. Also, the post-harvest supply chain may be significantly shorter, bringing higher-quality products to consumers at decreased price. Strawberry, a small herbaceous plant that produces a high-value fruit, is well-suited for vertical farms and controlled-environment agriculture. In fact, strawberry is likely the highest-value fruit, relative to total plant size and time required for productive yield. Thus, some fraction of the strawberry agricultural sector may be expected to develop towards controlled-environment 'plant factories' capable of providing high-value fruits within urban settings.

Adjustment of genetics will also bring new traits that should favourably reflect in economic sustainability as well as mitigate environmental impacts. Non-commercial wild strawberries and odd heirloom varieties retain genes long lost during intense selection and breeding, and they serve as a living repository for traits that may be re-introduced to modern varieties. The genes (or genomic regions) underlying the foundations of sensory quality and disease resistance are now known, and it is time to coalesce them into superior genetic backgrounds that also maintain the traditionally coveted traits of yield, fruit size and shipping quality. Modern breeding technologies such as gene editing have been initiated in strawberry. These technologies will be used to introduce functional allelic variants that cannot be moved into elite backgrounds via breeding. Traditional genetic engineering tricks may also be used to improve plants as these technologies become more accepted. Genomics-based tools have spurred the detection of new genetic loci associated with traits of interest. In these cases marker-assisted breeding can allow the rapid introgression of wild or heirloom traits into already improved varieties.

Traits will be controllable by the farmer. The next generation will use signals in the form of light signals or possibly small molecules to hasten flowering, trigger ripening, or help plants protect themselves from abiotic or biotic stress by pre-emptively triggering defence pathways, priming plant responses. These innovations are all alive and well in the laboratory today, and simply await an opportunity to thrive in the field.

Together the next frontiers in strawberry will lead to better fruits that tantalize the consumer's interest. Fruits with new colours, flavours, aromas and textures are alive in the breeder's mind, and in some cases in the greenhouse. These attributes will increase demand, leading to higher prices and a competitive edge for growers compared to traditional germplasm. The interest in environmental sustainability, plus the decreased availability of resources like water and phosphorus, will favour the development of efficient varieties that make the best use of resources provided. Next-generation plants will feature stacks of protective genes that allow them to grow in the presence of disease pressure without exogenous pesticides.

Optimal sustainability will be obtained at the intersection of new production techniques and advanced genetics. It is imperative that both be examined and optimized in parallel, with strong collaborative efforts to ensure that all new solutions are complementary and not being developed in isolation.

7 Where to look for further information

For professional growers as well as home gardeners, university extension programmes provide excellent informational resources for regional agriculture. For instance, the University of Florida provides an online repository of extension resources for strawberry, encompassing a variety of topics in strawberry agriculture. Principal topics include current cultural practices, common diseases and prevention, fertilization and nutrition, pest management, post-harvest and handling, common cultivated varieties and strategies for optimal water management. Growers, researchers and consumers may also review articles on DNA technologies used to create local varieties, and information for growing strawberries in the Florida home garden.

For researchers, the Genome Database for Rosaceae (GDR) maintains current genomic and molecular marker data for various strawberry species, including *Fragaria × ananassa* and *Fragaria vesca*. Genome assemblies and annotations may be accessed and downloaded via FTP or browsed directly on GDR. A comprehensive list of publically available raw sequencing data in *Fragaria* spp. is also maintained on GDR, and linked for download via the NCBI Short Read Archive. Basic tools for sequence comparison are also provided in-browser. Marker sequences and documentation for the Axiom genome-wide strawberry genotyping array are available on the Affymetrix website. Free and open-sourced tools for genome-wide association of traits, including GAPIT and FlexQTL, are available with documentation.

8 References

Aliasgarian, S., Ghassemzadeh, H. R., Moghaddam, M. and Ghaffari, H. 2015. Mechanical damage of strawberry during harvest and postharvest operations. *Acta Technologica Agriculturae*, 18, 1–5.

Antanaviciute, L., Šurbanovski, N., Harrison, N., McLeary, K. J., Simpson, D. W., Wilson, F., Sargent, D. J. and Harrison, R. J. 2015. Mapping QTL associated with *Verticillium dahliae* resistance in the cultivated strawberry (*Fragaria × ananassa*). *Horticulture Research*, 2, 15009.

Araus, J. L. and Cairns, J. E. 2014. Field high-throughput phenotyping: The new crop breeding frontier. *Trends in Plant Science*, 19, 52–61.

Asao, H. G., Nishizawa, Y., Arai, S., Sato, T., Hirai, M., Yoshida, K., Shinmyo, A. and Hibi, T. 1997. Enhanced resistance against a fungal pathogen *Sphaerotheca fumuli* in transgenic strawberry expressing a rice chitinase gene. *Plant Biotechnology*, 14, 145–9.

Bachman, P. M., Huizinga, K. M., Jensen, P. D., Mueller, G., Tan, J., Uffman, J. P. and Levine, S. L. 2016. Ecological risk assessment for DvSnf7 RNA: A plant-incorporated protectant with targeted activity against western corn rootworm. *Regulatory Toxicology and Pharmacology*, 81, 77–88.

Bolda, M., Tourte, L., Klonsky, K. and Bervejillo, J. E. 2003. Sample costs to produce organic strawberries: Central coast. *Santa Cruz and Monterey Counties University of California Cooperative Extension*.

Bolognesi, R., Ramaseshadri, P., Anderson, J., Bachman, P., Clinton, W., Flannagan, R., Ilagan, O., Lawrence, C., Levine, S., Moar, W., Mueller, G., Tan, J., Uffman, J., Wiggins, E., Heck, G. and Segers, G. 2012. Characterizing the mechanism of action of double-stranded RNA activity against western corn rootworm (*Diabrotica virgifera virgifera* LeConte). *PLoS ONE*, 7, e47534.

Bood, K. G. and Zabetakis, I. 2002. The biosynthesis of strawberry flavor (II): Biosynthetic and molecular biology studies. *Journal of Food Science*, 67, 2–8.

Booster, D. E. and Kirk, D. E. 1968. State of the art and future outlook for mechanical strawberry harvesting. Corvallis, OR: Agricultural Experiment Station, Oregon State University.

Braun, P. G. and Hildebrand, P. D. 2013. Effect of sugar alcohols, antioxidants and activators of systemically acquired resistance on severity of bacterial angular leaf spot (*Xanthomonas fragariae*) of strawberry in controlled environment conditions. *Canadian Journal of Plant Pathology*, 35, 20–6.

Calis, O., Cekic, C., Soylu, S. and Tör, M. 2015. Identification of new resistance sources from diploid wild strawberry against powdery mildew pathogen. *Pakistan Journal of Agricultural Sciences*, 52, 677–83.

Cantu, D., Vicente, A. R., Greve, L. C., Labavitch, J. M. and Powell, A. L. T. 2007. Genetic determinants of textural modifications in fruits and role of cell wall polysaccharides and defense proteins in the protection against pathogens. *Fresh Produce*, 1, 101–10.

Carrière, Y., Crickmore, N. and Tabashnik, B. E. 2015. Optimizing pyramided transgenic Bt crops for sustainable pest management. *Nature Biotechnology*, 33, 161–8.

Chalavi, V., Tabaeizadeh, Z. and Thibodeau, P. 2003. Enhanced resistance to *Verticillium dahliae* in transgenic strawberry plants expressing a *Lycopersicon chilense* chitinase gene. *Journal of the American Society for Horticultural Science*, 128, 747–53.

Chambers, A., Evans, S. and Folta, K. 2013. Methyl anthranilate and gamma decalactone inhibit strawberry pathogen growth and achene germination. *Journal of Agricultural and Food Chemistry*, 61, 12625–33.

Charles, D. 2017. [Interview]. NPR Radio. https://www.npr.org/sections/thesalt/2017/02/16/515380213/why-ditching-nafta-could-hurt-americas-farmers-more-than-mexicos

Chu, C.-C., Sun, W., Spencer, J. L., Pittendrigh, B. R. and Seufferheld, M. J. 2014. Differential effects of RNAi treatments on field populations of the western corn rootworm. *Pesticide Biochemistry and Physiology*, 110, 1–6.

Cools, H. J. and Hammond-Kosack, K. E. 2013. Exploitation of genomics in fungicide research: Current status and future perspectives. *Molecular Plant Pathology*, 14, 197–210.

Cordova, L. G., Amiri, A. and Peres, N. A. 2017. Effectiveness of fungicide treatments following the Strawberry Advisory System for control of Botrytis fruit rot in Florida. *Crop Protection*, 100, 163–7.

De Anda, J. and Shear, H. 2017. Potential of vertical hydroponic agriculture in Mexico. *Sustainability*, 9, 140.

Denisen, E. L. and Buchele, W. F. 1967. Mechanical harvesting of strawberries. *Proceedings of the American Society for Horticultural Science*, 91, 267–73.

Dianez, F., Santos, M., Blanco, R. and Tello, J. C. 2002. Fungicide resistance in *Botrytis cinerea* isolates from strawberry crops in Huelva (southwestern Spain). *Phytoparasitica*, 30, 529–34.

Elad, Y., Malathrakis, N. E. and Dik, A. J. 1996. Biological control of Botrytis-incited diseases and powdery mildews in greenhouse crops. *Crop Protection*, 15, 229–40.

Faedi, W., Mourgues, F. and Rosati, C. 2000. Strawberry breeding and varieties: Situation and perspectives. *Acta Horticulturae*, 567, 51–9.

Fang, X., Phillips, D., Verheyen, G., Li, H., Sivasithamparam, K. and Barbetti, M. J. 2012. Yields and resistance of strawberry cultivars to crown and root diseases in the field, and cultivar responses to pathogens under controlled environment conditions. *Phytopathologia Mediterranea*, 51, 69–84.

FAOSTAT. 2016. http://www.fao.org/faostat/en/#search/strawberries

Fernandez, G. E., Butler, L. M. and Louws, F. J. 2001. Strawberry growth and development in an annual plasticulture system. *HortScience*, 36, 1219–23.

Fernández-Cruz, M. L., Mansilla, M. L. and Tadeo, J. L. 2010. Mycotoxins in fruits and their processed products: Analysis, occurrence and health implications. *Journal of Advanced Research*, 1, 113–22.

Fernández-Ortuño, D., Chen, F. and Schnabel, G. 2012. Resistance to pyraclostrobin and boscalid in *Botrytis cinerea* isolates from strawberry fields in the Carolinas. *Plant disease*, 96, 1198–203.

Fernández-Ortuño, D., Grabke, A., Li, X. and Schnabel, G. 2015. Independent emergence of resistance to seven chemical classes of fungicides in *Botrytis cinerea*. *Phytopathology*, 105, 424–32.

Fletcher, S. 1917. *Strawberry Growing*. New York, NY: Macmillan Co.

Folta, K. M. and Davis, T. M. 2006. Strawberry genes and genomics. *Critical Reviews in Plant Sciences*, 25, 399–415.

Folta, K. M. and Klee, H. J. 2016. Sensory sacrifices when we mass-produce mass produce. *Horticulture Research*, 3. doi:10.1038/hortres.2016.32.

Forcelini, B. B., Seijo, T. E., Amiri, A. and Peres, N. A. 2016. Resistance in strawberry isolates of *Colletotrichum acutatum* from Florida to quinone-outside inhibitor fungicides. *Plant Disease*, 100, 2050–6.

Gatewood, B. 2017. Strawberries remain at top of pesticide list, report says (Online). http://www.cnn.com/2017/03/08/health/dirty-dozen-2017/index.html (accessed).

Giménez-Ferrer, R. M., Erb, W. A., Bishop, B. L. and Scheerens, J. C. 1994. Host-pest relationships between the two-spotted spider mite (Acari: Tetranychidae) and strawberry cultivars with differing levels of resistance. *Journal of Economic Entomology* 87(1), 168–75.

Girgenti, V., Peano, C., Baudino, C. and Tecco, N. 2014. From 'farm to fork' strawberry system: Current realities and potential innovative scenarios from life cycle assessment of non-renewable energy use and green house gas emissions. *Science of the Total Environment*, 473–4, 48–53.

Goff, S. A. and Klee, H. J. 2006. Plant volatile compounds: Sensory cues for health and nutritional value? *Science*, 311, 815–19.

Guan, Z., Wu, F., Roka, F. and Whidden, A. 2015. Agricultural labor and immigration reform. *Choices*, 30(4), 1–9.

Guidarelli, M., Zoli, L., Orlandini, A., Bertolini, P. and Baraldi, E. 2014. The *mannose-binding lectin* gene *FaMBL1* is involved in the resistance of unripe strawberry fruits to *Colletotrichum acutatum*. *Molecular Plant Pathology*, 15, 832–40.

Gunady, M. G. A., Biswas, W., Solah, V. A. and James, A. P. 2012. Evaluating the global warming potential of the fresh produce supply chain for strawberries, romaine/cos lettuces (*Lactuca sativa*), and button mushrooms (*Agaricus bisporus*) in Western Australia using life cycle assessment (LCA). *Journal of Cleaner Production*, 28, 81–7.

Guthman, J. 2016. Strawberry growers wavered over methyl iodide, feared public backlash. *California Agriculture*, 70, 124–9.

Hayashi, S., Shigematsu, K., Yamamoto, S., Kobayashi, K., Kohno, Y., Kamata, J. and Kurita, M. 2010. Evaluation of a strawberry-harvesting robot in a field test. *Biosystems Engineering*, 105, 160–71.

Haymes, K. M., Henken, B., Davis, T. M. and Vandeweg, W. E. 1997. Identification of RAPD markers linked to a *Phytophthora fragariae* resistance gene (*Rpf1*) in the cultivated strawberry. *Theoretical and Applied Genetics*, 94, 1097–101.

He, J. Q., Harrison, R. J. and Li, B. 2017. A novel 3D imaging system for strawberry phenotyping. *Plant Methods*, 13, 93.

Head, G. P., Carroll, M. W., Evans, S. P., Rule, D. M., Willse, A. R., Clark, T. L., Storer, N. P., Flannagan, R. D., Samuel, L. W. and Meinke, L. J. 2017. Evaluation of SmartStax and SmartStax PRO maize against western corn rootworm and northern corn rootworm: Efficacy and resistance management. *Pest Management Science*, 73, 1883–99.

Hernández-Muñoz, P., Almenar, E., Valle, V. D., Velez, D. and Gavara, R. 2008. Effect of chitosan coating combined with postharvest calcium treatment on strawberry (*Fragaria × ananassa*) quality during refrigerated storage. *Food Chemistry*, 110, 428–35.

Huang, G., Allen, R., Davis, E. L., Baum, T. J. and Hussey, R. S. 2006. Engineering broad root-knot resistance in transgenic plants by RNAi silencing of a conserved and essential root-knot nematode parasitism gene. *Proceedings of the National Academy of Sciences*, 103, 14302-6.

ISPRA (Istituto Superiore per la Protezione e la Ricerca Ambientale). 2013. Italian greenhouse gas inventory, 1990–2011 – National Inventory Report no. 177/2013. Available from: http://www.isprambiente.gov.it/files/pubbli-cazioni/rapporti/Rapporto_177_2013.pdf

Kanto, T., Matsuura, K., Ogawa, T., Yamada, M., Ishiwata, M., Usami, T. and Amemiya, Y. 2014. New UV-B lighting system controls powdery mildew of strawberry. *Acta Horticulturae*, 1049, 655–60.

Khoshnevisan, B., Rafiee, S. and Mousazadeh, H. 2013. Environmental impact assessment of open field and greenhouse strawberry production. *European Journal of Agronomy*, 50, 29–37.

Ku, V. V. V., Wills, R. B. H. and Ben-Yehoshua, S. 1999. 1-Methylcyclopropene can differentially affect the postharvest life of strawberries exposed to ethylene. *HortScience*, 34, 119-20.

Kudo, R. and Yamamoto, K. 2016. Induction of plant disease resistance and other physiological responses by green light illumination. In: Kozai, T., Fujiwara, K. and Runkle, E. S. (Eds), *LED Lighting for Urban Agriculture*. Singapore: Springer Singapore.

Leroch, M., Plesken, C., Weber, R. W. S., Kauff, F., Scalliet, G. and Hahn, M. 2013. Gray mold populations in German strawberry fields are resistant to multiple fungicides and dominated by a novel clade closely related to Botrytis cinerea. *Applied and Environmental Microbiology*, 79, 159–67.

Leroux, P., Gredt, M., Leroch, M. and Walker, A.-S. 2010. Exploring mechanisms of resistance to respiratory inhibitors in field strains of *Botrytis cinerea*, the causal agent of gray mold. *Applied and Environmental Microbiology*, 76, 6615–30.

Lieten, P. 2013. Advances in strawberry substrate culture during the last twenty years in the Netherlands and Belgium. *International Journal of Fruit Science*, 13, 84–90.

Lommen, A. 2009. MetAlign: Interface-driven, versatile metabolomics tool for hyphenated full-scan mass spectrometry data preprocessing. *Analytical Chemistry*, 81, 3079–86.

Maas, J. L. 2004. Strawberry disease management. In: Naqvi, S. A. M. H. (Eds), *Diseases of Fruits and Vegetables: Volume II*. Dordrecht, The Netherlands: Springer.

Machado, A. K., Brown, N. A., Urban, M., Kanyuka, K. and Hammond-Kosack, K. 2017. RNAi as an emerging approach to control Fusarium Head Blight disease and mycotoxin contamination in cereals. *Pest Management Science*, 74, 790–9.

Majumdar, R., Rajasekaran, K. and Cary, J. W. 2017. RNA Interference (RNAi) as a potential tool for control of mycotoxin contamination in crop plants: Concepts and considerations. *Frontiers in Plant Science*, 8, 200.

Mangandi, J., Verma, S., Osorio, L., Peres, N. A., Van De Weg, E. and Whitaker, V. M. 2017. Pedigree-based analysis in a multiparental population of octoploid strawberry reveals QTL alleles conferring resistance to *Phytophthora cactorum*. *G3: Genes, Genomes, Genetics*, 7, 1707–19.

Mazzola, M., Agostini, A. and Cohen, M. F. 2017. Incorporation of *Brassica* seed meal soil amendment and wheat cultivation for control of *Macrophomina phaseolina* in strawberry. *European Journal of Plant Pathology*, 149, 57–71.

Mercado, J. A., Barceló, M., Pliego, C., Rey, M., Caballero, J. L., Muñoz-Blanco, J., Ruano-Rosa, D., López-Herrera, C., De Los Santos, B., Romero-Muñoz, F. and Pliego-Alfaro, F. 2015. Expression of the β-1,3-glucanase gene *bgn13.1* from *Trichoderma harzianum* in strawberry increases tolerance to crown rot diseases but interferes with plant growth. *Transgenic Research*, 24, 979–89.

Mok, H.-F., Williamson, V. G., Grove, J. R., Burry, K., Barker, S. F. and Hamilton, A. J. 2014. Strawberry fields forever? Urban agriculture in developed countries: A review. *Agronomy for Sustainable Development*, 34, 21–43.

Mossler, M. 2015. Florida crop/pest management profiles: Strawberry. *University of Florida EDIS*, pp. 1–8.

Mossler, M. and Nesheim, O. 2007. Strawberry pest management strategic plan (PMSP). *University of Florida EDIS*, pp. 1–17.

National Academies of Sciences, Engineering, and Medicine. 2016. *Genetically Engineered Crops: Experiences and Prospects*. National Academies Press.

Noling, J. W. 1999. Nematode management in strawberries. University of Florida Cooperative Extension Service, Institute of Food and Agriculture Sciences, EDIS.

Olbricht, K., Grafe, C., Weiss, K. and Ulrich, D. 2008. Inheritance of aroma compounds in a model population of *Fragaria × ananassa* Duch. *Plant Breeding*, 127, 87–93.

Pavan, W., Fraisse, C. W. and Peres, N. A. 2011. Development of a web-based disease forecasting system for strawberries. *Computers and Electronics in Agriculture*, 75, 169–75.

Paynter, M. 2015. Breeding resistance in strawberry cultivars for *Fusarium oxysporum* f. sp. *fragariae*. MPhil Thesis, School of Agriculture and Food Sciences, The University of Queensland.

Pensala, O., Niskanen, A. and Lindroth, S. 1978. Aflatoxin production in black currant, blueberry and strawberry jams. *Journal of Food Protection*, 41, 344–7.

Peres, N. 2015. 2015 Florida plant disease management guide: Strawberry. *University of Florida IFAS EDIS Document* (Online).

Pillet, J., Chambers, A. H., Barbey, C., Bao, Z., Plotto, A., Bai, J., Schwieterman, M., Johnson, T., Harrison, B. and Whitaker, V. M. 2017. Identification of a methyltransferase catalyzing the final step of methyl anthranilate synthesis in cultivated strawberry. *BMC Plant Biology*, 17, 147.

Poling, E. B. 2005. An introductory guide to strawberry plasticulture. Department of Horticultural Science, North Carolina State, 30p.

Pradhan, P., Fischer, G., Van Velthuizen, H., Reusser, D. E. and Kropp, J. P. 2015. Closing yield gaps: How sustainable can we be? *PLoS ONE*, 10, e0129487.

Pérez, A. G., Sanz, C., Ríos, J. J., Olias, R. and Olías, J. M. 1999. Effects of ozone treatment on postharvest strawberry quality. *Journal of Agricultural and Food Chemistry*, 47, 1652–6.

Rabobank. 2015. US berries – increasing consumption is not a silver bullet. https://research.rabobank.com/far/en/sectors/regional-food-agri/us-berries.html

Razza, F., Degli Innocenti, F., Peano, C., Girgenti, V. and Bounous, M. 2012. Improving the efficiency of agricultural and food systems by supply chain: Experience from northern Italy. *Proceedings 8th International Conference on Life Cycle Assessment in the Agri-food Sector, INRA-Agrocampus Rennes*, p. 866.

Reddy, M. V. B., Belkacemi, K., Corcuff, R., Castaigne, F. and Arul, J. 2000. Effect of pre-harvest chitosan sprays on post-harvest infection by Botrytis cinerea and quality of strawberry fruit. *Postharvest Biology and Technology*, 20, 39–51.

Roach, J. A., Verma, S., Peres, N. A., Jamieson, A. R., Van De Weg, W. E., Bink, M. C. A. M., Bassil, N. V., Lee, S. and Whitaker, V. M. 2016. FaRXf1: A locus conferring resistance to angular leaf spot caused by *Xanthomonas fragariae* in octoploid strawberry. *Theoretical and Applied Genetics*, 129, 1191–201.

Romanazzi, G., Feliziani, E., Santini, M. and Landi, L. 2013. Effectiveness of postharvest treatment with chitosan and other resistance inducers in the control of storage decay of strawberry. *Postharvest Biology and Technology*, 75, 24–7.

Schestibratov, K. A. and Dolgov, S. V. 2005. Transgenic strawberry plants expressing a thaumatin II gene demonstrate enhanced resistance to *Botrytis cinerea*. *Scientia Horticulturae*, 106, 177–89.

Silva, K. J. P., Brunings, A., Peres, N. A., Mou, Z. and Folta, K. M. 2015. The Arabidopsis NPR1 gene confers broad-spectrum disease resistance in strawberry. *Transgenic Research*, 24, 693–704.

Silva, K. J. P., Brunings, A. M., Pereira, J. A., Peres, N. A., Folta, K. M. and Mou, Z. 2017. The Arabidopsis *ELP3/ELO3* and *ELP4/ELO1* genes enhance disease resistance in *Fragaria vesca* L. *BMC plant biology*, 17, 230.

Sivakumar, D. and Bautista-Baños, S. 2014. A review on the use of essential oils for postharvest decay control and maintenance of fruit quality during storage. *Crop Protection*, 64, 27–37.

Smith, L. E., Prendergast, A. J., Turner, P. C., Humphrey, J. H. and Stoltzfus, R. J. 2017. Aflatoxin exposure during pregnancy, maternal anemia, and adverse birth outcomes. *The American Journal of Tropical Medicine and Hygiene*, 96, 770–6.

Sogvar, O. B., Saba, M. K. and Emamifar, A. 2016. Aloe vera and ascorbic acid coatings maintain postharvest quality and reduce microbial load of strawberry fruit. *Postharvest Biology and Technology*, 114, 29–35.

Stegmeir, T. L., Finn, C. E., Warner, R. M. and Hancock, J. F. 2010. Performance of an elite strawberry population derived from wild germplasm of *Fragaria chiloensis* and *F. virginiana*. *HortScience*, 45, 1140–5.

Strand, L. L. 2008. *Integrated Pest Management for Strawberries*. UCANR Publications.

Suh, D. H., Guan, Z. and Khachatryan, H. 2017. The impact of Mexican competition on the US strawberry industry. *International Food and Agribusiness Management Review*, 20, 591–604.

Tabatabaie, S. M. H. and Murthy, G. S. 2016. Cradle to farm gate life cycle assessment of strawberry production in the United States. *Journal of Cleaner Production*, 127, 548–54.

Takagaki, M., Ozaki, M., Fujimoto, S. and Fukumoto, S. 2014. Development of a novel fungicide, pyribencarb. *Journal of Pesticide Science*, 39, 177–8.

Tan, J., Levine, S. L., Bachman, P. M., Jensen, P. D., Mueller, G. M., Uffman, J. P., Meng, C., Song, Z., Richards, K. B. and Beevers, M. H. 2016. No impact of DvSnf7 RNA on honey bee (*Apis mellifera* L.) adults and larvae in dietary feeding tests. *Environmental Toxicology and Chemistry*, 35, 287–94.

Tiezhong, Z. and Tianjuan, Z. 2004. Strawberry harvesting robot: I. Segmentation of strawberry image by BP neural network [J]. *Journal of China Agricultural University*, 4, 015.

Tikunov, Y. M., Laptenok, S., Hall, R. D., Bovy, A. and De Vos, R. C. H. 2012. MSClust: A tool for unsupervised mass spectra extraction of chromatography-mass spectrometry ion-wise aligned data. *Metabolomics*, 8, 714–18.

Tokarskyy, O., Schneider, K. R., Berry, A., Sargent, S. A. and Sreedharan, A. 2015. Sanitizer applicability in a laboratory model strawberry hydrocooling system. *Postharvest Biology and Technology*, 101, 103–6.

Tomazeli, V. N., Marchese, J. A., Danner, M. A., Perboni, A. T., Finatto, T. and Crisosto, C. H. 2016. Improved resistance to disease and mites in strawberry, through the use of acibenzolar-S-methyl and harpin to enhance photosynthesis and phenolic metabolism. *Theoretical and Experimental Plant Physiology*, 28, 287–96.

Ulrich, D. and Olbricht, K. 2014. Diversity of metabolite patterns and sensory characters in wild and cultivated strawberries. *Journal of Berry Research*, 4, 11–17.

USDA. 2018. Non-citrus fruits and nuts 2017 summary. National Agricultural Statistics Service. ISSN: 1948-2698.

Vandendriessche, T., Geerts, P., Membrebe, B. N., Keulemans, J., Nicolaï, B. M. and Hertog, M. L. A. T. M. 2013. Journeys through aroma space: A novel approach towards the selection of aroma-enriched strawberry cultivars in breeding programmes. *Plant Breeding*, 132, 217–23.

Vaughn, S. F., Spencer, G. F. and Shasha, B. S. 1993. Volatile compounds from raspberry and strawberry fruit inhibit postharvest decay fungi. *Journal of food Science*, 58, 793–6.

Vellicce, G. R., Ricci, J. C. D., Hernandez, L. and Castagnaro, A. P. 2006. Enhanced resistance to Botrytis cinerea mediated by the transgenic expression of the chitinase gene *ch5B* in strawberry. *Transgenic Research*, 15, 57–68.

Vining, K. J., Davis, T. M., Jamieson, A. R. and Mahoney, L. L. 2015. Germplasm resources for verticillium wilt resistance breeding and genetics in strawberry (*Fragaria*). *Journal of Berry Research*, 5, 183–95.

Wang, M. and Jin, H. 2017. Spray-induced gene silencing: A powerful innovative strategy for crop protection. *Trends in Microbiology*, 25, 4–6.

Wang, M., Weiberg, A., Lin, F.-M., Thomma, B. P. H. J., Huang, H.-D. and Jin, H. 2016. Bidirectional cross-kingdom RNAi and fungal uptake of external RNAs confer plant protection. *Nature Plants*, 2, 16151.

Whitaker, V. M. 2011. Applications of molecular markers in strawberry. *Journal of Berry Research*, 1, 115–27.

Whitaker, V. M., Lee, S., Osorio, L. F., Verma, S., Roach, J. A., Mangandi, J., Noh, Y. H., Gezan, S. and Peres, N. 2016. Advances in strawberry breeding at the University of Florida. *Acta Horticulturae*, 1156, 1–6.

Will, F. and Krüger, E. 1999. Fungicide residues in strawberry processing. *Journal of Agricultural and Food Chemistry*, 47, 858–61.

Williams, A., Pell, E., Webb, J., Moorhouse, E. and Audsley, E. 2008. Strawberry and tomato production for the UK compared between the UK and Spain. *Proceedings of the 6th International Conference on LCA in the Agri-food Sector*, pp. 254–62.

Wu, F. and Guan, Z. 2015. Modeling the interactions of strawberry commodity and labor markets in the US and Mexico. *Agricultural & Applied Economics Association and Western Agricultural Economics Association Annual Meeting, San Francisco, CA.* Retrieved from http://ageconsearch. umn. edu/bitstream/205887/2/Modeling. Vol. 20.

Xiong, L., Shen, Y.-Q., Jiang, L.-N., Zhu, X.-L., Yang, W.-C., Huang, W. and Yang, G.-F. 2015. Succinate dehydrogenase: An ideal target for fungicide discovery. In: *Discovery and Synthesis of Crop Protection Products.* American Chemical Society.

Xu, Y., Goodhue, R. E., Chalfant, J. A., Miller, T. and Fennimore, S. A. 2017. Economic viability of steam as an alternative to preplant soil fumigation in California strawberry production. *HortScience*, 52, 401–7.

Yoshida, H., Hikosaka, S., Goto, E., Takasuna, H. and Kudou, T. 2012. Effects of light quality and light period on flowering of everbearing strawberry in a closed plant production system. *Acta Horticulturae*, 956, 107–12.

Zhang, L., Gao, Z. and Vassalos, M. 2016. Sustainable consumer groups and their willingness to pay for tangible and intangible attributes of fresh strawberries. *Selected Paper presented at the Southern Agricultural Economics Association Annual Meeting*.

Ziedan, E.-S. H. and Farrag, E. S. H. 2016. Chemical fumigants as alternatives methyl bromide for soil disinfestation of plant integrated pest management. *International Journal of Agricultural Technology*, 12, 321–8.

Chapter 11

Strawberries: a case study of how evolving market expectations impact sustainability

M. P. Pritts, Cornell University, USA; and T. M. Sjulin, formerly Driscoll Strawberry Associates, USA

1 Introduction

Strawberry (*Fragaria* × *ananassa*) production in much of the world today would be unrecognizable to a farmer 150 years ago. Then, strawberry production was scattered across many states, and most towns had strawberry farms nearby. The highly perishable nature of the fruit required strawberries to be produced close to markets, and consumers were accustomed to eating the fruit for only a few weeks in spring. Centres of concentrated production began to develop, mostly along railroad lines that provided access to urban markets. For example, after Henry Plant extended a railway to Tampa, Florida, a new strawberry industry developed around Plant City (named in his honour) in the late 1800s for winter production of strawberries. These berries were shipped by express trains in iced 'pony refrigerators' up the eastern seaboard. Arkansas, Missouri, Louisiana, Tennessee, Virginia and Maryland also had access to rail shipping points to the major US cities, so concentrated strawberry production developed in these states. In the 1920s, over half of the US market production was delivered by rail in carload equivalents (Strowbridge, 1930). Shipments from California at that time were mainly within the state or to major west coast cities such as Portland and Seattle.

http://dx.doi.org/10.19103/AS.2018.0040.26

The California strawberry industry's development is an interesting case study of the environmental trade-offs that arise when market expectations evolve. In this particular example, increasing consumer demand and expectation for berries outside of traditional harvest times encouraged growers to introduce and reinforce many unsustainable practices.

The moderate climate along the central and south coasts of California is more favourable for strawberry production than for any other state in the United States (US), and yields have long been higher than in other production areas. Growers in California originally used a traditional perennial matted row production system similar to other areas of the US. In the early decades of the twentieth century, growers along the central California coast discovered that some short-day cultivars from the eastern US, such as 'Nich Ohmer' and 'Banner', had a prolonged flower initiation period that resulted in a longer production season and higher yields. Unfortunately, 'Nich Ohmer' plants were very susceptible to *Verticillium* wilt, and 'Banner' had poor runnering ability and was highly susceptible to the aphid-borne virus complex known as 'yellows' or 'xanthosis', but these two cultivars opened the door to the possibility of extended production seasons (Wilhelm and Sagen, 1974).

2 Development of an annual strawberry production system

In the 1920s, the University of California (UC) initiated pathology research and a breeding programme focused on strawberry production problems. In 1945, the breeding programme introduced its first set of cultivars (most notably 'Shasta' and 'Lassen') that combined the prolonged flower initiation response of 'Banner' and 'Nich Ohmer' with an improved level of virus tolerance. Strawberry nurseries moved out of the coastal area to the north end of the Sacramento River Valley to isolate plants from virus infected material, and to stimulate runnering from increased exposure to cooler temperatures. Improved virus detection and elimination techniques developed by UC Berkeley researchers supported clean stock development. Research by Steve Wilhelm and others led to the widespread use of soil fumigation to control *Verticillium* wilt and other soil-borne pests and diseases. Yields dramatically improved following fumigation, and it became the standard practice (Wilhelm and Sagen, 1974). Production methods shifted away from the traditional perennial matted row system (Fig. 1) to a summer-planted raised bed hill system of plants set at higher densities (Fig. 2) that were replaced annually (along the south coast) or after two to four years (along the central coast) (Darrow, 1966). Production research by Victor Voth at the UC South Coast Station near Irvine further improved the amounts and timing of production by using raised bed plasticulture and drip irrigation. During the 1950s and 1960s, both the UC and Driscoll's private breeding programs introduced cultivars better suited to

Figure 1 Matted row strawberry production system used in colder climates around the world.

Figure 2 Older summer-planting system used in California involves extensive runner removal and hand labour.

long-distance shipping. The industry capitalized on the development of the US interstate highway system, and shipments in refrigerated trucks to distant markets rapidly increased.

During this time, strawberry plants were propagated in low elevation nurseries, dug when dormant in January and stored until planting in summer ('frigo plants'). In the 1970s through 1980s, the California industry moved away from summer planting using cold stored plants to annual 'winter' plantings of freshly dug plants grown at high elevation nurseries in northern California where flower initiation had already occurred. These high elevation plants were the clonal offspring of low elevation nursery plants that were stimulated to produce runners under warm temperatures. After transplanting high elevation plants into production fields, plants continued to initiate flowers under the short days of winter and fruited earlier than frigo plants. Production volumes in the winter-planting system (Fig. 3) were similar to the summer-planting system, but the production peaks and valleys were less severe, and product flow from California became much more consistent, which assisted with marketing the fruit. Royce Bringhurst and Voth, the UC breeders at the time, introduced the first commercially successful 'day-neutral' cultivar Selva in 1983. This further smoothed the production peaks and allowed California growers to ship greater season-total volumes without increasing peak volumes (Sjulin, 2003). Researchers at Driscoll's introduced a new production system in the 1990s using cold-stored plants of 'everbearing' cultivars planted in mid-summer along the

Figure 3 Winter-planting system in California allows crops to be produced for most months of the year.

south coast. Fruit production in this system was concentrated around the mid-October through mid-December period, at a time when US fresh supplies were at their lowest. The result of these activities was a large increase in what was considered 'off-season' production, without pushing too much product into the market at peak periods (Welch et al., 1982).

Growers in Florida, where there was a long history of producing strawberries, adopted California growing practices to provide for the winter market before the timing of California production volumes reached high levels (Fig. 4). In the 1990s and 2000s, the University of Florida (UF) breeding program led by Craig Chandler introduced a series of improved cultivars that were earlier-yielding than the current industry standards and had better quality. Florida has no high elevation locations suitable for disease-free plant production. *Colletotrichum* infections are difficult to control there, so plant suppliers were established in the mountains of North Carolina, high elevation nurseries in California and in some Canadian provinces where disease pressure is less and the shorter days and cooler temperatures in autumn promote flower bud initiation. All Florida strawberry plants now originate from out-of-state nurseries.

3 Shifting markets

The overall growth in demand for California strawberries meant that winter supply from Florida could increase as well, as their production timing would

Figure 4 Winter-planting system (October planting) adapted to Florida conditions to target production during the winter months.

complement, rather than conflict, with that of California. In central Mexico, Driscoll researchers worked with strawberry growers in Jalisco and Michoacán states during the 1990s to develop a high tunnel system of winter production. This system, introduced around 2000, grew quickly in area and was copied by other grower-shipper entities. Today, the winter shipments from central Mexico are nearly equal to the Florida volume, further increasing US supplies in the December through March period (Sjulin, 2010).

During the last two decades, large supermarket chains have been acquiring smaller chains at a rapid rate and consolidating buying offices. This change favoured the larger shippers, as the supermarket chains wanted to deal directly with fewer shippers for supply and thereby reduce their purchases from terminal markets that had long been supplied, in part, by local producers. Point-of-purchase data collected by the chains indicated that there was a growing expectation and demand for fresh strawberries all year-round, driven in part by changes in the supply curve as well as consumer awareness of the health benefits of strawberry consumption.

4 Increasing inputs with expansion of annual production in favourable locations

The end result of the past century's changes is a radical shift from a geographically diverse, low-input perennial matted row production system with seasonal supplies, to year-round production from highly intensive annual production systems concentrated in three regions of North America. These annual production systems require large inputs of water, fertilizer, plastics and energy for water pumping, ground preparation, transporting plants from nurseries to production fields and transporting fruit to distant markets (Bolda et al., 2016). For example, fertility needs in matted row systems approach 112 kg/ha (100 lb/acre) nitrogen and, because the plants are small in stature and planted during cool weather in spring, water use is modest. In contrast, nitrogen needs in annual plasticulture systems approach 450 kg/ha (400 lb/acre) and, because transplanting occurs during a warm season, overhead irrigation often runs for up to 8 h/day for 2 weeks, resulting in the use of more than 4.7 million L/ha (500 000 gal/acre) of water under Florida conditions. Florida groundwater sources are rapidly becoming saline, rendering some traditional supplies unusable. Along the coast of California, growers rely on either overdrafts of groundwater or low-cost water diverted from government-funded supply systems.

In other regions of the world (e.g. Spain), the annual plasticulture system has been adopted with even further modification, such as the use of plastic tunnels to control climate and manage disease. A long fruiting season in a mild climate requires the use of frequent pesticide applications. In Florida, berries are treated with fungicides up to twice per week throughout the flowering and harvest season, depending on weather. Newer pests, such as the spotted winged drosophila (*Drosophila suzukii*), have forced growers to increase pesticide applications, especially in areas that produce late summer fruit or where fruit must be certified free of insect larvae.

In 1989, the Montreal Protocol went into effect. The goal of this international treaty was to reduce the use of halogenated hydrocarbons to allow the ozone layer above the Earth to recover. Methyl bromide, a common fumigant used in strawberry production, was identified as one of the major contributors to ozone depletion and its use was to be phased out by 2005. However, the strawberry industry made the case that methyl bromide fumigation was a critical use so should be exempted from the 2005 deadline. For each of the next 11 years, strawberry growers were granted an exemption to continue using the product. Even though the industry knew for 20 years that a phase-out was inevitable, fumigation practices mostly remained the same. Today, fumigation is still widely used in annual systems, but with alternative fumigants to methyl bromide.

5 Sustainability of the current model of strawberry production

The desire to have strawberries available throughout the year has been the driving force behind annual plasticulture production (Fig. 5), but unfortunately, this system has incorporated many unsustainable practices and the strawberry

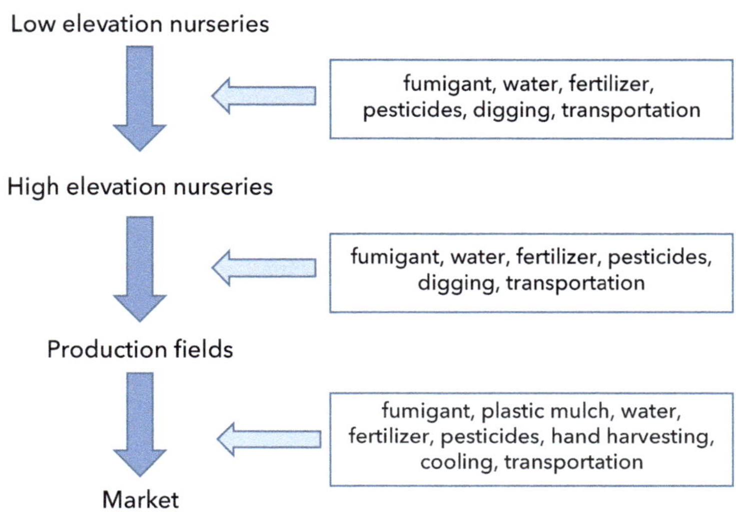

Figure 5 Steps and inputs required for production of strawberries under the annual hill system.

industry is reluctant to change to more sustainable growing methods. One could argue that annual strawberries are the most unsustainable fruit crop grown today, with a huge amount of energy expended on producing and distributing each berry. Despite high energy use, strawberries in the US now approximate the value of the fresh apple industry at $2.9 billion/year in sales (AgMRC, 2017).

Currently, California produces about 85% of the strawberries grown in the US, followed by Florida with 10%. Tabatabaie and Murthy (2016) assessed the energy used to produce strawberries from four regions of the US (California, Florida, North Carolina and Oregon) using life cycle analyses. They calculated three metrics of sustainability: nitrogen productivity, phosphorus productivity and fossil energy productivity. California had the best scores on these indices because yields were high relative to inputs. However, even under the best scenario (California), strawberry production emitted 1.75 kg of CO_2/kg of strawberry fruit. North Carolina producers emitted 5.48 kg of CO_2/kg fruit. A major contributor to the production of greenhouse gases is plastic mulch and poly tunnels, more so than agricultural machinery which has a long useful life. These numbers do not include the energy consumption of the nursery operation or transportation to point of sale, although Girgenti et al. (2014) estimated that nursery operations create an additional 6% of greenhouse gases. Peano et al. (2015) in Italy found that the greatest contributors to greenhouse gas emissions in the strawberry nursery are polystyrene cells, fertilizer and peat substrate. In the field, the major contributors are irrigation tubes and plastic mulch. The major post-harvest contributor is packaging. Together, their manufacture generates 0.55 kg CO_2/kg fruit. Williams et al. (2008) evaluated strawberry production in the United Kingdom and determined that 0.8 kg of CO_2 was released for every kg of strawberry produced. The value for Spain was lower due to greater yields per unit of input. Ingwersen (2012) compared the MJ (energy) input per serving of fruit for pineapples (*Ananas comosus*), apples (*Malus domestica*), oranges (*Citrus* spp.) and strawberries. Apples, oranges and strawberries were produced in multiple locations so there was some variation in values. However, strawberries grown in the United Kingdom used approximately 10× the energy (2.1 MJ/serving) of apples produced in either South Africa, New Zealand or the US, and the value for strawberries was also significantly higher than for oranges and pineapples. In contrast, many crops (potato [*Solanum tuberosum*], wheat [*Triticum aestivum*], maize [*Zea mays*], barley [*Hordeum vulgare*], chickpea [*Cicer arietinum*], lentil [*Lens culinaris*], sugarcane [*Saccharum officinarum*]) are carbon sinks (Chakrabarti et al., 2016).

Transporting strawberries from production areas to population centres also generates CO_2. A typical trailer will hold 14 pallets of strawberries, each containing 864 454 g (1 lb) 'clamshells', for a total of 5500 kg (12 096 lb) of fruit. Transporting these berries across the country from Watsonville, California, to

Hunts Point Terminal Market in New York City requires 1425 L (375 gal) of diesel fuel (3000 miles at 8 mpg) plus another 380 L (100 gal) for refrigeration (3.8 L/h for 100 h). The carbon dioxide equivalent of 1805 L (475 gal) of diesel fuel is 4836 kg of CO_2, indicating an almost 1:1 generation of CO_2/kg transported across the country. One kg of strawberry fruit contains enough sugar to generate about 176 g of CO_2 through respiration (assuming Brix of 12%), but 10× that amount of CO_2 is generated (1.75 kg) to produce each kg of berries (Tabatabaie and Murthy, 2016) and another 0.88 kg to transport them across the country.

Such a production and distribution system cannot be considered sustainable. Although some newer practices reduce the need for fumigation (e.g. solarization, biofumigation, anaerobic soil disinfestation [ASD]), pesticides (e.g. steam sterilization of plant material or steam weeding) and/or water (e.g. plugs rather than bare-rooted transplants), these often come with additional costs such as covering a field with a sheet of plastic to enhance solarization, building a structure to create plugs or using large amounts of energy to produce steam. In all cases, these modifications are dwarfed by the inherent unsustainable factors that characterize the current system that makes strawberries available to the consumer year-around. Questions about the sustainability of fruit production systems have been raised before (Merwin and Pritts, 1993), but the concerns involving strawberries appear to be greater than with most other fruit crops.

The strawberry industry is now facing the reality that it has been built on an unsustainable foundation. First, water is increasingly scarce in the regions where strawberries are grown. Aquifers continue to be depleted and land is subsiding in some areas as water is withdrawn. In Florida, while overall volume of water is not a concern, the quality of water has declined as saline water moves into aquifers. Second, methyl bromide fumigation has been phased out and less effective alternatives are being used. Border zones around fumigated fields are ever-expanding and it is only a matter of time before regulations further restrict the use of current fumigants in high population areas. Third, harvesting strawberries is based on migrant labour, of which about half is undocumented (National Agricultural Workers Survey, 2016). With restrictions on undocumented migrant labour increasing, finding labour to harvest the 1.36 billion kg (3 billion pounds) of strawberries per year (AgMRC, 2017) is becoming a major challenge. Fourth, land values are exceptionally high where strawberries are typically grown. Such high values not only increase costs, but they also deter beneficial crop rotations and increase pressure to use fumigants. Although current transportation costs are low, it is likely that this is a temporary situation since fossil fuels are a non-renewable resource. The production of 95% of the US strawberry crop in only three areas requires a significant reliance on fossil fuels to transport berries to market and to make the mulches, plastics and packaging used in production and distribution. Versions of the perennial matted row production system (that use less fertilizer, water and pesticides) persist in regions where the growing season is short and winters are cold, but this accounts for less than 5% of US production, and these perennial systems do not provide strawberries year around.

The National Strawberry Sustainability Initiative (NSSI, 2015), funded by the Walmart Foundation in 2013, supported 20 projects aimed at (1) improving the sustainability of strawberry growing, including increasing regional and local strawberry production, supply and availability; (2) reducing chemical and energy inputs in strawberry production; (3) conserving and preserving water resources; (4) improving soil quality and health in the production system; and (5) improving post-harvest value. Despite these and other efforts, there is still much work to be done to qualitatively impact the way that strawberries are grown so that resources are used more efficiently.

6 Future trends and conclusion

Are there ways forward that reduce reliance on fossil fuels and aquifers while still meeting consumer demand for high-quality berries available for much of the year? One step would be for supermarkets to follow the lead of stores like Wegmans (in the Northeast US) that work with local growers to provide high-quality berries for their stores. This, in turn, requires local growers to produce fruit over more than just 3–4 weeks as is currently done with perennial matted rows. Season extension technologies exist, but these also require plastic use for mulches and tunnels. An important step towards sustainability will be the development of biodegradable plastic/starch mulches and easy methods to recycle used plastics to allow local growers to extend their season. Girgenti et al. (2014) calculated that switching to biodegradable mulches and packaging would reduce greenhouse gas emissions in strawberry production by 20%. There is no universal solution to disposing of the volume of plastic used for mulch, fumigation and tunnelling. Most recyclers will not accept agricultural plastics because of the soil particles that adhere to the film, but eventual advances in technology should improve this situation.

Providing a several-month supply of local berries throughout the country will require growers to focus on day-neutral strawberries as a main crop that require regular attention and harvesting. Few strawberry growers outside the major production regions focus exclusively on strawberries, so a more dispersed system of production will require a cultural change among growers and perhaps some incentives to reduce the risks of growing a single crop. However, this change should reduce dependence on energy as production will be closer to markets and water will likely be more available.

Strawberry land in high-production areas will have to be rotated out of production to reduce reliance on fumigation, thereby decreasing overall supply from such areas. This is already happening to some extent due to competition with production areas in Mexico.

Breeders will need to focus on attributes other than size, yield and shelf life. While these are important, prioritizing soil disease resistance and perhaps a stronger root system to alleviate drought susceptibility would help reduce energy requirements and chemical inputs. If, at some future point, robotic harvesting (see below) becomes feasible, then plant architecture will have to be addressed. Fortunately, vast amounts of germplasm exist in North and South America, providing opportunities to exploit a multiplicity of traits. Currently, though, the modern strawberry is designed to be dependent on fertilizer inputs, chemical management of disease, regular availability of water and hand-harvest.

Traditionally, strawberry production economics has focused on per acre or per ha production costs, yields and returns, with little attention given to the effect of production system on labour costs per unit of fruit produced (Bolda et al., 2016). As labour becomes more expensive due to scarcity and/or labour law changes, harvest costs per unit of production have a greater impact on profitability, and thus sustainability. Newer production systems using day-neutral cultivars have markedly improved yields per season over short-day cultivars, but the harvest season is lengthened, and labour may not be utilized as efficiently during lower periods of production.

In Europe, where labour costs have long been higher than the US, the strawberry industry has developed a production system that improves harvest efficiency by improving fruit presentation, reducing stoop labour and increasing yields per lineal foot. This is the protected culture forcing system that utilizes substrates, crop support structures and short-duration crops from cultivars such as 'Elsanta' and 'Sonata'. Specialized nurseries develop plants with multiple crowns and large numbers of flower buds and place these into cold storage at −1°C to −2°C. Sixty or so days before a crop is desired, plants are set into plastic bags or pots filled with substrate. Before the first flowers open, the bags or pots are placed onto elevated support structures under plastic tunnels or in glasshouses. The harvest period is concentrated in the 60–90 day period from planting, and fruit is efficiently picked in a standing position from the well-presented trusses (Figs. 6 and 7). Multiple crops per year are completed using staggered planting dates, often under glass.

The protected culture forcing system has a higher fixed cost per unit area than open-field systems, but could provide a viable crop for an existing glasshouse or tunnel grower that is close to a large metropolitan market. Crops could be produced and harvested efficiently over longer periods of the year than the traditional in-ground system, and reduced transportation costs make

Figure 6 'Elsanta' forcing system in the Netherlands using troughs and tunnels.

the crop more sustainable. Certain diseases could be managed by manipulating the light source or adding UV-C rather than applying fungicides.

Success of the forcing system is critically dependent on using pathogen-free planting stock, high-quality water and substrates. Substrates in the protected culture forcing system function primarily as a support for plant roots, as plant fertility and water requirements are controlled by fertigation. Production with substrates, either on raised beds, trays or tabletops, comes with environmental impacts, particularly if the substrate is composed of a non-renewable resource such as peat or if it is shipped from long distances from manufacturer to grower

Figure 7 Forcing system in the Netherlands using pot culture, artificial substrate and tunnels.

such as vermiculite or perlite. Examples of renewable substrate components include coco coir, rice hulls and composted bark. Although coir is the most widely used substrate at present, alternatives such as aggregates could provide a longer-lasting substrate with less potential for disease infestation, less salt build-up and lower transportation costs. Fertigation water can be monitored, treated as needed with UV radiation or ozone, and re-used. Tunnels and glasshouse structures greatly reduce infection by fruit rot pathogens (e.g. *Botrytis cinerea*) and provide support for screening that can exclude arthropod pests such as spotted-wing drosophila.

Khoshnevisan et al. (2013) compared greenhouse to open-field production of strawberries in Iran and found that open-field production generated less CO_2 per unit of fruit (0.585 vs. 0.695 kg CO_2/kg fruit). It is unclear, however, if the greenhouses in this study used efficient state-of-the-art lighting, heating and irrigation systems. Even if the production of CO_2 per unit of fruit was similar, transportation costs presumably would be less with greenhouse production occurring closer to consumers, narrowing the gap.

Harvesting strawberries continues to be a major expense. Growers who only use documented migrant or local workers face potential shortages that can lead to crop abandonment and reduced acreage. Harvest aids such as solar-powered people carriers (e.g. Dole Solar Harvester, 2012) can assist the harvesters, but still require human input. Robotic harvesters are currently being developed (e.g. Agrobot, 2018) that can distinguish ripe from under ripe berries, but these are very slow. Genetic modification of the fruiting structure of the strawberry inflorescence from an indeterminate type to an upright determinant type could allow harvest of an entire truss instead of individual fruit, leading to greater efficiency. Increasing fruit size also increases harvest labour efficiency, but the size of strawberries may already be approaching the limit as to what consumers will accept. Growing strawberries on elevated benches is also less stressful for workers and improves harvest efficiency.

Packaging consumes a large percentage (30%) of the energy resources used to deliver strawberries to consumers, as most are sold in clear plastic protective clamshells (Peano et al., 2015). Meeting consumer expectations with less environmentally demanding packaging alternatives could significantly impact the carbon footprint of producing strawberries.

The strawberry industry faces many challenges that are currently overcome with considerable inputs. Some modest reductions of energy have been implemented and much improved integrated pest management (IPM) practices have been adopted by the industry, but these modifications have not fundamentally altered the unsustainable path of today's strawberry industry. Often new approaches to managing pests simply substitute one input for another without lowering the total amount of energy used. Significant movement towards sustainability will require re-localization of production, lower expectations of consumers for year-around strawberry availability and continued adoption of practices that lower the energy demand of strawberry production.

7 Where to look for further information

Research towards improving the sustainability of strawberry production must focus on reducing the energy inputs currently used. Developing methods to grow strawberries closer to the consumer and in regions with ample water supplies could reduce energy inputs considerably. Furthermore, as labour becomes scarce and unpredictable, methods to mechanically manage and harvest strawberries will be critical for economic survival. If the industry adopts soilless media for production of strawberries, then attention will need to be paid to nutrient and water management and securing media that is sustainably produced and recycled.

The North American Strawberry Growers Association and the California Strawberry Commission fund research and have websites that contain information for non-members. Several land-grant universities make available information about the details of strawberry production. Belgium and the Netherlands produce significant quantities of berries under glass, and Spain produces strawberries under low tunnels. While much information about strawberry growing is proprietary, sufficient knowledge is in the public domain to understand how issues of sustainability intersect with strawberry production around the world.

8 References

AgMRC. 2017. Agricultural marketing resource center. https://www.agmrc.org/commodities-products/fruits/strawberries (accessed 22 May 2018).

Agrobot. 2018. Agrobot robotic harvesters. http://www.agrobot.com/ (accessed 22 May 2018).

Bolda, M. P., L. Tourte, J. Murdock and D. A. Sumner. 2016. Sample costs to produce and harvest strawberries. University of California Agriculture and Natural Resources, Cooperative Extension and Agricultural Issues Center and UC Davis Department of Agricultural and Resource Economics, 20pp.

Chakrabarti, B., S. H. Kumar and H. Pathak. 2016. Carbon footprint of agricultural products. In: *The Carbon Footprint Handbook* (S. S. Muthu, ed.). CRC Press, New York, NY, pp. 431–54.

Darrow, G. M. 1966. *The Strawberry: History, Breeding and Physiology*. Holt, Rinehart and Winston, New York, NY, 447pp.

Dole Solar Harvester. 2012. https://www.youtube.com/watch?v=KeQOHdWQAPI (accessed 22 May 2018).

Girgenti, V., C. Peano, C. Baudino and N. Tecco. 2014. From 'farm to fork' strawberry system: Current realities and potential innovative scenarios from life cycle assessment of non-renewable energy use and greenhouse gas emissions. *Science of the Total Environment* 473: 48–53.

Ingwersen, W. W. 2012. Life cycle assessment of fresh pineapple from Costa Rica. *Journal of Cleaner Production* 35: 152–63.

Khoshnevisan, B., S. Rafiee and H. Mousazadeh. 2013. Environmental impact assessment of open field and greenhouse strawberry production. *European Journal of Agronomy* 50: 29–37.

Merwin, I. A. and M. P. Pritts. 1993. Are modern fruit production systems sustainable? *HortTechnology* 3: 128–36.

National Agricultural Workers Survey. 2016. Findings from the National Agricultural Workers Survey 2013–2014. https://www.doleta.gov/naws/pages/research/docs/NAWS_Research_Report_12.pdf (accessed 22 May 2018).

NSSI. 2015. The National Strawberry Sustainability Initiative. https://strawberry.uark.edu/ (accessed 22 May 2018).

Peano, C., C. Baudino, N. Tecco and V. Girgenti. 2015. Green marketing tools for fruit growers associated groups: Application of life cycle assessment (LCA) for strawberries and berry fruits ecobranding in northern Italy. *Journal of Cleaner Production* 104: 59–67.

Sjulin, T. M. 2003. The North American small fruit industry 1903–2003. II. Contributions of public and private research in the past 25 years and a view to the future. *HortScience* 38: 960–7.

Sjulin, T. M. 2010. Trends in North American strawberry production and supply. *3rd Intl. Strawberry Congress*. http://www.hoogstraten.eu/congress/data/congress-presentations/02_trends_in_north_american_strawberry_production_and_supply_tom_sjulin.pdf (accessed 22 May 2018).

Strowbridge, J. W. 1930. Origin and distribution of the commercial strawberry crop. USDA Tech. Bull. No. 180, 104pp.

Tabatabaie, S. M.H. and G. S. Murthy. 2016. Cradle to farm gate life cycle assessment of strawberry production in the United States. *Journal of Cleaner Production* 127: 548–54.

Welch, N. C., Bringhurst, R., A. S. Greathead, V. Voth, W. S. Seyman, N. F. McCalley and H. W. Otto. 1982. Strawberry production in California. Leaflet 2959, University of California Cooperative Extension, Berkeley, CA.

Wilhelm, S. and J. E. Sagen. 1974. *The History of the Strawberry*. University of California, Berkeley, CA, 298pp.

Williams, A., E. Pell, J. Webb, E. Moorhouse and E. Audsley. 2008. Strawberry and tomato production for the UK compared between the UK and Spain. In: *International Conference on LCA in the Agri-Food Sector*, Zurich.

Chapter 12

Advances and challenges in raspberry and blackberry breeding

Ramón Molina-Bravo, Universidad Nacional de Costa Rica, Costa Rica; Margaret Leigh Worthington, University of Arkansas, USA; and Gina E. Fernandez, North Carolina State University, USA

1 Introduction

The growing world population is driving higher demands for food, feed, land and energy in an increasingly resource-limited environment (Østergård et al., 2009). These concerns have contributed to renewed interest in sustainable, low-input agricultural systems (Lammerts van Bueren et al., 2008; Mezzetti, 2016). Modern agricultural systems are dependent on cheap fossil fuels and encourage high yields through short-term, high-input applications. These practices have had serious consequences on the environment, such as decline in agrobiodiversity and soil fertility (Østergård et al., 2009). A recent shift in paradigm envisages a more holistic view of crop production, rather than a system that mostly values economic outputs. Future crop production should be viewed in a broader sense as an environmentally, socially and economically sustainable complex system.

Irrespective of the farming system, plant breeding goals should focus on developing cultivars that increase productivity and quality, while simultaneously minimizing the environmental impact and the dependency on non-renewable resources. Although these objectives are not necessarily new, improved resource efficiency introduces a goal that is not commonly considered in a post-Green Revolution perspective (Dawson et al., 2007). Likewise, breeders must take into account adaptation to specific environments and farming systems when developing cultivars.

http://dx.doi.org/10.19103/AS.2018.0040.27

Fruit breeders must consider the aforementioned breeding objectives (Byrne, 2012) and face the added challenge of anticipating cultivar needs at least 10 years in advance due to the crops' long reproductive cycles. For this reason, the application and development of molecular tools are perhaps more important in perennial crops than in annual crops in meeting these objectives for sustainable cultivation.

Raspberries and blackberries (*Rubus* spp.), also called caneberries, are found in a wide variety of climates and on every continent, except Antarctica (Aguirre-Gutiérrez et al., 2017; Hummer, 2017). In modern commercial systems, individual canes (shoots) are annual or biennial and the plant's roots and crown are perennial with an expected life span between 5 and 40 years, depending on the cultivar and species (Strik and Bryla, 2015). Floricane-fruiting cultivars have biennial canes that are vegetative during the first season's growth and undergo bud dormancy that requires cessation of growth and sufficient chilling in order to flower the second season (Jennings, 1988). Primocane-fruiting cultivars have annual canes that have no chilling requirement and flower during the first season's growth (Strik and Finn, 2012).

Caneberries are important fruit crops that have gained substantial economic interest in the past several decades. In 2016, raspberry production was registered at 795 249 metric tonnes, a 2.3-fold increase since 1996 (Fig. 1; FAO, 2016). In 2005, there was an estimate of 20 035 ha of blackberries planted worldwide, a 45% increase since 1995 (Strik et al., 2007). The number of U.S. blackberry growers has since increased by 76% between 2008 and 2015 (Hall, 2017). Raspberries and blackberries were the third and fourth most economically important berry crops in the United States and accounted for US$914 million and US$549 million in sales, respectively, during 2016 (CA Strawberry Commission, 2018). The entire U.S. berry industry is expanding, and blackberries in particular have been driving this growth with an increase of 14.7% in sales between 2015 and 2016. This chapter focuses on the recent advances in molecular tools and challenges of *Rubus* breeding for achieving sustainable cultivation.

2 *Rubus* traits for sustainability

Improving sustainability in berry production is becoming a priority in Europe and North America. A European consortium, called the EUBerry project made up of 15 partners across the continent, is aiming to achieve sustainable improvement of European berries in a changing environment using strawberry, blueberry and raspberry as model crop species (Mezzetti, 2016). The funding priorities of the North American Bramble Growers Foundation, Inc. (NABGRF) include improvements in non-chemical strategies for pest management, improved cultivars for machine harvest and breeding for tolerance to abiotic

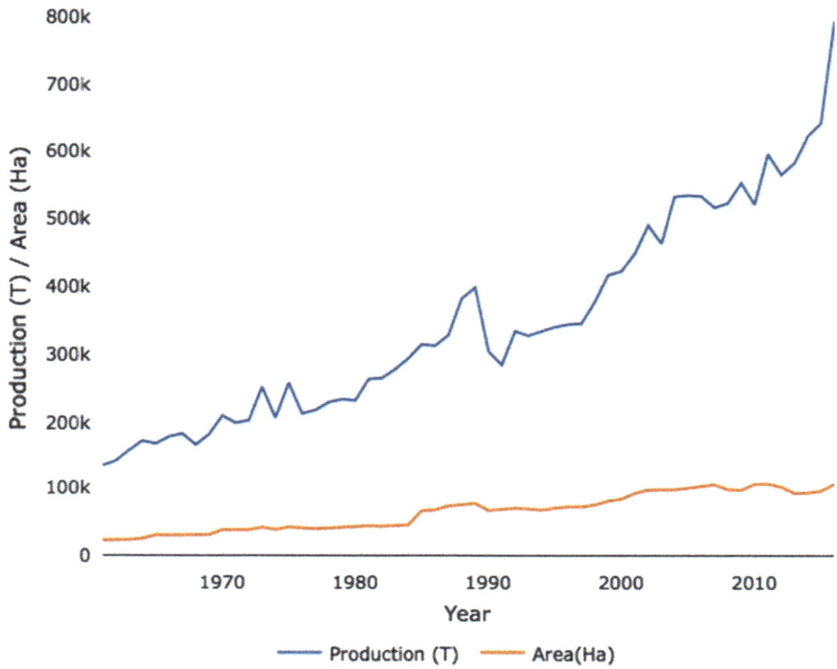

Figure 1 Red raspberry (*Rubus idaeus*) world production (metric tons; T) and area planted (ha), reported between 1961 and 2016 (FAO 2016).

stress (North American Raspberry and Blackberry Association (NARBA), 2019). There are several broad trends that can influence breeding for sustainability in *Rubus* crops and other fruit species. The following traits should be prioritized.

2.1 Nutrient efficiency

For many crops, there is considerable variation in nitrogen (N)-use efficiency (Fathi et al., 1997; Khoshgoftarmanesh et al., 2010; Muurinen et al., 2006; Sinebo et al., 2004; Singh et al., 1998). Lammerts van Bueren et al. (2011) suggest that breeding crops under conventional high-input fertilizer regimes may have resulted in cultivars that are dependent on readily available N. In caneberries, yield response to different N-fertilization rates varies among cultivars and fruiting types (floricane and primocane-fruiting) (Chaplin and Martin, 1980; Heiberg, 2002; Kowalenko, 1994, 2006; Mohadjer et al., 2001) and has a minimal effect on fruit quality and size (Alleyne and Clark, 1997; Nelson and Martin, 1986; Rempel et al., 2004). Fertilizer N recovery, as a percentage of traceable N added, is 32–45% in blackberry and 24–37% in red raspberry (Mohadjer et al., 2001; Naraguma et al., 1999; Rempel et al., 2004). However, traceable N studies might not provide a complete picture of N use *in planta*, as N from fertilizers can impact the allocation of other forms of plant-available N (Bañados et al., 2012). In floricane-fruiting red raspberry, if N in the form of fertilizer is applied when primocanes are less than 15 cm tall, N is taken up primarily by primocanes, fruiting laterals and fruit; but if fertilizer is applied when green fruit is present, most of the fertilizer N is taken up by the primocanes (Rempel et al., 2004). Similar results are observed in alternate-year, floricane-fruiting blackberry, where fertilizer N is an important source of N for fruiting laterals and fruit in the on-year (production year), and for primocanes in the off-year (Mohadjer et al., 2001). To date, there have been no studies examining the use of fertilizer in single or double cropping primocane-fruiting blackberry systems and growers are basing their fertility inputs on floricane-fruiting recommendations (Fernandez, pers. comm.).

Partly due to wide spacing between rows, small plant size and low yield, caneberries have relatively low N accumulation per area compared with other perennial crops (Dean et al., 2000; Mohadjer et al., 2001; Rempel et al., 2004; Strik, 2008). A significant portion of nutrients in *Rubus* crops are lost through fruiting, floricane removal, pruning or plant senescence. In trailing blackberry, nutrient loss in fruit and floricanes averages 33–79 kg/ha N, 4–12 kg/ha P and 36–84 kg/ha K, but varies among cultivars (Harkins et al., 2014). In red raspberry, losses average an annual total of 37 kg/ha N, 4 kg/ha P and 24 kg/ha K (Strik and Bryla, 2015). Fertilizer uptake should be further analysed, as this can vary with application method and product use in addition to cultivar and fruiting type (Fernandez-Salvador et al., 2015; Reickenberg and Pritts, 1996; Strik and Bryla, 2015).

2.2 Rhizosphere competence

Crop nutrient uptake can also be improved by bacterial and fungal communities that form rhizospheric associations. Ectomycorrhizal associations can facilitate transfer of amino N in woody plant systems (Abuzinadah and Read, 1989; Finlay et al., 1992; Taylor et al., 2004). More notably on the subject of breeding for sustainability, genes that control the ability to form mycorrhizal root symbiosis have been identified in tomato and wheat (Hetrick et al., 1995; Larkan et al., 2007), therefore, improving upon this trait can potentially be aided by marker-assisted selection (MAS) or other modern breeding methods. Other fungal associations also promote plant growth and protect from abiotic stress. Endophytic *Beauveria bassiana* and *Purpureocillium lilacinum* enhance growth in cultivated cotton (*Gossypium hirsutum*) (Lopez and Sword, 2015). The *Theobroma cacao* endophyte, *Trichoderma hamatum*, enhances seedling growth and delays the onset of drought stress (Bailey et al., 2011). In addition to improved nutrient uptake, rhizospheric associations can also protect plants from pests and diseases. Endophytic *B. bassiana* reduces the infestation of piercing-sucking insects in grapevine (Rondot and Reineke, 2018), and reduces the fitness and damage of the banana weevil (*Cosmopolites sordidus*) in *Musa* spp. (Akello et al., 2008). Little is known about *Rubus* endophytes or other rhizospheric associations, but Contreras et al. (2016) identified over 40 endophytic bacterial species in blackberry (*R. fruticosus*) roots. Nitrogen fixation has also been reported in *R. ellipticus* (Becking, 1979). Thus, there is potential in exploring traits that improve rhizosphere competence in *Rubus*.

2.3 Climate resilience

Crops are expected to experience earlier flowering, yield loss and growth reduction due to climate change. Climate models predict that the Earth's temperatures and solar radiation will increase (Lv et al., 2013; Stocker et al., 2013; Vose et al., 2005). Chilling models show that accumulated winter temperatures have decreased over time throughout the world (Luedeling and Brown, 2011; Luedeling et al., 2009, 2011). There are numerous traits that are involved in climate resilience. Many temperate-climate woody perennial crops, including *Rubus*, require exposure to prolonged hours between 0°C and 7°C in order to flower; this is known as the chilling requirement (Westwood, 1993). Therefore, developing cultivars with low-chilling requirements are of major interest. Low-chill cultivars are also important for expanding production areas. For example, blackberry production has greatly expanded in North America, particularly in Mexico, where most fruit produced is marketed for consumption in the United States and Europe (Clark and Finn, 2014; Strik et al., 2007). Thus, there has been interest in breeding low-chill cultivars similar to 'Brazos' and 'Tupy', with an estimated chilling requirement of 300 h (Clark and Finn, 2014).

Recently, blackberry production has expanded into climates where winters are colder as a result of new trellising systems that enable plants to be covered in the winter for cold protection (Takeda et al., 2013). Although cultural practices have expanded the regions where blackberries can be planted, cold hardiness is an important component for this type of system and breeding cultivars that tolerate extreme low temperatures will become more important as this industry continues to grow (Fernandez, pers. comm.).

Similarly, heat tolerance will become an increasingly important trait in the global climate change scenario. For example, typical raspberry cultivars are poorly adapted to warm temperatures during the growing season (Ballington and Fernandez, 2008; Kempler et al., 2011). The North Carolina *Rubus* Breeding Program expanded the growing regions throughout the state using germplasm from warm temperate regions of Asia (Ballington and Fernandez, 2008). The current breeding programme has some advanced raspberry selections that have improved heat tolerance; however, 'Dormanred' (*R. parvifolius* × 'Dorsett'), released in 1972, is the most widely adapted cultivar to the region. Climate resilience can also come about through alternative production practices.

Primocane fruiting (i.e. development of flower buds and fruit on first-season canes) is a trait that can extend the production season and allow growers to produce and market a summer and fall crop in their local area (Clark, 2008). Since the release of the first primocane-fruiting blackberry cultivars by the University of Arkansas Fruit Breeding Program in 2004, breeders have worked to combine this trait with other important attributes, including thornlessness (prickle-free canes), higher yield and fruit quality.

2.4 Health benefits and fruit quality

Breeding for high levels of health-promoting plant compounds contribute to sustainability from an economic standpoint. Health consciousness is one of the major driving forces of the market. Consumers associate a healthy diet with the prevention of disease, prevention of obesity and lowering cholesterol, along with other benefits (Dillard and German, 2000; Sloan, 2006). The perceived health benefits associated with dark-pigmented fruits is one of the major factors driving increased consumer demand for *Rubus* crops (Clark et al., 2002; Connor et al., 2005; Fan-Chiang and Wrolstad, 2005; Seeram, 2008). Blackberries, in particular, have significantly higher concentrations of anthocyanins and other antioxidants than most other fruit (Halvorsen et al., 2006; Machado et al., 2015). Blackberries are also rich in vitamin C, vitamin A, vitamin E, vitamin B6, folic acid, dietary fibre, potassium, phosphorous, magnesium, calcium and iron that promote skin, bone, heart and brain health (Lee, 2017). More recent reports have indicated that antioxidant capacity determined from *in vitro* studies are not necessarily important bioactive compounds in human systems (Haytowitz and Bhagwat, 2010). However, some of these compounds in food have a wide range of functions that may be associated with the absorbance of free radicals.

Rubus breeding for better postharvest quality has had great focus and has implications in sustainability in terms of storability and transport, since berries for the fresh market have poor durability. The quality of fruits for the fresh and processed market is determined by how a genotype responds to storage and handling practices from the moment of harvest until the fruit reaches consumers (Clark and Finn, 2008). Some of the characteristics that limit consumer appeal of fresh fruit are presence of decay, leakage of juice, mushiness of fruit and off-colour drupelets; and of frozen fruit are mushiness, perceived seediness, off-colour and poor flavour (especially high acidity) (Perkins-Veazie and Clark, 2005; Villamor et al., 2013; Yorgey and Finn, 2005). In raspberry, grey mould (caused by *Botrytis cinerea*) causes significant reduction in shelf life of harvested fruit (Kempler et al., 2011). Other factors that improve raspberry fruit quality such as firmness, small drupelet size and strong skin also tend to improve genotype resistance to fruit rot (Kempler et al., 2011). Methods for screening germplasm for resistance to postharvest rot and improved shelf life have been described by several authors (Daubeny and Pepin, 1974; Harshman et al., 2014; Stephens et al., 2002). Some genetic studies have analysed this trait indirectly by looking at fruit development stages (Simpson et al., 2017).

2.5 Plant architecture and morphology for simplified management

Sustainability goals across the globe have set the target to double agricultural productivity, specifically for small-scale food producers, by 2030, while maintaining ecological integrity (U.N. General Assembly, 2015). This challenge must be met by producing more in the same land space and, more than likely, with less input. Modifying plant size, growth or habit can have several benefits, not only for saving land space but also for reducing management and production costs. Certain traits in *Rubus*, such as thornlessness, can have dramatic effects on trellising, mechanization, hand harvesting and overall management of the plant (Clark and Finn, 2008). Machine-harvested berries are preferred in the processing industry and dislodged petioles are considered one of the most serious biological contaminants (Strik and Buller, 2002). Improving upon these traits can therefore increase productivity and simplify management. In *Rubus*, erect to semi-erect cane types are generally preferred over trailing types, as this eases trellising and other practices (Clark and Finn, 2008; Clark and Moore, 2008; Moore and Clark, 1989). Recently, *Rubus* breeding programmes have released a number of dwarf or semi-dwarf plants for ornamental purposes. Reduced plant stature may also ultimately have important implications for high-density production systems and substrate culture (Finn and Clark, 2017). Cultivars that are leafy, that have drooping laterals or that have tightly clustered fruit often show a higher incidence of fruit rot than those with open plant habit, upright laterals and widely spaced fruit (Hall et al., 2009).

2.6 Resistance to insects and diseases

Reducing the application of pesticides is crucial for achieving sustainability. Pesticides can accumulate in the environment, contaminating groundwater and entering the food chain (Bravo Durán et al., 2013; Cook, 2000; Gerhardson, 2002; Jiang et al., 2006; van Wendel de Joode et al., 2014, 2016). The loss of fumigants (e.g. Telone), increasing environmental regulations and lack of post-plant management options for soil-borne pathogens are major challenges for the *Rubus* industry (Walters et al., 2017). Root rot diseases are among the most important factors affecting production in raspberry (Kempler et al., 2011); blackberries are generally subjected to fewer diseases and insect problems

(Finn and Clark, 2011). In raspberry, root rot is usually caused by *Phytophthora rubi*, the most virulent form of *Phytophthora* (Wilcox, 1989), and the most effective method of control is through resistant cultivars (Kempler et al., 2011). Grey mould, caused by *B. cinerea*, is the most serious fruit rot disease, causing significant production losses in raspberry and blackberry, and can also infect canes (Finn and Clark, 2011; Kempler et al., 2011). Insect pests are also growing to be more problematic for the global *Rubus* industry. The invasive spotted wing drosophila (*Drosophila suzukii*) is the most significant pest in the United States (Burrack et al., 2013), while root boring insects and broad mites are also serious challenges for growers in multiple production regions. The development of disease- and pest-resistant cultivars is a crucial component of sustainable integrated pest management. Breeders should select healthy plants with minimal susceptibility to pests and diseases.

3 Molecular tools and resources for *Rubus*

Over the past quarter century, molecular markers have been applied to study and develop breeding tools for *Rubus*. Most of the molecular markers currently in use are based on polymerase chain reaction (PCR) assays, and have replaced isozyme markers and other marker systems because of their reproducibility, allele diversity and ease of use (Bus et al., 2009). PCR-based molecular markers have been used to study genetic diversity (Dossett et al., 2012; Graham and McNicol, 1995; Graham et al., 1997; Miyashita et al., 2015; Ryu et al., 2014), identify cultivars (Bassil et al., 2010, 2012, 2016; Parent et al., 1993; Stafne et al., 2003), construct genetic linkage maps (Bushakra et al., 2015a; Castro et al., 2013; Graham et al., 2004; Pattison et al., 2007; Ward et al., 2013) and associate traits through quantitative trait loci (QTL) analysis (Castro et al., 2013; Graham et al., 2009, 2006; Molina-Bravo et al., 2014; Sargent et al., 2007; Woodhead et al., 2013). Genetic linkage maps are available for red raspberry (*R. idaeus*) (Graham et al., 2004, 2006; Ward et al., 2013), blackberry (*Rubus* sp.) (Castro et al., 2013; Weber, 2014) and black raspberry (*R. occidentalis*) (Bushakra et al., 2012b, 2015a). Most of these maps have been constructed using PCR-based markers, but more recently constructed maps have used sequencing-based (such as genotype by sequencing) technology. PCR-free large parallel assays based on numerous markers that cover a large portion of the genome are the future for high-throughput genotyping technologies (Bus et al., 2009). Some PCR-free markers include single nucleotide polymorphisms (SNPs), single feature polymorphisms (SFPs) and restriction site-associated DNA (RAD) markers (Gupta et al., 2008). The implementation of next-generation sequencing (NGS) technologies in *Rubus* will be facilitated by the development of the first well-assembled *Rubus* genome from a highly homozygous accession of black raspberry (*R. occidentalis*) (Jibran et al., 2018; VanBuren et al., 2016). Along with these molecular advancements in *Rubus*, a comprehensive database of the genetic and genomic information is available at the Genome Database for Rosaceae (GDR, http://www.rosaceae.org). This database has been the long-standing repository and data mining resource for research in Rosaceae species (Jung et al., 2014, 2007). The database houses several annotated whole-genome sequences, thousands of unigenes, over 80 genetic linkage maps and over two million molecular marker sequences including SNPs, amplified fragment length polymorphisms (AFLPs), cleaved amplified polymorphic sequences (CAPs), short sequence repeats (microsatellites or SSRs), isozymes and sequence characterized amplified regions (SCARs), among others, from over 500 species. Although available information for *Rubus* is surpassed by other rosaceous species such as *Malus* and *Fragaria*, these species can serve as models for comparative analyses (Bushakra et al., 2012a; Decroocq et al., 2003; Dirlewanger et al., 2004; Ryu et al., 2017).

3.1 Molecular markers

Several molecular marker systems have been used to develop molecular tools for *Rubus*. Numerous studies have applied PCR-based markers such as randomly amplified polymorphisms (RAPDs), AFLPs, expressed sequence tags (ESTs) and SSRs (Bushakra et al., 2015b; Girichev et al., 2017; Graham and McNicol, 1995; Lewers et al., 2008; Marulanda et al., 2007; Miyashita et al., 2015; Woodhead et al., 2010); multiplexed fingerprinting assays (Bassil et al., 2012, 2016; Fernández-Fernández et al., 2011); and NGS technology for *de novo* marker development (Dossett et al., 2015; Li et al., 2016; Liu et al., 2017). Although microsatellite markers have been used in molecular applications for decades, they are still used because of their highly informative and polymorphic nature. SSRs are still used in fruit-breeding programmes for MAS and marker-assisted breeding (MAB) (Bus et al., 2009). The first SSR markers that were isolated and characterized for any *Rubus* species were published by Amsellem et al. (2001), in *R. alceifolius*, an invasive weed of Réunion Island. A year later, Graham et al. (2002) reported and isolated a set of ten SSR loci by screening a *Pst*I size-selected genomic library extracted from 'Glen Moy' red raspberry. Several reports of *Rubus*-extracted SSR markers followed, which included 66 SSRs and eight EST-SSRs in *R. idaeus* (Graham et al., 2004), 15 SSRs in *R. hochstetterorum* (Lopes et al., 2006), 25 EST-derived SSRs (EST-SSRs) in *R. idaeus* (Woodhead et al., 2008), 33 tested primer pairs to amplify EST-SSR markers from a 673 primer pair library in blackberry (*Rubus*) (Lewers et al., 2008), 13 SSRs in 'Meeker' red raspberry (*R. idaeus*) and 'Marion' blackberry (*Rubus* hybrid) (Castillo et al., 2010), 36 tested EST-SSR primer pairs from 131 designed primer pairs in red raspberry (*R. idaeus*) and 24 tested EST-SSR primer pairs from 288 designed primer pairs in black raspberry (*R. occidentalis*) (Bushakra et al., 2015b), 166 NGS-derived SSR markers in red

(*R. idaeus*) and black raspberry (*R. occidentalis*) (Dossett et al., 2015) and more recently through next-generation transcriptome sequence data mining, 3393 EST-SSRs identified in blackberry (*Rubus*) (Li et al., 2016) and 14210 EST-SSRs identified in red raspberry (*R. idaeus*) (Liu et al., 2017). SNPs are also available in red raspberry, black raspberry and blackberry. A SNP-based, dense genetic linkage map was developed by Ward et al. (2013) and a total of 4521 and 2391 SNP markers were mapped onto 'Heritage' and 'Tulameen' red raspberry parents, respectively, using GBS technology. Flanking nucleotide sequences of each of these SNPs are also available (Ward et al., 2013). GBS technology has also been applied in black raspberry, but due to the homozygous nature of *R. occidentalis*, only 399 SNPs were mapped (Bushakra et al., 2015a).

3.2 Linkage maps, MTLs and QTLs

Graham et al. (2004) published the first genetic linkage map for red raspberry using a full-sib family from a cross between the European cultivar Glen Moy (*R. idaeus* subs. *idaeus*) and the North American cultivar Latham (*R. idaeus* subs. *strigosus*). This map was initially constructed using 417 mapped markers, 60 SSRs, 8 EST-SSRs and 349 AFLP markers, and covered 465.8 cM with seven linkage groups. A final version was generated under more stringent criteria, which resulted in a map with nine linkage groups based on 273 markers with a total length of 788.1 cM. QTLs were identified for spines (prickle density), root sucker density and root sucker spread (diameter) (Graham et al., 2004). An updated version of this map was then published and comprised 349 markers of eight linkage groups, covering a total distance of 669 cM (Graham et al., 2006). Graham et al. (2006) identified several horticulturally important QTLs in this study, including resistance to cane botrytis (*B. cinerea*), spur blight (*Didymella applanata*), cane spot (*Elsinoë veneta*) and yellow rust (*Phragmidium rubi-idaei*). Gene *H*, which controls the presence of cane pubescence, was also mapped as a heritable phenotypic marker (MTL). A third red raspberry map was reported by Pattison et al. (2007) and was used to study the quantitative inheritance of *Phytophthora* root rot resistance. The map was constructed using a B_2 population from a 'Latham' × 'Titan' cross, where 'Latham' was the recurrent source of resistance. However, the map was constructed using only AFLP markers, and thus would require anchor markers based on conserved sequences to compare with other maps. The red raspberry map developed by Graham et al. underwent further marker enrichment and QTL analyses: developmental QTLs from bud break to ripe fruit and tallest cane height (Graham et al., 2009); QTL analysis of environmental and seasonal influences on anthocyanin contents (Kassim et al., 2009); QTLs and candidate genes influencing fruit colour (McCallum et al., 2010); QTLs, candidate genes and a potential marker associated with *Phytophthora* root rot resistance (Graham et al., 2011); over-seasons QTL analysis affecting total phenolic content and total anthocyanin content (Dobson et al., 2012); QTLs and candidate genes associated with environmental and seasonal influences on fruit volatiles (Paterson et al., 2013); QTLs involved in cane splitting (Woodhead et al., 2013); QTLs and candidate genes associated with the genetic and environmental regulation of plant architectural traits and their effect on arthropod pests (Graham et al., 2014); QTLs for 'crumbly' fruit analysed under field and polytunnel conditions over six seasons (Graham et al., 2015); and QTLs for several traits associated with fruit ripening and softening (Simpson et al., 2017). A fourth genetic map was constructed from an F1 population developed from a cross between an *R. occidentalis* selection and 'Latham' (*R. idaeus*) consisting of DNA sequence-based markers from across Rosaceae and was used for comparative genomic analysis (Bushakra et al., 2012b). On that map, 27 QTLs associated with cyanidin modification were identified (Bushakra et al., 2013). Castro et al. (2013) developed the first tetraploid blackberry genetic linkage map from a full-sib family of individuals segregating for thornlessness and primocane fruiting. A total of 119 SSR and SSR-EST *Rubus* markers, plus the thorny and floricane-fruiting traits were used to construct the map. In the same year, Ward et al. (2013) constructed a SNP-based saturated linkage map using GBS technology and genome-independent imputation. SSR markers cross-validated and linked this map to other published maps and spanned 462.7 cM in the 'Heritage' parent, and 376.6 cM in the 'Tulameen' parent. Molina-Bravo et al. (2014) published another genetic map from a full-sib family of individuals from a cross between (*R. parvifolius* × 'Tulameen') × 'Qualicum' and used AFLPs and several SSR anchor markers. The population segregated for a myriad of traits, and QTLs were found for chilling hour requirements, flower colour, thorn density and several fruit characteristics, plus, gene *s*, a gene for thornlessness, was mapped as an MTL (Molina-Bravo et al., 2014). A second tetraploid blackberry linkage map was constructed using restriction site-associated genomic DNA (RAD-Seq) markers (Weber, 2014). The research generated a high-density map for 'Chester Thornless' and 'Prime-Jim', consisting of 29 linkage groups spanning 1059.1 cM and 31 linkage groups spanning 1024.6 cM, respectively. Supporting evidence was found for the position of the thornless locus *S* (gene *s*) (Weber, 2014). The first full black raspberry (*R. occidentalis*) genetic linkage map was constructed using SNPs and transferable SSR markers (Bushakra et al., 2015a). The black raspberry map consisted of a total 399 GBS SNPs, 70 SSRs and 12 high-resolution melt (HRM) markers, and one MTL for aphid resistance (*Amphorophora agathonica*, Ag_4). In total, 91 traits have associated QTLs or MTLs mapped onto three red raspberry maps, one (tetraploid) blackberry map, one black raspberry by red raspberry map and one black raspberry map (Table 1).

Table 1 Summary of mapped quantitative trait loci (QTLs) and heritable phenotypic markers (MTLs) in *Rubus*

Trait	Marker	Map	Species	LG(#QTLs)	References
Root sucker spread (diameter)	QTL	Glen Moy × Latham	*R. idaeus* subs. *idaeus* × *R. idaeus* subs. *strigosus*	OLG8(2)[a]	Graham et al. (2004)
Root sucker density	QTL	Glen Moy × Latham	*R. idaeus* subs. *idaeus* × *R. idaeus* subs. *strigosus*	OLG8(1)[a]	Graham et al. (2004)
Thorn density (prickle/spine density)	QTL	Glen Moy × Latham	*R. idaeus* subs. *idaeus* × *R. idaeus* subs. *strigosus*	OLG2(2)[a]	Graham et al. (2004, 2006)
Cane botrytis (*B. cinerea*)	QTL	Glen Moy × Latham	*R. idaeus* subs. *idaeus* × *R. idaeus* subs. *strigosus*	1(1),3(3),5(1),6(1)	Graham et al. (2006)
Spur blight (*Didymella applanata*)	QTL	Glen Moy × Latham	*R. idaeus* subs. *idaeus* × *R. idaeus* subs. *strigosus*	1(1),3(3),5(1),6(1)	Graham et al. (2006)
Cane spot (*Elsinoë veneta*)	QTL	Glen Moy × Latham	*R. idaeus* subs. *idaeus* × *R. idaeus* subs. *strigosus*	3(1),6(1)	Graham et al. (2006)
Yellow rust (*Phragmidium rubi-idaei*)	QTL	Glen Moy × Latham	*R. idaeus* subs. *idaeus* × *R. idaeus* subs. *strigosus*	1(1),5(1)	Graham et al. (2006)
Cane pubescence (gene *H*)	MTL	Glen Moy × Latham	*R. idaeus* subs. *idaeus* × *R. idaeus* subs. *strigosus*	3	Graham et al. (2006)
Phytophthora root rot plant disease index	QTL	(Titan × Latham) × Titan	*R. idaeus*	1(1),5(1),7(1)	Pattison et al. (2007)
Phytophthora root rot stem lesion length	QTL	(Titan × Latham) × Titan	*R. idaeus*	5(1),7(1)	Pattison et al. (2007)
Phytophthora root rot incidence of petiole lesions	QTL	(Titan × Latham) × Titan	*R. idaeus*	1(1),5(1),7(1)	Pattison et al. (2007)
Phytophthora root rot plant root regeneration score	QTL	(Titan × Latham) × Titan	*R. idaeus*	1(1),5(1),7(1)	Pattison et al. (2007)
Resistance to *Amphoraphora idaei* (A1)	MTL	Malling Jewel × Malling Orion	*R. idaeus*	6	Sargent et al. (2007)

Dwarfing habit (sturdy, dw)	MTL	Malling Jewel × Malling Orion	*R. idaeus*	1	Sargent et al. (2007)
Tallest cane height	QTL	Glen Moy × Latham	*R. idaeus* subs. *idaeus* × *R. idaeus* subs. *strigosus*	5(1),7(1)	Graham et al. (2009)
Bud break to ripe fruit development	QTL	Glen Moy × Latham	*R. idaeus* subs. *idaeus* × *R. idaeus* subs. *strigosus*	1(3),3(4),5(5),6(15)	Graham et al. (2009)
Cyanidin-3-sophoroside (C3S)	QTL	Glen Moy × Latham	*R. idaeus* subs. *idaeus* × *R. idaeus* subs. *strigosus*	2(2),7(3)	Kassim et al. (2009)
Cyanidin 3-O-glucoside (C3G)	QTL	Glen Moy × Latham	*R. idaeus* subs. *idaeus* × *R. idaeus* subs. *strigosus*	2(2),7(3)	Kassim et al. (2009)
Cyanidin 3-*O*-glucosylrutinoside (C3GR)	QTL	Glen Moy × Latham	*R. idaeus* subs. *idaeus* × *R. idaeus* subs. *strigosus*	7(3)	Kassim et al. (2009)
Cyanidin-3-rutinoside (C3R)	QTL	Glen Moy × Latham	*R. idaeus* subs. *idaeus* × *R. idaeus* subs. *strigosus*	2(3),7(3)	Kassim et al. (2009)
Pelargonidin-3-sophoroside (P3S)	QTL	Glen Moy × Latham	*R. idaeus* subs. *idaeus* × *R. idaeus* subs. *strigosus*	2(1),7(3)	Kassim et al. (2009)
Pelargonidin-3-glucoside (P3G)	QTL	Glen Moy × Latham	*R. idaeus* subs. *idaeus* × *R. idaeus* subs. *strigosus*	2(3),7(3)	Kassim et al. (2009)
Pelargonidin-3-glucosylrutinoside (P3GR)	QTL	Glen Moy × Latham	*R. idaeus* subs. *idaeus* × *R. idaeus* subs. *strigosus*	2(1),7(3)	Kassim et al. (2009)
Pelargonidin-3-rutinoside (P3R)	QTL	Glen Moy × Latham	*R. idaeus* subs. *idaeus* × *R. idaeus* subs. *strigosus*	2(1),7(3)	Kassim et al. (2009)
Fruit colour	QTL	Glen Moy × Latham	*R. idaeus* subs. *idaeus* × *R. idaeus* subs. *strigosus*	1(14),2(5),3(16),6(13)	McCallum et al. (2010)
Phytophthora root rot (field scores)	QTL	Glen Moy × Latham	*R. idaeus* subs. *idaeus* × *R. idaeus* subs. *strigosus*	1(3),6(3)	Graham et al. (2011)

(Continued)

Table 1 (*Continued*)

Trait	Marker	Map	Species	LG(#QTLs)	References
Phytophthora root rot (greenhouse scores)	QTL	Glen Moy × Latham	*R. idaeus* subs. *idaeus* × *R. idaeus* subs. *strigosus*	1(2)	Graham et al. (2011)
Phytophthora root rot (root density and diameter, field)	QTL	Glen Moy × Latham	*R. idaeus* subs. *idaeus* × *R. idaeus* subs. *strigosus*	1(1),6(7)	Graham et al. (2011)
Phytophthora root rot (height at infection site)	QTL	Glen Moy × Latham	*R. idaeus* subs. *idaeus* × *R. idaeus* subs. *strigosus*	5(2)	Graham et al. (2011)
Total phenol content	QTL	Glen Moy × Latham	*R. idaeus* subs. *idaeus* × *R. idaeus* subs. *strigosus*	1(1),5(1),6(1)	Dobson et al. (2012)
Total anthocyanin content	QTL	Glen Moy × Latham	*R. idaeus* subs. *idaeus* × *R. idaeus* subs. *strigosus*	1(1),3(1)	Dobson et al. (2012)
Antioxidant capacity (FRAP, TEAC)	QTL	Glen Moy × Latham	*R. idaeus* subs. *idaeus* × *R. idaeus* subs. *strigosus*	1(2),5(2),6(2)	Dobson et al. (2012)
Cyanidin 3-*O*-sophoroside (C3S9)	QTL	96395S1 × Latham	*R. idaeus* subs. *idaeus* × *R. idaeus* subs. *strigosus*	2(3),3(4),6(2),7(2)	Bushakra et al. (2013)
Cyanidin 3-*O*-rutinoside (C3XR_C3R)	QTL	96395S1 × Latham	*R. idaeus* subs. *idaeus* × *R. idaeus* subs. *strigosus*	2(3),3(1),5(1),6(3),7(1)	Bushakra et al. (2013)
Cyanidin 3-*O*-sambubioside (C3G_C3Sb)	QTL	96395S1 × Latham	*R. idaeus* subs. *idaeus* × *R. idaeus* subs. *strigosus*	1(1),2(1),3(3),6(1),7(1)	Bushakra et al. (2013)
Cyanidin 3-*O*-glucoside (C3G)	QTL	96395S1 × Latham	*R. idaeus* subs. *idaeus* × *R. idaeus* subs. *strigosus*	3(2),6(1),7(1)	Bushakra et al. (2013)
Cyanidin 3-*O*-glucosylrutinoside (C3GR)	QTL	96395S1 × Latham	*R. idaeus* subs. *idaeus* × *R. idaeus* subs. *strigosus*	2(3),3(5),6(1),7(1)	Bushakra et al. (2013)
Pelargonodin 3-*O*-rutinoside (P3R)	QTL	96395S1 × Latham	*R. idaeus* subs. *idaeus* × *R. idaeus* subs. *strigosus*	2(2),5(3)	Bushakra et al. (2013)
Primocane fruiting	MTL	Prime-Jim × Arapaho	*Rubus* × *Rubus*	7	Castro et al. (2013)

Thornlessness (gene *s*)	MTL	Prime-Jim × Arapaho	*Rubus* × *Rubus*	4	Castro et al. (2013)
α-Ionol	QTL	Glen Moy × Latham	*R. idaeus* subs. *idaeus* × *R. idaeus* subs. *strigosus*	5(3),6(2)	Paterson et al. (2013)
α-Ionone	QTL	Glen Moy × Latham	*R. idaeus* subs. *idaeus* × *R. idaeus* subs. *strigosus*	6(2),7(2)	Paterson et al. (2013)
β-Ionone	QTL	Glen Moy × Latham	*R. idaeus* subs. *idaeus* × *R. idaeus* subs. *strigosus*	1(2),5(2),6(2)	Paterson et al. (2013)
β-Demascenone	QTL	Glen Moy × Latham	*R. idaeus* subs. *idaeus* × *R. idaeus* subs. *strigosus*	5(2),6(2),7(2)	Paterson et al. (2013)
Linalool	QTL	Glen Moy × Latham	*R. idaeus* subs. *idaeus* × *R. idaeus* subs. *strigosus*	1(3),7(2)	Paterson et al. (2013)
Geraniol	QTL	Glen Moy × Latham	*R. idaeus* subs. *idaeus* × *R. idaeus* subs. *strigosus*	3(1),5(1)	Paterson et al. (2013)
Benzyl alcohol	QTL	Glen Moy × Latham	*R. idaeus* subs. *idaeus* × *R. idaeus* subs. *strigosus*	6(2)	Paterson et al. (2013)
Acetic acid	QTL	Glen Moy × Latham	*R. idaeus* subs. *idaeus* × *R. idaeus* subs. *strigosus*	4(1)	Paterson et al. (2013)
Hexanoic acid	QTL	Glen Moy × Latham	*R. idaeus* subs. *idaeus* × *R. idaeus* subs. *strigosus*	3(1),6(1)	Paterson et al. (2013)
Hexanol	QTL	Glen Moy × Latham	*R. idaeus* subs. *idaeus* × *R. idaeus* subs. *strigosus*	6(1)	Paterson et al. (2013)
Raspberry ketone	QTL	Glen Moy × Latham	*R. idaeus* subs. *idaeus* × *R. idaeus* subs. *strigosus*	3(1),6(1)	Paterson et al. (2013)
Cane splitting	QTL	Glen Moy × Latham	*R. idaeus* subs. *idaeus* × *R. idaeus* subs. *strigosus*	1(3),3(2),5(2),6(5)	Woodhead et al. (2013)

(*Continued*)

Table 1 (*Continued*)

Trait	Marker	Map	Species	LG(#QTLs)	References
Leaf trichomes (hairs)	QTL	Glen Moy × Latham	*R. idaeus* subs. *idaeus* × *R. idaeus* subs. *strigosus*	2(1),3(1),6(2)	Graham et al. (2014)
Leaf density	QTL	Glen Moy × Latham	*R. idaeus* subs. *idaeus* × *R. idaeus* subs. *strigosus*	1(2),6(3)	Graham et al. (2014)
Cane number (density)	QTL	Glen Moy × Latham	*R. idaeus* subs. *idaeus* × *R. idaeus* subs. *strigosus*	6(3),7(1)	Graham et al. (2014)
Bush density	QTL	Glen Moy × Latham	*R. idaeus* subs. *idaeus* × *R. idaeus* subs. *strigosus*	1(1),6(1)	Graham et al. (2014)
Overall density	QTL	Glen Moy × Latham	*R. idaeus* subs. *idaeus* × *R. idaeus* subs. *strigosus*	2(1),6(4)	Graham et al. (2014)
Lateral length and number	QTL	Glen Moy × Latham	*R. idaeus* subs. *idaeus* × *R. idaeus* subs. *strigosus*	5(1),6(2)	Graham et al. (2014)
Thorn density (prickle/spine density)	QTL	(*R. parvifolius* × Tulameen) × Qualicum	(*R. parvifolius* × *R. idaeus*) × *R. Idaeus*	4(2),6(2)	Molina-Bravo et al. (2014)
Thornlessness (gene s)	MTL	(*R. parvifolius* × Tulameen) × Qualicum	(*R. parvifolius* × *R. idaeus*) × *R. Idaeus*	4	Molina-Bravo et al. (2014)
Chilling requirement	QTL	(*R. parvifolius* × Tulameen) × Qualicum	(*R. parvifolius* × *R. idaeus*) × *R. Idaeus*	1(2),4(2),5(1),6(2)	Molina-Bravo et al. (2014)
Fruit size	QTL	(*R. parvifolius* × Tulameen) × Qualicum	(*R. parvifolius* × *R. idaeus*) × *R. Idaeus*	5(2)	Molina-Bravo et al. (2014)
Fruit shape	QTL	(*R. parvifolius* × Tulameen) × Qualicum	(*R. parvifolius* × *R. idaeus*) × *R. Idaeus*	3(2)	Molina-Bravo et al. (2014)
Flower colour	QTL	(*R. parvifolius* × Tulameen) × Qualicum	(*R. parvifolius* × *R. idaeus*) × *R. Idaeus*	5(2),6(2)	Molina-Bravo et al. (2014)

Fruit colour	QTL	(*R. parvifolius* × Tulameen) × Qualicum	(*R. parvifolius* × *R. idaeus*) × *R. Idaeus*	1(2),5(2)	Molina-Bravo et al. (2014)
Aphid resistance (*Amphorophora agathonica*, Ag4)	MTL	ORUS3021-2 × ORUS4153-1	*R. occidentalis* × *R. occidentalis*	6	Bushakra et al. (2015a)
Crumbly fruit	QTL	Glen Moy × Latham	*R. idaeus* subs. *idaeus* × *R. idaeus* subs. *strigosus*	6(2),7(2)	Graham et al. (2015)
Fruit firmness	QTL	Glen Moy × Latham	*R. idaeus* subs. *idaeus* × *R. idaeus* subs. *strigosus*	6(4),7(1)	Simpson et al. (2017)
Fruit length	QTL	Glen Moy × Latham	*R. idaeus* subs. *idaeus* × *R. idaeus* subs. *strigosus*	7(1)	Simpson et al. (2017)
Fruit mass	QTL	Glen Moy × Latham	*R. idaeus* subs. *idaeus* × *R. idaeus* subs. *strigosus*	5(2),6(1),7(1)	Simpson et al. (2017)
Fruit resistance to applied force (force/mass)	QTL	Glen Moy × Latham	*R. idaeus* subs. *idaeus* × *R. idaeus* subs. *strigosus*	6(2)	Simpson et al. (2017)
Fruit firmness peak stress (2 * final load/pi * diameter * length)	QTL	Glen Moy × Latham	*R. idaeus* subs. *idaeus* × *R. idaeus* subs. *strigosus*	6(2)	Simpson et al. (2017)
Fruit firmness final load (force)	QTL	Glen Moy × Latham	*R. idaeus* subs. *idaeus* × *R. idaeus* subs. *strigosus*	5(1),6(3)	Simpson et al. (2017)

[a] Linkage group numbers only specific to that publication.
All linkage group numbers assigned were proposed by Bushakra et al. (2012b) based on the *Fragaria* (*F. vesca* × *F. bucharica*) reference map (Sargent et al., 2006).

*3.3 The black raspberry (***R. occidentalis***) and red raspberry (***R. idaeus***) genomes*

VanBuren et al. (2016) first reported a draft assembly of the black raspberry genome (*R. occidentalis*) using Illumina paired-end libraries. The draft assembly included 2226 scaffolds spanning 243 Mb, which corresponded to 83% of the estimated genome size, and the quality of the assembly was similar to that of other Illumina-based Rosaceae draft genome assemblies, including pear and Chinese plum. Additionally, chloroplast and mitochondrial genomes were assembled in that study. The draft genome contained widely dispersed centromeric satellite repeats and was largely collinear to the woodland strawberry genome (*Fragaria vesca*). Genes that may be related to key agronomic traits were also identified.

Most NGS-based approaches yield low-cost genome data, but the short-read lengths produce low-quality assemblies compared to plant genome assemblies developed through Sanger technology (Michael and VanBuren, 2015). Recent advances in whole-genome sequencing and bioinformatic methods have been used to overcome these limitations, and have been applied to improve upon the black raspberry genome assembly. Jibran et al. (2018) scaffolded the draft genome into a chromosome-scale assembly. High-throughput chromatin conformation capture (Hi-C) and proximity-guided assembly (PGA) were used to cluster and order 9650 out of 11 936 contigs into seven pseudo-chromosomes that covered over 97% of the total contig length (~223.8 Mb). Jibran et al. were able to locate existing genetic markers on the physical map that resolved multiple discrepancies in marker order on the genetic map. Centromeric regions were also inferred from recombination frequencies of genetic markers.

VanBuren et al. (2018) released a third version of the published black raspberry genome, utilizing long-read PacBio sequencing and Hi-C to finish and re-annotate the genome and includes thousands of new genes. The updated V3 assembly has a contig N50 of 5.1 Mb, representing approximately a 200-fold improvement over the previous Illumina-based version. Each of the 235 contigs is anchored and oriented into seven chromosomes. The newest chromosome-scale assembly contains 47 Mb of new sequences, including large pericentromeric regions and thousands of previously unannotated protein-coding genes related to agronomic traits such as disease resistance and secondary metabolite biosynthesis.

More recently, Wight et al. (2019) released a red raspberry (*R. idaeus*) draft genome assembly and annotation. Using a hybrid sequence assembly approach that included data from both long and short sequence reads, the *de novo* assembled genome consisted of 2145 scaffolds with a 95.3% completeness, while 80.1% of the genome was anchored onto seven chromosomes. 35 566 protein-coding genes were annotated using over 1 billion paired-end RNAseq reads, which represented a transcriptome completeness of 97.2%.

4 Future trends and conclusion

Despite the significant advances in molecular marker development, linkage mapping and QTL mapping, fewer molecular tools have been developed for raspberries and blackberries compared to other crops in Rosaceae; molecular markers are not used routinely for selecting seedlings or planning crosses in most active cultivar development programmes. While *Rubus* breeders have made significant progress in improving the productivity, quality and stress tolerance using classical phenotypic selection, efficiency could be improved with molecular breeding. The delay in development of useful molecular tools in *Rubus* crops is due, in part, to complications presented by polyploidy. Red and black raspberries are diploids (2n = 2x = 18), while cultivated blackberries are bred at multiple ploidy levels with the bulk of the cultivars at the tetraploid (2n = 4x = 28) and hexaploid levels (2n = 6x = 42) (Finn and Clark, 2017). The majority of the previously described advances in molecular breeding of *Rubus* species have been achieved at the diploid level. Most of the new analytical tools to support SNP calling, linkage and QTL mapping, association analyses, genomic selection and genome assembly have been designed for diploid species. Furthermore, the interpretation of markers in polyploid species is complicated by the problem of estimating the allelic dosage (e.g. AAAT, AATT, ATTT) in heterozygous individuals. Fortunately, major advances have been achieved in economically important autopolyploid species like potato (*Solanum tuberosum*) using new software for dosage calling (Carley et al., 2017; Voorrips et al., 2011), linkage and QTL mapping (Hackett et al., 2017), and genomic selection (Endelman et al., 2018) in autopolyploids. Blackberry breeders will be able to build on the advances made in other polyploid crops and use synteny with diploid raspberry and other diploid Rosaceae crops to their advantage as they begin to implement molecular breeding.

Molecular breeding strategies will likely facilitate breeding for traits related to sustainable cultivation, including nutrient efficiency, rhizosphere competence, climate resilience, health benefits and fruit quality, plant architecture and morphology for simplified management, and resistance to insect and diseases. Wild and unadapted germplasm are important sources of trait variation for many of these attributes. These genetic resources can be accessed at the U.S. National Clonal Germplasm Repository and other gene banks and repositories around the world. These germplasm accessions have been characterized with molecular markers and phenotyped for various traits of interest, but much more work remains to identify and characterize sources of stress resistance and other attributes of interest. Unfortunately, using unadapted germplasm in breeding programmes poses challenges and can be a difficult and time-consuming process to break linkage drag and introgress the trait of interest into an elite cultivar. For instance, primocane fruiting in blackberries was introgressed from 'Hillquist', a wild diploid *R. argutus* plant, and 37 years have transpired between the initial 'Brazos' × 'Hillquist' cross and the release of the first primocane-fruiting cultivars, 'Prime-Jim®' and 'Prime-Jan®' in 2005 (Clark, 2008). Molecular markers linked to target traits identified through QTL or association mapping may facilitate more rapid introgression of traits from wild relatives into adapted germplasm.

QTL studies require phenotyping numerous individuals within study populations in multi-year evaluations. Although current genotyping technology can process large sample sizes and generate vast amounts of data, phenotyping large populations for complex traits is still a limitation. High-throughput phenotyping methods can aid assessing complex traits. Nanoparticle and nanotechnology is an area of research that has gained overwhelming recognition and has expanded to many fields and applications, including clinical diagnosis, environmental monitoring and food analysis (Fanali et al., 2011; Sapsford et al., 2013; Sattler, 2011; Suginta et al., 2013; Wu et al., 2007; Zhang et al., 2003). One area of particular interest is the application of biosensors to quantify important plant molecules. Biosensors are carbon nanotubes or metal nanoparticles in suspension linked to a specific probe (Sattler, 2011). These probes range from detection enzymes, such as horseradish peroxidases (Xiao et al., 1999), to oligonucleotide probes that function as electrochemical DNA sensors (Bai et al., 2007). Despite numerous applications, biosensors have not been extensively used (if at all) for phenotyping breeding populations. One example where biosensor technology can aid the field of phenomics and QTL discovery is through the use of quantum dots (QD) for quantifying phenolic compounds. QD are colloidal nanocrystalline semiconductors that possess unique spectral properties (Akshath et al., 2014). These structures are used as enzymatic biosensors that function as reporters through the transfer of electrons that modulate their fluorescent properties, that is quenching of the QDs' fluorescence (Clarke et al., 2006). Akshath et al. (2014) have successfully used immobilized laccase-QD technology to convert polyphenols into quinones as an ultrasensitive optical label to detect polyphenols by way of 15-min assays. Quantifying specific phenolic compounds in a relatively short time can have great applicability, as antioxidants and other phenolic compounds are very important in *Rubus* crops. Other potential applications include antibody bound nanoparticles that can detect proteins of interest, such defence response proteins or viruses in plants. Huang et al. (2012) developed a modified graphene paste electrode with gold nanoparticles and a Nafion-L-cysteine composite film as an electrochemical immunosensor for the detection of hepatitis B surface antigen. Further advancements in this technology allow us to envisage QD assays, nanoparticle devices or nanotube sensors that can detect molecules of interest in a relatively short time, and in multi-well assays to evaluate large mapping populations.

5 Where to look for further information

Integrating molecular tools into conventional breeding within Rosaceae is a topic of interest. Peace (2017) discusses the future prospects and perspectives of the applications of DNA-informed breeding for rosaceous crops in detail. The USDA-funded project RosBREED 2 addresses the need to develop and validate modern genomics and genetics tools for implementation in breeding programmes of rosaceous crops (Iezzoni et al., 2017); blackberry is on the list of target crops. More information can be found on the RosBREED website: https://www.rosbreed.org/home.

On the subject of breeding and genetic resources:

- Worldwide raspberry breeding programmes (Finn and Hancock, 2008)
- Genomic and breeding resources in raspberry (Graham and Woodhead 2010; Hall et al. 2009)
- Cultivar release and intellectual property in raspberry (Carew et al., 2009)
- Genetic diversity and erosion of berries (Debnath, 2015; Finn, 2008)
- Exhaustive review of breeding goals, breeding techniques and regions of domestication of raspberry and blackberry (Clark et al., 2007; Finn, 2008; Finn and Hancock, 2008; Finn and Clark, 2011; Kempler et al., 2011)
- Developments in raspberry breeding and molecular marker applications in Scotland (Jennings et al., 2016) and cultivar development in Montenegro (Galić and Vidaković, 2012)

Williams et al. (2017) explore further advancements in high-throughput phenotyping using spectral imaging in raspberry. In the context of global climate change and sustainable in fruit production, Fernandez et al. (2018) review what is known about heat tolerance in *Rubus* crops.

6 References

Abuzinadah, R. A. and Read, D. J. 1989. The role of proteins in the nitrogen nutrition of ectomycorrhizal plants. IV. The utilization of peptides by birch (*Betula pendula* L.) infected with different mycorrhizal fungi. *New Phytologist*. 112(1), 55–60.

Aguirre-Gutiérrez, J., van Treuren, R., Hoekstra, R. and van Hintum, T. J. L. 2017. Crop wild relatives range shifts and conservation in Europe under climate change. *Diversity and Distributions* 23(7), 739–50.

Akello, J., Dubois, T., Coyne, D. and Kyamanywa, S. 2008. Endophytic *Beauveria bassiana* in banana (*Musa* spp.) reduces banana weevil (*Cosmopolites sordidus*) fitness and damage. *Crop Protection* 27(11), 1437–41. doi:10.1016/j.cropro.2008.07.003.

Akshath, U. S., Shubha, L. R., Bhatt, P. and Thakur, M. S. 2014. Quantum dots as optical labels for ultrasensitive detection of polyphenols. *Biosensors and Bioelectronics* 57, 317–23. doi:10.1016/j.bios.2014.01.038.

Alleyne, V. and Clark, J. R. 1997. Fruit composition of 'Arapaho' blackberry following nitrogen fertilization. *HortScience* 32(2), 282–3.

Amsellem, L., Dutech, C. and Billotte, N. 2001. Isolation and characterization of polymorphic microsatellite loci in *Rubus alceifolius* Poir. (Rosaceae), an invasive weed in La Réunion island. *Molecular Ecology Notes* 1(1-2), 33–5. doi:10.1046/j.1471-8278.2000.00013.x.

Bai, S.-L., Zhong, X., Ma, L., Zheng, W., Fan, L.-M., Wei, N. and Deng, X. W. 2007. A simple and reliable assay for detecting specific nucleotide sequences in plants using optical thin-film biosensor chips. *The Plant Journal* 49(2), 354–66. doi:10.1111/j.1365-313X.2006.02951.x.

Bailey, B. A., Bae, H., Melnick, R. and Crozier, J. 2011. The endophytic *Trichoderma hamatum* isolate DIS 219b enhances seedling growth and delays the onset of drought stress in Theobroma cacao. In: Pirttilä, A. M. and Frank, A. C. (Eds), *Endophytes of Forest Trees*. Springer, Dordrecht, The Netherlands.

Ballington, J. R. and Fernandez, G. E. 2008. Breeding raspberries to warm humid climates with fluctuating temperatures in winter. *Acta Horticulturae* 777(777), 87–90. doi:10.17660/ActaHortic.2008.777.9.

Bañados, M. P., Strik, B. C., Bryla, D. R. and Righetti, T. L. 2012. Response of highbush blueberry to nitrogen fertilizer during field establishment, I: accumulation and allocation of fertilizer nitrogen and biomass. *HortScience* 47(5), 648–55.

Bassil, N. V., Muminova, M. and Njuguna, W. 2010. Microsatellite-based fingerprinting of western blackberries from plants, IQF berries and puree. *Acta Horticulturae* 859(859), 73–80. doi:10.17660/ActaHortic.2010.859.6.

Bassil, N. V., Nyberg, A., Hummer, K. E., Graham, J., Dossett, M. and Finn, C. E. 2012. A universal fingerprinting set for red raspberry. *Acta Horticulturae* 946(946), 83–7. doi:10.17660/ActaHortic.2012.946.10.

Bassil, N. V., Nyberg, A. M., Finn, C. E., Clark, J. R., Peace, C. P. and Iezzoni, A. 2016. Development of a multiplexed fingerprinting set in blackberry. *Acta Horticulturae* 1133(1133), 89–96. doi:10.17660/ActaHortic.2016.1133.14.

Becking, J. H. 1979. Nitrogen fixation by *Rubus ellipticus* J. E. Smith. *Plant and Soil* 53(4), 541–5.

Bravo Durán, V., la Cruz Malavassi de, E., Herrera Ledezma, G. and Ramírez-Muñoz, F. 2013. Uso de plaguicidas en cultivos agrícolas como herramienta para el monitoreo de peligros en salud. *Uniciencia* 27(1), 351–76.

Burrack, H. J., Fernandez, G. E., Spivey, T. and Kraus, D. A. 2013. Variation in selection and utilization of host crops in the field and laboratory by *Drosophila suzukii* Matsumara (Diptera: Drosophilidae), an invasive frugivore. *Pest Management Science* 69(10), 1173–80.

Bus, V., Esmenjaud, D., Buck, E. and Laurens, F. 2009. Application of genetic markers in rosaceous crops. In: Folta, K. M. and Gardiner, S. E. (Eds), *Genetics and Genomics of Rosaceae*. Springer, New York, NY, pp. 563–99.

Bushakra, J. M., Sargent, D. J., Cabrera, A., Crowhurst, R., Girona, E. L., Velasco, R., Symonds, V. V., van der Knaap, E., Troggio, M., Gardiner, S. E., et al. 2012a. Rosaceae conserved orthologous set (RosCOS) markers as a tool to assess genome synteny between *Malus* and *Fragaria*. *Tree Genetics and Genomes* 8(4), 643–58. doi:10.1007/s11295-011-0450-y.

Bushakra, J. M., Stephens, M. J., Atmadjaja, A. N., Lewers, K. S., Symonds, V. V., Udall, J. A., Chagné, D., Buck, E. J. and Gardiner, S. E. 2012b. Construction of black (*Rubus occidentalis*) and red (*R. idaeus*) raspberry linkage maps and their comparison to the genomes of strawberry, apple, and peach. *Theoretical and Applied Genetics* 125(2), 311-27. doi:10.1007/s00122-012-1835-5.

Bushakra, J. M., Krieger, C., Deng, D., Stephens, M. J., Allan, A. C., Storey, R., Symonds, V. V., Stevenson, D., McGhie, T., Chagné, D., et al. 2013. QTL involved in the modification of cyanidin compounds in black and red raspberry fruit. *Theoretical and Applied Genetics* 126, 847-65. doi:10.1007/s00122-012-2022-4.

Bushakra, J. M., Bryant, D. W., Dossett, M., Vining, K. J., VanBuren, R., Gilmore, B. S., Lee, J., Mockler, T. C., Finn, C. E. and Bassil, N. V. 2015a. A genetic linkage map of black raspberry (*Rubus occidentalis*) and the mapping of Ag4 conferring resistance to the aphid *Amphorophora agathonica*. *Theoretical and Applied Genetics* 128(8), 1631-46.

Bushakra, J. M., Lewers, K. S., Staton, M. E., Zhebentyayeva, T. and Saski, C. A. 2015b. Developing expressed sequence tag libraries and the discovery of simple sequence repeat markers for two species of raspberry (*Rubus* L.). *BMC Plant Biology* 15(1), 258. doi:10.1186/s12870-015-0629-8.

Byrne, D. H. 2012. Trends in fruit breeding. In: Badenes, M. L. and Byrne, D. H. (Eds), *Fruit Breeding*. Springer, Boston, MA, pp. 3-36.

California Strawberry Commission. 2018. U.S. retail category trends. California Strawberry Commission, pp. 1-8. Available at: http://www.calstrawberry.com/en-us/market-data/retail-category-trends (accessed 11 September 2018).

Carew, R., Kempler, C., Moore, P. and Walters, T. 2009. Developments in raspberry production, cultivar releases, and intellectual property rights: a comparative study of British Columbia and Washington State. *International Journal of Fruit Science* 9(1), 54-77. doi:10.1080/15538360902801338.

Carley, C. A. S., Coombs, J. J., Douches, D. S., Bethke, P. C., Palta, J. P., Novy, R. G. and Endelman, J. B. 2017. Automated tetraploid genotype calling by hierarchical clustering. *Theoretical and Applied Genetics* 130(4), 717-26.

Castillo, N. R. F., Reed, B. M., Graham, J., Fernández-Fernández, F. and Bassil, N. V. 2010. Microsatellite markers for raspberry and blackberry. *Journal of the American Society for Horticultural Science* 135(3), 271-8. doi:10.21273/JASHS.135.3.271.

Castro, P., Stafne, E. T., Clark, J. R. and Lewers, K. S. 2013. Genetic map of the primocane-fruiting and thornless traits of tetraploid blackberry. *Theoretical and Applied Genetics* 126(10), 2521-32.

Chaplin, M. H. and Martin, L. W. 1980. The effect of nitrogen and boron fertilizer applications on leaf levels, yield and fruit size of the red raspberry. *Communications in Soil Science and Plant Analysis* 11(6), 547-56. doi:10.1080/00103628009367062.

Clark, J. R. 2008. Primocane-fruiting blackberry breeding. *HortScience* 43(6), 1637-9.

Clark, J. R. and Finn, C. E. 2008. New trends in blackberry breeding. *Acta Horticulturae* 777(777), 41-8. doi:10.17660/ActaHortic.2008.777.2.

Clark, J. R. and Finn, C. E. 2014. Blackberry cultivation in the world. *Revista Brasileira de Fruticultura* 36(1), 46-57. doi:10.1590/0100-2945-445/13.

Clark, J. R. and Moore, J. N. 2008. 'Natchez' thornless blackberry. *HortScience* 43(6), 1897-9.

Clark, J. R., Howard, L. and Talcott, S. 2002. Antioxidant activity of blackberry genotypes. *Acta Horticulturae* 585(585), 475-80. doi:10.17660/ActaHortic.2002.585.76.

Clark, J. R., Stafne, E. T., Hall, H. K. and Finn, C. E. 2007. Blackberry breeding and genetics. In: Janick, J. (Ed.), *Plant Breeding Reviews* (vol. 29). John Wiley & Sons, Inc.

Clarke, S. J., Hollmann, C. A., Zhang, Z., Suffern, D., Bradforth, S. E., Dimitrijevic, N. M., Minarik, W. G. and Nadeau, J. L. 2006. Photophysics of dopamine-modified quantum dots and effects on biological systems. *Nature Materials* 5(5), 409–17. doi:10.1038/nmat1631.

Connor, A. M., Finn, C. E. and Alspach, P. A. 2005. Genotypic and environmental variation in antioxidant activity and total phenolic content among blackberry and hybridberry cultivars. *Journal of the American Society for Horticultural Science* 130(4), 527–33. doi:10.21273/JASHS.130.4.527.

Contreras, M., Loeza, P. D., Villegas, J., Farias, R. and Santoyo, G. 2016. A glimpse of the endophytic bacterial diversity in roots of blackberry plants (*Rubus fruticosus*). *Genetics and Molecular Research* 15(3), 1–11. doi:10.4238/gmr.15038542.

Cook, R. J. 2000. Advances in plant health management in the twentieth century. *Annual Review of Phytopathology* 38(1), 95–116. doi:10.1146/annurev.phyto.38.1.95.

Daubeny, H. A. and Pepin, H. S. 1974. Variations among red raspberry cultivars and selections in susceptibility to the fruit rot causal organisms *Botrytis cinerea* and *Rhizopus* spp. *Canadian Journal of Plant Science* 54(3), 511–6. doi:10.4141/cjps74-086.

Dawson, J. C., Murphy, K. M. and Jones, S. S. 2007. Decentralized selection and participatory approaches in plant breeding for low-input systems. *Euphytica* 160(2), 143–54.

Dean, D. M., Zebarth, B. J., Kowalenko, C. G., Paul, J. W. and Chipperfield, K. 2000. Poultry manure effects on soil nitrogen processes and nitrogen accumulation in red raspberry. *Canadian Journal of Plant Science* 80(4), 849–60. doi:10.4141/P99-136.

Debnath, S. C. 2015. Genetic diversity and erosion in berries. In: Nandwani, D. (Ed.), *Organic Farming for Sustainable Agriculture*. Springer International Publishing, Cham, pp. 75–129.

Decroocq, V., Favé, M. G., Hagen, L., Bordenave, L. and Decroocq, S. 2003. Development and transferability of apricot and grape EST microsatellite markers across taxa. *Theoretical and Applied Genetics* 106(5), 912–22. doi:10.1007/s00122-002-1158-z.

Dillard, C. J. and German, J. B. 2000. Phytochemicals: nutraceuticals and human health. *Journal of the Science of Food and Agriculture* 80(12), 1744–56. doi:10.1002/1097-0010(20000915)80:12<1744::AID-JSFA725>3.0.CO;2-W.

Dirlewanger, E., Graziano, E., Joobeur, T., Garriga-Calderé, F., Cosson, P., Howad, W. and Arús, P. 2004. Comparative mapping and marker-assisted selection in Rosaceae fruit crops. *Proceedings of the National Academy of Sciences* 101(26), 9891–6. doi:10.1073/pnas.0307937101.

Dobson, P., Graham, J., Stewart, D., Brennan, R., Hackett, C. A. and McDougall, G. J. 2012. Over-seasons analysis of quantitative trait loci affecting phenolic content and antioxidant capacity in raspberry. *Journal of Agricultural and Food Chemistry* 60(21), 5360–6. doi:10.1021/jf3005178.

Dossett, M., Bassil, N. V., Lewers, K. S. and Finn, C. E. 2012. Genetic diversity in wild and cultivated black raspberry (*Rubus occidentalis* L.) evaluated by simple sequence repeat markers. *Genetic Resources and Crop Evolution* 59(8), 1849–65. doi:10.1007/s10722-012-9808-8.

Dossett, M., Bushakra, J. M., Gilmore, B., Koch, C. A., Kempler, C., Finn, C. E. and Bassil, N. V. 2015. Development and transferability of black and red raspberry microsatellite markers from short-read sequences. *Journal of the American Society for Horticultural Science* 140(3), 243–52. doi:10.21273/JASHS.140.3.243.

Endelman, J. B., Carley, C. A. S., Bethke, P. C., Coombs, J. J., Clough, M. E., da Silva, W. L., De Jong, W. S., Douches, D. S., Frederick, C. M., Haynes, K. G., et al. 2018. Genetic variance partitioning and genome-wide prediction with allele dosage information in autotetraploid potato. *Genetics* 209(1), 77–87. doi:10.1534/genetics.118.300685.

Fan-Chiang, H.-J. and Wrolstad, R. E. 2005. Anthocyanin pigment composition of blackberries. *Journal of Food Science* 70(3), C198–202. doi:10.1111/j.1365-2621.2005.tb07125.x.

Fanali, C., Dugo, L., D'Orazio, G., Lirangi, M., Dachà, M., Dugo, P. and Mondello, L. 2011. Analysis of anthocyanins in commercial fruit juices by using nano-liquid chromatography-electrospray-mass spectrometry and high-performance liquid chromatography with UV-vis detector. *Journal of Separation Science* 34(2), 150–9. doi:10.1002/jssc.201000665.

FAO. 2016. Food and agriculture data (FAOSTAT). Available at: http://www.fao.org/faostat/en/#home (accessed on 23 April 2018).

Fathi, G., McDonald, G. K. and Lance, R. C. M. 1997. Responsiveness of barley cultivars to nitrogen fertiliser. *Australian Journal of Experimental Agriculture* 37(2), 199–211. doi:10.1071/EA96052.

Fernandez, G. E., Molina-Bravo, R. and Takeda, F. 2018. What we know about heat stress in *Rubus*. In: Graham, J. and Brennan, R. (Eds), *Raspberry*. Springer International Publishing, Cham, pp. 29–40. doi:10.1007/978-3-319-99031-6_3.

Fernández-Fernández, F., Antanaviciute, L., Govan, C. L. and Sargent, D. J. 2011. Development of a multiplexed microsatellite set for fingerprinting red raspberry (*Rubus idaeus*) germplasm and its transferability to other *Rubus* species. *Journal of Berry Research* 1, 177–87.

Fernandez-Salvador, J., Strik, B. C. and Bryla, D. R. 2015. Liquid corn and fish fertilizers are good options for fertigation in blackberry cultivars grown in an organic production system. *HortScience* 50(2), 225–33.

Finlay, R. D., Frostegård, Å and Sonnerfeldt, A.-M. 1992. Utilization of organic and inorganic nitrogen sources by ectomycorrhizal fungi in pure culture and in symbiosis with *Pinus contorta* Dougl. ex Loud. *New Phytologist* 120(1), 105–15. doi:10.1111/j.1469-8137.1992.tb01063.x.

Finn, C. E. 2008. Blackberries. In: Hancock, J. F. (Ed.), *Temperate Fruit Crop Breeding*. Springer, Dordrecht, pp. 83–114. doi:10.1007/978-1-4419-0763-9_5.

Finn, C. E. and Clark, J. R. 2011. Blackberry. In: Badenes, M. L. and Byrne, D. H. (Eds), *Fruit Breeding*. Springer US, Boston, MA, pp. 151–90. doi:10.1007/978-1-4419-0763-9_5.

Finn, C. E. and Clark, J. R. 2017. Cultivar development and selection. In: Hall, H. K. and Funt, R. C. (Eds), *Blackberries and Their Hybrids*. CABI Publishing, Boston, MA, pp. 63–92.

Finn, C. E. and Hancock, J. F. 2008. Raspberries. In: Hancock, J. F. (Ed.), *Temperate Fruit Crop Breeding*. Springer Netherlands, Dordrecht, pp. 359–92. doi:10.1007/978-1-4020-6907-9_12.

Galić, D. and Vidaković, Z. 2012. Raspberry breeding in Montenegro. *Acta Horticulturae* 946, 151–6. doi:10.17660/ActaHortic.2012.946.22.

Gerhardson, B. 2002. Biological substitutes for pesticides. *Trends in Biotechnology* 20(8), 338–43. doi:10.1016/S0167-7799(02)02021-8.

Girichev, V., Hanke, M.-V., Peil, A. and Flachowsky, H. 2017. SSR fingerprinting of a German *Rubus* collection and pedigree based evaluation on trueness-to-type. *Genetic Resources and Crop Evolution* 64(1), 189–203. doi:10.1007/s10722-015-0345-0.

Graham, J. and McNicol, R. J. 1995. An examination of the ability of RAPD markers to determine the relationships within and between Rubus species. *Theoretical and Applied Genetics* 90(7–8), 1128–32. doi:10.1007/BF00222932.

Graham, J. and Woodhead, M. 2010. *Rubus*. In: Kole, C. (Ed.), *Wild Crop Relatives: Genomic and Breeding Resources*. Springer, Berlin Heidelberg, pp. 179–96. doi:10.1007/978-3-642-16057-8_9.

Graham, J., Squire, G. R., Marshall, B. and Harrison, R. E. 1997. Spatially dependent genetic diversity within and between colonies of wild raspberry *Rubus idaeus* detected using RAPD markers. *Molecular Ecology* 6(11), 1001–8. doi:10.1046/j.1365-294X.1997.00272.x.

Graham, J., Smith, K., Woodhead, M. and Russell, J. 2002. Development and use of simple sequence repeat SSR markers in *Rubus* species. *Molecular Ecology Notes* 2(3), 250–2. doi:10.1046/j.1471-8286.2002.00203.x.

Graham, J., Smith, K., MacKenzie, K., Jorgenson, L., Hackett, C. and Powell, W. 2004. The construction of a genetic linkage map of red raspberry (*Rubus idaeus* subsp. *idaeus*) based on AFLPs, genomic-SSR and EST-SSR markers. *Theoretical and Applied Genetics* 109(4), 740–9. doi:10.1007/s00122-004-1687-8.

Graham, J., Smith, K., Tierney, I., MacKenzie, K. and Hackett, C. A. 2006. Mapping gene *H* controlling cane pubescence in raspberry and its association with resistance to cane Botrytis and spur blight, rust and cane spot. *Theoretical and Applied Genetics* 112(5), 818–31. doi:10.1007/s00122-005-0184-z.

Graham, J., Hackett, C. A., Smith, K., Woodhead, M., Hein, I. and McCallum, S. 2009. Mapping QTLs for developmental traits in raspberry from bud break to ripe fruit. *Theoretical and Applied Genetics* 118(6), 1143–55. doi:10.1007/s00122-009-0969-6.

Graham, J., Hackett, C. A., Smith, K., Woodhead, M., MacKenzie, K., Tierney, I., Cooke, D., Bayer, M. and Jennings, N. 2011. Towards an understanding of the nature of resistance to *Phytophthora* root rot in red raspberry. *Theoretical and Applied Genetics* 123(4), 585–601. doi:10.1007/s00122-011-1609-5.

Graham, J., Hackett, C. A., Smith, K., Karley, A. J., Mitchell, C., Roberts, H. and O'Neill, T. 2014. Genetic and environmental regulation of plant architectural traits and opportunities for pest control in raspberry. *Annals of Applied Biology* 165(3), 318–28.

Graham, J., Smith, K., McCallum, S., Hedley, P. E., Cullen, D. W., Dolan, A., Milne, L., McNicol, J. W. and Hackett, C. A. 2015. Towards an understanding of the control of 'crumbly' fruit in red raspberry 4(1), 223. doi:10.1186/s40064-015-1010-y.

Gupta, P. K., Rustgi, S. and Mir, R. R. 2008. Array-based high-throughput DNA markers for crop improvement. *Heredity* 101(1), 5–18. doi:10.1038/hdy.2008.35.

Hackett, C. A., Boskamp, B., Vogogias, A., Preedy, K. F. and Milne, I. 2017. TetraploidSNPMap: software for linkage analysis and QTL mapping in autotetraploid populations using SNP dosage data (3rd edition). *Journal of Heredity* 108(4), 438–42.

Hall, H. K. 2017. World blackberry production. In: Hall, H. K. and Funt, R. C. (Eds), *Blackberries and Their Hybrids*. CABI Publishing, Boston, MA, pp. 308–14.

Hall, H., Hummer, K. E., Jamieson, A. R., Jennings, S. N. and Weber, C. A. 2009. Raspberry breeding and genetics. *Plant Breeding Reviews* 32, 39–382.

Halvorsen, B. L., Carlsen, M. H., Phillips, K. M., Bøhn, S. K., Holte, K., Jacobs, D. R. J. and Blomhoff, R. 2006. Content of redox-active compounds (ie, antioxidants) in foods consumed in the United States. *The American Journal of Clinical Nutrition* 84(1), 95–135. doi:10.1093/ajcn/84.1.95.

Harkins, R. H., Strik, B. C. and Bryla, D. R. 2014. Weed management practices for organic production of trailing blackberry, II. Accumulation and loss of biomass and nutrients. *HortScience* 49(1), 35–43.

Harshman, J. M., Lewers, K. S., Jurick, W. M. and Walsh, C. S. 2014. Breeding selection efficiency for raspberry postharvest shelf life affected by storage temperature and harvest season. *Euphytica* 199(3), 283–92. doi:10.1007/s10681-014-1123-3.

Haytowitz, D. B. and Bhagwat, S. 2010. USDA database for the oxygen radical absorbance capacity (ORAC) of selected foods, Release 2. US Department of Agriculture, pp. 10–48.

Heiberg, N. 2002. Effect of vegetation control and nitrogen fertilization in red raspberry. *Acta Horticulturae* 585(585), 579–83. doi:10.17660/ActaHortic.2002.585.93.

Hetrick, B. A. D., Wilson, G. W. T., Gil, B. S. and Cox, T. S. 1995. Chromosome location of mycorrhizal responsive genes in wheat. *Canadian Journal of Botany* 73(6), 891–7. doi:10.1139/b95-097.

Huang, K.-J., Li, J., Liu, Y.-M., Cao, X., Yu, S. and Yu, M. 2012. Disposable immunoassay for hepatitis B surface antigen based on a graphene paste electrode functionalized with gold nanoparticles and a nafion-cysteine conjugate. *Microchimica Acta* 177(3–4), 419–26. doi:10.1007/s00604-012-0805-6.

Hummer, K. E. 2017. Blackberries: an introduction. In: Hall, H. K. and Funt, R. C. (Eds), *Blackberries and Their Hybrids*. CABI Publishing, Boston, MA, pp. 1–16.

Iezzoni, A., Peace, C., Main, D., Bassil, N., Coe, M., Finn, C., Gasic, K., Luby, J., Hokanson, S., McFerson, J., et al. 2017. RosBREED2: progress and future plans to enable DNA-informed breeding in the *Rosaceae*. *Acta Horticulturae* 1172, 115–18. doi:10.17660/ActaHortic.2017.1172.20

Jennings, D. 1988. *Raspberries and Blackberries: Their Breeding, Disease and Growth*. Academic Press, London.

Jennings, S. N., Graham, J., Ferguson, L. and Young, V. 2016. New developments in raspberry breeding in Scotland. *Acta Horticulturae* 1133, 23–8. doi:10.17660/ActaHortic.2016.1133.4.

Jiang, Y., Yuan, D., Xie, S., Li, L., Zhang, G. and He, R. 2006. Groundwater quality and land use change in a typical karst agricultural region: a case study of Xiaojiang watershed, Yunnan. *Journal of Geographical Sciences* 16(4), 405–14. doi:10.1007/s11442-006-0403-9.

Jibran, R., Dzierzon, H., Bassil, N., Bushakra, J. M., Edger, P. P., Sullivan, S., Finn, C. E., Dossett, M., Vining, K. J., VanBuren, R., et al. 2018. Chromosome-scale scaffolding of the black raspberry (*Rubus occidentalis* L.) genome based on chromatin interaction data. *Horticulture Research* 5(1), 1–11.

Jung, S., Staton, M., Lee, T., Blenda, A., Svancara, R., Abbott, A. and Main, D. 2007. GDR (Genome Database for Rosaceae): integrated web-database for Rosaceae genomics and genetics data. *Nucleic Acids Research* 36(Database), D1034–40. doi:10.1093/nar/gkm803.

Jung, S., Ficklin, S. P., Lee, T., Cheng, C. H., Blenda, A., Zheng, P., Yu, J., Bombarely, A., Cho, I., Ru, S., et al. 2014. The Genome Database for Rosaceae (GDR): year 10 update. *Nucleic Acids Research* 42(Database issue), D1237–44. doi:10.1093/nar/gkt1012.

Kassim, A., Poette, J., Paterson, A., Zait, D., McCallum, S., Woodhead, M., Smith, K., Hackett, C. and Graham, J. 2009. Environmental and seasonal influences on red raspberry anthocyanin antioxidant contents and identification of quantitative traits loci (QTL). *Molecular Nutrition and Food Research* 53(5), 625–34. doi:10.1002/mnfr.200800174.

Kempler, C., Hall, H. and Finn, C. E. 2011. Raspberry. In: Badenes, M. L. and Byrne, D. H. (Eds), *Fruit Breeding*. Springer US, Boston, MA, pp. 263–304. doi:10.1007/978-1-4419-0763-9_8.

Khoshgoftarmanesh, A. H., Schulin, R., Chaney, R. L., Daneshbakhsh, B. and Afyuni, M. 2010. Micronutrient-efficient genotypes for crop yield and nutritional quality in sustainable agriculture. A review. *Agronomy for Sustainable Development* 30(1), 83–107.

Kowalenko, C. G. 1994. Growing season changes in the concentration and distribution of macroelements in Willamette red raspberry plant parts. *Canadian Journal of Plant Science* 74(4), 833–9. doi:10.4141/cjps94-151.

Kowalenko, C. G. 2006. The effect of nitrogen and boron fertilizer applications on Willamette red raspberry growth, and on applied and other nutrients in the plant and soil over two growing seasons. *Canadian Journal of Plant Science* 86(1), 213–25. doi:10.4141/P04-036.

Lammerts van Bueren, E. T., Østergård, H., Goldringer, I. and Scholten, O. 2008. Plant breeding for organic and sustainable, low-input agriculture: dealing with genotype-environment interactions. *Euphytica* 163(3), 321–2. doi:10.1007/s10681-008-9731-4.

Lammerts van Bueren, E. T., Jones, S. S., Tamm, L., Murphy, K. M., Myers, J. R., Leifert, C. and Messmer, M. M. 2011. The need to breed crop varieties suitable for organic farming, using wheat, tomato and broccoli as examples: a review. *NJAS - Wageningen Journal of Life Sciences* 58(3-4), 193–205. doi:10.1016/j.njas.2010.04.001.

Larkan, N. J., Smith, S. E. and Barker, S. J. 2007. Position of the reduced mycorrhizal colonisation (Rmc) locus on the tomato genome map. *Mycorrhiza* 17(4), 311–8. doi:10.1007/s00572-007-0106-9.

Lee, J. 2017. Blackberry fruit quality components, composition, and potential health benefits. In: Hall, H. K. and Funt, R. C. (Eds), *Blackberries and Their Hybrids*. CABI Publishing, Boston, MA.

Lewers, K. S., Saski, C. A., Cuthbertson, B. J., Henry, D. C., Staton, M. E., Main, D. S., Dhanaraj, A. L., Rowland, L. J. and Tomkins, J. P. 2008. A blackberry (*Rubus* L.) expressed sequence tag library for the development of simple sequence repeat markers. *BMC Plant Biology* 8(1), 69. doi:10.1186/1471-2229-8-69.

Li, W. L., Zhang, C. H., Yang, H. Y., Wu, W. L. and Lyu, L. F. 2016. SSR marker information mining in blackberry from transcriptome sequences. *Acta Horticulturae* 1133(1133), 97–102. doi:10.17660/ActaHortic.2016.1133.15.

Liu, J., Yang, B., Ming, Y., Zhang, Y. and Cheng, Y. 2017. Identification of expressed sequence tag-simple sequence repeat markers from the de novo transcriptome sequence of red raspberry (*Rubus idaeus* L.). *HortScience* 52(4), 554–9.

Lopes, M. S., Maciel, G. B., Mendonca, D., Gil, F. S. and Da Câmara Machado, A. 2006. Isolation and characterization of simple sequence repeat loci in *Rubus hochstetterorum* and their use in other species from the Rosaceae family. *Molecular Ecology Notes* 6(3), 750–2. doi:10.1111/j.1471-8286.2006.01329.x.

Lopez, D. C. and Sword, G. A. 2015. The endophytic fungal entomopathogens *Beauveria bassiana* and *Purpureocillium lilacinum* enhance the growth of cultivated cotton (*Gossypium hirsutum*) and negatively affect survival of the cotton bollworm (*Helicoverpa zea*). *Biological Control* 89, 53–60. doi:10.1016/j.biocontrol.2015.03.010.

Luedeling, E. and Brown, P. H. 2011. A global analysis of the comparability of winter chill models for fruit and nut trees. *International Journal of Biometeorology* 55(3), 411–21. doi:10.1007/s00484-010-0352-y.

Luedeling, E., Gebauer, J. and Buerkert, A. 2009. Climate change effects on winter chill for tree crops with chilling requirements on the Arabian Peninsula. *Climatic Change* 96(1-2), 219-37. doi:10.1007/s10584-009-9581-7.

Luedeling, E., Girvetz, E. H., Semenov, M. A. and Brown, P. H. 2011. Climate change affects winter chill for temperate fruit and nut trees. *PLoS ONE* 6(5), e20155. doi:10.1371/journal.pone.0020155.

Lv, Z., Liu, X., Cao, W. and Zhu, Y. 2013. Climate change impacts on regional winter wheat production in main wheat production regions of China. *Agricultural and Forest Meteorology* 171-172, 234-48. doi:10.1016/j.agrformet.2012.12.008.

Machado, A. P. D. F., Pasquel-Reátegui, J. L., Barbero, G. F. and Martínez, J. 2015. Pressurized liquid extraction of bioactive compounds from blackberry (*Rubus fruticosus* L.) residues: a comparison with conventional methods. *Food Research International* 77, 675-83. doi:10.1016/j.foodres.2014.12.042.

Marulanda, M. L., López, A. M. and Aguilar, S. B. 2007. Genetic diversity of wild and cultivated Rubus species in Colombia using AFLP and SSR markers. *Crop Breeding and Applied Biotechnology* 7(3), 242-52. doi:10.12702/1984-7033.v07n03a03.

McCallum, S., Woodhead, M., Hackett, C. A., Kassim, A., Paterson, A. and Graham, J. 2010. Genetic and environmental effects influencing fruit colour and QTL analysis in raspberry. *Theoretical and Applied Genetics* 121(4), 611-27. doi:10.1007/s00122-010-1334-5.

Mezzetti, B. 2016. The sustainable improvement of European berry production, quality and nutritional value in a changing environment: strawberries, currants, blackberries, blueberries and raspberries - the EUBerry project. *Acta Horticulturae* 1117, 309-14.

Michael, T. P. and VanBuren, R. 2015. Progress, challenges and the future of crop genomes. *Current Opinion in Plant Biology* 24, 71-81. doi:10.1016/j.pbi.2015.02.002.

Miyashita, T., Kunitake, H., Yotsukura, N. and Hoshino, Y. 2015. Assessment of genetic relationships among cultivated and wild *Rubus* accessions using AFLP markers. *Scientia Horticulturae* 193, 165-73.

Mohadjer, P., Strik, B. C., Zebarth, B. J. and Righetti, T. L. 2001. Nitrogen uptake, partitioning and remobilization in 'Kotata' blackberry in alternate-year production. *The Journal of Horticultural Science and Biotechnology* 76(6), 700-8. doi:10.1080/14620316.2001.11511434.

Molina-Bravo, R., Fernandez, G. E. and Sosinski, B. R. 2014. Quantitative trait locus analysis of tolerance to temperature fluctuations in winter, fruit characteristics, flower color, and prickle-free canes in raspberry. *Molecular Breeding* 33(2), 267-80. doi:10.1007/s11032-013-9947-4.

Moore, J. N. and Clark, J. R. 1989. 'Navaho' erect thornless blackberry. *HortScience* 24(5), 863-5.

Muurinen, S., Slafer, G. A. and Peltonen-Sainio, P. 2006. Breeding effects on nitrogen use efficiency of spring cereals under northern conditions. *Crop Science* 46(2), 561-8. doi:10.2135/cropsci2005-05-0046.

Naraguma, J., Clark, J. R., Norman, R. J. and McNew, R. W. 1999. Nitrogen uptake and allocation by field-grown 'Arapaho' Thornless blackberry. *Journal of Plant Nutrition* 22(4-5), 753-68. doi:10.1080/01904169909365669.

Nelson, E. and Martin, L. W. 1986. The relationship of soil-applied N and K to yield and quality of 'Thornless Evergreen' blackberry. *HortScience* 21, 1153-4.

North American Raspberry and Blackberry Association (NARBA). 2019. Request for proposals. Available at: http://www.raspberryblackberry.com (accessed on 14 January 2019).

Østergård, H., Finckh, M. R., Fontaine, L., Goldringer, I., Hoad, S. P., Kristensen, K., Lammerts van Bueren, E. T., Mascher, F., Munk, L. and Wolfe, M. S. 2009. Time for a shift in crop production: embracing complexity through diversity at all levels. *Journal of the Science of Food and Agriculture* 89(9), 1439–45. doi:10.1002/jsfa.3615.

Parent, J. G., Fortin, M. G. and Page, D. 1993. Identification of raspberry cultivars by random amplified polymorphic DNA (RAPD) analysis. *Canadian Journal of Plant Science* 73, 1115–22.

Paterson, A., Kassim, A., McCallum, S., Woodhead, M., Smith, K., Zait, D. and Graham, J. 2013. Environmental and seasonal influences on red raspberry flavour volatiles and identification of quantitative trait loci (QTL) and candidate genes. *Theoretical and Applied Genetics* 126(1), 33–48. doi:10.1007/s00122-012-1957-9.

Pattison, J. A., Samuelian, S. K. and Weber, C. A. 2007. Inheritance of Phytophthora root rot resistance in red raspberry determined by generation means and molecular linkage analysis. *Theoretical and Applied Genetics* 115(2), pp. 225–36.

Peace, C. P. 2017. DNA-informed breeding of rosaceous crops: promises, progress and prospects. *Horticulture Research* 4, 17006. doi:10.1038/hortres.2017.6

Perkins-Veazie, P. and Clark, J. R. 2005. Blackberry research in Arkansas and Oklahoma. *Proceedings of the North American Bramble Growers Association Annual Meeting*, pp. 39–42.

Reickenberg, R. L. and Pritts, M. P. 1996. Dynamics of nutrient uptake from foliar fertilizers in red raspberry (*Rubus idaeus* L.). *Journal of the American Society for Horticultural Science* 121(1), 158–63.

Rempel, H. G., Strik, B. C. and Righetti, T. L. 2004. Uptake, partitioning, and storage of fertilizer nitrogen in red raspberry as affected by rate and timing of application. *Journal of the American Society for Horticultural Science* 129(3), 439–48.

Rondot, Y. and Reineke, A. 2018. Endophytic *Beauveria bassiana* in grapevine *Vitis vinifera* (L.) reduces infestation with piercing-sucking insects. *Biological Control* 116, 82–9. doi:10.1016/j.biocontrol.2016.10.006.

Ryu, J., Ha, B.-K., Kim, Ds, Kim, J.-B., Kim, S. H., Ahn, J.-W., Jeong, I. Y., Jo, H.-J., Kim, E.-Y. and Kang, S.-Y. 2014. Genetic diversity and relationship assessment based on AFLP analysis in blackberry (*Rubus fructicosus* L.) mutant lines. *Plant Breeding and Biotechnology* 2(4), 386–95. doi:10.9787/PBB.2014.2.4.386.

Ryu, J., Kim, W. J., Im, J., Kim, S. H., Oh, S. C., Cho, L., Kang, S.-Y. and Ha, B.-K. 2017. Study of transferability of *Rubus* microsatellite markers to hybrid boysenberry. *Plant Breeding and Biotechnology* 5(4), 253–60. doi:10.9787/PBB.2017.5.4.253.

Sapsford, K. E., Algar, W. R., Berti, L., Gemmill, K. B., Casey, B. J., Oh, E., Stewart, M. H. and Medintz, I. L. 2013. Functionalizing nanoparticles with biological molecules: developing chemistries that facilitate nanotechnology. *Chemical Reviews* 113(3), 1904–2074. doi:10.1021/cr300143v.

Sargent, D. J., Clarke, J., Simpson, D. W., Tobutt, K. R., Arús, P., Monfort, A., Vilanova, S., Denoyes-Rothan, B., Rousseau, M., Folta, K. M., et al. 2006. An enhanced microsatellite map of diploid Fragaria. *Theoretical and Applied Genetics* 112(7), 1349–59.

Sargent, D. J., Fernández-Fernández, F., Rys, A., Knight, V. H., Simpson, D. W. and Tobutt, K. R. 2007. Mapping of A1 conferring resistance to the aphid *Amphorophora idaei* and dw (dwarfing habit) in red raspberry (*Rubus idaeus* L.) using AFLP and microsatellite markers. *BMC Plant Biology* 7(1), 15. doi:10.1186/1471-2229-7-15.

Sattler, K. D. 2011. *Handbook of Nanophysics*. CRC Press, Boca Raton, FL, pp. 1–718.

Seeram, N. P. 2008. Berry fruits for cancer prevention: current status and future prospects. *Journal of Agricultural and Food Chemistry* 56(3), 630–5. doi:10.1021/jf072504n.

Simpson, C. G., Cullen, D. W., Hackett, C. A., Smith, K., Hallett, P. D., McNicol, J., Woodhead, M. and Graham, J. 2017. Mapping and expression of genes associated with raspberry fruit ripening and softening. *Theoretical and Applied Genetics* 130(3), 557–72.

Sinebo, W., Gretzmacher, R. and Edelbauer, A. 2004. Genotypic variation for nitrogen use efficiency in Ethiopian barley. *Field Crops Research* 85(1), 43–60. doi:10.1016/S0378-4290(03)00135-7.

Singh, U., Ladha, J. K. and Castillo, E. G., Punzalan, G., Tirol-Padre, A. and Duqueza, M. 1998. Genotypic variation in nitrogen use efficiency in medium-and long-duration rice. *Studies in Mycology* 58(1), 35–53.

Sloan, A. E. 2006. Top 10 functional food trends. *Food Technology* 60(4), 22–40.

Stafne, E. T., Clark, J. R. and Szalanski, A. L. 2003. Nuclear ribosomal ITS region sequences for differentiation of *Rubus* genotypes. *Journal of the Arkansas Academy of Science* 57, 176–80.

Stephens, M. J., Alspach, P., Hall, H. K. and Kempler, C. 2002. Red raspberry-grey mould resistance from *Rubus* species. *Acta Horticulturae* 585(585), 349–53. doi:10.17660/ActaHortic.2002.585.58.

Stocker, T. F., Qin, D., Plattner, G. K., Tignor, M. and Allen, S. K. 2013. Technical summary. In: Stocker, T. F., Qin, D., Plattner, G. K., Tignor, M. and Allen, S. K. (Eds), *IPCC, 2013: Climate Change 2013*. Cambridge University Press, pp. 33–115.

Strik, B. C. 2008. A review of nitrogen nutrition of *Rubus*. *Acta Horticulturae* 777(777), 403–10. doi:10.17660/ActaHortic.2008.777.61.

Strik, B. C. and Bryla, D. R. 2015. Uptake and partitioning of nutrients in blackberry and raspberry and evaluating plant nutrient status for accurate assessment of fertilizer requirements. *HortTechnology* 25(4), 452–9.

Strik, B. C. and Buller, G. 2002. Reducing thorn contamination in machine harvested 'Marion' blackberry. *Acta Horticulturae* 585(585), 677–81. doi:10.17660/ActaHortic.2002.585.112.

Strik, B. C. and Finn, C. E. 2012. Blackberry production systems-a worldwide perspective. *Acta Horticulturae* 946(946), 341–7. doi:10.17660/ActaHortic.2012.946.56.

Strik, B. C., Clark, J. R., Finn, C. E. and Bañados, M. P. 2007. Worldwide blackberry production. *HortTechnology* 17(2), 205–13.

Suginta, W., Khunkaewla, P. and Schulte, A. 2013. Electrochemical biosensor applications of polysaccharides chitin and chitosan. *Chemical Reviews* 113(7), 5458–79.

Takeda, F., Glenn, D. M. and Tworkoski, T. 2013. Rotating cross-arm trellis technology for blackberry production. *Journal of Berry Research* 3(1), 25–40.

Taylor, A. F. S., Gebauer, G. and Read, D. J. 2004. Uptake of nitrogen and carbon from double-labelled (15N and 13C) glycine by mycorrhizal pine seedlings. *New Phytologist* 164(2), 383–8. doi:10.1111/j.1469-8137.2004.01164.x.

United Nations General Assembly. 2015. Transforming our world: the 2030 agenda for sustainable development. A/RES/70/1. Available at: http://www.un.org/ga/search/view:doc.asp?symbol=A/RES/70/1&Lang=E (accessed on 14 January 2019).

van Wendel de Joode, B., Mora, A. M., Córdoba, L., Cano, J. C., Quesada, R., Faniband, M., Wesseling, C., Ruepert, C., Öberg, M., Eskenazi, B., et al. 2014. Aerial application of mancozeb and urinary ethylene thiourea (ETU) concentrations among pregnant women in Costa Rica: the Infants' Environmental Health Study (ISA). *Environmental Health Perspectives* 122(12), 1321–8. doi:10.1289/ehp.1307679.

van Wendel de Joode, B., Barbeau, B., Bouchard, M. F., Mora, A. M., Skytt, Å., Córdoba, L., Quesada, R., Lundh, T., Lindh, C. H. and Mergler, D. 2016. Manganese concentrations in drinking water from villages near banana plantations with aerial mancozeb spraying in Costa Rica: results from the Infants' Environmental Health Study (ISA). *Environmental Pollution* 215, 247–57. doi:10.1016/j.envpol.2016.04.015.

VanBuren, R., Bryant, D., Bushakra, J. M., Vining, K. J., Edger, P. P., Rowley, E. R., Priest, H. D., Michael, T. P., Lyons, E., Filichkin, S. A., et al. 2016. The genome of black raspberry (*Rubus occidentalis*). *The Plant Journal* 87(6), 535–47.

VanBuren, R., Wai, C. M., Colle, M., Wang, J., Sullivan, S., Bushakra, J. M., Liachko, I., Vining, K. J., Dossett, M., Finn, C. E., et al. 2018. A near complete, chromosome-scale assembly of the black raspberry (*Rubus occidentalis*) genome. *GigaScience*. Oxford University Press 7(8), 71–9.

Villamor, R. R., Daniels, C. H., Moore, P. P. and Ross, C. F. 2013. Preference mapping of frozen and fresh raspberries. *Journal of Food Science* 78(6), S911–9. doi:10.1111/1750-3841.12125.

Voorrips, R. E., Gort, G. and Vosman, B. 2011. Genotype calling in tetraploid species from bi-allelic marker data using mixture models. *BMC Bioinformatics* 12(1), 172.

Vose, R. S., Easterling, D. R. and Gleason, B. 2005. Maximum and minimum temperature trends for the globe: an update through 2004. *Geophysical Research Letters* 32(23), L23822. doi:10.1029/2005GL024379.

Walters, T. W., Bolda, M. and Zasada, I. A. 2017. Alternatives to current fumigation practices in western states raspberry. *Plant Health Progress* 18(2), 104–11. doi:10.1094/PHP-RS-16-0068.

Ward, J. A., Bhangoo, J., Fernández-Fernández, F., Moore, P., Swanson, J. D., Viola, R., Velasco, R., Bassil, N., Weber, C. A. and Sargent, D. J. 2013. Saturated linkage map construction in *Rubus idaeus* using genotyping by sequencing and genome-independent imputation. *BMC Genomics* 14(1), 2.

Weber, D. E. 2014. Linkage mapping in tetraploid blackberry (*Rubus* spp.) using high-throughput genomic sequencing and restriction-site associated DNA sequencing (RAD-Seq). Ph.D. Dissertation. University of Illinois at Urbana-Champaign, pp. 1–259.

Westwood, M. N. 1993. *Temperate-Zone Pomology: Physiology and Culture* (3rd edn.). Timber Press Inc, Portland, OR.

Wight, H., Zhou, J., Li, M., Hannenhalli, S., Mount, S. and Liu, Z. 2019. Draft genome assembly and annotation of red raspberry *Rubus idaeus*. *bioRxiv* 546135. doi:10.1101/546135.

Wilcox, W. F. 1989. Identity, virulence, and isolation frequency of seven *Phytophthora* spp. causing root rot of raspberry in New York. *Phytopathology* 79(1), 93–101. doi:10.1094/Phyto-79-93.

Williams, D., Britten, A., McCallum, S., Jones, H., Aitkenhead, M., Karley, A., Loades, K., Prashar, A. and Graham, J. 2017. A method for automatic segmentation and splitting of hyperspectral images of raspberry plants collected in field conditions. *Plant Methods* 13(1), 1–12. doi:10.1186/s13007-017-0226-y.

Woodhead, M., McCallum, S., Smith, K., Cardle, L., Mazzitelli, L. and Graham, J. 2008. Identification, characterisation and mapping of simple sequence repeat (SSR) markers from raspberry root and bud ESTs. *Molecular Breeding* 22(4), 555–63.

Woodhead, M., Weir, A., Smith, K., McCallum, S., MacKenzie, K. and Graham, J. 2010. Functional markers for red raspberry. *Journal of the American Society for Horticultural Science* 135(5), 418–27. doi:10.21273/JASHS.135.5.418.

Woodhead, M., Williamson, S., Smith, K., McCallum, S., Jennings, S. N., Hackett, C. and Graham, J. 2013. Identification of quantitative trait loci for cane splitting in red raspberry (*Rubus idaeus*). *Molecular Breeding* 31(1), 111–22.

Wu, J., Yan, F., Tang, J., Zhai, C. and Ju, H. 2007. A disposable multianalyte electrochemical immunosensor array for automated simultaneous determination of tumor markers. *Clinical Chemistry* 53(8), 1495–502. doi:10.1373/clinchem.2007.086975.

Xiao, Y., Ju, H. X. and Chen, H. Y. 1999. Hydrogen peroxide sensor based on horseradish peroxidase-labeled Au colloids immobilized on gold electrode surface by cysteamine monolayer. *Analytica Chimica Acta* 391(1), 73–82.

Yorgey, B. M. and Finn, C. E. 2005. Comparison of 'Marion' to thornless blackberry genotypes as individually quick-frozen and puree products. *HortScience* 40(3), 513–5.

Zhang, H., Zhou, Z., Yang, B. and Gao, M. 2003. The influence of carboxyl groups on the photoluminescence of mercaptocarboxylic acid-stabilized CdTe nanoparticles. *The Journal of Physical Chemistry B* 107(1), 8–13. doi:10.1021/jp025910c.

Chapter 13

Advances and challenges in sustainable raspberry/blackberry cultivation

Julie Graham, Alison Karley, Alison Dolan, Dominic Williams and Nikki Jennings, James Hutton Institute, UK

1 Introduction

Rubus crops, particularly red raspberry (*R. idaeus* and *R. strigosis*) and blackberries (*R. plicatus*), have seen a significant increase in consumer demand, while at the same time facing considerable challenges in cultivation. Consumer interest in environmentally sustainable food will require an increase in local production if these challenges are to be met. Numerous biotic and abiotic stresses such as pest and disease burdens, lack of or too much water and suboptimal temperatures restrict production of these crops. Additionally, the overall negative effects of greater seasonal variability and longer-term climate change on productivity and quality are predicted to lead to serious crop yield declines in many countries. Given that farming in general suffers from unstable yields in high-yielding environments, this raises serious concerns about the sustainability of existing production systems (Mackay et al., 2011). Raspberry growers are already aware of seasonal- and climatically induced issues such as the altered developmental schedule of varieties, the reduced winter chill affecting bud break and an increased incidence of unwanted fruit phenotypes. These challenges need to be overcome if the increasing demand for raspberry is to be met. This chapter reviews the major pest and disease challenges in bramble production, followed by environmental impacts on the crop as well as environmental interactions with pests and diseases that may be exacerbated by climate change.

http://dx.doi.org/10.19103/AS.2018.0040.28

Possible solutions for continued growth of raspberry and other *Rubus* crops production, through resilience and adaptation to these environmental challenges, are discussed using altered or new cultivation practices and potential breeding approaches that will become essential for future sustainability.

2 Major pest and disease stresses that affect *Rubus* production

In fruit plantations, there is now greater emphasis on reducing pesticide inputs and using integrated pest management (IPM) and other more environmentally benign and sustainable practices. Breeding objectives reflect this change. The incorporation of novel resistance to, or tolerance of, pests and diseases is regarded as essential for the development of varieties suitable for growth with IPM systems. Sources of pest and disease resistance in diverse *Rubus* spp. have been identified and exploited in conventional cross-breeding (Keep et al., 1977; Jones, 1987; Jennings, 1988; Knight, 1991; Williamson and Jennings, 1992; Birch et al., 2002; Jones et al., 2001). Pests and diseases of raspberry in Europe have been reviewed extensively (Gordon and Williamson, 2004; Gordon et al., 2006; Jennings and Dolan, 2014; Parikka et al., 2016). The main pests and diseases are briefly mentioned below along with breeding or molecular breeding strategies for control.

2.1 Root rot

Raspberry root rot, caused by *Phytophthora* species, is a devastating soil-borne disease that has caused serious problems in Europe, North America and Australia for many years. Screening for resistance has been ongoing since the late 1970s (Barritt et al., 1979; Duncan et al., 1987; Bristow et al., 1988; Knight and Fernández-Fernández, 2008). Several breeding programmes list resistance to root rot as a major breeding target. Identifying robust and transferable molecular markers, to reliably select genotypes with high levels of resistance, would be very useful for breeders. Pattison et al. (2007) presented a two gene hypothesis for resistance, and Graham et al. (2011) subsequently identified two QTLs for resistance; a root rot resistance marker has since been deployed for breeding purposes and selections with the marker have been shown to be resistant. In terms of coping with this disease until resistant varieties become available, field growing on ridges has been carried out with some limited success by ensuring plants are not rooted into wet soil. The fungicides available for use on raspberry have limited effect, as they were designed originally for use with *Phytophthora* species and diseases of other crops. These control measures have proved partially successful, and therefore a large proportion of growers have moved production from the soil into containers with substrate, most commonly coir. Substrate production of raspberries is associated with high costs, and requires higher outlays for pots, substrate and irrigation/fertigation equipment. Coir is derived from coconut and is sourced and processed in the southern hemisphere, and coir import therefore involves high shipping costs. Further, these production systems are associated with increased plastic waste. Offering varieties which combine root rot resistance with fruit quality and yields that are acceptable to the supply chain, combined with only growing clean certified stock, will give growers the option to return to soil production for raspberries with lower production costs and reduced waste.

2.2 Large raspberry aphids

The large raspberry aphid (*Amphorophora idaei* in Europe and *A. agathonica* in North America) is a vector for several viruses including Raspberry leaf mottle virus (RLMV), Black raspberry necrosis virus (BRNV) and *Rubus* yellow net virus (RYNV). Breeding for aphid resistance is therefore a goal of many breeding programmes. Knight et al. (1960) and Keep and Knight (1967) studied the genetic control of *A. idaei* resistance derived from different sources and attributed it to a number of single dominant genes (A_1-A_{10}, A_{L518}, A_{K4a}). Resistance is strongly dependent upon the *A. idaei* biotype, with some genes providing resistance to certain biotypes but not others (Sargent et al., 2007). The widespread growing of cultivars carrying A_1 resistance imposed strong selection pressure on aphid populations, leading to predominance of biotypes that could overcome this resistance (Gordon et al., 1997). Resistance to the most common biotypes can be conferred by one of three genes: A_{10} from *R. occidentalis*, A_{L518} from *R. idaeus* subsp. *strigosus* or A_{K4a} from a German clone of *R. idaeus* subsp. *vulgatus*. However, an A_{10} resistance-breaking aphid biotype has also been reported (Gordon et al., 1997), making it important to pyramid several aphid resistance genes in breeding lines to provide a robust and durable vector resistance. Suitable molecular markers are essential to determine how many resistance genes are carried by resistant selections, as the equivalent effects of many of the reported resistance genes make them phenotypically indistinguishable and possibly synonymous. Thus, the development of maps of progenies carrying resistance genes, and the identification of molecular markers linked to these genes, is a key tool in differentiating reported genes, identifying their presence in modern hybrid materials and in managing strategies for pyramiding. A_1 has been mapped to LG3 (Sargent et al., 2007) and linked markers to the other genes are needed to enable efficient gene pyramiding. Recently, three sources of aphid resistance have been identified in black raspberry: Ag (4) from Ontario (ON), Ag (5) from Maine (ME) and a third source from Michigan (MI) with no formal designation (Bushakra et al., 2015, 2018). ON was placed on a genetic linkage map on *Rubus* linkage group (RLG) 6 which corresponds to

LG 3 of Sargent et al. (2007). Segregation ratios in populations with single and combined resistance sources, and linkage mapping in two populations (ORUS 4304 and ORUS 4812) segregating for the Ag (5) and MI sources, respectively, indicated that these three sources of resistance are each conferred by single dominant genes/alleles that are linked on RLG6.

To complement resistance-breeding efforts, other work has focussed on testing naturally bio-pesticidal products and natural enemies to control aphid infestations. Recent research has shown that an extract derived from neem (*Azadirachta indica*) and a terpenoid blend suppressed aphid abundance in tunnel-grown raspberry crops infested with large raspberry aphid and potato aphid (*Macrosiphum euphorbiae*), particularly when applied with augmented release of parasitoid wasps (O'Neill et al., 2014). The use of 'banker' plants is also being tested as a sustainable system to supply a continuous source of natural enemies that disperse into the crop (Balzan, 2017; Miller and Rebek, 2018). Typically, banker plants consist of cereal plants infested with cereal aphids (which do not infest the target crop) as a constant supply of aphid hosts for parasitoid wasps to reproduce, thus acting as a source of parasitoid wasps for pest control in the crop. These could be grown alongside *Rubus* tunnels to augment raspberry aphid control throughout the growing season. Other methods that could be used to promote the abundance and activity of natural enemies of aphids in *Rubus* include introduction of plants that attract natural enemies, such as buckwheat (*Fagopyrum esculentum*) for promoting hoverflies (Berndt et al., 2006).

2.3 Cane and leaf diseases

Cane botrytis (*Botrytis cinerea*) and spur blight (*Didymella applanata*) infect mature or senescent leaves on raspberry primocanes. The infection spreads to the buds where the disease lesions develop. The following spring, buds from infected nodes will break later or not at all, constituting a major cause of yield loss, particularly in the case of cane botrytis. Williamson and Jennings (1986) proposed a common resistance mechanism to both diseases and Graham et al. (2006) identified two major QTLs associated with both resistances in LG2 and LG3 of the 'Latham' × 'Glen Moy' (L × GM) reference mapping population (Graham et al., 2004), suggesting multiple gene control of the trait. Interestingly, the QTL in LG2 co-located with Gene *H*. The Gene *H* marker is now used routinely in the James Hutton Limited breeding programme (www.huttonltd.com/plant-varieties/berries).

Rain shelters and white synthetic reflective groundcovers have been assessed for their independent and combined effects on cane *Botrytis* and spur blight (as well as on crop productivity and effects on ground arthropods involved in natural pest control). Rain shelters decreased the severity of spur blight infections by ca. 50% and reduced the rate of post-harvest raspberry degradation caused by *Botrytis* in comparison with the control treatment. Additionally, no detrimental effects of the structures were detected on ground beetle communities. Based on this, it is suggested that the combination of these two structures constitutes a promising approach to counter certain adverse effects caused by untimely precipitation while maintaining a productive cropping system (Comeau et al., 2012).

Cane spot or anthracnose (*Elsinoë veneta*) can develop in most raspberry tissues, but it is most recognizable in the second year's canes where it produces deep lesions that can lead to vascular damage, and therefore reduce yields. The resistance to this pathogen has been associated with the presence of hairy cane (*H*) in European red raspberry, but not in North American cultivars (Graham et al., 2006). Genetic control of the trait has not yet been firmly established, although Graham et al. (2006) identified two QTLs in LG2 and LG4 of the L × GM progeny associated with response to the disease which could be investigated further.

Yellow rust (*Phragmidium rubi-idaei*) has increased in prevalence in recent years, associated with the more common and widespread cultivation under tunnels of susceptible cultivars (e.g. 'Glen Ample' and 'Tulameen'). A major resistance gene (*Yr*) from 'Latham' was identified by Anthony et al. (1986), and Graham et al. (2006) postulated this gene to be located on LG3 of the L × GM progeny. Anthony et al. (1986) studied the inheritance of complete and incomplete resistance to rust in a half diallel cross including 'Boyne', which derives complete resistance from 'Latham'. They found that crosses of 'Boyne' to susceptible varieties all segregated for complete resistance and proposed that 'Boyne' was heterozygous for a single resistance gene, designated *Yr*, which was derived from 'Latham'. As the L × GM population also segregates for resistance to rust, Graham et al. (2006) proposed that 'Latham' is also heterozygous for *Yr*. Anthony et al. (1986) also found variation in the degree of susceptibility among offsprings of 'Boyne' without complete resistance and concluded 'Boyne' to also be a source of incomplete resistance, making breeding a possible solution. In plantations, thinning canes to improve air circulation, use of effective weed control and removal of the first flush of primocanes by hand or with herbicide are recommended control measures (Dolan et al., 2018).

2.4 Spotted wing drosophila (SWD)

Drosophila suzukii is a vinegar fly native to Asia; however, over the last decade it has been detected in many countries worldwide including North America (2008), Italy (2009), England (2012) and Scotland (2014). The fly infests both cultivated and wild hosts, and when found in significant numbers, it can cause considerable economic damage. The host range is broad and includes both soft and stone fruit, with *Rubus* spp. being one of its preferred hosts. It is thought that most species of vinegar fly do not cause economic losses as they infest fruit that is overripe and already unmarketable. *D. suzukii* differs in that the female has a serrated ovipositor which can cut through the skin of developing fruit, laying her eggs under the skin causing the fruit to be malformed and rendering it inedible. The number of generations per year and length of each life cycle is temperature dependent. The pest prefers a moderate climate with temperatures around 20°C, but it can survive in colder conditions and overwinter as an adult in temperatures below zero.

The pest is an active flyer, but it is more probable that it spreads over long distances due to the movement of infested fruit. Monitoring for the pest using baited traps enables a better understanding of the density and possible establishment of *D. suzukii* in a new region, and where and when to adopt control measures. An invasion of the pest inevitably has an effect on IPM measures already in place, since strategic broad-spectrum products often are recommended to control this pest. As with any addition to a spray programme, it is important to consider the impact on beneficial organisms, the potential increase in chemical residues on the fruit and possible build-up of resistance by the pest to the chemicals available for its control.

A recent study in Northern Italy (Del Fava et al., 2017) assessed the impact of SWD on the soft fruit industry (strawberries, raspberries, blackberries and blueberries). Here a conventional IPM strategy based on insecticide, mass trapping and cultural measures was compared with an upgraded IPM strategy based on exclusion netting. In a period of low pest pressure, the conventional IPM strategy would be more profitable than no management. However, with higher pest pressure levels, the upgraded IPM strategy would become increasingly profitable compared with conventional IPM as pest damage levels increase. Exclusion netting, despite being more expensive, will be more profitable, in terms of its effectiveness in reducing pest pressure and through not having to increase spraying of target crops.

2.5 Vine weevil

Rubus plantations are at risk from infestation with vine weevil, *Otiorhynchus sulcatus* (Coleoptera: Curculionidae), which is a highly polyphagous pest that feeds on a large number of horticultural and garden plant species (Smith, 1932). Adults are nocturnal and feed on leaves, causing classic 'notching' of the leaf margins. However, damage from adult insects is relatively minor in comparison with the significant damage caused by the larval stages, which feed below ground on roots for an extended period between late summer and the following spring. At present, there are no known genetic targets for breeding resistance to this pest in *Rubus*, although *Rubus* genotypes with improved root vigour might better tolerate root infestations (Mitchell et al., 2016). Soil conditioners and beneficial microbes also could be incorporated into the soil to promote plant vigour (e.g. Orhan et al., 2006), although this approach has yet to be tested for *Rubus*. Potential biocontrol organisms for vine weevil include entomopathogenic fungi, such as *Beauveria bassiana* and *Metarhizium anisopliae* (e.g. Hirsch and Reineke, 2014), and entomophagous nematodes. These organisms are available commercially and can be effective at reducing weevil survival independently or in combination (Bruck, 2007; Ansari et al., 2008), although temperature differentially affects the efficacy of entomopathogenic fungi compared to nematodes (Bennison et al., 2014). Lure traps supplied with plant-derived volatiles that attract weevils (van Tol et al., 2012) could be used to monitor adult vine weevil abundance and to supplement other weevil control methods.

2.6 Viral diseases

Virus problems often do not receive the same level of attention as other pest and disease problems. For a review of viruses detected in *Rubus* species, see Martin et al. (2013) who listed 30 viruses that have been reported to infect red raspberry, black raspberry (*Rubus occidentalis*) and blackberry. There are currently some viruses that are detected only in the United States (e.g. Blackberry virus Y) or Europe (e.g. Raspberry leaf blotch virus). These differences may be real, caused by propagation of different plant stocks in the United States and Europe, or may be artificial, caused by biases in the testing regimes employed in these continents (Dolan et al., 2018). Similarly, whether red raspberry viruses are capable of infecting black raspberry or blackberry plants, and vice versa, is often not known, and with the tendency now for growers to produce crops using a number of different *Rubus* spp., this may become more important. Viruses tend to be managed through vector control when possible. Raspberry bushy dwarf virus is the most important pollen-transmitted virus in raspberry. It affects the development of the fruit and leads to reduced drupelet numbers and increased size of these drupelets, resulting in unmarketable fruit (Dolan et al., 2018).

In addition to the methods of pest and disease control considered above, there has been recent interest in plant physical or structural characteristics that contribute to pest and disease resistance or tolerance in *Rubus* breeding programmes (Graham et al., 2015; Karley et al., 2016; Mitchell et al., 2016). Plant traits conferring resistance include structural traits (e.g. trichomes, spinescence, waxy cuticles, sclerophylly, cane pubescence) that act as a physical barrier. Epicuticular waxes form a slippery film or form crystals which prevent pests from attaching to the plant surface (White and Eigenbrode, 2000) or ovipositing (Voigt and Gorb, 2010). Glandular trichomes deter pests through repellence or toxicity (Figueiredo et al., 2013) while non-glandular trichomes prevent insect attachment, limit pest movement or oviposition (Tian et al., 2012). In raspberry, it was shown that the presence of cane hairs is associated with resistance to *Botrytis* and spur blight (Jennings, 1982a; Jennings and Brydon, 1989; Graham et al., 2006). Cane hairs are easy to select for in crossing programmes where breeders rarely have the resources to include a primary screen for each fungal disease. The cane hairs may affect the ability of fungi to adhere to and infect tissues (Jennings, 1962). Jennings (1982b) also found that water run-off was more rapid on hairy canes than non-hairy ones and suggested that, consequently, bud infection occurred less often on canes with hairs due to reduced residence time of fungal spores.

Oviposition by the generalist phytophagous mite, *Tetranychus uticae*, was significantly reduced on raspberry genotypes with high leaf trichome densities (Karley et al., 2016), and with identification of underlying genetic markers, this trait has potential utility in mite control (Graham et al., 2014).

Cane splitting in raspberry is another example of a physical trait that affects pest and disease burdens. Cane splitting is a normal feature of raspberry growth, and severe splitting can lead to plant infestation by cane midge (*Resseliella theobaldi* (Barnes)) followed by fungal infection by many pathogens. This can lead to a disease complex called cane blight, with losses in yield of up to 50% if left untreated (Jennings, 1988). Raspberry genotypes differ in the degree of cane splitting, and regions of the raspberry genome associated with cane splitting have been identified (Woodhead et al., 2013). A correlation between cane splitting and cane height has been shown, with shorter caned genotypes exhibiting less cane splitting than longer caned genotypes. Loci accounting for 49% of height variation have been identified (Graham et al., 2009a; Woodhead et al., 2013), thus allowing breeding directly or indirectly for genotypes with reduced propensity to splitting.

Root vigour has been shown to have an impact on the ability of the plant to resist *Phytophthora* root rot infection (Graham et al., 2011). The demonstration of a highly significant correlation between root sucker parameters and root rot resistance is of great value to breeders, as it provides a simple visual screen for identifying germplasm with some level of resistance/tolerance to *Phytophthora* root rot (Graham et al., 2011). Plant vigour also could allow plants to tolerate herbivory (Price, 1991), and QTL approaches have been used to identify genetic markers for vigour in raspberry (e.g. root vigour: Graham et al., 2011) that could be targeted for *Rubus* breeding.

3 Environmental impacts on crop development

In *Rubus* crops, understanding how climate affects the developmental process will be crucial for sustaining the industry against a background of increasingly aggressive biotic stresses. Disrupted developmental processes caused by greater seasonal variability lead to altered timings of the ripening period and to fruit developmental issues. For a review of the developmental transitions to fruiting in raspberry, see Graham and Simpson (2018). The timing of flowering is a key component of plant adaptation and there is evidence that flowering time has shifted in response to changes in climate (Fitter and Fitter, 2002; Amano et al., 2010). Recent reports on the effect of greater seasonal variability/ climate change impacting *Rubus* crops include unpredicted phenotypic traits or variation in key transitional timings, which are apparent to breeders selecting new genotypes and to growers who have altered fruit harvest times. The unpredictability in timing of key developmental processes is altering the production scheduling of varieties which could traditionally be classified as early, mid- and late season. Also evident is a lack of evenness in expression of key developmental stages, such as bud break, which results in variety failure upon release to grower test plots due to unpredictability in behaviour (Jennings, *pers. comm.*).

Spatial variations also are observed at the plant, row and field level. Modification of primocane and biennial variety behaviour has been observed, and increased expression of fruit disorders, particularly crumbly fruit, is further challenging growers (Jennings, *pers. comm.*; Graham et al., 2015). Natural variation between varieties and across seasons does occur in flowering, fruiting time and fruit quality (www.huttonltd.com/plant-varieties/berries; Graham et al., 2009a) but, increasingly, growers are seeing unpredicted problems in commercial production. Wild populations also exhibit differences in developmental transitions across a spatial scale. As these wild raspberry populations often exist near cultivated raspberry crops, the potential for gene flow has encouraged the study of the interactions among wild and cultivated forms of this species (Graham et al., 1997, 2003; Marshall et al., 2001). These studies revealed that wild raspberry exists as a series of local populations which are differentiated genetically and physiologically, and that the physiological differences are maintained in common environments (Marshall et al., 2001; Graham et al., 2003, 2009b). Recent studies in these wild populations have shown an increase in fruit disorders and shifts in developmental timings (Graham, *unpublished data*).

The biology of this crop, in terms of differing life cycles, means both annual and biennial types are available to growers. Most raspberry and blackberry plants develop short-lived woody shoots on a long-lived perennial root system, bearing both juvenile and mature shoots (canes) simultaneously on an individual plant. In biennial fruiting cultivars, the canes have a two-year life cycle responding to photoperiod and temperature, while primocane cultivars complete the cycle of vegetative growth, flowering and fruiting in a single season, and are effectively day-length and temperature neutral (Jennings, 1988).

An improved understanding of the genetic, environmental and epigenetic factors impacting flowering and fruiting, along with the development of breeding strategies and/or novel cultivation practices to mitigate unwanted developmental and fruit quality phenotypes, are therefore crucial (along with managing other key factors such as biotic stresses) to ensure the continued future success of cane fruit. Seasonal and environmental factors influence key crop development processes such that multiple environmental and physiological cues can regulate the transition to flowering (Fornara et al., 2010; Pin and Nilson, 2012; Song et al., 2012). The effect of a changing climate on temporal and spatial variation in these factors and developmental processes is largely unknown. Photoperiod and temperature play a major role in controlling flowering time in many plants (Fornara et al., 2010). Plants keep track of photoperiod with an endogenous clock (Green et al., 2002; Dodd et al., 2006; Brachi et al., 2010). Activity of CONSTANS (CO), a key component in leaves of the photoperiodic pathway, accumulates in long day conditions and activates transcription of FLOWERING LOCUS T (FT) (Simon et al., 2015) which interacts with bZIP transcription factors (Abe et al., 2005; Cao et al., 2015, 2016) activating downstream genes leading to flowering. This basic flowering process is influenced by several autonomous and stress-related signals. In terms of temperature regulation, the MADS box FLOWERING LOCUS C (FLC) is central. FLC and short vegetative phase proteins (SVP) form a complex to repress flowering until the plant is exposed to the appropriate level of cold. Age is a factor in vernalization, as the activity of miRNA156 (Wang, 2014), which acts by repressing members of the SQUAMOSA PROMOTER BINDING LIKE (SPL) family of transcription factors (Wu et al., 2009), is reduced as the plant ages. The plant thus requires downregulation of both miRNA156 and FLC for flowering to occur.

We therefore need to characterize the effects of environmental signals on the downstream gene regulation of phenotype to understand the impact of climate change on irregular phenotype. In a biennial raspberry population (Graham et al., 2009a), QTLs were identified for various developmental stage timings from bud break through to ripe fruit and for ripening as a continuous process. QTL mapping has also identified underlying genes of importance for a number of other developmental- and ripening-related traits (Kassim et al., 2009; McCallum et al., 2010; Paterson et al., 2013; Simpson et al., 2017). In the ripening study, RiMADS_01 was identified as a potential candidate gene affecting vernalization (Graham et al., 2009a). The gene is similar to SVP, modulating the timing of the developmental transition to flowering phase in response to temperature (Lee et al., 2007). In a colder season, RiMADS_01 was associated with earlier flowering. SPL was also identified in raspberry on linkage group 5 associated with a QTL for floral transition. Another gene, Gene *H*, previously associated with cane morphology (Graham et al., 2006) was shown to be associated with a reduced rate of ripening across all stages (Graham et al., 2009a). Sequencing of the Gene *H* region (McKenzie et al., 2015) identified a DIVIA-like Myb transcription factor (Werewolf) which has been shown to be a posttranscriptional regulator of FT (Seo et al., 2011). Recently these developmental data have been re-analysed on a new, marker-dense, raspberry genotyping by sequencing (GbS) map to enhance QTL identification (Hackett et al., 2018). This GbS map is linked to the genome sequence of 'Glen Moy' and has highlighted numerous genes of interest which can be studied for their role in the ripening process. This has implications for modulating flowering time through breeding for favourable alleles in selections in response to changing environments.

Fruit disorders such as crumbly fruit, a serious and unpredictable disorder, have become more prevalent and problematic over the last few seasons (Graham et al., 2015), and initial findings suggest these may also be associated with timings of key developmental transitions. Understanding regulation of key developmental processes at the molecular level and identification of allelic variation would allow breeding strategies to focus on producing crop varieties with increased tolerance to this seasonal variability, thereby improving environmental resilience.

Other cultivation practices can also be utilized to mitigate seasonal unpredictability. There has been a move to protected cropping with much of the world's raspberry and other soft and stone fruits grown in tunnels. These tunnels modify the environment and have enabled growers to evolve production systems that extend the capability of production beyond the traditional cropping season (for in-depth discussion and production guidelines, see Pritts et al., 2017). These tunnel systems offer protection from the weather, tend to produce better quality fruit and higher yield, extend plantation life and offer some protection against certain pest and diseases. These tunnel systems work well for both soil-grown and pot-grown crops. Frost protection can be provided by sprinklers or portable heaters protecting plants where the tunnels are left over the entire season. In the fruiting season tunnel doors can be opened and ventilation provided to cool the environment. Regardless of growth method, irrigation (and feed if in pots) is essential with the water source providing almost all the moisture input for the tunnel. Although the costs of tunnel establishment are high and there are risks from weather that cause significant damage these systems have revitalized soft fruit growing.

An increasing number of growers are moving raspberry production out of the soil and into containers with various types of substrate. Initially this practice was adopted to avoid soil-borne diseases, such as *Phytophthora* root rot, but it has progressed into a system to extend the cropping season and manipulate harvest timings of varieties to several months beyond the traditional 4- to 6-week production window of field-grown plantations. This is commonly practised with long-cane production, which involves pre-propagated, year-old canes which have received adequate chilling and are ready to break dormancy (Carew et al., 2000; Gillespie et al., 1999). Planting dates are manipulated and staggered via the cold storage of canes so production either precedes the main season or is delayed. This method can achieve a prolonged fruiting season over several months and allows growers to displace imported fruit and avoid oversupply during the season to maintain a higher return on investment.

A shift in growing conditions created by an extended season, or seasonal or climatic differences in temperature, can impose stress on the plants, and varieties vary in their ability to adapt to these new growing conditions. Understanding a variety's response to the environment enables growers to match varieties to specific commercial needs through more focussed decision-making and manipulation of crop timings and yields. The amount of chilling required to break dormancy and responses to the environment in raspberry are poorly understood. The use of cold storage to manipulate dormancy and chilling requirements underpins key production techniques such as long-cane production. An understanding of the dormancy and chilling requirements also supports breeders in selecting varieties which are more resilient to future climate change, since increases in global temperatures may reduce the length and timing of the chilling period of winter chilling (Atkinson et al., 2013). Further understanding of the genetic differences in dormancy response will help the industry manipulate varieties to maximize production as well as enable future breeding strategies to select varieties which show a high degree of resilience to environmental change.

4 Effects of climate change on bramble crop management and phenology

In raspberry, the main season summer-fruiting crop is usually supported on a post-and-wire system designed to carry the weight of fruits and to protect canes from excessive damage due to wind, harvesting and cultivation. Primocane-fruiting types also require support and modern systems have evolved into an adjustable horizontal system, for example using *Chrysanthemum* netting, that can be raised as plants grow to support the canes and subsequent flowers and fruit. In the main season crop, primocanes are produced in numbers excessive to requirements for cropping in the following season, so many must be removed by pruning in winter and early spring to reduce inter-cane competition and create an open crop canopy for efficient light capture. Old dead fruiting canes must also be removed by pruning after harvest. These pruning operations remove sources of fungal inoculum from the plantation and are important for the long-term health of the crop. This management will become even more important with increased pest and disease burdens as primocanes and fruiting canes are in close proximity, and so the plant provides spatial and temporal continuity for pests and pathogens (Willmer et al., 1996). The complex nature of the plant architecture also creates a barrier of foliage that impedes spray penetration of plant protectant chemicals, requiring specialized chemical application equipment (Gordon and Williamson, 1988).

There is great uncertainty about the extent to which climate change will exacerbate pest and disease problems (Deutsch et al., 2008). What is expected is that with increased mean temperatures, the number of insects and the period when infestation or infection can occur will increase, as will the amount of plant material insects have consumed. In berry crops where a shift to protected cultivation has occurred (and is now the norm in some countries), some insight into the effects of warming have been obtained (Birch et al., 2011). The use of tunnels to extend the crop growing season is thought to be likely to lead to greater pest problems due to elevated temperatures and a longer period over which pest infestations can persist (Birch et al., 2011). Elevated temperatures may increase the number of insect generations produced each season (e.g. De Lucia et al., 2012; Johnson et al., 2012; Caffarra et al., 2012; Juszczak et al., 2013; Buckley et al., 2017). Research is needed to address knowledge gaps, including for climate change, to deliver successful IPM (Birch et al., 2011) and allow growers to maintain crop yields in spite of increasing pest and disease burdens. Climate change may lead to a range of weather events, some of which may even be beneficial under some circumstances. For example, problems like raspberry root rot, which prefers cool wet conditions, may become less severe where warmer drier conditions are experienced.

Climate change is altering plant phenology. It is not known if bees and other pollinators have developed strategies to mitigate fitness losses following temporal mismatches. A recent study examined how the shift of apple (*Malus domestica*) tree phenology affected the abundance and diversity of pollinators, generalist and specialist herbivores and predatory arthropods (Korosi et al., 2018). While arthropod abundances either declined or increased, species diversity tended to be lower on apple trees with shifted phenology. A study of herbarium specimens of blackberry confirmed a trend towards earlier dates of first flower, full flower and first fruit as global temperatures increased (Diskin et al., 2012). The interaction of these changes in *Rubus* phenology with temperature-driven changes in the abundance and distribution of pollinator populations (Memmott et al., 2007) is likely to be complex and the outcomes for *Rubus* yields difficult to predict. These insects also are affected by more subtle changes, for example, in humidity which is different in tunnel-grown crops. The impacts of abiotic factors (mainly temperature and relative humidity) on honeybee activities have been reviewed (Abou-Shaara et al., 2017) and as temperatures increase, humidity levels become more critical.

5 Environmental stresses

Water relations can be an additional stress on berry crop growth. Water stress effects will depend on the nature of the stress (continuous drought or waterlogging vs. intermittent/inundation events) and interaction with specific pests or diseases (e.g. Huberty and Denno, 2004). Berry crops have a high requirement for irrigation, particularly with increasing temperatures both under protective covers and in the field. Studies on more efficient yield with the minimum level of irrigation may allow reductions in water use in areas where water is an issue. In a number of crops, such research has demonstrated that water reductions are possible. For example, see the studies in apple (El Jaouhari et al., 2018), grapevine (*Vitis vinifera*) (Wenter et al., 2018) and cucumber (*Cucumis sativus*) (Abd El-Mageed et al., 2018). In strawberry (*Fragaria × ananassa*), a recent study examined the economics of irrigated UK strawberry production and the importance of a reliable water supply (Morris et al., 2017). Climate change and growing pressures on water resources require a greater interest in irrigation economics in the soft fruit sector, especially with respect to restrictions on summer extraction and rising competition and charges for using the public water supply.

Heat stress limits productivity in cultivated raspberry (Stafne et al., 2000; Ballington and Fernandez, 2008; Jennings, 1988). High temperature stress experienced in recent years usually coincides with the flowering of primocane raspberries and negatively affects fruit quality parameters. Tipping young primocanes (Strick et al., 2012) has been used to mitigate plant response to these conditions. Zorenc et al. (2017) carried out a study to examine the effect of this practice on raspberry fruit characteristics and concluded that primocane tipping was a good cultivation practice for delaying the production season to avoid periods of high temperature stress, with negligible differences in fruit quality.

In some regions, low temperature stress must be managed. In the Nordic regions (Nestby and Takeda, 2015) red raspberry floricanes can suffer 20–30% injury due to low temperatures. Management practices such as bending and covering the raspberry floricanes have been examined for their effects on the extent of winter injury and yield. The results showed that bending canes to avoid freezing injury was possible without much visible cane injury. However, fruit yield was less than with standard trellised erect canes. Another study (Mettler and Hatterman-Valenti, 2016) examined the effect of four winter covers (black plastic with corn stover, black plastic with wheat straw, thermal blanket with corn stover and thermal blanket without mulch) on ten different blackberry cultivars. Here all treatments except the thermal blanket alone moderated minimum and maximum temperatures compared to the outside air temperature. Higher bud survival and total live buds occurred when the thermal blanket with corn stover was used, and this was attributed to greater temperature moderation.

In a recent study in apple, bloom date was modelled to advance towards early April when the risk of spring frosts may actually increase in a warmer climate due to a stronger connection between blocking (a stable high-pressure system blocking the regular atmospheric circulation and favouring advection of cold air from the north) and cold spells in early spring (Unterberger et al., 2018). Increased use of frost protection methods such as sprinklers and improved air circulation may need to be supplemented by artificial heating methods.

Regarding other strategies to mitigate climate stress, it is known that arbuscular mycorrhizal fungi (AMF) can improve the growth and nutritional quality of crops and the maintenance of host plant fitness under stressful environments (e.g. Barros et al., 2018; Chen et al., 2018; Ouledali et al., 2018). The future scenarios of climate change may also modify the diversity and the growth of AMF in soils as well as the functionality of the mycorrhizal symbiosis. A recent paper addressed the potential role of AMF for maintaining fruit quality in future climate change scenarios (Torres et al., 2018).

5.1 Combined stress studies

Plants, being sessile, need to adapt to multiple ongoing stresses and have developed a multitude of defence responses that allow them to adapt and survive under these conditions. While many studies have been carried out examining individual stresses, few have examined the control of complex plant regulatory networks in response to multiple stress factors. Climate change is predicted to alter pest and pathogen burdens as well as abiotic stresses, so understanding these networks and decisions at the plant level will be essential to assist breeders in the development of resilient varieties (Dowd and Johnson, 2018).

Growers are increasingly using precision monitoring equipment and services to make production systems more efficient and detect environmental stress early. Soil and substrate moisture, temperature, humidity and electrical conductivity sensors are deployed by growers alongside local, mobile weather stations to predict applications of irrigation and fertigation, and to estimate conditions for fungal diseases, for example powdery mildew, so chemical application can be targeted and focussed.

5.2 Phenotyping and plantation stress management

Plant adaptation to changing environmental conditions is known to involve an array of physiological processes and signalling events (Moshou et al., 2005). Among these, regulation of stomatal aperture is essential for stabilizing plant water relations and maintaining nutrient uptake, controlling canopy temperature and optimizing photosynthetic capacity (Jones et al., 2002, 2004, 2014a; Prashar et al., 2013). These effects can be monitored using non-destructive imaging technologies that provide a robust indicator of plant physiological status. Furthermore, environmental stresses can induce changes in the leaf surface and internal structure, cause the accumulation of secondary metabolites or lead to the breakdown of photosynthetic pigments, with adverse effects on photosynthesis and ultimately crop yield (Merlot et al., 2002; Jones et al., 2009; Jones, 2014).

Imaging methodologies offer exciting opportunities for large-scale visualization of plant populations in controlled and field conditions, allowing rapid collection of light signals from the plant surface across a wide spectrum of wavelengths ranging between visible and infrared (Fahlgren et al., 2015; Jones, 2004; Jones and Vaughan, 2010). The main advantage of imaging technologies is the speed at which large-scale plant trials can be investigated. Spectral signatures extracted from imaging datasets can provide information about canopy temperature, pigment composition and water status, which can be informative regarding concomitant changes in plant performance (Fahlgren et al., 2015). The potential application of different imaging methods in marker-assisted breeding has been discussed comprehensively by Singh and Singh (2015). In raspberry, imaging has focussed initially on hyperspectral imaging (Williams et al., 2017), which is increasingly used to detect canopy characteristics (Fahlgren et al., 2015). Spectral reflectance of the plant canopy can be used to acquire information on its biochemistry, detecting the presence of chlorophyll, carotenoids and xanthophyll as well as water-soluble carbohydrates. Many hyperspectral indices have been developed to detect different physiological processes. These work by combining a limited number of wavelength bands to measure absorption by a particular chemical. The best known of these are normalized difference vegetation index (NDVI), which can be used to measure canopy coverage, and the photochemical reflectance index (PRI), which has been used as a proxy for rate of photosynthesis (Garbulsky et al., 2011). Recently, a high-throughput field phenotyping platform has been developed for cane and bush crops utilizing hyperspectral imaging (Williams et al., 2017) with the aim of linking spectral signatures to desired phenotypes for use in breeding and research. Other work has used machine learning techniques (such as support vector machines and artificial neural networks) to model responses using the entire spectrum. Use of hyperspectral imaging can provide a non-destructive measure of plant responses to environmental conditions, and there is significant interest in using imaging techniques to phenotype plant responses to insect pests (Goggin et al., 2015).

6 Conclusion and future trends

The future sustained success of *Rubus* crops production should be achievable using a combination of horticultural tools and approaches, including plant breeding based on knowledge of complex trait regulation using plant genetics and genomics resources, coupled with high-throughput phenotyping under differing environmental stresses. This would provide a greater understanding of the regulation of traits for resilience, combined with active plant management practices such as substrate production in protected environments, and control of canopy structure and irrigation, to minimize biotic and abiotic stresses.

7 References

Abd El-Mageed, T. A., Semida, W. M., Taha, R. S. and Rady, M. M. 2018. Effect of summer-fall deficit irrigation on morpho-physiological, anatomical responses, fruit yield and water use efficiency of cucumber under salt affected soil. *Scientia Horticulturae* 237, 148–55. doi:10.1016/j.scienta.2018.04.014.

Abe, M., Kobayashi, Y., Yamamoto, S., Daimon, Y., Yamaguchi, A., Ikeda, Y., Ichinoki, H., Notaguchi, M., Goto, K. and Araki, T. 2005. FD, a bZIP protein mediating signals from the floral pathway integrator FT at the shoot apex. *Science* 309(5737), 1052–6. doi:10.1126/science.1115983.

Abou-Shaara, H. F., Owayss, A. A., Ibrahim, Y. Y. and Basuny, N. K. 2017. A review of impacts of temperature and relative humidity on various activities of honey bees. *Insectes Sociaux* 64, 455–63.

Amano, T., Smithers, R. J., Sparke, T. H. and Sutherland, W. J. 2010. A 250-year index of first flowering dates and its response to temperature changes. *Proceedings of the Royal Society B: Biological Sciences* 277(1693), 2451–7. doi:10.1098/rspb.2010.0291.

Ansari, M. A., Shah, F. A. and Butt, T. M. 2008. Combined use of entomopathogenic nematodes and *Metarhizium anisopliae* as a new approach for black vine weevil, *Otiorhynchus sulcatus*, control. *Entomologia Experimentalis et Applicata* 129(3), 340–7. doi:10.1111/j.1570-7458.2008.00783.x.

Anthony, V. M., Williamson, B., Jennings, D. L. and Shattock, R. C. 1986. Inheritance of resistance to yellow rust (Phragmidium rubi-idaei) in red raspberry. *Annals of Applied Biology* 109(2), 365–74. doi:10.1111/j.1744-7348.1986.tb05328.x.

Atkinson, C. J., Brennan, R. M. and Jones, H. G. 2013. Declining chilling and it impact on temperate perennial crops. *Environmental and Experimental Botany* 91, 48–62. doi:10.1016/j.envexpbot.2013.02.004.

Ballington, J. R. and Fernandez, G. E. 2008. Breeding raspberries adapted to warm humid climates with fluctuating temperatures in winter. *Acta Horticulturae* 777(777), 87–90. doi:10.17660/ActaHortic.2008.777.9.

Balzan, M. V. 2017. Flowering banker plants for the delivery of multiple agroecosystem services. *Arthropod-Plant Interactions* 11(6), 743–54. doi:10.1007/s11829-017-9544-2.

Barritt, B. H., Crandall, P. C. and Bristow, P. R. 1979. Breeding for root rot resistance in red raspberry. *Journal of the American Society of Horticultural Science* 104, 92–4.

Barros, V., Frosi, G., Santos, M., Ramos, D. G., Falcao, H. M. and Santos, M. G. 2018. Arbuscular mycorrhizal fungi improve photosynthetic energy use efficiency and decrease foliar construction cost under recurrent water deficit in woody evergreen species. *Plant Physiology and Biochemistry: PPB* 127, 469–77. doi:10.1016/j.plaphy.2018.04.016.

Bennison, J., Chandler, D., Prince, G., Pope, T., Atwood, J., Talbot, D., Roberts, H. and Creed, C. 2014. A review of vine weevil knowledge in order to design best-practice IPM protocols suitable for implementation in UK horticulture. HDC Project CP111 Final report.

Berndt, L. A., Wratten, S. D. and Scarratt, S. L. 2006. The influence of floral resource subsidies on parasitism rates of leafrollers (Lepidoptera: Tortricidae) in New Zealand vineyards. *Biological Control* 37(1), 50–5. doi:10.1016/j.biocontrol.2005.12.005.

Birch, A. N. E., Jones, A. T., Fenton, B., Malloch, G., Geoghegan, I., Gordon, S. C., Hillier, J. and Begg, G. 2002. Resistance-breaking raspberry aphid biotypes: Constraints to sustainable control through plant breeding. *Acta Horticulturae* 585(585), 315–7. doi:10.17660/ActaHortic.2002.585.51.

Birch, A. N. E., Begg, G. S. and Squire, G. R. 2011. How agro-ecological research helps to address food security issues under new IPM and pesticide reduction policies for global crop production systems. *Journal of Experimental Botany* 62(10), 3251–61. doi:10.1093/jxb/err064.

Brachi, B., Faure, N., Horton, M., Flahauw, E., Vazquez, A., Nordborg, M., Bergelson, J., Cuguen, J. and Roux, F. 2010. Linkage and association mapping of *Arabidopsis thaliana* flowering time in nature. *PLoS Genetics* 6(5), e1000940. doi:10.1371/journal.pgen.1000940.

Bristow, P. R., Daubeny, H. A., Sjulin, T. M., Pepin, H. S., Nestby, R. and Windom, G. E. 1988. Evaluation of *Rubus* germplasm for reaction to root-rot caused by *Phytophthora-erythroseptica*. *Journal of the American Society for Horticultural Science* 113(4), 588–91.

Bruck, D. J. 2007. Efficacy of *Metarhizium anisopliae* as a curative application for Black Vine Weevil (*Otiorhynchus sulcatus*) infesting container-grown nursery crops. *Journal of Environmental Horticulture* 25(3), 150–6.

Buckley, L. B., Arakaki, A. J., Cannistra, A. F., Kharouba, H. M. and Kingsolver, J. G. 2017. Insect development, thermal plasticity and fitness implications in changing, seasonal environments. *Integrative and Comparative Biology* 57(5), 988–98. doi:10.1093/icb/icx032.

Bushakra, J. M., Bryant, D. W., Dossett, M., Vining, K. J., VanBuren, R., Gilmore, B. S., Lee, J., Mockler, T. C., Finn, C. E., et al. 2015. A genetic linkage map of black raspberry (*Rubus occidentalis*) and the mapping of Ag (4) conferring resistance to the aphid *Amphorophora agathonica*. *Theoretical and Applied Genetics* 128(8), 1631–46.

Bushakra, J. M., Dossett, M., Carter, K. A., Vining, K. J., Lee, J. C., Bryant, D. W., VanBuren, R., Lee, J., Mockler, T. C., Finn, C. E., et al. 2018. Characterization of aphid resistance loci in black raspberry (*Rubus occidentalis* L.). *Molecular Breeding* 38(7), 7. doi:10.1007/s11032-018-0839-5.

Caffarra, A., Rinaldi, M., Eccel, E., Rossi, V. and Pertot, I. 2012. Modelling the impact of climate change on the interaction between grapevine and its pests and pathogens: European grapevine moth and powdery mildew. *Agriculture, Ecosystems and Environment* 148, 89–101. doi:10.1016/j.agee.2011.11.017.

Cao, K., Cui, L., Zhou, X., Ye, L., Zou, Z. and Deng, S. 2015. Four tomato FLOWERING LOCUS T-like proteins act antagonistically to regulate floral initiation. *Frontiers in Plant Science* 6, 1213.

Cao, Y. S., Xiao, Y., Huang, H. Q., Xu, J. C., Hu, W. H. and Wang, N. 2016. Simulated warming shifts the flowering phenology and sexual reproduction of *Cardamine hirsuta* under different planting densities. *Scientific Reports* 6, 27835. doi:10.1038/srep27835.

Carew, J. G., Gillespie, T., White, J., Wainwright, H., Brennan, R. and Battey, N. H. 2000. Techniques for the manipulation of the annual growth cycle in raspberry. *Journal of Horticultural Science and Biotechnology* 75, 504–9.

Chen, M., Arato, M., Borghi, L., Nouri, E. and Reinhardt, D. 2018. Beneficial services of arbuscular mycorrhizal fungi - From ecology to application. *Frontiers in Plant Science* 9(1270). doi:10.3389/fpls.2018.01270.

Comeau, C., Prive, J. P. and Moreau, G. 2012. Beneficial impacts of the combined use of rain shelters and reflective groundcovers in an organic raspberry cropping system. *Agriculture, Ecosystems and Environment* 155, 117–23. doi:10.1016/j.agee.2012.04.007.

De Lucia, E. H., Nabity, P. D., Zavala, J. A. and Berenbaum, M. R. 2012. Climate change: Resetting plant-insect interactions. *Plant Physiology* 160(4), 1677–85. doi:10.1104/pp.112.204750.

Del Fava, E., Ioriatti, C. and Melegaro, A. 2017. Cost-benefit analysis of controlling the spotted wing drosophila (*Drosophila suzukii* (Matsumura)) spread and infestation of soft fruits in Trentino, Northern Italy. *Pest Management Science* 73(11), 2318–27. doi:10.1002/ps.4618.

Deutsch, C. A., Tewksbury, J. J., Huey, R. B., Sheldon, K. S., Ghalambor, C. K., Haak, D. C. and Martin, P. R. 2008. Impacts of climate warming on terrestrial ectotherms across latitude. *Proceedings of the National Academy of Sciences* 105(18), 6668–72. doi:10.1073/pnas.0709472105.

Diskin, E., Proctor, H., Jebb, M., Sparks, T. and Donnelly, A. 2012. The phenology of *Rubus fruticosus* in Ireland: Herbarium specimens provide evidence for the response of phenophases to temperature, with implications for climate warming. *International Journal of Biometeorology* 56(6), 1103–11. doi:10.1007/s00484-012-0524-z.

Dodd, A. N., Jakobsen, M. K., Baker, A. J., Telzerow, A., Hou, S. W., Laplaze, L., Barrot, L., Poethig, R. S., Haseloff, J. and Webb, A. A. R. 2006. Time of day modulates low-temperature Ca2+ signals in Arabidopsis. *Plant Journal* 48(6), 962–73.

Dolan, A., McFarlane, S. and Jennings, S. N. 2018. Pathogens in raspberry and other *Rubus* spp. In: Graham, J. and Brennan, R. (Eds), *Raspberry: Breeding, Challenges and Advances*. Cham: Springer, pp. 41–61.

Dowd, P. F. and Johnson, E. T. 2018. Insect damage influences heat and water stress resistance gene expression in field-grown popcorn: Implications in developing crop varieties adapted to climate change. *Mitigation and Adaptation Strategies for Global Change* 23(7), 1063–81. doi:10.1007/s11027-017-9772-x.

Duncan, J. M., Kennedy, D. M. and Seemüller, E. 1987. Identities and pathogenicities of *Phytophthora* spp. causing root rot of red raspberry. *Plant Pathology* 36(3), 276–89. doi:10.1111/j.1365-3059.1987.tb02235.x.

El Jaouhari, N., Abouabdillah, A., Bouabid, R., Bourioug, M., Aleya, L. and Chaoui, M. 2018. Assessment of sustainable deficit irrigation in a Moroccan apple orchard as a climate change adaptation strategy. *Science of the Total Environment* 642, 574–81. doi:10.1016/j.scitotenv.2018.06.108.

Fahlgren, N., Gehan, M. A. and Baxter, I. 2015. Lights, camera, action: High-throughput plant phenotyping is ready for a close-up. *Current Opinion in Plant Biology* 24, 93–9. doi:10.1016/j.pbi.2015.02.006.

Figueiredo, A. S. T., Resende, J. T. V., Morales, R. G. F., Gonçalves, A. P. S. and Da Silva, P. R. 2013. The role of glandular and non-glandular trichomes in the negative interactions between strawberry cultivars and spider mite. *Arthropod-Plant Interactions* 7(1), 53-8. doi:10.1007/s11829-012-9218-z.

Fitter, A. H. and Fitter, R. S. R. 2002. Rapid changes in flowering time in British plants. *Science* 296(5573), 1689–91. doi:10.1126/science.1071617.

Fornara, F., de Montaigu, A. and Coupland, G. 2010. Snapshot control of flowering in Arabidopsis. *Cell* 141, 550. doi:10.1016/j.cell.2010.04.024.

Garbulsky, M. F., Penuelas, J., Gamon, J., Inoue, Y. and Filella, I. 2011. The photochemical reflectance index (PRI) and the remote sensing of leaf, canopy and ecosystem radiation use efficiencies: A review and meta-analysis. *Remote Sensing of Environment* 115: 281-97.

Gillespie, T., Brennan, R., Raffle, S. and McNicol, R. J. 1999. Cultivar responses to long-cane fruit production in raspberry. Scottish Crop Research Institute Annual Report for 1999, pp. 105–9.

Goggin, F. L., Lorence, A. and Topp, C. N. 2015. Applying high-throughput phenotyping to plant-insect interactions: Picturing more resistant crops. *Current Opinion in Insect Science* 9: 69-76.

Gordon, S. C. and Williamson, B. 1988. Comparison of an air-assisted cross-flow sprayer with a conventional hydraulic sprayer for control of raspberry aphids by fenitrothion. *Crop Protection* 7(2), 106–11. doi:10.1016/0261-2194(88)90020-8.

Gordon, S. C. and Williamson, B. 2004. Review of current knowledge of disease and pest control trials in cane and bush fruits in Europe. *Acta Horticulturae* 649, 283–7.

Gordon, S. C., Woodford, J. A. T. and Birch, A. N. E. 1997. Arthropod pests of *Rubus* in Europe: Pest status, current and future control strategies. *Journal of Horticultural Science* 72(6), 831-62. doi:10.1080/14620316.1997.11515577.

Gordon, S. C., Williamson, B. and Graham, J. 2006. Current and future control strategies for major arthropod pests and fungal diseases of red raspberry (*Rubus idaeus*) in Europe. In: Ramdane, D. (Ed.), *Crops: Quality, Growth and Biotechnology*. Helsinki: WFL Publisher, pp. 925–50.

Graham, J. and Simpson, C. 2018. Developmental transitions to fruiting in red raspberry. In: Hytonen, T., Graham, J. and Harrison, R. (Eds), *The Genomes of Rosaceous Berries and their Wild Relatives*. Compendium of Plant Genomes. Cham: Springer, pp. 199–212. doi:10.1007/978-3-319-76020-9_14.

Graham, J., Squire, G. R., Marshall, B. and Harrison, R. E. 1997. Spatially dependent genetic diversity within and between colonies of wild raspberry *Rubus idaeus* detected using RAPD markers. *Molecular Ecology* 6(11), 1001–8. doi:10.1046/j.1365-294X.1997.00272.x.

Graham, J., Marshall, B. and Squire, G. R. 2003. Genetic differentiation over a spatial environmental gradient in wild *Rubus idaeus* populations. *New Phytologist* 157(3), 667–75. doi:10.1046/j.1469-8137.2003.00693.x.

Graham, J., Smith, K., MacKenzie, K., Jorgensen, L., Hackett, C. A. and Powell, W. 2004. The construction of a genetic linkage map of red raspberry (*Rubus idaeus* subsp. *idaeus*) based on AFLPs, genomic-SSR and EST-SSR markers. *Theoretical and Applied Genetics* 109(4), 740-9. doi:10.1007/s00122-004-1687-8.

Graham, J., Smith, K., Tierney, I., MacKenzie, K. and Hackett, C. 2006. Mapping gene H controlling cane pubescence in raspberry and its association with resistance to cane botyritis and spur blight, rust and cane spot. *Theoretical and Applied Genetics* 112, 818–31.

Graham, J., Hackett, C. A., Smith, K., Woodhead, M., Hein, I. and McCallum, S. 2009a. Mapping QTLs for developmental traits in raspberry from bud break to ripe fruit. *Theoretical and Applied Genetics* 118(6), 1143–55. doi:10.1007/s00122-009-0969-6.

Graham, J., Woodhead, M., Smith, K., Russell, J. R., Marshall, B., Ramsay, G. and Squire, G. R. 2009b. New insight into wild red raspberry populations using simple sequence repeat markers. *Journal of the American Society of Horticultural Science* 134, 109–19.

Graham, J., Hackett, C. A., Smith, K., Woodhead, M., MacKenzie, K., Tierney, I., Cooke, D. E. L., Bayer, M. and Jennings, N. 2011. Towards an understanding of the nature of resistance to Phytophthora root rot in red raspberry: Is it mainly root vigour? *Theoretical and Applied Genetics* 123(4), 585–601. doi:10.1007/s00122-011-1609-5.

Graham, J., Hackett, C. A., Smith, K., Karley, A. J., Mitchell, C., Roberts, H. and O'Neill, T. 2014. Genetic and environmental regulation of plant architectural traits and opportunities for pest control in raspberry. *Annals of Applied Biology* 165(3), 318–28. doi:10.1111/aab.12134.

Graham, J., Smith, K., McCallum, S., Hedley, P. E., Cullen, D. W., Dolan, A., Milne, L., McNicol, J. W. and Hackett, C. A. 2015. Towards an understanding of crumbly fruit in red raspberry. *SpringerPlus* 4, 223. doi:10.1186/s40064-015-1010-y.

Green, R. M., Tingay, S., Wang, Z. Y. and Tobin, E. M. 2002. Circadian rhythms confer a higher level of fitness to Arabidopsis plants. *Plant Physiology* 129(2), 576–84. doi:10.1104/pp.004374.

Hackett, C., Milne, L., Smith, K., Hedley, P., Morris, J., Simpson, C. G., Preedy, K. and Graham, J. 2018. Enhancement of Glen Moy × Latham raspberry linkage map using GbS to further understand control of developmental processes leading to fruit ripening. *BMC Genetics* 19, 59.

Hirsch, J. and Reineke, A. 2014. Efficiency of commercial entomopathogenic fungal species against different members of the genus *Otiorhynchus* (Coleoptera: Curculionidae) under laboratory and semi-field conditions. *Journal of Plant Diseases and Protection* 121(5), 211–8. doi:10.1007/BF03356513.

Huberty, A. F. and Denno, R. F. 2004. Plant water stress and its consequences for herbivorous insects: A new synthesis. *Ecology* 85(5), 1383–98. doi:10.1890/03-0352.

Jennings, D. L. 1962. Some evidence on the influence of the morphology of raspberry canes upon their ability to be attacked by certain fungi. *Horticulture Research* 1, 100–11.

Jennings, D. L. 1982a. Resistance to *Didymella applanata* in red raspberry and some related species. *Annals of Applied Biology* 101(2), 331–7. doi:10.1111/j.1744-7348.1982.tb00829.x.

Jennings, D. L. 1982b. Further evidence on the effects of Gene-H, which confers cane hairiness, on resistance to raspberry diseases. *Euphytica* 31(3), 953–6. doi:10.1007/BF00039234.

Jennings, D. L. 1988. *Raspberries and Blackberries: Their Breeding, Diseases and Growth*. London: Academic Press.

Jennings, D. L. and Brydon, E. 1989. Further-studies on breeding for resistance to *Botrytis-cinerea* in red raspberry canes. *Annals of Applied Biology* 115(3), 507–13. doi:10.1111/j.1744-7348.1989.tb06571.x.

Jennings, N. and Dolan, A. 2014. Rubus: Blackberry, raspberry and loganberry. In: George, R. A. T. and Fox, R. T. V. (Eds), *Diseases of Temperate Horticultural Plants*. Oxfordshire, UK: CABI, pp. 21–52.

Johnson, S. N., Young, M. W. and Karley, A. J. 2012. Protected raspberry production alters aphid-plant interactions but not aphid population size. *Agricultural and Forest Entomology* 14(2), 217–24. doi:10.1111/j.1461-9563.2011.00561.x.

Jones, A. T. 1987. Control of virus-infection in crop plants through vector resistance – A review of achievements, prospects and problems. *Annals of Applied Biology* 111(3), 745–72. doi:10.1111/j.1744-7348.1987.tb02033.x.

Jones, H. G. 2004. Application of thermal imaging and infrared sensing in plant physiology and ecophysiology. *Advances in Botanical Research* 41, 107–63. doi:10.1016/S0065-2296(04)41003-9.

Jones, H. G. 2014. *Plants and Microclimate: A Quantitative Approach to Environmental Plant Physiology* (3rd edn.). Cambridge: Cambridge University Press.

Jones, H. G. and Vaughan, R. A. 2010. *Remote Sensing of Vegetation: Principles, Techniques, and Applications*. Oxford: Oxford University Press.

Jones, A. T., McGavin, W. J. and Birch, A. N. E. 2001. Some factors influencing the effectiveness of resistance genes to the aphid virus-vector, *Amphorophora idaei* Borner, in raspberry. *Acta Horticulturae* 551, 39–44.

Jones, H. G., Stoll, M., Santos, T., de Sousa, C., Chaves, M. M. and Grant, O. M. 2002. Use of infrared thermography for monitoring stomatal closure in the field: Application to grapevine. *Journal of Experimental Botany* 53(378), 2249–60. doi:10.1093/jxb/erf083.

Jones, H. G., Serraj, R., Loveys, B. R., Xiong, L. Z., Wheaton, A. and Price, A. H. 2009. Thermal infrared imaging of crop canopies for the remote diagnosis and quantification of plant responses to water stress in the field. *Functional Plant Biology* 36(11), 978–89. doi:10.1071/FP09123.

Juszczak, R., Kuchar, L., Lesny, J. and Olejnik, J. 2013. Climate change impact on development rates of the codling moth (*Cydia pomonella* L.) in the Wielkopolska region, Poland. *International Journal of Biometeorology* 57(1), 31–44. doi:10.1007/s00484-012-0531-0.

Karley, A. J., Mitchell, C., Brookes, C., McNicol, J., O'Neill, T., Roberts, H., Graham, J. and Johnson, S. N. 2016. Exploiting physical defence traits for crop protection: Leaf trichomes of *Rubus idaeus* have deterrent effects on spider mites but not aphids. *Annals of Applied Biology* 168(2), 159–72. doi:10.1111/aab.12252.

Kassim, A., Poette, J., Paterson, A., Zait, D., McCallum, S., Woodhead, M., Smith, K., Hackett, C. A. and Graham, J. 2009. Environmental and seasonal influences on red raspberry anthocyanin antioxidant contents and identification of quantitative traits loci (QTL). *Molecular Nutrition and Food Research* 53(5), 625–34. doi:10.1002/mnfr.200800174.

Keep, E. and Knight, R. L. 1967. A new gene from *Rubus occidentalis* l for resistance to strains 1,2 and 3 of rubus aphid *Amphorophora rubi* Kalt. *Euphytica* 16(2), 209.

Keep, E., Knight, V. H. and Parker, J. H. 1977. *Rubus coreanus* as donor of resistance to cane diseases and mildew in red raspberry breeding. *Euphytica* 26(2), 505–10. doi:10.1007/BF00027019.

Knight, V. H. 1991. Use of the salmonberry, *Rubus spectabilis* Pursh, in red raspberry breeding. *Journal of Horticultural Science* 66(5), 575–81. doi:10.1080/00221589.1991.11516186.

Knight, R. L., Briggs, J. B. and Keep, E. 1960. Genetics of resistance to *Amphorophora-rubi* (Kalt) in the raspberry. 2. The genes a2-a7 from the American variety, Chief. *Genetical Research* 1(2), 319–31. doi:10.1017/S0016672300000288.

Knight, V. H. and Fernández Fernández, F. 2008. Screening for resistance to *Phytophthora fragariae* var. *rubi* in *Rubus germplasm* at East Malling. *Acta Horticulturae* 777(777), 353–60. doi:10.17660/ActaHortic.2008.777.53.

Korosi, Á., Marko, V., Kovacs-Hostyanszki, A., Somay, L., Varga, Á., Elek, Z., Boreux, V., Klein, A. M., Foldesi, R. and Baldi, A. 2018. Climate-induced phenological shift of apple trees has diverse effects on pollinators, herbivores and natural enemies. *Peer Journal* 6, e5269. doi:10.7717/peerj.5269.

Lee, J. H., Yoo, S. J., Park, S. H., Hwang, I., Lee, J. S. and Ahn, J. H. 2007. Role of SVP in the control of flowering time by ambient temperature in Arabidopsis. *Genes and Development* 21(4), 397–402. doi:10.1101/gad.1518407.

Mackay, I., Horwell, A., Garner, J., White, J., McKee, J. and Philpott, H. 2011. Reanalyses of the historical series of UK variety trials to quantify the contributions of genetic and environmental factors to trends and variability in yield over time. *Theoretical and Applied Genetics* 122(1), 225–38. doi:10.1007/s00122-010-1438-y.

Marshall, B., Harrison, R. E., Graham, J., McNicol, J. W., Wright, G. and Squire, G. R. 2001. Spatial trends of phenotypic diversity between colonies of wild raspberry *Rubus idaeus*. *New Phytologist* 151(3), 671–82. doi:10.1046/j.0028-646x.2001.00220.x.

Martin, R. R., MacFarlane, S., Sabanadzovic, S., Quito, D., Poudel, B. and Tzanetakis, I. E. 2013. Viruses and virus diseases of Rubus. *Plant Disease* 97(2), 168–82. doi:10.1094/PDIS-04-12-0362-FE.

McCallum, S., Woodhead, M., Hackett, C. A., Kassim, A., Paterson, A. and Graham, J. 2010. Genetic and environmental effects influencing fruit colour and QTL analysis in raspberry. *Theoretical and Applied Genetics* 121(4), 611-27. doi:10.1007/s00122-010-1334-5.

McKenzie, K., Williamson, S., Smith, K., Woodhead, M., McCallum, S. and Graham, J. 2015. Characterisation of the Gene H region in red raspberry: Exploring its role in cane morphology, disease resistance, and timing of fruit ripening. *Journal of Horticulture* 2, 3.

Memmott, J., Craze, P. G., Nickolas, M., Waser, N. M. and Price, M. V. 2007. Global warming and the disruption of plant-pollinator interactions. *Ecology Letters* 10(8), 710–7.

Merlot, S., Mustilli, A. C., Genty, B., North, H., Lefebvre, V., Sotta, B., Vavasseur, A. and Giraudat, J. 2002. Use of infrared thermal imaging to isolate Arabidopsis mutants defective in stomatal regulation. *The Plant Journal: For Cell and Molecular Biology* 30(5), 601–9. doi:10.1046/j.1365-313X.2002.01322.x.

Mettler, D. and Hatterman-Valenti, H. 2016. Cultivar comparison and winter row covers for blackberry production in North Dakota. *Acta Horticulturae* 1133(1133), 217–22. doi:10.17660/ActaHortic.2016.1133.33.

Miller, T. L. P. and Rebek, E. J. 2018. Banker plants for aphid biological control in greenhouses. *Journal of Integrated Pest Management* 9(1).

Mitchell, C., Brennan, R. M., Graham, J. and Karley, A. J. 2016. Plant defense against herbivorous pests: Exploiting resistance and tolerance traits for sustainable crop protection. *Frontiers in Plant Science* 7, 1132. doi:10.3389/fpls.2016.01132.

Morris, J., Else, M. A., El Chami, D., Daccache, A., Rey, D. and Knox, J. W. 2017. Essential irrigation and the economics of strawberries in a temperate climate. *Agricultural Water Management* 194, 90–9. doi:10.1016/j.agwat.2017.09.004.

Moshou, D., Bravo, C., Oberti, R., West, J., Bodria, L., McCartney, A. and Ramon, H. 2005. Plant disease detection based on data fusion of hyper-spectral and multi-spectral fluorescence imaging using Kohonen maps. *Real-Time Imaging* 11(2), 75–83. doi:10.1016/j.rti.2005.03.003.

Nestby, R. and Takeda, F. 2015. Method to reduce low temperature stress (LTS) in red raspberry (*Rubus idaeus* L.). *Journal of Berry Research* 5(4), 219–30. doi:10.3233/JBR-150104.

O'Neill, T., Gladders, P., Wedgewood, E., Huckle, A., Collier, R., Jukes, A., Richardson, A., Knott, K., Berrie, A., Saville, R., et al. 2014. Sustainable crop and environment protection – Target research for edibles (SCEPTRE). HDC Project CP 077 Final Report.

Orhan, E., Esitken, A., Ercisli, S., Turan, M. and Sahin, F. 2006. Effects of plant growth promoting rhizobacteria (PGPR) on yield, growth and nutrient contents in organically growing raspberry. *Scientia Horticulturae* 111(1), 38–43. doi:10.1016/j.scienta.2006.09.002.

Ouledali, S., Ennajeh, M., Zrig, A., Gianinazzi, S. and Khemira, H. 2018. Estimating the contribution of arbuscular mycorrhizal fungi to drought tolerance of potted olive trees (*Olea europaea*). *Acta Physiologiae Plantarum* 40(5). doi:10.1007/s11738-018-2656-1.

Parikka, P., Tuovinen, T. and Lemmetty, A. 2016. Challenges for plant protection of berry crops in northern Europe. *Acta Horticulturae* 1117(1117), 95–102. doi:10.17660/ActaHortic.2016.1117.17.

Paterson, A., Kassim, A., McCallum, S., Woodhead, M., Smith, K., Zait, D. and Graham, J. 2013. Environmental and seasonal influences on red raspberry flavour volatiles and identification of quantitative trait loci (QTL) and candidate genes. *Theoretical and Applied Genetics* 126(1), 33–48. doi:10.1007/s00122-012-1957-9.

Pattison, J. A., Samuelian, S. K. and Weber, C. A. 2007. Inheritance of Phytophthora root rot resistance in red raspberry determined by generation means and molecular linkage analysis. *Theoretical and Applied Genetics* 115(2), 225–36. doi:10.1007/s00122-007-0558-5.

Pin, P. A. and Nilsson, O. 2012. The multifaceted roles of *FLOWERING LOCUS T* in plant development. *Plant, Cell and Environment* 35(10), 1742–55. doi:10.1111/j.1365-3040.2012.02558.x.

Prashar, A., Yildiz, J., McNicol, J. W., Bryan, G. J. and Jones, H. G. 2013. Infra-red thermography for high throughput field phenotyping in *Solanum tuberosum*. *PLoS ONE* 8(6), e65816. doi:10.1371/journal.pone.0065816.

Price, P. W. 1991. The plant vigor hypothesis and herbivore attack. *Oikos* 62, 244–51.

Pritts, M., McDermott, L., Demchak, K., Hanson, E., Weber, C., Both, A. J., Loeb, G. and Heidenreich, C. 2017. High tunnel raspberries and blackberries. Available at: http://www.hort.cornell.edu/fruit/pdfs/high-tunnel-brambles.pdf.

Sargent, D. J., Fernandez-Fernandez, F., Rys, A., Knight, V. H., Simpson, D. W. and Tobutt, K. R. 2007. Mapping of A1 conferring resistance to the aphid *Amphorophora idaei* and dw (dwarfing habit) in red raspberry (*Rubus idaeus* L.) using AFLP and microsatellite markers. *BMC Plant Biology* 7, 15.

Seo, E., Yu, J., Ryu, K. H., Lee, M. M. and Lee, I. 2011. WEREWOLF, a regulator of root hair pattern formation, controls flowering time through the regulation of FT mRNA stability. *Plant Physiology* 156(4), 1867–77. doi:10.1104/pp.111.176685.

Simon, S., Rühl, M., de Montaigu, A., Wötzel, S. and Coupland, G. 2015. Evolution of CONSTANS regulation and function after gene duplication produced a photoperiodic flowering switch in the *Brassicaceae*. *Molecular Biology and Evolution* 32(9), 2284–301. doi:10.1093/molbev/msv110.

Simpson, C. G., Cullen, D. W., Hackett, C. A., Smith, K., Hallett, P. D., McNicol, J., Woodhead, M. and Graham, J. 2017. Mapping and expression of genes associated with raspberry fruit ripening and softening. *Theoretical and Applied Genetics* 130(3), 557–72. doi:10.1007/s00122-016-2835-7.

Singh, B. D. and Singh, A. K. 2015. Marker assisted plant breeding: Principles and practices. In: Sing, B. D. and Sing, A. K. (Eds), *Phenomics*. New Delhi, India: Springer, pp. 431–62. Chapter 15.

Smith, F. F. 1932. Biology and control of the black vine weevil. US Dept Agric Tech Bull 325, 45pp.

Song, J., Angel, A., Howard, M. and Dean, C. 2012. Vernalisation-a cold induced epigenetic switch. *Journal of Cell Science* 125, 3723–31.

Stafne, E. T., Clark, J. R. and Rom, C. R. 2000. Leaf gas exchange characteristics of red raspberry germplasm in a hot environment. *HortScience* 35(2), 278–80.

Strick, B. C., Clark, J. R., Finn, C. E. and Buller, G. 2012. Management of primocane-fruiting blackberry: Impacts on yield, fruiting, season and cane architecture. *HortScience* 47(5), 593–8.

Tian, D. L., Tooker, J., Peiffer, M., Chung, S. H. and Felton, G. W. 2012. Role of trichomes in defense against herbivores: Comparison of herbivore response to woolly and hairless trichome mutants in tomato (*Solanum lycopersicum*). *Planta* 236(4), 1053–66. doi:10.1007/s00425-012-1651-9.

Torres, N., Antolin, M. C. and Goicoechea, N. 2018. Arbuscular mycorrhizal symbiosis as a promising resource for improving berry quality in grapevines Under changing environments. *Frontiers in Plant Science* 9, 897. doi:10.3389/fpls.2018.00897.

Unterberger, C., Brunner, L., Nabernegg, S., Steininger, K. W., Steiner, A. K., Stabentheiner, E., Monschein, S. and Truhetz, H. 2018. Spring frost risk for regional apple production under a warmer climate. *PLoS ONE* 13(7). doi:10.1371/journal.pone.0200201.

van Tol, R. W. H. M., Bruck, D. J., Griepink, F. C. and de Kogel, W. J. 2012. Field attraction of the vine weevil *Otiorhynchus sulcatus* to kairomones. *Journal of Economic Entomology* 105(1), 169–75. doi:10.1603/EC11248.

Voigt, D. and Gorb, S. 2010. Egg attachment of the asparagus beetle *Crioceris asparagi* to the crystalline waxy surface of *Asparagus officinalis*. *Proceedings of the Royal Society B: Biological Sciences* 277(1683), 895–903. doi:10.1098/rspb.2009.1706.

Wang, J. W. 2014. Regulation of flowering time by the miR156-mediated age pathway. *Journal of Experimental Botany* 65(17), 4723–30. doi:10.1093/jxb/eru246.

Wenter, A., Zanotelli, D., Montagnani, L., Tagliavini, M. and Andreotti, C. 2018. Effect of different timings and intensities of water stress on yield and berry composition of grapevine (cv. Sauvignon blanc) in a mountain environment. *Scientia Horticulturae* 236, 137–45. doi:10.1016/j.scienta.2018.03.037.

White, C. and Eigenbrode, S. D. 2000. Effects of surface wax variation in *Pisum sativum* on herbivorous and entomophagous insects in the field. *Environmental Entomology* 29(4), 773–80. doi:10.1603/0046-225X-29.4.773.

Williams, D., Britten, A., McCallum, S., Jones, H., Aitkenhead, M., Karley, A., Loades, K., Prashar, A. and Graham, J. 2017. A method for automatic segmentation and splitting of hyperspectral images of raspberry plants collected in field conditions. *Plant Methods* 13(1). doi:10.1186/s13007-017-0226-y.

Williamson, B. and Jennings, D. L. 1986. Common resistance in red raspberry to *Botrytis-cinerea* and *Didymella-applanata*, 2 pathogens occupying the same ecological niche. *Annals of Applied Biology* 109(3), 581–93. doi:10.1111/j.1744-7348.1986.tb03215.x.

Williamson, B. and Jennings, D. L. 1992. Resistance to cane and foliar diseases in red raspberry (*Rubus-idaeus*) and related species. *Euphytica* 63(1-2), 59–70. doi:10.1007/BF00023912.

Willmer, P. G., Hughes, J. P., Woodford, J. A. T. and Gordon, S. C. 1996. The effects of crop microclimate and associated physiological constraints on the seasonal and diurnal distribution patterns of raspberry beetle (*Byturus tomentosus*) on the host plant *Rubus idaeus*. *Ecological Entomology* 21(1), 87–97. doi:10.1111/j.1365-2311.1996.tb00270.x.

Woodhead, M., Williamson, S., Smith, K., McCallum, S., Jennings, N., Hackett, C. and Graham, J. 2013. Identification of quantitative trait loci for cane splitting in red raspberry (*Rubus idaeus*). *Molecular Breeding* 31(1), 111–22. doi:10.1007/s11032-012-9775-y.

Wu, G., Park, M. Y., Conway, S. R., Wang, J. W., Weigel, D. and Poethig, R. S. 2009. The sequential action of miR156 and miR172 regulates developmental timing in Arabidopsis. *Cell* 138, 750–9.

Zorenc, Z., Veberic, R., Koron, D. and Mikulic-Petkovsek, M. 2017. Impact of raspberry (*Rubus idaeus* L.) primocane tipping on fruit yield and quality. *Notulae Botanicae Horti Agrobotanici Cluj-Napoca* 45(2), 417–24.

Chapter 14

Advances and challenges in blueberry breeding

Susan McCallum, James Hutton Institute, UK

1 Introduction

Consumer demand for blueberries (*Vaccinium corymbosum*) is at an all-time high and continues to rise year on year, having recently become the second-most important berry behind strawberry (*Fragaria* x *ananassa*) (Giongo et al., 2013). This rise in demand is due, in part, to the many recognised health benefits and nutritional qualities of blueberries (Cardeñosa et al., 2016), in addition to increased worldwide production. Blueberry fruit are rich in phenolic compounds, which have been associated with improved night vision, prevention of macular degeneration, anti-cancer activity, inhibition of pro-inflammatory molecules and reduced risk of heart disease (Johnson and Arjmandi, 2013; Kalt et al., 2007). Blueberry polyphenols also have been shown to increase endothelium-dependent vasodilation as assessed by flow-mediated dilation leading to an improvement in overall vascular function (Rodriguez-Mateos et al., 2013). Pterostilbene, the primary antioxidant component in blueberries, has been shown to have both preventive and therapeutic effects on neurological, cardiovascular, metabolic and haematological disorders (Ness and Powles, 1997; Rimando et al., 2004).

http://dx.doi.org/10.19103/AS.2018.0040.29

In the United Kingdom (UK), the high consumer demand for blueberries, combined with the lack of appropriate high-quality cultivars suitable for UK climatic conditions, has resulted in a real need by the industry for the rapid development of new blueberry cultivars with high fruit and nutritional quality, and an expanded fruiting season to meet the demand for home-grown fruit. External quality, such as fruit colour and appearance, will influence initial sales, but internal quality such as flavour and texture will determine likelihood of repeat purchase. Fruit flavour depends on a combination of taste (sweetness, sourness, low astringency), aroma (concentration, ratio and interaction of volatile organic compounds [VOCs]) and texture (nice crunch when bitten). As far as biochemical drivers of the sensory experience are concerned, sugars and acids remain the principal contributors to fruit taste (Kader, 1991). Citric acid is the predominant organic acid found in blueberry, followed by malic and quinic acid, while glucose and fructose are the main sugars (Forney et al., 2012). While low levels of citric acid can produce less tart fruit, the absence of acid altogether results in a bland, tasteless fruit (Kader, 1991). Although hundreds of VOCs may be detected, it is often only a few which are present in high enough quantities to impact flavour and generate the key flavour signals contributing to the overall 'blueberry' experience (Baldwin et al., 2000). Rich and fruity aromas are most desirable for consumer purchase, as this signifies a fresh and ripe product. A lack of odour, or off notes such as mould or fermented fruit, has a negative effect and suggests spoiled, unripe or poor-quality fruit (Forney, 2001).

Most breeding programmes have focused on commercial traits such as yield and size due to the complexity of defining and quantifying fruit flavour (Klee and Tieman, 2018). A current challenge is breeding blueberries with greater levels of genetic diversity which can withstand slight climatic changes that meet both grower and consumer expectations in terms of yield and size, but without compromising on sensory quality.

2 Breeding challenges today

The main reasons given for the general decline in flavour of fresh produce are genetics, harvest maturity and post-harvest handling (Baldwin et al., 2007). The starting genetic material is therefore an appropriate place to begin to ensure an inherent flavour potential becomes an integral part of any breeding programme (Baldwin et al., 2007). Blueberry genotype has been confirmed as a major factor relating to the VOCs detected and profiles recorded (Forney, 2001; Farnetti et al., 2017). Interestingly, in terms of breeding potential, considerable variation between cultivated and wild *Vaccinium* species has also been noted in both flavonoid and phenolic profiles. The penetration level for blueberry purchase in UK households in 2017 was 47%, with less than 15% of these buyers likely to be regular purchasers. Increasing this market penetration and recruiting regular buyers will be an important driver for continued expansion (Gilbert et al., 2014). The goal for breeders, therefore, will be to convert these occasional buyers into regular buyers, and a big influence on this will be through product satisfaction. In current blueberry germplasm, problems with a lack of variation in flavour intensity have been attributed to a lack of genetic diversity in flavour attributes. It also has been suggested that sensory panellists involved in blueberry trials may exhibit a poor understanding of flavour (as opposed to taste) in a crop which is renowned for being less aromatic than other berries (Gilbert et al., 2015). Volatile concentrations in blueberry are reported at 0.5–0.75 mg kg^{-1} compared with 2–6 mg kg^{-1} for strawberry and up to 62 mg kg^{-1} in wild raspberries (Forney, 2001).

Recent developments in both genetics and genomics have advanced research in most berry species. Molecular markers that can detect genome-wide variability in both coding and non-coding protein regions have enabled genetic mapping studies to move beyond simple linkages between morphological traits (Jennings, 1988, 1967; Ourecky, 1975; Crane and Lawrence, 1931; Keep, 1968) to the development of saturated linkage maps containing numerous markers which can be utilised in marker-assisted breeding. Until recently, mapping in blueberry species has trailed that of other berries due to their more complex genetic make-up and lesser economic importance.

The production of highbush blueberry (*Vaccinium corymbosum* L.) has continued to expand worldwide over the last 30 years. There have, however, been reported cases across Europe and some American states indicating that many have experienced significant instability in blueberry yield across seasons. This yield instability could be the result of sudden and largely unpredictable catastrophic events, such as late spring frosts, that occur during vulnerable stages of fruit development (Warmund et al., 2008; Rowland et al., 2013). On the other hand, yield instability also could be the result of subtle year-to-year environmental variation that causes a disruption in carbon assimilation, storage or partitioning (Petridis et al., 2018). Photosynthesis and whole plant carbohydrate partitioning are known to impact greatly on fruit yield, so any factor which disrupts these processes can lead to fruit abortion, resulting in economical yield losses (Teo et al., 2006; Gauci et al., 2009; Osorio et al., 2014).

For the last 30 years, blueberry breeding efforts have been focused on improving yield and fruit size, but as we have seen in tomato (*Solanum lycopersicum*) and strawberry breeding, this often has a detrimental effect on flavour quality (Ulrich et al., 2007; Ruiz et al., 2005). A recent study identified 13 flavour chemicals in tomato which were all significantly reduced in commercial varieties that had been described as tasting 'bland' (Tieman et al., 2017). Consumers are becoming more driven by health benefits of fresh fruit and are becoming concerned with 'food miles' and sustainability. Poor quality issues have been raised in blueberries that have been imported, at times spending several weeks in a controlled atmosphere en route to the UK, and for some consumers, this experience is their first, and often last, taste of blueberries. In a product that has low levels of volatiles when fresh, any quality factor, such as picking before fully ripe or poor or prolonged post-harvest treatment, will undoubtedly have a detrimental effect on sensory experience (Miszczak et al., 1995). If increased consumption of blueberries is set to continue, then a new generation of cultivars with appropriate seasonal and fruit quality attributes will be essential. Identification of appropriate character-impacting volatiles for sweet and fruity blueberry flavour (such as [*E*]-2-hexenal, linalool and ethyl acetate) could help breeders to bridge the current gap in new cultivar development to consistently deliver superior external and internal qualities.

New varieties, high in quality, capable of tolerating environmental stresses, pests and diseases, are essential for maintaining productivity in current and future production environments. These abiotic and biotic stresses can limit crop productivity, and new superior varieties are essential for survival, growth and sustainability of the UK berry industry. Advances in genetics, for example, use of marker-assisted breeding, as well as metabolomic resources, have brought many improvements in breeding techniques, but the evaluation of the performance of new varieties is currently a bottleneck in the breeding process. The precision and speed at which the response of new varieties to stress can be detected is still currently limited. There is a great opportunity to fast track breeding by improving the speed and precision of phenotypic selection using imaging sensor and agricultural engineering systems, thereby reducing the cost of years of field assessment.

3 Genetic material and their uses in breeding programmes

Blueberry belongs to the genus *Vaccinium* comprising around 450 species, many of which are native to North America. There are three species predominantly cultivated: the wild American lowbush (*Vaccinium angustifolium* Ait.), highbush (*Vaccinium corymbosum* L.) and rabbiteye (*Vaccinium ashei* Reade) blueberries. The highbush blueberries can be divided further into Northern and Southern types depending on their winter hardiness and chilling requirements, with the Northern species being more suitable for colder climates and are the only species currently grown commercially in the UK (Hancock et al., 2008). There are three main ploidy levels found in *Vaccinium* species, $2n = 2X = 24$, $4X = 48$ and $6X = 72$ (Vander Kloet, 1988). These diploid, tetraploid and hexaploid genomes indicate that extensive genome rearrangements are likely to have occurred during their evolution, resulting in these complex genomes making genetic studies difficult (Die and Rowland, 2013).

Current breeding efforts in *Vaccinium* depend primarily on the location of the breeding programme as well as the target market. The breeding use of several different species has facilitated the adaption of blueberry into warmer and more humid climates of the southeastern United States. Low chilling requirement has been introduced through hybrids with species such as *V. darrowii*, *V. virgatum* (ashei) and *V. tenellum*. 'Florida 4B' is a selection from the diploid species *V. darrowii* that not only conveys desirable fruit and chilling characteristics but can also allow breeders to introduce germplasm from additional diploid species which themselves may not produce unreduced gametes (Draper and Hancock, 2003). Southern highbush breeding programmes (SHB), currently located in Florida, Australia, Chile, Spain, North Carolina and Georgia, are focusing on berry firmness, flavour, yield and machine harvestability. Here the hybrid between *V. darrowii* and *V. corymbosum* 'US75' has been instrumental in the development of low-chilling Southern highbush cultivars such as 'Cooper', 'O'Neal' and 'Sierra'. Several northern highbush breeding programmes (NHB) are in Oregon, Michigan, New Jersey, Chile, New Zealand and the UK, with priorities focused on flavour, firmness, shelf life and expanding harvest dates. Plants for machine harvesting are also of interest to all breeding programmes, providing premium fruit is achievable. Rabbiteye and other hybrid programmes exist in Georgia, North Carolina, Florida and Mississippi, and on a smaller scale, the UK. Here shelf life, size, flavour and yield are the primary targets.

V. angustifolium is a popular choice for conferring several desirable traits, including winter hardiness, early ripening (a short fruit development period), frost tolerance and a small picking scar. *V. angustifolium* has also proven valuable as a source of resistance to stem canker (*Botryosphaeria* spp.) (Prodorutti et al., 2007). *V. constablaei* can pass on winter hardiness, high chilling and a nice light colouring. *V. ashei* is often used to increase winter hardiness, low chilling, firmness and resistance to *Phytophthora* root rot, while it can also be a source of genes that contribute to faster de-acclimation (Arora et al., 2004). *V. darrowii* can confer low chilling, heat tolerance, late ripening and superior flavour as well as a tolerance to high pH. *V. elliottii* can confer drought tolerance, low chilling, late flowering and good flavour. *V. myrsinites* contributes genes for low chilling, good firmness and a desirable low acidity (Ballington et al., 1984). Global blueberry production continues to rise, and this is in part due to the successful breeding efforts for developing cultivars better adapted to more challenging growing environments which now include subtropical regions such as Africa, Asia and New Zealand. Breeding programmes focusing on these regions can either select adapted wild germplasm in which to cross with or use elite breeding lines from temperate zones which can be adapted to suit localised conditions (Lyrene, 2005). While some cultivars, like 'Bluecrop', have survived the test of time and continue to be grown in many regions 70 years after their release, it remains to be seen how many of the newer cultivars will continue to be selected 10 years or more after their launch. Failure of cultivars to be continued by growers can be for several reasons including poor shelf life so unable to export fruit, susceptibility to local pest and disease pressures as well as other limiting factors such as light levels or day length affecting yield. Some of the key cultivars currently under production today range from 70 years old to <5 years old (Table 1).

Many native species can be introduced into breeding programmes to enhance both agronomic and quality traits (Hancock et al., 2008). Although crossing between species can present difficulties, particularly where differences between ploidy levels exist, intersectional crosses can improve fruit quality. Some of the most promising native species with the highest levels of soluble solids are those from *V. angustifolium*, *V. pallidum* Ait. and *V. stamineum*. Low titratable acidity can be found in *V. ashei*, *V. darrowii* and *V. myrsinites*, while large fruit size and firmness are present in *V. corymbosum*, *V. ashei* and *V. darrowii*, and *V. ashei*, respectively (Ballington et al., 1984; Hancock et al., 2008). One great success to come from interspecific crosses was the southern highbush blueberries developed in the 1980s following the introduction of native material from Florida into *V. corymbosum*. While specifically bred to reduce chilling requirement, allowing production in hotter and dryer climates, another SHB result was an improvement in some flavour and texture attributes (Lobos and Hancock, 2015; Gilbert et al., 2015; Du and Rouseff, 2014). Aldehydes have

Table 1 Key highbush blueberry cultivars currently under production

Cultivars	Advantages	Disadvantages
Duke	Popular early variety with consistent fruit size.	Flavour can be bland and challenging to maintain vigour over time.
Bluecrop	Standard cultivar for many years and still widely grown. Consistent yield, size and flavour.	Can be tart if picked too early. Needs time on the bush after turning blue.
Draper	Excellent shelf life and flavour. Easy to pick by hand.	Challenging to grow in some areas.
Liberty	Great flavour and yield potential. Excellent shelf life.	Low tolerance to high heat resulting in soft fruit.
Aurora	Large fruit and good yield.	Can be tart if picked too early.
Chandler	Large fruit with good flavour. Long ripening season.	Can be soft which reduces shelf life.
Last call	High yield, firmness and good flavour.	Prone to cracking, needs aggressive pruning, tart if picked too early.
Cargo	Good flavour, size and firmness.	Can lack vigour if not pruned and managed appropriately. Occasional problems with poor fruit set.
Ozarkblue	Large fruit, good flavour.	Can be difficult to establish in some regions.
Jewel	High yield and good fruit size, low chilling requirement.	Flavour can be a bit tart.

been found to be the most abundant aroma group in SHB, contributing fresh, grassy and fruity notes, with an ester composition more similar to lowbush than to NHB blueberry levels (Du and Rouseff, 2014).

Hirvi and Honkanen (1983) found that most aroma components, as well as sugar content, in highbush blueberries were largely inherited by the progenies. Commercial varieties developed and bred in America have been known to fail to establish and grow across some European countries due to factors such as frost damage or disease susceptibility. Attempts to develop interspecific crosses between blueberry species have gained interest to produce germplasm better adapted to local conditions. The bog blueberry (*V. uliginosum* L.) that grows in the wild across Northern Europe has been successfully crossed with *V. corymbosum*, creating a hybrid which has reached commercial cultivation with combined traits found to be superior to both parents. As well as characteristics such as fruit yield, size and bush architecture, it was found that aroma compounds such as trans-2-hexenal, geraniol and trans-2-hexen-1-ol were also inherited.

Of particular interest to UK breeding programmes is the potential to cross highbush blueberries with native wild European blueberries (*V. myrtillus*). European blueberry occurs throughout the UK, being most abundant in the Highlands, particularly around spruce- and pine-dominated heath forests (Nestby et al., 2011). *V. myrtillus* is not commercially grown due to the small size of fruit and low plant yields; they do, however, have high anthocyanin content and are coloured throughout the berry and not just the skin, unlike highbush blueberries, and are often described by consumers as having a desirable blueberry flavour (Rohloff et al., 2009).

4 Advances in technology to aid phenotyping

Selecting potential new varieties from a breeding programme requires extensive and repeated phenotypic evaluations in order to assess and compare how germplasm performs year on year. This can often be complicated due to complex genotypic × environmental interactions, with evaluations differing significantly depending on year or location. Breeders, therefore, require robust and high-throughput phenotypic tools that are capable of screening large numbers of plants, whether for the mapping of quantitative trait loci (QTLs) or for direct selection as part of a breeding programme (Williams et al., 2017). Physiological traits provide excellent indicators of many desirable crop characteristics. For example, the different ways varieties respond to stress by closing stomata, stopping water loss through leaves, provide a robust indicator of plant physiological status. Infrared thermography can be used to easily measure leaf temperature, which is related to stomatal opening. Spectral reflectance of leaves is influenced by leaf structure and chemical composition, and spectral signatures can be linked to concentrations of chlorophyll and other compounds to provide additional information on abiotic and biotic stress responses. Hyperspectral imaging measures these spectral signatures. Research is currently underway to develop a field-based phenotyping platform that can function both in controlled glasshouse conditions and, importantly, under field conditions that use automated imaging techniques (infrared thermography and hyperspectral) in order to link phenotyping to genetics (Williams et al., 2017).

Spectral signatures detected in the form of wavelengths from plants often can be linked to specific plant conditions or characteristics (Araus and Cairns, 2014). These techniques have been used previously to estimate yield, stress responses, leaf area index, gas exchange rates and photosynthetic capacity, amongst others. Plant temperature also has been widely used to evaluate a number of key plant traits, and with the development of cheaper and more portable spectrometry and thermography devices, this technology should become more readily available to growers and breeders.

5 Marker-assisted breeding

Marker-assisted breeding has been used successfully in other crop species, resulting in the release of new cultivars in a faster timescale. Based on DNA markers and screening protocols, potential breeding parents or progeny carrying desirable traits, particularly those which are difficult to screen such as chilling hours, can be detected. More practical at times, however, is the identification of individuals with undesirable traits which can then be eliminated from programmes prior to long and expensive field evaluations or through the ability to break linkages between desirable (e.g. large fruit) and undesirable (e.g. high acidity) traits (Rowland et al., 2014). This ability to look directly at a genotype's DNA to screen for desirable traits without the need to wait for these traits to be expressed can only be pursued by the development of molecular markers capable of discriminating between alleles linked to traits of interest.

Utilising a range of markers, as they have been developed, has led to the generation of linkage maps which provide a framework for increased efficiency of selection, where markers linked to the gene(s) or quantitative trait loci (QTL) underlying the trait(s) of interest can be developed further for use in breeding programmes, as well as a research tool for assisting in the understanding of the genetic control of desirable phenotypes. Genome sequencing of blueberry has estimated the haploid genome size to be around 500–600 Mb. This puts the blueberry genome at 1.5- to 3-fold larger than other woody plants currently sequenced such as grape (*Vitis vinifera*), peach (*Prunus persica*) and raspberry (*Rubus idaeus*) (Die and Rowland, 2013). A systematic approach that can utilise global molecular approaches alongside integrated bioinformatics will therefore be required to meet new challenges.

The first diploid blueberry map was published by Rowland and Levi (1994) and was based on a population found to segregate for chilling requirement derived from an interspecific hybrid (*V. darrowii* × *V. elliottii*) × *V. darrowii*, which consisted of 12 linkage groups covering 950 cM with 72 markers. Progress has since been made with a recently updated diploid map (*V. darrowii* 'Fla4B' × *V. corymbosum* 'W85-20') × *V. corymbosum* 'W85-23', covering 1740 cM with 265 markers (Rowland et al., 2014), with QTLs identified for both chilling requirement and cold hardiness. Other advances have included the development of a tetraploid blueberry linkage map covering 1619 cM and consisting of 12 linkage groups and 536 markers (McCallum et al., 2016). This mapping population was an F_1 population generated from a cross between two tetraploid cultivars: the northern highbush 'Draper' released by Michigan State University and the southern highbush 'Jewel' released by the University of Florida. Other *Vaccinium* maps including cranberry (*Vaccinium macrocarpon*) with 573 markers covering the 12 linkage groups (Schlautman et al., 2015) are also available with the diploid genome sequence, transcriptome data and various expression datasets all available to access and search online (Bian et al., 2014; Rowland et al., 2003, 2014).

6 Review of how research can improve blueberry flavour

Flavour, described as sweetness with a trace of acidity in a 'crunchy' textured fruit, is what consumers find most appealing in blueberries, along with a light blue colouring with a good wax coating combined with the perceived health benefits. In terms of production priorities, growers prioritise yield, shelf life, firmness (which is associated with shelf life), flavour that attracts repeat purchases and a small picking scar. Generally, however, high acid fruit tends to have a longer storage life than the more desirable low acidity fruit. While genetic improvement has already been made through conventional breeding regarding many of these traits, flavour, aroma and antioxidant capacity have not made as much progress, perhaps due to the complexity and environmental variability of these traits. Conventional breeding of *Vaccinium* species through successive germplasm selections or hybridisation approaches can be both time-consuming and labour intensive. Not all traits of interest can be easily identified within natural germplasm pools, which leads to the benefits of making interspecific crosses. This method can also be time-consuming and unpredictable, however, with the potential to transfer not only desirable but also undesirable traits (Song and Hancock, 2012).

While blueberry breeders successfully used traditional approaches, the incorporation of marker-assisted techniques could result in more cost-effective cultivar development strategies, principally by allowing breeders to improve selection of parent plants for crosses and eliminate offspring with undesirable traits before they are planted in the field for long-term evaluation. The chilling requirement and cold tolerance of progeny genotypes, for example, could be more accurately predicted with the allocation of appropriate markers, allowing breeders to evaluate germplasm in appropriate environments. Marker-assisted breeding could significantly aid in genetic improvement, particularly when combining certain traits like climatic adaptation and season extension with other important traits like fruit quality. Woody perennials, like blueberry, are especially suitable for improvement via marker-assisted breeding because of their long generation times, high heterozygosity, self-incompatibility, inbreeding depression and polyploidy of commercial types.

Breeding priorities will need to start focusing on developing products that consumers will actively seek out and this will depend on improving available flavour attributes and availability. The primary flavour compounds identified in blueberry include aldehydes, alcohols, ketones, terpenoids and esters which are all perceived on a sensory level (Gilbert et al., 2015) and are therefore preferred targets for breeding programmes. It has generally been accepted that most fruit breeding programmes have primarily focused on improving external fruit quality with selection based on size, colour and shelf life. Many of these traits, however, may have unintentional side effects leading to a negative sensory experience as well as reducing nutritional qualities (Goff and Klee, 2006). Phenotyping for flavour remains expensive and time-consuming, and this will be out of the realms of possibility for most breeding programmes for deployment on large-scale advanced selections. It may, however, be feasible to screen small numbers of hybrid samples from interspecific crosses in order to better select parental material with favourable flavour profiles.

Many of these VOC concentrations coincide with fruit ripening and are therefore considered crucial to the overall aroma profile from their ecological perspective of serving as attractants to both human and animals. It has been found that the low molecular weight volatiles such as (*E*)-2-hexenol and (*Z*)-2-hexenol (green and grassy aroma) tend to decrease as the fruit ripens, whereas high molecular weight volatiles including linalool and geraniol (floral aromas) increase (Horvat and Senter, 1985). Care must also be given to post-harvest conditions, as these can have either a positive or negative effect on fruit quality (Forney, 2001). In strawberries, for example, off flavours have been reported due to the generation of acetaldehyde, ethanol and ethyl acetate due to packaging under high concentrations of CO_2 and O_2 resulting in the beginning of the fermentation process (Ke et al., 1994). High CO_2 can also have a negative impact on blueberries (Rosenfeld et al., 1999).

7 Summary

Environmental and experimental factors affecting quality measurements must be considered prior to genetic intervention strategies, particularly in cases for which unfavourable traits may be associated with favourable traits. This was demonstrated in an experiment by Conner et al. (2002) in which fruit size was found to be negatively correlated with total phenolic content. Further study showed that the genetic correlations between these traits were not significant. Where unfavourable traits are found to be genetically controlled, novel germplasm containing alternative biological or physiological processes will be required for these to be overcome (Luby and Shaw, 2008).

Considerable financial outlays are required to successfully grow blueberries in the UK due to the need for protective polytunnels, irrigation and fertigation regimes, as well as the cost of hand-harvest. Grower priorities have recently focused on flavour to achieve premium supermarket status. Unlike other berries, there is a minor processed market for blueberries and due to production in polytunnels, machine harvesting is impractical, so high-quality fruit is essential for profitability. Fresh locally produced fruit picked at the appropriate stage of ripeness and carefully transported directly to the point of sale will likely provide the highest quality product. That is, of course, providing the starting plant material is of the highest genetic quality in terms of internal and external traits. Making more efficient use of natural resources not only saves money but can help to lower risks and benefit society. Utilising innovative products or services as well as developing efficient fruit production and processing techniques are key for this to be achieved. The marketing and promotion of locally grown, high-quality blueberries could give growers a sustainable competitive advantage which does not harm the environment and provides fair wages to farmers.

If increased consumption of blueberries is to continue, a new generation of cultivars with appropriate seasonal and fruit quality attributes will be essential. Identification of appropriate character-impacting volatiles for sweet and fruity blueberry flavour (such as [*E*]-2-hexenal, linalool and ethyl acetate) could help breeders to bridge the current gap in new cultivar development and deliver consistently superior external and internal qualities. As to whether the perfect blueberry exists reflects not only the plant genetics, but also the human genetics of those consuming the fruit. With approximately 400 genes found to encode olfactory receptors in humans, with almost 6000 polymorphisms currently identified, there will be high allelic diversity present which will result in variations in sensory perception (Gilbert et al., 2015; Niimura, 2012; Olender et al., 2012). Current conventional breeding efforts take around 10–15 years between initial crossing to cultivar release, which essentially means that breeders today are developing germplasm for a market almost two decades in the future (Gilbert et al., 2014).

8 Future trends in research

There are many factors that contribute to fruit flavour, from sugar and acid content to volatile and flavonoid interactions, making this a difficult trait to measure on both a chemical and sensory level. Variations between the flavour profiles of the primary blueberry species have been reported which would allow this trait to be selectively crossed and bred for. Bioactive fruit composition can therefore be enhanced through selective breeding, should target phenolic- and volatile-rich cultivars be identified. Environmental interactions will, however, remain a challenge, as will the potential to influence compound activity via certain agronomic practices such as irrigation, fertigation and climatic conditions (Finn et al., 2003). Identifying hybrid genotypes whereby biochemical components are stable across environmental conditions, resulting in reliable and consistently high fruit quality, would undoubtedly change the face of breeding programmes and allow the consistent production and marketing of premium quality blueberries with excellent texture, flavour and the ideal sweet and sour balance (Gilbert et al., 2014).

Flavonoid content relating to fruit colour and antioxidant properties has been studied extensively in blueberries by many groups (Conner et al., 2002; Bravo, 1998; Kalt and Dufour, 1997), while volatile and aroma composition contributing to flavour have attracted less attention. Volatile compounds are responsible for plant protection from external stress through their antioxidant activities as well as imparting desirable berry flavour. Volatile and aroma compounds whose regulation are controlled more by genotype than by environment will be the most desirable targets for breeding to achieve consistent performance of cultivars across growing conditions (Gilbert et al., 2015). Recent efforts in the development of higher resolution genotyping by sequencing (GBS) maps, in combination with genome sequencing, have increased the utility and accuracy by which traits can be located and linked to underlying genes.

9 Where to look for further information

Many comprehensive and extensive studies have been conducted covering blueberry research and breeding from around the world, and some of these have been reviewed by Luby et al. (1991) and Ballington (2001), amongst others.

Bioinformatics is an important tool to manage data effectively and provide researchers and breeders with the ability to mine and utilise available datasets. There are many websites available now which are rich with blueberry sequences and annotated transcripts. The genome database for *Vaccinium* is available online (http://www.vaccinium.org/) and contains a wide range of genome sequences as well as annotations of transcripts and markers. Another resource is The Blueberry Genomics Database (BBGD) available at http://bioinformatics.towson.edu/BBGD/ which contains both EST and microarray data available for analysis. The Plant Genome Network at PGN (http://www.pgn.cornell.edu/) focuses on EST processing and annotation through a general-purpose analysis pipeline (Die and Rowland, 2013; Alkharouf et al., 2007; Albert et al., 2005).

10 References

Albert, V. A., Soltis, D. E., Carlson, J. E., Farmerie, W. G., Wall, P. K., Ilut, D. C., Solow, T. M., Mueller, L. A., Landherr, L. L., Hu, Y., et al. 2005. Floral gene resources from basal angiosperms for comparative genomics research. *BMC Plant Biology* 5(1), 5. doi:10.1186/1471-2229-5-5.

Alkharouf, N. W., Dhanaraj, A. L., Naik, D., Overall, C., Matthews, B. F. and Rowland, L. J. 2007. BBGD: An online database for blueberry genomics data. *BMC Plant Biology* 7(1), 5. doi:10.1186/1471-2229-7-5.

Araus, J. L. and Cairns, J. E. 2014. Field high-throughput phenotyping: The new crop breeding frontier. *Trends in Plant Science* 19(1), 52–61. doi:10.1016/j.tplants.2013.09.008.

Arora, R., Rowland, L. J., Ogden, E. L., Dhanaraj, A. L., Marian, C. O., Ehlenfeldt, M. K. and Vinyard, B. 2004. Rate of dehardening, stage of bud opening, and changes in dehydrin metabolism in blueberry (*Vaccinium* spp.) cultivars during deacclimation at constant, warm temperatures. *Journal of the American Society for Horticultural Science* 129, 667–74.

Baldwin, E. A., Scott, J. W., Shewmaker, C. K. and Schuch, W. 2000. Flavour trivia and tomato aroma: Biochemistry and possible mechanisms for control of important aroma components. *HortScience* 35(6), 1013–22.

Baldwin, E. A., Plotto, A. and Goodner, K. 2007. Shelf-life versus flavour-life for fruits and vegetables: How to evaluate this complex trait. *Stewart Postharvest Review* 1, 3.

Ballington, J. R. 2001. Collection, utilization and preservation of genetic resources in *Vaccinium. HortScience* 36, 213–20.

Ballington, J. R., Ballinger, W. E., Swallow, W. H., Galleta, G. J. and Kushman, L. J. 1984. Fruit quality characterization of 11 *Vaccinium* species. *Journal of the American Society for HortScience* 109(5), 684–9.

Bian, Y., Ballington, J., Raja, A., Brouwer, C., Reid, R., Burke, M., Wang, X., Rowland, L. J., Bassil, N. and Brown, A. 2014. Patterns of simple sequence repeats in cultivated blueberries (*Vaccinium* section *Cyanococcus* spp.) and their use in revealing genetic diversity and population structure. *Molecular Breeding* 34(2), 675–89. doi:10.1007/s11032-014-0066-7.

Bravo, L. 1998. Polyphenols: Chemistry, dietary sources, metabolism, and nutritional significance. *Nutrition Reviews* 56(11), 317–33. doi:10.1111/j.1753-4887.1998.tb01670.x.

Cardeñosa, V., Girones-Vilplana, A., Muriel, J. L., Moreno, D. A. and Moreno-Rojas, J. M. 2016. Influence of genotype, cultivation system and irrigation regime on antioxidant capacity and selected phenolics of blueberries (*Vaccinium corymbosum* L.). *Food Chemistry* 202, 276–83. doi:10.1016/j.foodchem.2016.01.118.

Conner, A. M., Luby, J. J., Hancock, J. F., Berkheimer, S. and Hanson, E. J. 2002. Changes in fruit antioxidant activity among blueberry cultivars during cold-temperature storage. *Journal of Agricultural and Food Chemistry* 50(4), 893–8. doi:10.1021/jf011212y.

Crane, M. B. and Lawrence, W. J. C. 1931. Inheritance of sex, colour and hairiness in the raspberry, *Rubus idaeus* L. *Journal of Genetics* 24(2), 243–55. doi:10.1007/BF02983857.

Die, J. V. and Rowland, L. J. 2013. Advent of genomics in blueberry. *Molecular Breeding* 32(3), 493–504. doi:10.1007/s11032-013-9893-1.

Draper, A. and Hancock, J. 2003. Florida 4B: Native blueberry with exceptional breeding value. *Journal of American Pomological Society* 57, 138–41.

Du, X. and Rouseff, R. 2014. Aroma active volatiles in four Southern highbush blueberry cultivars determined by gas chromatography-olfactometry (GC-O) and gas chromatography-mass spectrometry (GC-MS). *Journal of Agricultural and Food Chemistry* 62(20), 4537–43. doi:10.1021/jf500315t.

Farnetti, B., Khomenko, I., Grisenti, M., Ajelli, M., Betta, E., Algarra, A. A., Cappellin, L., Aprea, E., Gasperi, F., Biasioli, F., et al. 2017. Exploring blueberry aroma complexity by chromatographic and direct-injection spectrometric techniques. *Frontiers in Plant Science* 8(617). doi:10.3389/fpls.2017.00617.

Finn, C. E., Hancock, J. F., Mackey, T. and Serçe, S. 2003. Genotype × Environment interactions in highbush blueberry (*Vaccinium* sp. L.) families grown in Michigan and Oregon. *Journal of the American Society of Horticultural Science* 128(2), 196–200.

Forney, C. F. 2001. Horticultural and other factors affecting aroma volatile composition of small fruit. *HortTechnology* 11(4), 529–38. doi:10.21273/HORTTECH.11.4.529.

Forney, C. F., Kalt, W., Jordan, M. A., Vinqvist-Tymchuk, M. R. and Fillmore, S. A. E. 2012. Blueberry and Cranberry fruit composition during development. *Journal of Berry Research* 2, 169–77.

Gauci, R., Otrysko, B., Catford, J. G. and Lapointe, L. 2009. Carbon allocation during fruiting in *Rubus chamaemorus*. *Annals of Botany* 104(4), 703–13. doi:10.1093/aob/mcp142.

Gilbert, J. L., Olmstead, J. W., Colquhoun, T. A., Levin, L. A., Clark, D. G. and Moskowitz, H. R. 2014. Consumer assisted selection of blueberry fruit quality traits. *HortScience* 49(7), 864–73.

Gilbert, J. L., Guthart, M. J., Gezan, S. A., Pisaroglo de Carvalho, M., Schwieterman, M. L., Colquhoun, T. A., Bartoshuk, L. M., Sims, C. A., Clark, D. G. and Olmstead, J. W. 2015. Identifying breeding priorities for blueberry flavor using biochemical, sensory, and genotype by environment analysis. *PLoS One* 10(9), e0138494. doi:10.1371/journal.pone.0138494.

Giongo, L., Poncetta, P., Loretti, P. and Costa, F. 2013. Texture profiling of blueberries (*Vaccinium* spp.) during fruit development, ripening and storage. *Postharvest Biology and Technology* 76, 34–9. doi:10.1016/j.postharvbio.2012.09.004.

Goff, S. A. and Klee, H. J. 2006. Plant volatile compounds: Sensory cues for health and nutritional value? *Science* 311(5762), 815–9. doi:10.1126/science.1112614.

Hancock, J. F., Lyrene, P., Vorsa, N. and Lobos, G. A. 2008. Blueberry and cranberry. In: Hancock, J. F. (Ed.), *Temperate Fruit Crop Breeding: Germplasm to Genomics*. Dordrecht, The Netherlands: Springer, Kluwer Academic Publishers, pp. 115–49.

Hirvi, T. and Honkanen, E. 1983. The aroma of some hybrids between high-bush blueberry (*Vaccinium corymbosum*, L.) and bog blueberry (*Vaccinium ulginosum*, L.). *Zeitschrift Lebensmittel-Untersuchung und -Forschung* 176, 346–9.

Horvat, R. J. and Senter, S. D. 1985. Comparison of the volatile constituents from rabbiteye blueberries (*Vaccinium ashei*) during ripening. *Journal of Food Science* 50(2), 429-31. doi:10.1111/j.1365-2621.1985.tb13419.x.

Jennings, D. L. 1967. Balanced lethals and polymorphism in *Rubus idaeus*. *Heredity* 22(4), 465–79. doi:10.1038/hdy.1967.64.

Jennings, D. L. 1988. *Raspberries and Blackberries: Their Breeding, Diseases and Growth*. London: Academic Press.

Johnson, S. A. and Arjmandi, B. H. 2013. Evidence for anti-cancer properties of blueberries: A mini review. *Anti-Cancer Agents in Medicinal Chemistry* 13(8), 1142–8. doi:10.2174/18715206113139990137.

Kader, A. A. 1991. Quality and its maintenance in relation to the postharvest physiology of strawberry. In: Dale, A. and Luby, J. J. (Eds), *The Strawberry Into the 21st Century*. Portland, OR: Timber Press, pp. 145–52.

Kalt, W. and Dufour, D. 1997. Health functionality of blueberries. *HortTechnology* 7(3), 216–21. doi:10.21273/HORTTECH.7.3.216.

Kalt, W., Joseph, J. A. and Shukitt-Hale, B. 2007. Blueberries and human health: A review of the current research. *Journal of the American Pomological Society* 61, 151–60.

Ke, D., Zhou, L. and Kadar, A. A. 1994. Mode of oxygen and carbon dioxide action on strawberry ester biosynthesis. *Journal of the American Society for Horticultural Science* 119, 971–5.

Keep, E. 1968. Inheritance of resistance to powdery mildew *Sphaerotheca macularis* (Fr.) Jaczewski in the red raspberry *Rubus idaeus* L. *Euphytica* 17, 417–38.

Klee, H. J. and Tieman, D. M. 2018. The genetics of fruit flavour preferences. *Nature Reviews Genetics* 19(6), 347–56. doi:10.1038/s41576-018-0002-5.

Lobos, G. A. and Hancock, J. F. 2015. Breeding blueberries for a changing global environment: A review. *Frontiers in Plant Science* 6, Art. 782. doi:10.3389/fpls.2015.00782.

Luby, J. J. and Shaw, D. V. 2008. Plant breeders' perspectives on improving yield and quality traits in horticultural food crops. *HortScience* 44(1), 20–2.

Luby, J. J., Ballington, J. R., Draper, A. D., Pliszka, K. and Austin, M. E. 1991. Blueberries and cranberries (*Vaccinium*). *Acta Horticulturae* 290, 393–458. doi:10.17660/ActaHortic.1991.290.9.

Lyrene, P. 2005. Breeding low-chill blueberries and peaches for subtropical areas. *HortScience* 40(7), 1947–9.

McCallum, S., Graham, J., Jorgensen, A., Rowland, L. J., Bassil, N. V., Hancock, J. F., Wheeler, E. J., Vining, K., Poland, J. A., Olmstead, J. W., et al. 2016. Construction of a SNP and SSR linkage map in autotetraploid blueberry using genotyping by sequencing. *Molecular Breeding* 36, Article No.4.

Miszczak, A., Forney, C. F. and Prange, R. K. 1995. Development of aroma volatiles and color during postharvest ripening of 'Kent' strawberries. *Journal of American Society of HortScience* 120(4), 650–5.

Ness, A. R. and Powles, J. W. 1997. Fruit and vegetables, and cardiovascular disease: A review. *International Journal of Epidemiology* 26(1), 1–13. doi:10.1093/ije/26.1.1.

Nestby, R., Percival, D., Martinussen, I., Opstad, N. and Rohloff, J. 2011. The European blueberry (*Vaccinium myrtillus* L.) and the potential for cultivation: A review. *The European Journal of Plant Science and Biotechnology* 5(1), 5–16.

Niimura, Y. 2012. Olfactory receptor multigene family in vertebrates: From the viewpoint of evolutionary genomics. *Current Genomics* 13(2), 103–14. doi:10.2174/138920212799860706.

Olender, T., Waszak, S. M., Viavant, M., Khen, M., Ben-Asher, E., Reyes, A., Nativ, N., Wysocki, C. J., Ge, D. and Lancet, D. 2012. Personal receptor repertoires: Olfaction as a model. *BMC Genomics* 13, 414. doi:10.1186/1471-2164-13-414.

Osorio, S., Ruan, Y. L. and Fernie, A. R. 2014. An update on source-to-sink carbon partitioning in tomato. *Frontiers in Plant Science* 5, 516. doi:10.3389/fpls.2014.00516.

Ourecky, D. K. 1975. *Advances in Fruit Breeding*. West Lafayette, IN: Purdue University Press.

Petridis, A., Van der Kaay, J., Chrysanthou, E., McCallum, S., Graham, J. and Hancock, R. D. 2018. Photosynthetic limitation as a factor influencing yield in highbush blueberries (*Vaccinium corymbosum*) grown in a northern European environment. *Journal of Experimental Botany* 69(12), 3069–80. doi:10.1093/jxb/ery118.

Prodorutti, D., Pertot, I., Giongo, L. and Gessler, C. 2007. Highbush blueberry: Cultivation, protection, breeding and biotechnology. *European Journal of Plant Science and Biotechnology* 1, 44–56.

Rimando, A. M., Kalt, W., Magee, J. B., Dewey, J. and Ballington, J. R. 2004. Resveratrol, pterostilbene, and piceatannol in *Vaccinium* berries. *Journal of Agricultural and Food Chemistry* 52(15), 4713–9. doi:10.1021/jf040095e.

Rodriguez-Mateos, A., Rendeiro, C., Bergillos-Meca, T., Tabatabaee, S., George, T. W., Heiss, C. and Spencer, J. P. E. 2013. Intake and time dependence of blueberry flavonoid-induced improvements in vascular function: A randomized, controlled, double-blind, crossover intervention study with mechanistic insights into biological activity. *The American Journal of Clinical Nutrition* 98(5), 1179–91. doi:10.3945/ajcn.113.066639.

Rohloff, J., Nestby, R., Nes, A. and Martinussen, I. 2009. Volatile profiles of European blueberry: Few major players, but complex aroma patterns. *Latvian Journal of Agronomy* 12, 98–103.

Rosenfeld, H. J., Meberg, K. R., Haffner, K. and Sundell, H. A. 1999. MAP of highbush blueberries: Sensory quality in relation to storage temperature, film type and initial high oxygen atmosphere. *Postharvest Biology and Technology* 16(1), 27–36. doi:10.1016/S0925-5214(98)00102-1.

Rowland, L. J. and Levi, A. 1994. RAPD-based genetic linkage map of blueberry derived from a cross between diploid species (*Vaccinium darrowii* and *V elliottii*). *Theoretical and Applied Genetics* 87(7), 863–8. doi:10.1007/BF00221139.

Rowland, L. J., Mehra, S. and Arora, R. 2003. Identification of molecular markers associated with cold tolerance in blueberry. *Acta Horticulturae* 625(625), 59–69. doi:10.17660/ActaHortic.2003.625.5.

Rowland, L. J., Ogden, E. L., Takeda, F., Glenn, D. M., Ehlenfeldt, M. K. and Vinyard, B. T. 2013. Variation among highbush blueberry cultivars for frost tolerance of open flowers. *HortScience* 48, 692–5.

Rowland, L. J., Ogden, E. L., Bassil, N., Buck, E. J., McCallum, S., Graham, J., Brown, A., Wiedow, C., Campbell, A. M., Haynes, K. G., et al. 2014. Construction of a genetic linkage map of an interspecific diploid blueberry population and identification of QTL for chilling requirement and cold hardiness. *Molecular Breeding* 34(4), 2033–48. doi:10.1007/s11032-014-0161-9.

Ruiz, J. J., Alonso, A., Garcia-Martinez, S., Valero, M., Blasco, P. and Ruiz-Bevia, F. 2005. Quantitative analysis of flavour volatiles detects differences among closely related traditional cultivars of tomato. *Journal of the Science of Food and Agriculture* 85(1), 54–60. doi:10.1002/jsfa.1879.

Schlautman, B., Covarrubias-Pazaran, G., Diaz-Garcia, L. A., Johnson-Cicalese, J., Iorrizo, M., Rodriguez-Bonilla, L., Bougie, T., Bougie, T., Wiesman, E., Steffan, S., et al. 2015. Development of a high-density cranberry SSR linkage map for comparative genetic analysis and trait detection. *Molecular Breeding* 35(8), 177. doi:10.1007/s11032-015-0367-5.

Song, G. and Hancock, J. F. 2012. Recent advances in blueberry transformation. *International Journal of Fruit Science* 12(1–3), 316–32. doi:10.1080/15538362.2011.623083.

Teo, G., Suzuki, Y., Uratsu, S. L., Lampinen, B., Ormonde, N., Hu, W. K., DeJong, T. M. and Dandekar, A. M. 2006. Silencing leaf sorbitol synthesis alters long-distance partitioning and apple fruit quality. *Proceedings of the National Academy of Sciences of the United States of America* 103(49), 18842–7. doi:10.1073/pnas.0605873103.

Tieman, D., Zhu, G., Resende Jr., M. F., Lin, T., Nguyen, C., Bies, D., Rambla, J. L., Beltran, K. S., Taylor, M., Zhang, B., et al. 2017. A chemical genetic roadmap to improve tomato flavor. *Science* 355(6323), 391–4. doi:10.1126/science.aal1556.

Ulrich, D., Komes, D., Olbricht, K. and Hoberg, E. 2007. Diversity of aroma patterns in wild and cultivated *Fragaria* accessions. *Genetic Resources and Crop Evolution* 54(6), 1185–96. doi:10.1007/s10722-006-9009-4.

Vander Kloet, S. P. 1988. *The Genus Vaccinium in North America*. Ottawa Centre, Ontario: Canadian Government Publication.

Warmund, M. R., Guinan, P. and Fernandez, G. 2008. Temperatures and cold damage to small fruit crops across the eastern United States associated with the April 2007 freeze. *HortScience* 43, 1643–7.

Williams, D., Britten, A., McCallum, S., Jones, H., Aitkenhead, M., Karley, A., Loades, K., Prashar, A. and Graham, J. 2017. A method for automatic segmentation and splitting of hyperspectral images of raspberry plants collected in field conditions. *Plant Methods* 13, 74. doi:10.1186/s13007-017-0226-y.

Index